Enterprise and Entrepreneurship for Agri-Business Management and Planning

The Global Perspective

The Editor

Prof. M.M. Adhikary, the author, is an erudite personality in the academic world of Social Science and related disciplines, who is presently offering his invaluable services as Professor, Department of Agricultural Extension, Faculty of Agriculture, Bidhan Chandra Krishi Viswavidyalaya. This distinct personality has been attached with the profession of teaching, research and extension for about thirty four years. His immense potentiality of academic activities and capability to work with the people, mettle of research and extension works helped him serve as Dean, Faculty of Agriculture, Registrar, Bidhan Chandra Krishi Viswavidyalaya, Vice-Chancellor, Bidhan Chandra Krishi Viswavidyalaya and Director, Directorate of Extension Education in the same University as well as the Head of the Department of Agricultural Extension. His hard work and the devotion in the domain of extension science affiliated him as Master Trainers of Experiential learning Cycle in ICAR-USAID programme in 1991. He is the expert of different specialized fields and skills like Group dynamics and leadership, Training of Trainers, Project Evaluation, Participatory Project Management, Value added Agriculture, Media Management etc. He has conducted fourteen research and extension projects as investigator. To his credit, there are 89 scientific research papers published in the National and International Journals, a good score of chapter contribution and authorship to 16 number of books in related areas. He has been widely traveled on academic mission covering China, Japan, USA, Australia, Istanbul, Bangladesh and Thailand etc. He has been the Director of six ICAR sponsored Winter/ Summer Schools in a row and successfully 11 Ph.D. scholars to evince his elegance of research and extension pursuits. To these, entire he has also been the chairman of KVK site selection committee of ICAR, a coveted position to exert at the National Level.

Enterprise and Entrepreneurship for Agri-Business Management and Planning

The Global Perspective

— Editor —

M. M. Adhikary

Professor of Agricultural Extension,
Faculty of Agriculture,
BCKV, Mohanpur, Nadia, W.B.

2014

Daya Publishing House

A Division of

Astral International Pvt. Ltd.

New Delhi – 110 002

ISBN 9789351302544

Published by : **Daya Publishing House®**
A Division of
Astral International Pvt. Ltd.
– ISO 9001:2008 Certified Company –
4760-61/23, Ansari Road, Darya Ganj
New Delhi-110 002
Ph. 011-43549197, 23278134
E-mail: info@astralint.com
Website: www.astralint.com

Laser Typesetting : **Classic Computer Services**, Delhi - 110 035

Printed at : **Replika Press Pvt. Ltd.**

PRINTED IN INDIA

Foreword

Indian Council of Agricultural Research, has envisaged the need for comprehensive reforms in agricultural education in India through State Agricultural Universities, Central Agricultural Universities or ICAR Institutes. The technocrats produced from agricultural education systems must elicit a demand in the job market as well as academic organization in terms of capacity, mettle and relevance. Accordingly, the text on enterprise, agri-business and agricultural marketing has been adequately highlighted in social science course module. In achieving the stated objective, quality text books, research manuals and referential publications are essential. The changing scenario of Indian agriculture alongside the transformation in global trade has encompassed this necessity, rightly and aptly.

Today, we are not only concerned about production and productivity of crops, we contemplate more about enhancing livelihood, per capita value of return from the same unit of area. So, we need to revamp and restructure our agriculture into a roaring entrepreneurship to usher in a much-needed step change in the rate of growth of the Indian economy.

How to transform the farmers with small resources/capital into an entrepreneur or to train the timid and traditional farmwomen into a determined woman entrepreneur, generating agri-business is the challenge of the day. All these are essential interventions to improve India's contribution to the value added global agricultural market from meager 1 – 1.5 per cent now, of the total global market.

Unleashing entrepreneurial motivation itself is a synergistic process, integrating both science and skills. A holistic approach has to encompass contemporary challenges like global warming, climate change, natural resource management, food and nutritional security, inclusive growth and gender empowerment.

I hope that the book, entitled "*Enterprise and Entrepreneurship for Agri-Business Management and Planning: The Global Perspective*" shall encourage a clear focus on improving income and, livelihood, nutrition and environment management. It should enthuse the entrepreneur to imbibe human values, philanthropic pursuits and methodological innovations. I congratulate Professor Manas Mohan Adhikary for painstakingly bringing out this publication.

The purpose of the book has well been dovetailed with the objectives set by the ICAR and accordingly, the chapters have been well thought out and well arranged. This book is the reflection of the need for creating upcoming managerial skills, entrepreneurial leadership, motivational pursuits, planning pedagogy, resource conservation and ICT interventions in Indian agriculture. I strongly believe that this book will be a worthy collection and useful to the professional managers, scholars and policy makers, striving hard and relentlessly to ensure that Indian agriculture be a major player in the global market of twenty first century.

Pradip K Mazumdar

Advisor to the Hon'ble Chief Minister,
Government of West Bengal on Agriculture and Allied Sectors

Contents

List of Contributors

1. **Dr. P. Das**, Former Deputy Director General (AE), Indian Council of Agricultural Research, New Delhi-12.
2. **Mr. P. K. Mazumdar**, Advisor to Hon'ble Chief Minister on Agriculture, Govt. of West Bengal, Writers Buildings, Kolkata-700001
3. **Prof. S. K. T. Nasar**, Former Director of Research, BCKV, Kalyani, Nadia
4. **Dr. S. K. Samanta**, Former Joint Director of Research, BCKV, Kalyani, Nadia
5. **Prof. Nilimesh Rochaudhury**, Rtd. Professor, Faculty of Horticulture, BCKV
6. **Dr. C. Satapathy**, Director AMITY, Bhubaneswar, Odisha.
7. **Dr. Kaushik Pradhan**, Head, Agril. Extension, F/Ag., UBKV, Coochbehar, WB
8. **Dr. S. D. Mukhopadhyay**, Associate Professor, PSB, Deptt of EEE, Visva Bharati, Santiniketan, Birbhum, WB
9. **Dr. Raktima Mukhopadhyay**, Executive Director, IBARD, Kolkata
10. **Prof. S. Pan**, Professor, West Bengal University of Animal and Fishery Sciences, Kolkata
11. **Prof. Dipak De**, Professor and Head, Department of Agricultural Extension, Banaras Hindu University, Varanasi, Uttar Pradesh
12. **Dr. Asitava Sur**, General Manager, Metro Diary, Barasat, 24-Parganas (North), WB

13. **Prof. Sisir Kr. Mitra**, Professor, Department of Fruits and Orchard Management, Faculty of Horticulture, BCKV
14. **Prof. Anandamoy Puste**, Professor of Agronomy, Faculty of Agriculture BCKV, Mohanpur, Nadia
15. **Prof. Tapan Kumar Maity**, Professor, Deptt. of Vegetable Crops,Faculty of Horticulture, BCKV, Mohanpur, Nadia.
16. **Dr. Amit Baran Sarengi**, Associate Professor, Department of Spices and Plantation Crops, Faculty of Horticulture, BCKV
17. **Dr. Kaushik Brahmachari**, Associate Professor, Deptt. of Agronomy, Faculty of Agriculture BCKV, Mohanpur, Nadia
18. **Dr. Amitava Biswas**, Associate Professor, Deptt. of Agril. Extension, Faculty of Agriculture BCKV, Mohanpur, Nadia
19. **Ms Manashi Saha**, Ph.D. Research Student, Deptt. of Agril. Extension, Faculty of Agriculture BCKV, Mohanpur, Nadia
20. **Prof. Asit Kumar Basu**, Professor, Deptt. of Seed Science and Technology, Faculty of Agriculture BCKV, Mohanpur, Nadia
21. **Prof. Santanu Jha**, Professor, Deptt. of Agril. Entomology, Faculty of Agriculture BCKV, Mohanpur, Nadia
22. **Prof. Anjan Bhattacharya**, Professor and Head, Deptt. of Agril. Chemicals, Faculty of Agriculture BCKV, Mohanpur, Nadia
23. **Dr. Ivy Chakraborty**, Associate Professor, Deptt. of Post Harvest Technology of Horticultural Crops, Faculty of Horticulture, BCKV
24. **Prof. Abu Hassain**, Professor, Department of Fruits and Orchard Management Faculty of Horticulture, BCKV
25. **Prof. Jayanta Kumar Das**, Professor and Head, Deptt. of Agril. Extension, Faculty of Agriculture BCKV, Mohanpur, Nadia
26. **Prof. Sankar Kumar Acharya**, Professor of Agricultural Extension, Faculty of Agriculture BCKV, Mohanpur, Nadia
27. **Prof. M.M. Adhikary**, Professor of Agricultural Extension, Faculty of Agriculture, BCKV, Mohanpur, Nadia.

– *Part I* –

Definition, Concept and Strategy of Enterprise and Entrepreneurship

Chapter 1

Enterprise, Entrepreneurship and Agri-Business Management

Manas Mohan Adhikary and Sankar Kumar Acharya

What is Entrepreneurship?

Entrepreneurship is the act and art of being an entrepreneur or "'one' who undertakes innovations or introducing new things, finance and business acumen in an effort to transform innovations into economic goods". *This may result in new organizations or may be part of revitalizing mature organizations in response to a perceived opportunity. The most obvious form of entrepreneurship is that of starting new businesses (referred as Startup Company); however, in recent years, the term has been extended to include social and political forms of entrepreneurial activity. When entrepreneurship is describing activities within a firm or large organization it is referred to as intra-preneurship and may include corporate venturing, when large entities spin-off organizations.*

Entrepreneurship is the process of exploring the opportunities in the market place and arranging resources required to exploit these opportunities for long term gain. It is the process of planning, organising, opportunities and assuming. Thus it is a risk of business enterprise. It may be distinguished as an ability to take risk independently to make utmost earnings in the market. It is a creative and innovative skill and adapting response to environment.

According to Paul Reynolds, entrepreneurship scholar and creator of the Global Entrepreneurship Monitor, "by the time they reach their retirement years, half of all working men in the United States probably have a period of self-employment of one or more years; one in four may have engaged in self-employment for six or more years. Participating in a new business creation is a common activity among U.S. workers over the course of their careers." And in recent years this has been documented by scholars such as David Audretsch to be a major driver of economic growth in both the

United States and Western Europe. "As well, entrepreneurship may be defined as the pursuit of opportunity without regard to resources currently controlled (Stevenson, 1983)" Entrepreneurship can be defined by describing what entrepreneurs do. For example: *"Entrepreneurs use personal initiative, and engage in calculated risk-taking, to create new business ventures by raising resources to apply innovative new ideas that solve problems, meet challenges, or satisfy the needs of a clearly defined market."*

But as the following definitions state, entrepreneurship is not restricted to business and profit: *"Entrepreneurship involves bringing about change to achieve some benefit. This benefit may be financial but it also involves the satisfaction of knowing you have changed something for the better.* (*Entrepreneurship: Creating a Venture* by Lily Kretchman *et al.*, Toronto: Wiley, 1991) *"Entrepreneurship is essentially the act of creation requiring the ability to recognize an opportunity, shape a goal, and take advantage of a situation. Entrepreneurs plan, persuade, raise resources, and give birth to new ventures."* (*Entrepreneurship: The Spirit of Adventure* by Richard Bodell *et al.*, Toronto: Harcourt Brace Jovanovich, 1991.)

Entrepreneurship can be described as a process of action an entrepreneur undertakes to establish his enterprise. Entrepreneurship is a creative activity. It is the ability to create and build something from practically nothing. It is a knack of sensing opportunity where others see chaos, contradiction and confusion. Entrepreneurship is the attitude of mind to seek opportunities, take calculated risks and derive benefits by setting up a venture. It comprises of numerous activities involved in conception, creation and running an enterprise.

According to Peter Drucker Entrepreneurship is defined as 'a systematic innovation, which consists in the purposeful and organized search for changes, and it is the systematic analysis of the opportunities such changes might offer for economic and social innovation.' 7 Entrepreneurship is a discipline with a knowledge base theory. It is an outcome of complex socio-economic, psychological, technological, legal and other factors. It is a dynamic and risky process. It involves a fusion of capital, technology and human talent. Entrepreneurship is equally applicable to big and small businesses, to economic and non-economic activities. Different entrepreneurs might have some common traits but all of them will have some different and unique features. If we just concentrate on the entrepreneurs then there will be as many models as there are ventures and we will not be able to predict or plan, how and where, and when these entrepreneurs will start their ventures. Entrepreneurship is a process. It is not a combination of some stray incidents. It is the purposeful and organized search for change, conducted after systematic analysis of opportunities in the environment. Entrepreneurship is a philosophy- it is the way one thinks, one acts and therefore it can exist in any situation be it business or government or in the field of education, science and technology or poverty alleviation or any others.

Difference between Business and Entrepreneurship

What is the difference between a business and an entrepreneurship? How about a business owner from an entrepreneur? These terms are commonly used interchangeably. However, these terms have different meaning – and those differences are what we will try to find out. To illustrate and understand simply and clearly the difference between business and entrepreneurship, we will have a face-off between a businessman and an entrepreneur. So the showdown begins.

Businessman/Business	*Entrepreneur/Entrepreneurship*
Own an enterprise or venture	Own an enterprise or venture
Usually a profit oriented	Costumer oriented
Usually plays safe	Bold and ambitious
May be purchased, donated or inherited	Creates his own idea and realize it as a business
Generally traditional	An innovator
Works for the Company	The Company works for him
Usually don't have time for their families and love ones	Shares enough time with their families and love ones
Usually distressed and experiences sleepless nights	Always a happy and enthusiastic businessman
Generally hire people to contribute profit	Hire people to make their lives better
Commonly commit tax evasion to save taxes and increase profit	Practice tax avoidance to save taxes and increase profit

As a conclusion, a businessman is similar to an entrepreneur because an entrepreneur is a businessman. Entrepreneur is a businessman which is not traditional but is an innovative and creative type of a businessman. An entrepreneur differs a lot from a usual businessman. His entrepreneurship is created out of his own innovative idea and creation which is realized to be placed in business or to be transformed as a good business. Entrepreneurs are also more concerned on the quality of their products and on the satisfaction of their customers rather than being too much concerned to their profits.

Entrepreneurship and Agri-preneurship

Agripreneurs and Agrirepreneurship Development

There is no single definition of the term "Agrirepreneur". However, an agrirepreneur may be thought of someone who undertakes a variety of activities in agricultural to be agrirepreneur. For example an agrirepreneur may start an agri-business, change a business's direction, acquire a business or be involved in innovatory activity in agricultural value addition. Thus an agrirepreneur may undertake a range of different activities which have a common link - the perception of an opportunity and the willingness to do something to take advantage of it. Explicit in this is that the agrirepreneur is a risk-taker and has the opportunity to initiate and to implement decisions which deal with the uncertain agricultural business environment within which the firm operates. The business environment consists of many complex interrelated elements including: customers, competition, economic factors, social and demographic trends, government policy (both macroeconomic and microeconomic) and technology. In this paper I have identify five frameworks within which to analyze entrepreneurial activity. First, the Intuitive which stresses the personal psychological makeup of the individual to avail the opportunity; second, the Classical which highlights the importance of recognizing opportunities; third, the Leadership which concentrates on a person's ability to motivate and direct people for success; fourth, the Management which looks at the importance of organizing businesses by such means as developing strategies and improved decision-making; fifth, the Intrapreneurship looks at innovation within organizations. Within these frameworks of thought, the entrepreneur is responsible for identifying and meeting market needs. Entrepreneurs are often discussed in terms of starting a business. Research identifies a number of strategies necessary for successful entrepreneurial start-ups which include identifying unmet demand, developing new products which meet changing market conditions, producing marketing and financial plans, weighing up risks and rewards and having the necessary resources to launch the business. Agripreneurship requires more than just the identification of new opportunities. Additional qualities such as imagination, commitment, decisiveness and self confidence are also important elements of the entrepreneurial process. Thus the agrirepreneur introduces change into the market place but must also be aware of the outside pressures for change which are being experienced as the result of other entrepreneurial activity.

The following table is adapted from Edward DeBono's lateral thinking principles (1990) and is applied to highlight the difference between the "Agricultural" of current approaches to Agripreneurship education and the "art" of agrirepreneurship education which is needed. The poor farmers have failed in agriculture but agripreneurs are becoming very successful by doing business in Agri Export Zones (AEZ) of India.

1. Agricultural is selective, whereas the Art of Agrirepreneurship is generative.
2. Agricultural moves only if there is a direction in which to move, whereas the art of Agrirepreneurship moves in order to generate direction.
3. Agricultural is mechanical, whereas the Art of Agrirepreneurship is provocative.
4. Agricultural is sequential, whereas the Art of Agrirepreneurship can make jumps.
5. Agricultural requires each step to be correct, whereas with the Art of Agrirepreneurship this does not need to be so.
6. Agricultural uses the negative to block off certain pathways, whereas with the Art of Agrirepreneurship there is no negative.
7. Agricultural excludes irrelevant information, whereas the Art of Entrepreneurship welcomes chance intrusions.
8. Agricultural has fixed categories, classifications and labels, whereas the Art of
9. Agrirepreneurship there is no such constraints.

Agri-Entrepreneurship

Amidst the changing paradigms and demanding global structure, India, in order to remain a front-runner needs to primarily focus on the agriculture sector, the backbone of the economy. This specialization is developing agripreneurs with distinct traits and skills to exploit opportunities galore in the field of agriculture. Among the various strategies to promote planned growth in this sector, focus on promoting viable enterprises is certainly helping to exploit its operational efficiency to the hilt.

Agri-business Development: Private Sector Role

Agriculture being vital for Indian economy, Government has very long played an important role. The time has come for private sector to come forward. Today, even Government is encouraging private sector participation by inviting trade bodies to suggest necessary reforms in Indian agriculture.

The Task Force highlighted the potential for global opportunities and India's competitive advantages in the context of WTO and Globalization. Somehow, this initiative while preparing a document to completely transform Indian agriculture got new light through new policies and regulations prevalent in the country. In this document, many success stories have been presented, and final solution come with clear cut action plans for the future. It also started private sector role that could also complement and supplement the government efforts.

1. Many individuals, small business houses and large corporate were looking for new investment opportunities. In many sectors of the economy, not much scope was available. With the reducing interest rates, business plans with long gestation periods have acquired new relevance for the private sector. Agriculture Export Zones has become a new business opportunity for their entry at any investment level.
2. Agriculture being top priority, its business environment in terms of regulations, is also conducive. It attracts minimum taxes and gets maximum support. Reforms in agriculture have become top agenda for the government in the country.
3. Many private sector players engaged in agriculture related business activities have synergy in either backward or forward business linkages related to AEZ's
4. Increasing demand for quality and convenience food by the world consumers has created new level of business opportunities.
5. Many interest groups and development agencies etc. have also identified Agriculture Export Zones as a new agenda that will create conducive Agri environment for healthy development in future.

Why India promoting Agripreneurship: A Quick Perspective

In India, 52 per cent of total land is cultivable as against 11 per cent in the world. All 15 major climates of the world, snow bound Himalayas to hot humid southern peninsula; Thar Desert to heavy rain areas all exist in India. There are 20 agro-climatic regions and nearly 46 out of 60 soil types in the country. Sunshine hours and day length are ideally suited for round the year cultivation of crops. India is the centre for biodiversity in plants, animals, insects, micro-organism and accounts for 17 per cent animal, 12 per cent plants and 10 per cent fish genetic resources of the globe. In the live stock sector, India has 16 per cent of cattle, 57 per cent of buffalo, 17 per cent of goats and 5 per cent of sheep population of the world. Agriculture contributes 24.2 per cent to GDP, 15.2 per cent of total exports and provides employment to 58.4 per cent of country's work force.

- ☆ Agriculture remains a key sector of the Indian economy accounting for 25 per cent share in the gross domestic product (GDP) and about 13 per cent of the total export earnings.
- ☆ India is the second largest producer of rice and wheat in the world; first in pulses and fourth in coarse grains. India is also one of the largest producers of cotton, sugar, sugarcane, peanuts, jute, tea and an assortment of spices
- ☆ In terms of the real value added, the Indian agriculture sector ranks third, after China and the United States.
- ☆ The share of agriculture in the total value added to the economy, at around 25 per cent, is still quite high. This implies that agriculture is likely to remain a priority, both for policy makers as well as businesses, in the foreseeable future and any move to ramp up the sector calls for a multi-pronged strategy.
- ☆ In recent years, there has been a considerable emphasis on crop diversification towards horticulture (fruits, vegetables, ornamental crops, medicinal and aromatic plants and spices), plantation crops (coconut, cashew nuts and cocoa) and allied activities.
- ☆ Creation of critical infrastructure for cold storage, refrigerated transportation, rapid transit, grading, processing, packaging and quality control measures open major opportunities for investment.
- ☆ Second highest fruit and vegetable producer in the world (134.5 million tones) with cold storage facilities available only for 10 per cent of the produce.
- ☆ Second highest producer of milk with a cold storage capacity of 70,000 tonne.
- ☆ Fifth largest producer of eggs. Investments in cold chain required to store 20 per cent of surplus of meat and poultry products during 10th plan requires Rs 500 Crore (US$100M)
- ☆ Sixth largest producer of fish with harvesting volumes of 5.2 million tones.

In a fast changing International Trade Environment, with a view to providing remunerative return to the Agripreneurs community in a sustained manner, efforts will be made to provide improved access to the produce/products of the agriculture and allied sectors in the International markets by setting up Agri Export Zones. The AEZ is expected to give a focus and direction for exports of key agricultural produce with potential from the country. It involves a detailed action plan for the development of a specified geographic area/s for effecting systematically greater exports of a specific produce.

Agri Export Zone

With the primary objective of boosting agricultural exports from India, in March 2001, Government of India announced a policy of setting up of Agri Export Zones (AEZs) across the country. The Central Government has sanctioned 60 AEZs comprising about 40 agricultural commodities till March 2005. These AEZs are spread across 20 states in the country.

Objectives

The objective of setting up AEZs is to converge the efforts made, hitherto, by various central and state government departments for increasing exports of agricultural commodities from India. The AEZ takes a comprehensive view of a particular produce/product located in a geographically contiguous area for the purpose of developing and sourcing raw materials, their processing/packaging, and leading to final exports.

The major components of this comprehensive concept are

- Cluster approach of identifying the potential products and geographical region in which these products are grown.
- Adopting an end-to-end approach of integrating the entire process, right from the stage of production till it reaches the consumption stage.
- Integration of the activities of various agencies connected with the department of the product.

Agri Export Zones (AEZs)

AEZ are identified by the State Government, who would evolve a comprehensive package of services provided by all State Government agencies, State agriculture universities and all institutions and agencies of the Union Government for intensive delivery in these zones to develop agripreneurship. Private sector with proven credentials would be encouraged through entrepreneurship development to sponsor new agri export zone or take over already notified agri export zone or part of such zones for boosting agri exports from the zones. Services which would be managed and co-coordinated by State Government/Private sector and would include provision of pre/post harvest treatment and operations, plant protection, processing, packaging, storage and related research and development etc. Nodal agency will supplement, within its schemes and provisions, efforts of State Governments for facilitating such exports.

The concept of the Agri Export Zone attempts to take a comprehensive look at a particular produce/products located in a contiguous area for the purpose of developing and sourcing the raw materials, their processing and packaging, finally exporting them. It is a part of agripreneurship development Thus, the entire effort is centered on a cluster approach of identifying the potential products, the geographical region in which these are grown and adopting an end to end approach of integrating the entire process, right from the stage of production till it reaches the market. The government helps agripreneurs in sourcing for raw materials, the setting up of processing facilities, providing finance at low interest rates and even matching with international buyers.

The export zones mooted by the Agricultural and Processed Food Products Export Development Authority (APEDA) to increase international trade in agri-commodities are an attempt to take a holistic approach for encouraging trade in specific commodities located in contiguous areas.

The concept of AEZ can be summarized as "to produce what could be sold rather sell what is produced". It visualizes identification of interventions in a project mode. By making all central, state governments and local agencies to work under a single umbrella and comprehensively address all the issues relating to each stage of the entire value chain from farm to the ultimate consumer we can penetrate in to new markets in international scenario. Under AEZ India has successfully gone ahead to a large extend private entrepreneurs. Traceability of the produce from the farmer to the end consumer is also very important in global trading.

Agri Export zones are concerned with A to Z of Agri – exports. The emphasis is on partnership on various agencies/systems and convergence of interventions of various agencies like APEDA, Ministry of Food Processing Industries (MFPI), National Horticulture Board (NHB) etc. The focus will be on increasing exports of identified commodities, which is very important factor for entrepreneurship development.

The Concept of AEZ's which aims to give fillip to agriculture exports through entrepreneurship development, comprises the following

- ☆ Identifying a potential zone based on agro-climatic requirements for a particular crop.
- ☆ Integrating various assistance programmes of Central and State Government agencies and providing fiscal incentives to exporters.
- ☆ Implementing the same through involvement of private and public partnership and
- ☆ Integrating all the activities till the produce reaches the market.

Measures Envisaged to Promote Exports from Such Zones by Agripreneurs

1) Financial Assistance to Motivate Entrepreneur

Both central as well as state government and their agencies are providing a variety of financial assistance to various agri export related activities. These extend from providing financial assistance for training and extension, R and D, quality up gradation, infrastructure, marketing etc. All these facilities have to be dovetailed and extended to promote agri exports from the proposed zones in a coordinated manner to promote agripreneurship.

2) Fiscal Incentives for the Entrepreneurial Support

The benefits under Export Promotion of Capital Goods Scheme, which were hitherto available to direct exporters, have now been extended to service exporters in the Agri Export Zones. Thus even service provided to ultimate exporters will be eligible for import of capital goods at a concessional duty for setting up common facilities.

They shall fulfill their export obligation through receipt of foreign exchange from ultimate exporters who will make payments from their account.

Exporters of value added agri products will be eligible for sourcing duty free fuel for generation of power, provided cost component of power in the ultimate product is 10 per cent or more and the input out norms are fixed by the advance licensing committee of the Director General Foreign Trade. In view of the power intensive nature of most of the value addition, almost all the exporters of the value added agriculture produce will become eligible for such facility.

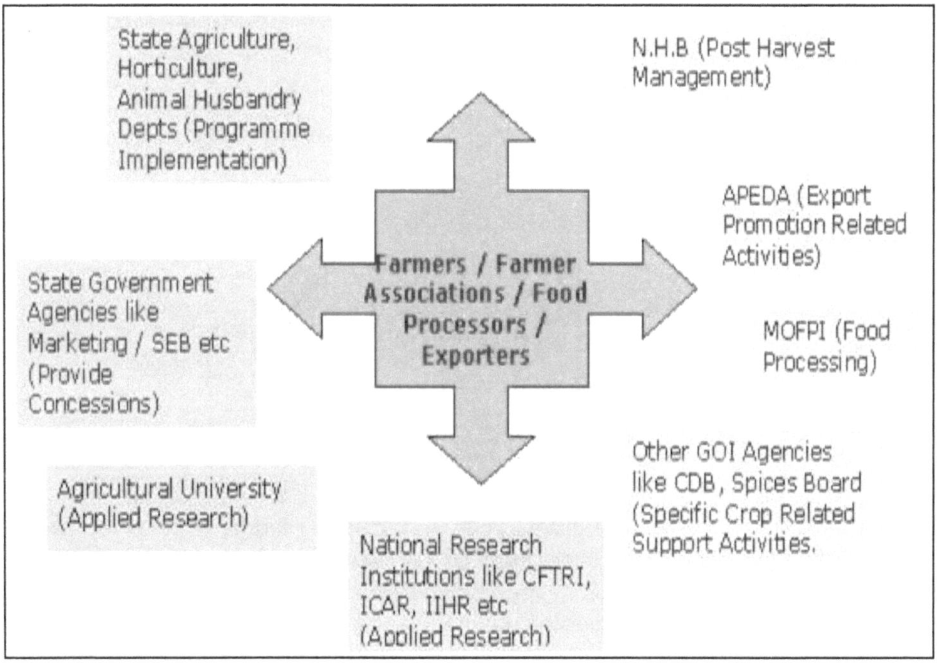

Figure 1.1: Agencies Involved in Sixty Agri Export Zones Set Up (AEZ) in India for Agripreneurship Development.

Need or Rational of Export Market

Investment required is estimated to be Rs 350 Crore (US$ 70M) In spite of the vast natural resources and abundant agricultural produce India ranks below 10th in the export of food products. Conservative estimates put processing levels in the fruits and vegetables sector at 2 per cent, meat and poultry at 2 per cent, milk by way of modern dairies at 14 per cent, fish at 4 per cent, bulk meat de-boning is to the tune of 21 per cent. Currently, the food processing sector, though in the nascent stage, constitutes 14 per cent of manufacturing GDP amounting to products value of Rs.2, 80,000 Crores. It employs 130 lakh persons (13 million) and is supposed to increase at an annual rate of 7 per cent.

Although Agri export zone will boost export but later on they will help to cater domestic market in a cluster form. Food chain clusters (Figure 1.2) are formed with

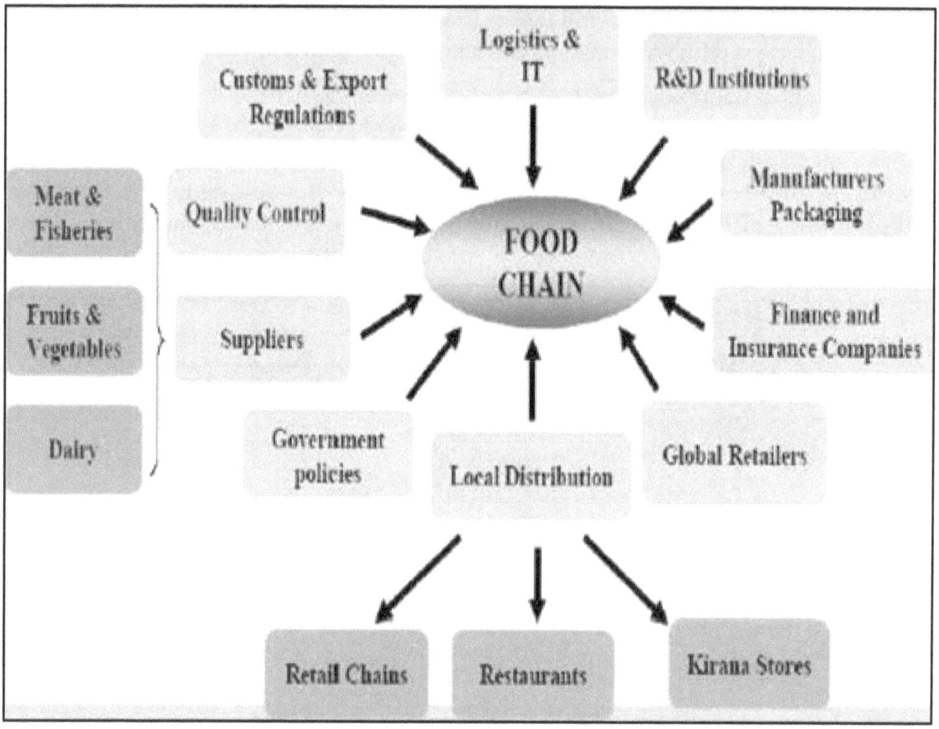

Figure 1.2: Food Supply Chain Cluster.

the participation of all stake holders such as farmers, seed growers, merchants, transporters, wholesalers, retailers, financial institutions, and insurance companies. Information sharing is essential for generating the efficiencies. The Internet and mobile communications are used to enable information and financial transfer between the stake holders. Also, recent advances in RFID technology will have tremendous impact in the management of the food chain particularly for source identification and tracking and also in providing supply chain visibility.

Present Status of India

State Governments identified product specific Agri-export zone for export of specific products from a geographically contiguous area and evolve a comprehensive package of services for delivery in these zones. Such services would include provision of pre/post harvest treatment and operations, plant protection, processing, packaging, storage and related research and development, etc. The service providers as well as the setting up of common infrastructural facilities like sorting, grading, polishing, packaging, cold storage, transport equipment/refrigerated vans, vapour and heat treatment plants, X-ray screening facility, etc.,(Such activities are also promoting entrepreneurship in these sectors) shall be entitled for EPCG Scheme. Agri exporters shall be entitled for recognition as Export House/Trading House/Star Trading House/Super Star Trading House on achieving a third of the threshold limit prescribed for exporters of goods.

Agribusiness v/s Agripreneurship Unlocking the Potential of India's Agribusiness through Agripreneurship

For India to achieve 8-10 per cent growth, agriculture must improve its growth rate to over 5 per cent. If more than 7 per cent is not reached, no serious reduction of poverty can be achieved. India has major advantages including diverse agro climatic regions. On the supply side, India has a large pool of research manpower, large marketable surpluses and abundant raw materials for agribusiness. On the demand side, there is a huge growing market with consumption growth predicted to increase to 7.8 per cent. The problems of the agricultural sector include low levels of productivity, infrastructure and investment. There are weak linkages between research and extension, low seed replacement rates and an imbalanced use of fertilizers. Limited access to credit, insufficient risk cover, and quarantine and sanitary and phytosanitary standard infrastructure are other inadequacies of the system. The government is trying to improve the environment for investment. It intends to spend Rs 15,000 crores in the perishable sector on the backward and forward production linkages in horticulture. Priority sector lending and credit are expected to double in three years. The government is bringing 10 million hectares of additional land under irrigation. India is on track to becoming the world's leading food producer through a number of cross-sectoral initiatives and investment opportunities. Percy Barnevik, Honorary Chairman, Sandvik, Sweden, stated that the biggest potential in India is to mobilize 350 million people who live on less than one dollar a day. Mobilizing them into entrepreneurship will make a tremendous difference to India's economy. An NGO in Tamil Nadu has helped 80,000 women organize themselves into self-help groups. Thirty-five per cent of these 5,000 projects are agri-projects and 10 per cent are agri-related industrial activities. This programme, when replicated, may reach 1.3 million women and help 4 per cent of the state's population. Such projects provide social and economic empowerment to women and will also delay the migration to slum areas. Reema Nanavaty, Director, Rural Development, Self- Employed Women's Association (SEWA), India, said that farmers' contributions need to be recognized and should convert into agripreneurship. Agriculture is seen in a traditional manner and is not treated as an industry. Therefore, it is not attracting enough investment. There is a need for collective and joint cooperation among farmers to enter into farming and marketing contracts. Through the Agri-export Zone network, such initiatives reach 700,000 households. It is necessary to have agriculture stabilization inputs and funds, a mechanism that provides solutions to marginal farmers in order to absorb such shocks as erratic rainfall. Farmers need to be given direct market access through collectives and cooperatives.

Results produce by Agripreneurs in sixty Agri-Export Zones all over India. These zones are set up on the basis of produce.

Results of Last Five Years Affords

1. Actual Export done by Agripreneurs In the year of 2005-2006 Rs. 2107.00 Cr. (210.00 Million US Dollar) from Agri Export Zones
2. Actual investment made by only Agripreneurs in the same period Rs.197.30 Cr. (19.3 Million US Dollar) in Agri Export Zones

The 60 AEZs sanctioned by the Government envisaged an investment of Rs. 1717.95 crores and export of Rs. 11821.47 crores over a period of 5 years. Against these projections, these AEZs could mobilize an investment of Rs. 449.90 crores and exports of Rs. 2768.00 crores over the last three and a half years ending 31 March 2005.

The encouraging results of goal-oriented "Green Revolution", "White Revolution", "Yellow Revolution" etc. enthuse the agricultural fraternity of the country to set a new goal for "Agri-Export Revolution" which is not only the need of the hour but also a compulsion to strengthen and revitalize the economy of the country. While India holds an important position in the export market for a set of traditional agricultural commodities, new areas and new commodities are likely to emerge such as live animals and animal products, fruits, vegetables, floriculture, medicinal plants and processed agricultural products. In order to understand the emerging challenges and to identify ways and means of placing our country as a major player in the global export market for agricultural products

Solutions for Agri-Entrepreneurs

Over the years, problems of unemployment, underemployment and disguised unemployment have plagued the country, especially the rural populace. Considering that two-thirds of the Indian population is employed (either directly or indirectly) in the agriculture sector, providing viable and sustainable business opportunities in Indian agribusiness is imperative for generating employment in the country. Sustainable agribusiness ventures can promote job-led economic growth in rural areas through harnessing science and technology for environmentally sustainable and socially equitable development. Entrepreneurship in sustainable agriculture is a key factor which has the potential to transform the face of rural India, the prime driver of our economy.

Holistic Approach

A holistic approach for cultivating entrepreneurship amongst rural communities at large is essential. This can be done through sustainable livelihoods that can be generated in the agriculture sector by providing business solutions that envelop natural resource management and innovative eco-technological principles. The 'Sustainable Livelihoods' approach, values wide range of activities people do to make a living, as well as assets and capabilities, they possess. In other words, the root of human development and economic growth is "livelihoods" rather than "jobs." While entrepreneurship has widely been viewed by policy makers and educators alike as the creative act of an independent businessperson, we are now learning that the roots of entrepreneurship extend deep into a community's civic structure.

Enterprise Environment

Entrepreneurship Principles

"The Seven Principles of Entrepreneurship":

1. Ski with your knees bent.

2. Refine the skill of falling down.
3. Get comfortable with "close enough."
4. Be happy with a "conditional yes."
5. Remember that business model innovation is often as important as tech innovation.
6. Think small.
7. Strive to understand and mitigate risk.

Let's examine these principles in detail:

Ski with Your Knees Bent

Those of you who enjoy downhill skiing know that one of the first principle of survival is to keep your knees bent and flexible and your center of gravity low. This style enables you to adjust to change – on the slope, in surface conditions or with obstacles or other skiers – and still achieve your goal of gracefully traversing the hill. Conversely, skiing with locked knees, a rigid posture and a fixed gaze is a formula for disaster.

It's too easy, when working in an established organization, to develop certain rigidity in how one approaches decision-making and day-to-day operations. Most times, you can get away with it, standing upright, knees locked, eyes trained straight ahead. Why? Because established businesses necessarily develop standard operating procedures, and many times little changes day-to-day. The rigidity can creep up on you. Sameness and predictability are comforting, and it's human nature to embrace and standardize behaviour that succeeded in the past.

By contrast, entrepreneurship is, metaphorically, a bit like skiing moguls (big, scary, unpredictable bumps)... blindfolded. If you're an entrepreneur, you have to keep your knees bent. You have to stay loose. You know full well that things will change, probably dramatically, and that you'll experience dramatic shocks; you just don't know exactly what those shocks will be, where they'll come from or when they'll occur.

Keeping loose and with a low center of gravity helps business managers absorb change and keep the business on its feet. Entrepreneurs have always operated this way as a matter of course. And this sort of flexibility and adjustability can be a crucial advantage for corporate executives as well, whether in accommodating change in existing markets or tackling new business initiatives.

Refine the Skill of Falling Down

To continue the skiing metaphor, one of the first things a ski instructor teaches novices is how to fall. Why? Because it's an inevitable part of the sport, and it's the primary way of getting hurt, but good skiers fall gracefully and bounce back quickly.

Similarly, successful entrepreneurship requires getting comfortable with the idea of falling down repeatedly and springing back up each time. Startup business is all about expecting, gracefully accommodating and learning from failure. After all, even with the best-thought-through venture, it's reasonable to expect that 50 per cent of the original business plan will prove to be wrong. Worse yet, you won't know which 50 per cent until you get into it – until you point your skis down the hill and go.

Understanding this phenomenon is why venture investors often prefer to invest in entrepreneurs who've experienced failure. It's also why many startups prefer to hire, as key managers, individuals who have experienced the good and the bad of a previous startup or two. A previous fall or two is not considered a scarlet letter of failure on a person's career, but rather an indication of maturity and a willingness to take calculated risks.

For established organizations to successfully grow through innovation, they must delve into less certain and more ambiguous environments. Therefore, they need to take more calculated risks without being paralyzed by fear of failure. They need to refine the skill of falling down.

Get Comfortable with "Close Enough"

The vast majority of corporate innovations never see the light of day because they're "killed in committee." Why do so many die that way? Because innovative ventures and business initiatives almost always have too many unknowns for many people's comfort, and the powers that be in established companies – often groups or committees – possess the power to say "No" based on that uncertainty. Just as, in the old days, IT executives knew they'd never get fired by buying IBM, with corporate innovation it's nearly always a safer bet to say "No" to something new.

Meanwhile, successful entrepreneurship – or corporate venturing and new-business-development – requires operating in a highly uncertain, ambiguous environment. It's a bit like trying to solve an algebraic equation with seven variables and six unknowns. Technically, it can't be done, so the "correct" answer is, "We can't do it." The equation can't be solved without taking intelligent guesses and trying out different combinations based on inadequate information, approximations and instinct.

But the perfect is often the enemy of the good. Remember, in entrepreneurship: The best decision is the perfect decision (which you'll never have sufficient information or time to divine). The next best decision is "close enough" and get moving – you can always adjust course as you go (*i.e.*, ski with your knees bent). And the worst decision of all is to continue to study, or form a committee (which so often translates to the "safe no," and therefore doing nothing).

The entrepreneurial approach accepts "close enough": Roll up your sleeves and work with customers from the start. Get something in customers' hands, even if it's not finished. Experiment, and don't be afraid to adjust, occasionally fall down and get back up. Do it, try it, fix it... and repeat.

Be Happy with a "Conditional Yes"

The tendency in big organizations is to seek budget approval for an entire multiyear project upfront. After all, nobody wants to launch into building, say, a $275-million plant when they only have corporate funding approved for the first $30 million for planning and site prep. The problem is when we see corporate new-business-development folks trying to apply this upfront approval formula to venturing.

In the entrepreneurial world, nobody expects to receive 100 per cent funding upfront; it just doesn't work that way. With independent ventures, investors believe in "milestone investing," progressively meting out capital sufficient to fund the next 9 to 18 months of activity and the achievement of the next crucial value-building milestones. For instance, it's not uncommon for a venture requires a total of $15 million in investment capital in order to reach self-sustaining profitability to seek seed funding of only a million dollars or less to build a prototype and do some preliminary testing. A subsequent "A" round may be for just a few million dollars to enable the venture to build a team, productize the technology and sign up the first few customers. And so on. Typically, early-stage investors are keenly interested in continuing to participate in subsequent rounds; they just like to see incremental progress along the way.

Internal corporate ventures – and here we're referring to risky ones entailing new technologies, new business models and/or new markets, not capacity-expansion projects and the like – should approach funding with the same venture-funding mentality. Remember the principle we mentioned in our last issue: that 50 per cent of a new venture's business plan will inevitably prove to be wrong, you just don't know which 50. If that holds true, it only makes sense for the parent company (playing the role of venture capitalist) and the internal startup team to agree on funding increments and associated milestones rather than the all-in approach. Less capital is committed, inevitable mistakes or discoveries are less costly and more easily accommodated, and the new business remains nimble.

Remember that Business Model Innovation is often as Important as Tech Innovation

We've never seen any statistics or studies in this regard, but it sure seems that the majority of shareholder value created over the last half century had a lot more to do with companies innovating around their business model than around technology. Think of eBay with online auctions. Store brands and generic drugs. Amazon cutting out the retail middleman. Manufacturers asking suppliers to co-locate. Dell building PCs to order. Social networking typified by sites such as MySpace and Facebook. The way HMOs and PPOs fused insurance and healthcare delivery. Sure, in many cases, technology was involved, but technology was not the strategic driver that created shareholder value. Instead, it was creativity applied to the business model (product/service mix, value proposition, channels, pricing) that made the difference. This kind of thinking needs to be applied not only by entrepreneurs but by corporate new-business professionals as well.

Think Small

An executive in a tech startup, recently hired away from a Fortune 500 company, unfortunately brought his big-company thinking with him. Inheriting management responsibility for a professional services operation of about 50 people growing at over 50 per cent annually, he saw a crying need for more coherent project management. His solution? Call in the vendor who'd provided similar software and services to his last employer, and get a quote. The result? A half-million-dollar expense where cloud-

or PC-based project management software and rigorous management communication would have sufficed nicely; worse yet, the expensive "solution" never worked. The manager was fired and the system scrapped for a simpler approach.

Too often, we see established organizations trying to innovate and getting caught in this big-company, go-big-or-go-home mentality. I observed one internal corporate venture of a multinational tech company spend millions on PR – because "That's how we do things at XYZ Corp." – before they'd even fully defined their product, value proposition, positioning and go-to-market strategy. Bizarre. Entrepreneurship, even if it's taking place under the corporate umbrella, calls for small, inexpensive, rapid-turnaround experiments and trials. "Thinking small" doesn't mean that you don't have big aspirations for your new venture. (Indeed, I tend not to think of startup ventures as small businesses; we think of them as global enterprises that happen to be young.) But by iteratively discovering what works and what doesn't, you'd be surprised how far you can get on how little capital.

Strive to Understand and Mitigate Risk

Contrary to popular belief, entrepreneurs and venture investors are not risk-seeking nuts, the business equivalent of helmet-free bungee jumpers. In fact, the best ones are remarkably risk-averse, skilled at identifying and mitigating venture risk. Whether they do it intuitively or explicitly, A-list venture folks are constantly working to wring risk – whether it's product risk or risk of a market, financial or management nature – out of their startups. There's a method to their madness that corporate startups need to apply.

"The Seven Principles of Entrepreneurship" are enumerated here to stimulate and challenge the thinking of corporate managers. In many cases, entrepreneurial behaviors that may, on the surface, seem to be inappropriately risky turn out, on closer examination, to be worth emulating if done in a thoughtful manner. And emulating certain entrepreneurial behaviour can help corporate executives excel, particularly when it comes to launching new initiatives or product lines, trying out new business models or entering new markets.

Entrepreneurship Leadership

Rarely can entrepreneurs make a company succeed by themselves. This is much like the fact that the greatest athlete doesn't ensure that their team will win if the other players cannot perform. As such, entrepreneurs need to be able to identify staffing needs, expertly fill them, and lead the team to success. Leadership is the process through which an entrepreneur is able to influence employees to achieve the objectives of the organization. To be an effective leader, an entrepreneur must build trust and confidence among employees.

Communicate Effectively with them

Leaders can build trust in many ways. They can achieve it by working hard, maintaining a constant message and/or being available to solve employees' problems. By showing employees that they are fully committed in achieving the vision, entrepreneurial leaders build trust and confidence in employees. This in turn yields high employee satisfaction and commitment.

Effective communication is equally critical to successful leadership. If employees are unclear about the company's vision, and/or receive mixed messages over time, they will be unable to focus all of their efforts on achieving the company's goals. Conversely, by delineating the company's vision and goals, and reinforcing them over time with the same message, the company's goals become engrained in its employees. In addition to building trust and effective communications, other keys to entrepreneurial leadership include the following:

- **Seeking self-improvement:** A great leader always seeks to become even better.
- **Possessing technical skills:** While the leader may not need to have the greatest technical skills in their organizations, they need to be savvy enough to lead the team.
- **Accepting responsibility for actions:** Leaders and companies always make mistakes. Great leaders don't place blame on others.
- **Making decisions:** Good leaders must make good and timely decisions.
- **Being a role model:** A leader must set an example to his employees and guide them to excel.

The ability to lead effectively is a crucial factor in the success, or lack thereof, in entrepreneurial ventures. By understanding and embodying what it takes to lead effectively, entrepreneurs can maximize their chances of success.

Entrepreneurship Motivation

We believe that the study of the entrepreneurial process is important for several reasons. First, entrepreneurship drives innovation and technical change, and therefore generates economic growth (Schumpeter, 1934). Second, as the Austrian economists have explained, entrepreneurial action is the process through which supply and demand are equilibrated (Kirzner, 1997). Third, entrepreneurship is an important process by which new knowledge is converted into products and services (Shane and Venkataraman, 2000). Fourth, entrepreneurship has become an important vocation and we need to understand its role in the development of human and intellectual capital (Zahra and Dess, 2001).Here, we adopt Shane and Venkataraman's (2000, p. 218) definition of entrepreneurship as the process by which "opportunities to create future goods and services are discovered, evaluated, and exploited." As these authors have explained, this definition does not require viewing entrepreneurs as the founders of new organizations.

1. An options trader can be an entrepreneur, as can a corporate salesman who discovers and pursues opportunities for the creation of new products.
2. Moreover, this definition shows that entrepreneurship is a creative process. By rearranging resources in a new way, entrepreneurs engage in creative activity. However, the degree of creativity involved in entrepreneurship varies across the types of resource recombination that occurs. For example,

the creation of business to undertake space tourism may be more creative than the creation of a new restaurant in a strip mall.

Entrepreneurship involves human agency. The entrepreneurial process occurs because people act to pursue opportunities. People differ in their willingness and abilities to act on these opportunities because they are different from each other. We argue that the variation among people in their willingness and ability to act has important effects on the entrepreneurial process.

1. By this definition, we do not imply that the study of firm formation is unimportant. We believe that firm founding entrepreneurship is an interesting and important special case of entrepreneurial activity and one that is in need of greater study.
2. Interested readers might note that this definition is consistent with several classical definitions of entrepreneurship, including those of Kirzner (1973) and McClelland (1961) who argue that people other than business owners or founders can be entrepreneurs.

   S. Shane *et al.*/Human Resource Management Review 13 (2003) 257–279 259. Recent research has been relatively accepting the arguments that people vary in their willingness and ability to engage in the entrepreneurial process because of non-motivational individual differences. Researchers have shown that the willingness of people to pursue entrepreneurial opportunities depends on such things as their opportunity cost (Amit, Meuller, and Cockburn, 1995), their stocks of financial capital (Evans and Leighton, 1989), their social ties to investors (Aldrich and Zimmer, 1986), and their career experience (Carroll and Mosakowski, 1987; Cooper, Woo, and Dunkleberg, 1989).

   We argue that motivational differences also influence the entrepreneurial process. For example, such things as variation across people in their perceptions of risk and opportunity influence entrepreneurial decisions (Shane and Venkataraman, 2000). People vary in how they view the risk of expending resources before knowing the distribution of outcomes (Palich and Bagby, 1995). Similarly, the probability of success at the entrepreneurial process is low, and those people who are willing to proceed despite these odds might be more optimistic or higher in self-efficacy than people deterred by these odds. We consider the incorporation of this individual-level variation in motivation to be important to the entrepreneurial process. To isolate the effects of entrepreneurial motivation, other factors that could have a causal effect on the process and outcome of entrepreneurship need to be controlled. One category of control variables is the external environment. These would include such things as (1) political factors (*e.g.*, legal restrictions, quality of law enforcement, political stability, and currency stability); (2) market forces (*e.g.*, structure of the industry, technology regime, potential barriers to entry, market size, and population demographics); and (3) resources (*e.g.*, availability of investment capital, labour market including skill availability, transportation infrastructure, and complementary technology). Most researchers either explicitly or implicitly agree that these categories of factors influence the entrepreneurial process and need to be controlled to measure the effect of motivations on the entrepreneurial process.

3. While it is clear why the above factors need to be controlled if we are to fully understand how motivation is related to entrepreneurship, it is less clear how opportunities affect this relationship.

Agri-preneurship and India

Agri-business and Human Resources

Emerging Trend: Why Agriculture to Agri-business?

- ☆ Increasing demand for organic/quality food both in India as well as Abroad. Market growth of around 15-25 per cent per year.
- ☆ Competitive advantages for many primary production activities in agriculture. Rain-fed farming, tropical fruits and vegetables, livestock, animal husbandry, aquaculture, wild craft, etc. are produced through real low cost production methods.
- ☆ Private sector is willing to enter into agri-businesses at all levels of operations. Changing consumer demand and retail revolution have open the doors for investments by private sector in agri-agribusinesses like Reliance, Bharati, Pantaloon, Carrefour, etc.

Key Issues: Agriculture to Agri-business

- ☆ Policy issues: From absolute control and management of agriculture by Government, today it is being opened to public-private partnerships.
- ☆ Production technology issues: From input/s oriented technology development for increasing production, today practice based value addition is being promoted.
- ☆ Quality and certification issues: Demands from consumer/s for better quality has forced Government/s to establish regulatory mechanisms for quality certification.
- ☆ Logistics and supply chain issues: Modern retail formats requires efficient and dedicated supply chain management facilities.
- ☆ Human resource issues: Lack of appropriately trained human resource is today considered as the biggest constraint in conversion of agriculture to agri-business.

Key Challenges for HR Skill Development in Transforming Agriculture to Agri-business

- ☆ Agriculture to agri-business potential has neither been recognized by the academic/s nor is being promoted as a career making opportunity among practitioners.

- ✰ Region-location specific agri-business opportunities need to be identified and accordingly suitable human resource training programs should be launched.
- ✰ Learning processes for human resource to be engaged in agri-business enterprises should be designed to suit the candidates as per the job requirements.
- ✰ The focus to upgrade the skills of conventional practitioners of agriculture into agri-business human resource will provide the missing links.

HR Skills - Present Status of Availability: Agriculture to Agri-business - Case Study

Presently Complete Agri-business Sector is Starved of Appropriately

Sl.No.	*HR SkillLevels*	*Export House*	*Processing Unit*	*Aggregator*
1.	**Top Management**	Mostly managed by the owner/s.		
2.	**Senior Management**	Mostly senior managers with qualifications and experiences in marketing, finance, logistics and distribution management are engaged in agri-businesses		
3.	**Middle Management**	Mostly managers with hands-on experiences in respective functional areas are engaged by the agri-business enterprises.		
4.	**Junior Management**	Professionally qualified fresh graduates and post graduates in management are recruited	Graduates and post graduates with science background are recruited	Persons having some relevant experiences are employed
5.	**Senior Supervisory Management**	Any person seeking a job is generally recruited and trained on the job.		
6.	**Junior Supervisory Management**	Only temporary and stop-gap human resource is engaged on as and when required basis		
7.	**Entry Level Positions**	Candidates with some educational background but mostly without any relevant skills are engaged		

Trained Human Resource

Trained Man Power - Demand Assessment: Agriculture to Agri-business

Sl.No.	*HR SkillLevels*	*Small Enterprises*	*Medium Enterprises*	*Large Corporation*
1.	**Top Management**	Every year about 100,000 small entrepreneurial units are expected to come into existence. They all need appropriate training and skill Development support through formal and informal courses.	Presently employed Senior Management Staff (about 1000 persons every year) needs executive Development programs.	
2.	**Senior Management**			
3.	**Middle Management**		Presently employed and to be employed management staff (about 5000 persons every year) would require both structured class room as well as distance learning programs.	
4.	**Junior Management**			

Contd...

Table–*Contd...*

Sl.No.	*HR SkillLevels*	*Small Enterprises*	*Medium Enterprises*	*Large Corporation*
5.	**Senior Supervisory Management**	3-6 months full time courses for 20,000 higher secondary plus candidates every year	6-12 months full time courses for 50,000 higher secondary plus candidates every year	
6.	**Junior Supervisory Management**	6-12 months full time courses for 100,000 secondary plus candidates every year.		
7.	**Entry Level Positions**	3-6 months full time courses for 100,000 candidates from farming background (irrespective of their education qualifications) every year.		

There is an urgent need for professional training courses to meet this ever increasing demand for trained man power

Status of Present Day Education in Agriculture

- ☆ Education is based on scientific developments carried out from 1950s to till date. Mostly candidates are trained in input based technologies and extension works.
- ☆ Focus on external inputs for increasing productivity. The skills for enabling value addition in agriculture are not imparted.
- ☆ Focus on government jobs (extension and research). Managerial skills are not included in the curriculum/s.
- ☆ Absence of value addition and supply chain management subjects. Even trained candidates do not learn the skills to manage agri-business enterprises.
- ☆ Accessible to careerists. For very long candidates joining these courses have mainly been attracted because of government jobs with total exclusion of practitioners wanting to create their own agri-business enterprises.

Moorpark Foundation's Initiatives: Agriculture to Agri-business

- ☆ Developed implementable models for conversion of agriculture to agri-businesses covering complete value chain. Over 70 different types of agri-business enterprises has been created.
- ☆ Evolved methodologies for entrepreneurial initiatives for agri-business enterprises. Established linkages for resource mobilization with banks and marketing channels.
- ☆ Transformed conventionally educated human resource into agri-business managers. Both structured and unstructured learning methods have been developed and implemented.
- ☆ Innovated human resource skill development programmes for HR needs in agri-businesses. Currently running formal courses for five different needs of trained man power.

Morarka Foundation's Initiatives: Innovations in HR Skill Development

- Initiated entrepreneurship in agriculture: Over 1000 small entrepreneurs have been assisted in 15 States of the country to set up sericulture units in last eight years.
- Implemented AC and ABC scheme: Trained 265 agri-graduates to set up their own small agri-business enterprises.
- Collaborated with MPUAT for Post Graduate Diploma in Organic Agriculture: Completed three Batches (56 candidates) with 100 per cent placements record.
- Launched Certificate Course with Rajasthan Vidyapeeth: Completed three batches (90 candidates) with 100 per cent placements record.
- Joined hands with National Productivity Council: Two Batches (53 candidates) are undergoing training.

Offers from Moorpark Foundation: Structured and Unstructured Learning

- Moorpark Foundation is positioning itself as a capacity building institution to meet the growing demand for trained human resource by the agri-businesses by offering short duration courses (through both structured and unstructured learning) comprising of formal classroom training as well as on the job practical training.
- Moorpark Foundation has joined hands with academic/s to provide course material/s, faculty training/s, practical-hands on work experience, for candidates enrolled by them for conversion of young human resource into agri-business professionals.

Components of HR Course/s: Agriculture to Agri-business

1. Basic Courses

- Organic/Quality Agriculture Practices
- Agriculture Product Marketing
- Food Quality Management
- International Marketing
- Technologies for Organic/Quality Production
- Value Addition in Agriculture
- Agripreneurship Development and Management
- Logistics and Supply Chain Management
- Organic/Quality Standards and Certification

2. Business Skill Courses

- ☆ Basic Language Skills
- ☆ Effective Communication
- ☆ Personality Development
- ☆ Business Soft Skills
- ☆ IT-ICT Application

Chapter 2

Entrepreneurship Development and Management: Operational Concept and Dimension

Sankar Kumar Acharya and Manas Mohan Adhikary

What is an Enterprise?

An enterprise is a continuous process of transforming any activity or investment into an innovative venture generating relentlessly both jobs and ideas, resource and value.

What is a Business?

A business is any kind of transition with in a stipulated goal and objective.

Innovation and Enterprise, the prime mover for any enterprise building is innovation. Otherwise, enterprise is getting anything done newly or innovatively.

The word enterprise is, at present, much used in a variety of context and with a wide range of meanings. Within this range there are narrow meanings of word specifically related to business, and there are wider meanings indicating a way of behaving that can apply in a variety of context, including business. The narrower meaning are closely linked to entrepreneurship, and in term, the concepts of enterprise and entrepreneurship embraced much that would be considered to be expression of small business activity.

Indeed, the words enterprise, entrepreneurship and small business often appeared to be used interchangeably. The emerging enterprise domain is all set to embrace the following humane and managerial skill.

- ☆ Communication Skill

The Role of Innovation

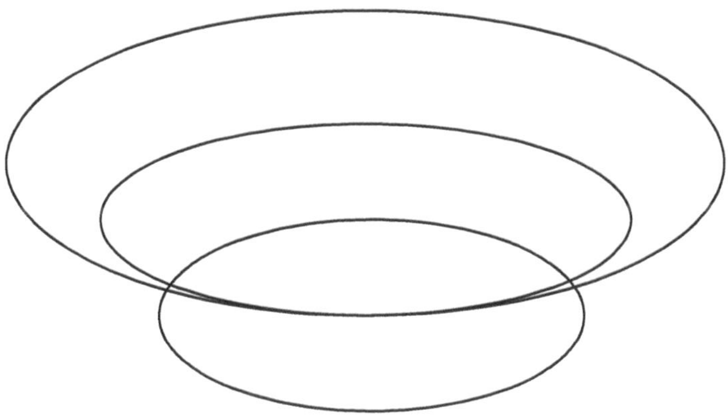

- ✰ Group work skill
- ✰ Personal Skill
- ✰ Organizational skill
- ✰ Interpersonal skill
- ✰ Problem solving
- ✰ Social and Community Awareness
- ✰ Resource Management Skill

"One can be enterprising both by making a million before one's fortieth but they and by shepherding passengers out of burning Aeroplane" Coffield and Mcdonald (1991)

"Enterprise is having ideas and imagination and using them" Marks and Spencers.

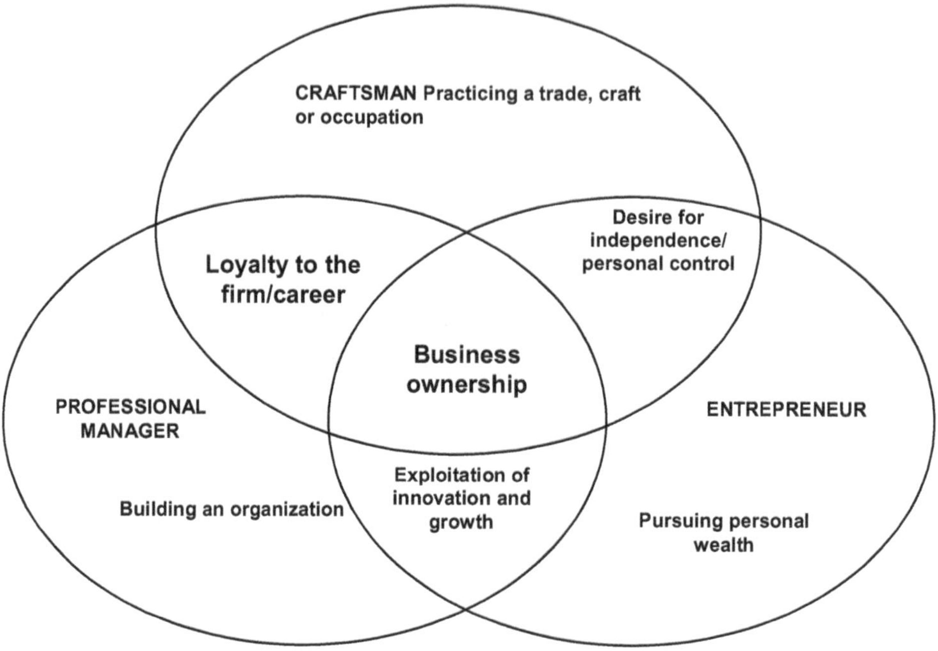

The Networking of Enterprise Personality

Approaches to Enterprise

Entrepreneurial Model	*Focus*	*Assumption*	*Skill*
Great person school	The entrepreneur has an intuitive ability	Without inborn in intuition the individual would be like other mortals	Intuitions, vigour, energy, self esteem
Management School	Organizers of an economic venture	Entrepreneur trained in technical function of management	Planning, organizing, capitalization, budgeting etc.
Leadership School	Individuals are leaders of people	An entrepreneur can not accomplish his/her goals alone, depends on other	Motivating, directing and leading
Intrapreneurship School	Entrepreneuranal skills can be use full in complex organization; intrapreneurship is the development of independent unit to create market and expand services	Organization needs to adopt to survive; entrepreneurial activity leads to organizational building an entrepre-neurs becoming managers	Alertness to opportunity, maximizing decision

Entrepreneurial or Enterprising Attributes

- ☆ Autonomy/independence
- ☆ Belief in control of one's own destiny
- ☆ Creativity
- ☆ Determination
- ☆ Flexibility
- ☆ Goal orientation
- ☆ Hard work
- ☆ Imagination
- ☆ Initiative
- ☆ Leadership
- ☆ Risk taking
- ☆ Achievement motivation
- ☆ Optimism
- ☆ Perseverance
- ☆ Persuasion
- ☆ Problem solving ability

If enterprise, or entrepreneurship, is the exercise of a set of attributes or behaviours then those attributes can also be illustrated by the process by which they might, or might not, be developed. Gibb has analyzed and described the relevant attributes, ideas, ambitions and culture and their development utilizing a number of diagram.

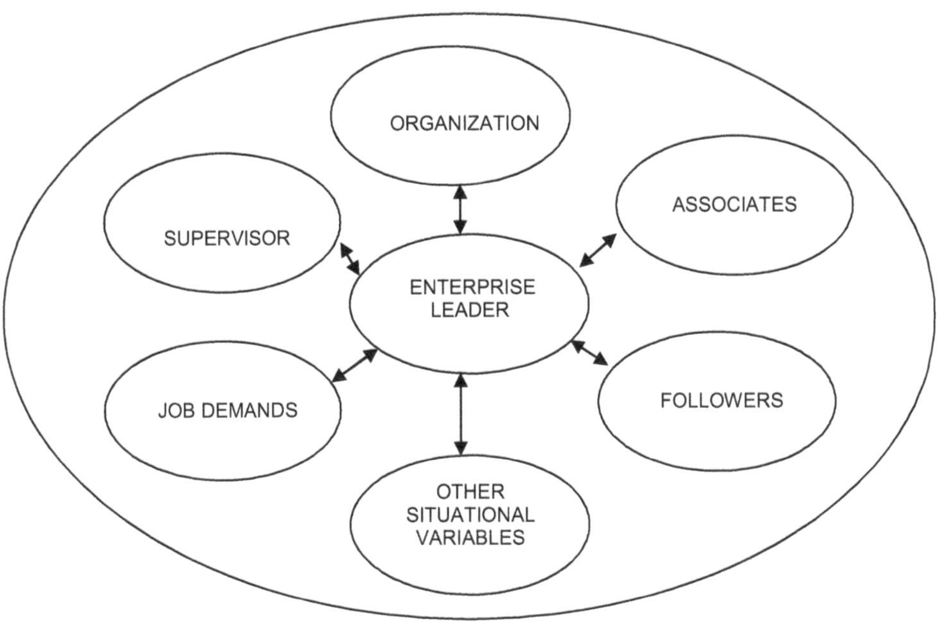

There are considerable variations in the levels of enterprise and the rate of emergence of entrepreneurs found in different countries and cultures. The factors which have been identified as influencing enterprise are:

- National and Social Culture
- Economic conditions such as market economy
- Political conditions such as Governance, administrative and regulatory regimes
- Demand side factors, such as technological development, globalization, economic development and industrial structure
- Supply side factors, such as population growth and density, age structure, immigrations, participation rates, income level and unemployment
- Intervention by Government and Other agencies
- Equilibrium rates of enterprise
- Other influences, including social and cultural enterprise

Integrated theories are advanced to explain the influence of such factors, and those internal factors. A significant development has also been the work of global enterprise monitor which has explore the extend of entrepreneurship in many different countries and its relationship with national economic development and growth.

Collectivism to Individualism: The Unique Entrepreneurial Traits

Collectivist	*Individualist*
Collective interests prevail over individual interests	Individual interests prevail over collective interest
Private life is invaded by groups	Everyone has a right to privacy
Opinions are predetermined by group membership	Everyone is expected to have a private opinion
Economy based on collective interests	Economy based on individual interest
Ideologies of equality prevail over ideologies of individual freedom	Ideologies of individual freedom prevail over ideologies of equality
Harmony and consensus in society are ultimate goals	Self-actualization by every individual is an ultimate goal

References

R. Burrows: Deciphering the Enterprise Culture; Entrepreneurship, Petty Capitalism and the restructuring of Britain (London 1991).

A. Morrison: Entrepreneurship, an International Perspective.

T. Garrison: International Business culture.

Bridge O'Neill and Cromie: Understanding Enterprise, entrepreneurship and Small Business.

Chapter 3

Quality Issues in Horticultural Production for both Domestic and International Market

Sisir Kumar Mitra

Concepts of Biodiversity

"Biodiversity" is often defined as the variety of all forms of life, from genes to species, through to the broad scale of ecosystems (Gaston, 1996). "Biodiversity" was coined as a contraction of "biological diversity" in 1985, but the new term arguably has taken on a meaning and import all its own. While the history of this term is relatively short (compared to other terms covered in the *Biological Abstracts*), it already has raised important, distinctive, philosophical issues. Overall, the major issue for biodiversity is how its conservation may be integrated with other needs of society. In fact, "biodiversity" is now used sometimes to mean "life" or "wilderness" or other conservation values. "Biodiversity" also has served on occasion as a catch-all for "conservation" itself.

The nub of the problem of defining biodiversity is that it is hard to exclude anything from a concept that is taken so easily to mean "everything". Callicott *et al.* (1999) examined "biodiversity" as one of the current normative concepts in conservation. They concluded that it remains ill-defined, and that distinctions can be made between "functional" and "compositional" perspectives in approaching biodiversity. "Functional" refers to a primarily concern with ecosystem and evolutionary processes, while "compositional" sees organisms as aggregated into populations, species, higher taxa, communities, and other categories.

There is a parallel between the debates about the definitions/concepts of species and of biodiversity. Despite a wide range of usage, biodiversity emerges as a concept linked primarily to the idea of biological variation that is largely unknown in its extent, and its future values. Any "calculus" of biodiversity providing quantitative estimates of this unknown variation automatically provides at the same time a measure of its central values - option values and intrinsic values. Such values broadly reflect values of elements of biodiversity having unknown present value. These quantified values typically will not be in conventional units (*e.g.* dollars), but nevertheless can be balanced with other values of society. Decision making (for example, deciding whether we should invest in conservation of area A or area B) may require only estimates of *relative* gains in represented variation offered by different places (their "complementarily" values). Complementarily helps integrate biodiversity option values with other values attributed to biodiversity, and with values of society more generally. This integrative process, together with processes for the growth of knowledge about components of biodiversity, provide an alternative to the "post-positivism" perspective that sees biodiversity conservation as predominantly value-laden. The perspective that biodiversity reflects option and intrinsic values, to be balanced with other values, appear to be compatible with the broader discipline of conservation biology: "the field is rooted in a philosophy of stewardship rather than one of utilitarianism or consumption. The latter has been the basis of traditional resource conservation that is, conserving resources solely for their economic use and human consumption" (Meffe, 2000).

Agricultural/horticultural biodiversity refers to the variety and variability of plants, animals, microorganisms on the earth that are important to food and agricultural, which results the interaction between the environment, genetics resources, management systems and practices used by mankind. Biodiversity takes into account not only genetics species, but also cultural diversity which influences human interactions at all levels. It intrigues production of food, fodder, fibre, fuel and pharmaceuticals for the nutritional security and over all human development (Mathewkutty, 2004).

Values of Biodiversity vis-à-vis Horticulture

World Food Day theme for 2004 was "Biodiversity for Food Security", reflects the richness of our planet as well as tribute to biodiversity's role in ensuring that people have sustainable access to enough high quality food to lead active and healthy lives.

Food Supply

Virtually 100 per cent of the protein from domesticated animals consumed by people comes from nine species (cattle, pigs, sheep, goats, water buffalo, chickens, ducks, geese and turkeys), while fish are becoming a new kind of domesticated animal through the development of aquaculture techniques. Only a very small proportion of the world's plants have been used for food on a large scale. About 10-50,000 are thought to be edible, but only about 150 are used as human food. As economies have become more global, man has concentrated on fewer species so that today, 90 per cent of the world's food comes from 15 species. Three of them - wheat, corn, and rice - supply two-thirds of this amount. Although there are over 10,000 species of cereals, no new ones have been brought into cultivation during the past 2000 years.

India is a land of nearly 33 lakh sq. km with more than one billion people, a coastline of 7616 km, 14 different types of climatic forests covering 650,000 sq. km, 17,000 species of flowering plants, 20,000 species of fungi, 50,000 insects, 65,000 species of fauna including 2,000 species of fishes, 1200 of birds, 350 of mammals and 420 of reptiles. There are 70,000 wetlands and 28,000 km of rivers. The "Great Irish Potato Famine" is the historical evidence of the dangers of relying on too few crops. Potatoes were introduced into Ireland from the New World in about 1600 and eventually most of the Irish people became dependent on this one crop. But in 1845-1847, the wind-borne potato blight fungus spread throughout the country and caused almost complete failure of the potato crop which took a massive tool of 1 million people died of starvation. Similarly, the banana and plantains is a staple food crop of West and Central Africa but the rich source of diversity of wild species are present in the Asia and Pacific region where the genus is originated.

Fortunately, the vast store of species in the tropics is the obvious place to look for new potential crop plants. At least 1650 known tropical forest plants have potential as vegetable crops. The botanist N. Vavilov identified the major areas of high diversity of crops and of the origins of food crops around the world. Most of them are in developing tropical countries. Some new tropical plant products (from tropical America) are gradually making their way on to the market, at least in health-food stores *viz.*, Uvilla (fruit), Lulo (fruit), Pupunha (fruit) and Guanabana (fruit).

Source of Gene Resources

Wild plants are also important sources of genes that can confer useful properties on our conventional crops. These genes from wild species usually have to be brought into the crop by hybridizing it with the wild species, then selectively breeding the hybrids. Wild plant species usually have a great deal of genetic variability, so that markedly different strains can be developed by selective breeding from one species. This is an important reason for conserving not only species, but a good sampling of the genetic variability within species *i.e.* samples from different locations, different subspecies, etc. some important gene sources have been found:

- ✰ A wild relative of the potato was found in Peru, and when it was hybridized with the standard crop plant a variety was obtained that was resistant to potato blight.
- ✰ A wild barley plant from Ethiopia provided a gene that protects the $160 million California barley crop from lethal yellow dwarf virus.
- ✰ Rice grown in Asia is protected from the four main rice diseases by genes brought in from a wild species from India.
- ✰ In both India and Africa, yields of cassava (tapioca) - one of the most important root crops throughout the tropics - were increased 18-fold because of disease resistance brought in from wild Brazilian cassava.
- ✰ The sugar cane industry in the U.S. was saved from collapse by disease-resistance genes brought in from wild Asian species.
- ✰ A wild tomato discovered in the Andes has been used to increase the sugar content of cultivated varieties, increasing their commercial value by $5-8 million per year.

About 2.5 million entries are kept in 700 seed banks world wide, but there has been only limited success in using these wild species for crop improvement. Most of the successes have been in disease resistance, and this is because disease resistance is controlled by one or a few genes, while the other characteristics, especially quantitative ones like annual yield or nutritional quality, are controlled by many genetic differences.

Availability rich gene source may likely to have greater impacts in horticulture, where even minor changes such as in colour, aroma, quality and post harvest behaviour would make significant commercial impacts. The newly developed technique of Genetic Modification (Genetic Engineering) makes it possible to introduce the desired genes into a species. The desired gene first inserted into an individual plant cell, and this cell is then grown into a clone in tissue culture, and the clone can be used to regenerate an entire transgenic plant. As of February 1999, at least 64 GM crop varieties had been approved in the USA and Canada, 20 in Japan but only eight in Europe. Many kinds of GM crops are being produced such as 'insecticidal crops'(resistant to insect pests), 'round-up ready crops' (resistant to herbicides), 'plants that immunize' (produce vaccine) and GM crops with 'terminator technology' (sterile seed transgenics).

Source of Biological Control Agents

On many occasions, man has introduced exotic animals or plants that have become serious pests. Usually, the best control methods have been to find the natural antagonists of the species in its original home. Some well known examples are:

- *Rodolia cardinalis* for controlling *Icerya purchasi* in citrus
- *Baculovirus* for controlling *Oryctes rhinoceros* in coconut
- *Trichogramma brasiliensis* for controlling *Helicoverpa armigera* in tomato

An agro-ecosystem very rich biodiversity ensures a perfect balance of all forms of living entity and no one selectively is allowed to dominate in the system. A number of bio-control agents have identified for management insect pests and diseases of horticultural crops.

Source of Pesticides

Many tropical plants produce chemicals that deter herbivores. Tropical indigenous people have discovered many of these plants, used them as poisons or medicines which later led to the development of different pesticides of today :

- Calabar bean was traditionally used as a poison in West Africa. Chemical studies of this plant led to the development of methyl carbamate insecticides.
- Daisy plants (*Chrysanthemum cinerariaefolium*) were first used centuries ago as a lice remedy in the Middle East, and this led to the discovery of pyrethrum insecticides.
- In South America, the natives use an extract of a forest vine to stun fish; this led to the discovery of rotenone, a biodegradable insecticide.

- ☆ The bacterium *Bacillus thuringiensis* produces toxic proteins that kill certain insects but are apparently harmless to humans. These are being produced and marketed as biopesticides.
- ☆ The Indian Neem tree has been found to be a source of the insecticide azadirachtin, as well as fungicides, spermicide, and agents potentially valuable in birth control.

Other tropical plants have been found to be toxic to leafcutter ants, mosquitoes, and other insects, and could lead to the discovery of other pesticides.

Source of Medicines

The potential for discovering medicinal compounds in wild organisms is enormous, and provides one of the most powerful arguments for conservation of biological diversity. This is especially true of tropical forests. The pharmaceutical industry is much more dependent on natural products than is generally realized. About 121 prescription drugs are derived from higher plants. Of the 121 drugs, 74 per cent were identified through native folklore. These include morphine, codeine, quinine, atropine, and digitalis (Table 3.1). Yet less than 1 per cent of rainforest plants have been tested.

Source of Anti-Cancer Drugs

In 1986, the National Cancer Institute started a new, extensive plant collection and screening program and has tested (on cultured cells) 35,000 species of higher plants as well as other organisms for anti-AIDS and anti-cancer activity. As of 1991, over 800 had shown some anti-HIV activity and 60 had shown anti-cancer activity (the HIV screens have been given higher priority). NCI is also supporting botanical exploration, research and training to increase the number of species to be tested. The collections have identified more than 40 new species of plants.

Table 3.1: Source of Medicines from Diverse Horticultural Crops

Purpose	*Drug*	*Source*	*Traditional Use*
Painkillers	Aspirin	Willow	+
	Codeine	Opium poppy	+
	Morphine	Opium poppy	+
	Cocaine	*Erthroxylum coca*	+
	Tetrodotoxin	Central American frog	+ (trance-inducer)
Antimalarial	Quinine	*Cinchona* (coffee family)	+ (Indian fever bark)
Amebicide	Emetine	*Cephaelis ipecacuanha*	+
Heart stimulants	Digitalis	Foxglove	+
	Ouabain	*Strophanthus gratus*	+ (arrow poison)
Pupil dilator	Atropine	Nightshade (Belladonna)	+
Antispasmodics for intestinal disorders	Scopolamine, Hyoscyamine	Nightshade (Belladonna)	+
Muscle relaxant	Tubocurarine	*Chondrodendron tomentosum*	+ (arrow poison)
High Blood pressure	Reserpine	*Rauwolfia serpentina*	+
Glaucoma	Pilocarpine	*Pilocarpus jaborandi*	+

For example (Table 3.2), the 'rosy periwinkle' was used in Cuba, Philippines and South Africa for the treatment of inflammation, rheumatism, and diabetes (*ethnobotanical knowledge*). In the late 1950s, vincristine and vinblastine were isolated from the periwinkle plant by Eli Lilly scientists and these chemicals were shown to have anti-cancer effects. Treatment with these drugs has increased the chances of remission to 99 per cent in childhood leukemia and to 70 per cent in Hodgkin's disease. Global sales of vincristine and vinblastine earn the Eli Lilly Company about $100 million each year.

Table 3.2: Source of Anti-cancer Drugs from Horticultural Plants

Drug	*Source*	*Traditional Use*
Cantharidin	Chinese Blister beetle	+ (abortifacient)
Etoposide, podophyllotixin, teniposide	*Podophyllum peltatum*	+ (snake bites, weakness, condyloma, lymphadenopathy, tumors)
Monocrotaline	*Crotalaria spectabilis*	+ (skin cancer)
Vincristine, Vinblastine	Rosy periwinkle	+ (diabetes)
Taxol	Yew (*Taxus brevifolia*)	_

Source of Fertilizers and other Useful Materials

Natural fixation of nitrogen by diverse species of leguminous plants, bacteria is a useful substitute of chemical fertilizers for crop plants. Fibre from silkworm, shellac from lac insects, adhesive, biopolymers and bio-diesel are some of the useful materials derived from natural source of diverse organisms. About 3 million tons of vegetable fats and oils are already used annually to supplement non-renewable fossil fuels. Further development of bio-fuel industries could help in reducing our dependence on non-renewable fossil fuels.

Source of Bio-fuel

- Jatropha – Oil from seeds is mixed with diesel (bio-diesel) and used for running engines.
- *Pongamia pinnata* - Oil from the seed of this tree is being used as a substitute for diesel in electricity generation in rural areas of the country.
- Patuá palm - The fruit produces an oil that is almost identical to olive oil, and is rich in protein as well.
- Babassú palm - Produces a coconut-like fruit rich in oil and protein. The tree grows well in deforested areas.
- Fevillea vine - Seeds have an extremely high oil content.
- Jojoba - A desert plant of the southwestern United States and northern Mexico that produces a "liquid wax" (chemically different from an oil), is very useful as a heat-resistant lubricant.
- Guayule – A desert plant, has a very high content of natural rubber and could help to relieve the dependence on rubber, grown in Southeast Asia.

Nurture Biometic Materials Science

Biometic materials science is the name given to attempts to mimic the materials found in nature. Much of it is funded by the Navy and the Air Force, because of the potential importance of lightweight but strong materials in aerospace engineering, and of underwater glues and coatings in marine engineering. It is necessary to understand not only the chemical makeup of the biological materials, but also the organization of different components within the material (*e.g.* calcite and proteins in sea urchin skeleton, or carbohydrate and protein in cuticle), which in nature is controlled by the cells producing the material.

Environmental Services

Table 3.3: Diverse Living Forms Carry Out Thousands of Environmental Services

Services	*Organism*
Pollination	Bees, bats
Biodegradation	Micro-organisms
Soil aeration	Earthworms
Fertilization	Soil bacteria
CO_2 - O_2 exchange	Plants
Water storage	Plants

Wild organisms carry out many functions in the environment that are vital to us, and that would be very difficult to do ourselves. Bees pollinate the blossoms and help to set fruits. Bats pollinate wild bananas (not the cultivated ones that are parthenocarpic), breadfruit, guava and durian (Malaysian fruit). Wild micro-organisms biodegrade much of our garbage as well as fallen leaves and other dead animal and plant matter. Earthworms turn over soil and keep it aerated. Soil bacteria turn nitrogen into nitrate fertilizer. Plants use up carbon dioxide and produce oxygen, thereby slowing global warming due to CO_2. All of these services (Table 3.3), and many more besides, are provided free of charge and usually taken for granted until they stop. In a paper published in *Nature* in May 1997, a group of 13 ecologists, geographers and economists estimated the economic value of these environmental services at between $16 and $54 trillion per year.

Phytoremediation (use of plants to clean up toxic wastes) is potentially much more effective and less expensive than current methods which consist mainly of excavation and reburial. Some plant species that live naturally on soils that are rich in heavy metals have evolved biochemical mechanisms for extracting those metals from the soil and accumulating them to very high levels in their tissues. They are called hyperaccumulators. Plants in the mustard family are especially promising. Some of the metals that are toxic at high levels (*e.g.*, selenium) are actually required by the body at low levels. So hyperaccumulating plants may also be useful in providing essential metals in the diet at the appropriate levels. Other sites where this technique is being used include abandoned mines (zinc and lead); military sites (lead and cadmium); municipal waste dumps (copper, mercury, lead); and sewage dumps (all of these metals).

Source of Beauty, Love and Inspiration

Wildlife is worth conserving because it is interesting, beautiful, spectacular, or contributes to landscapes that are interesting, beautiful, or spectacular. Wild animals and plants provide inspiration not only to biologists but also to millions of naturalists, explorers, painters, photographers, writers, poets and musicians. Enjoying wildlife is far from being restricted to poets, however. According to a survey conducted for the U.S. Fish and Wildlife Service, 77 million Americans participated in wildlife-related recreation in 1996. During that time, they spent $108 billion compared to only $81 billion on cars. Preserving the environment is healthy for the economy as well as for the soul.

Future Options

We do not know what our value systems will be in the future, or what the value systems of our successors will be. Perhaps they will need vast quantities of some species that we now consider insignificant or even harmful. Many of the natural sources of medicines are, in fact, poisonous. Nobody could have predicted that the Madagascar periwinkle would be a source of an anti-leukemic drug, or that a heat-loving microbe living in a hot spring at Yellowstone National Park would provide a key ingredient in the DNA fingerprinting.

Diversified Horticultural Crops for Farm-Enterprise Diversification

Crop diversification in present context, is only option to face new challenges of International and domestic markets. It's the necessary change in the conventional cropping system to bridge the gap between increasing demands and supply of cereals, pulses, oilseeds, fibre, fuel, fruits, vegetables etc. and to make balance cropping system with the aim of improvement in soil health and agro-ecosystem. Higher yields and diversification from cereals to high value, labour intensive horticulture, organic farming, commercial crops, medicinal and aromatic plants, spices and plantation crops, agro-processing etc. are the high potential areas to face the challenges in WTO context.

Species Diversity in Horticultural Crops

Eight agro-ecological regions have been recognized in India based on physiography, climate, soil and associated factors wherein the entire native diversity in plant genetic resource is distributed. These regions are the Western Himalayas, Indus Plains, Eastern Himalayas, Eastern Peninsular Region, North Eastern Himalayas, Western Peninsular Region, Gangetic Plains and Island Regions. The prominent horticultural crops with wide diversity grown in the above agro-ecological regions are given in Table 3.4. Substantial genetic diversity is found accumulated in the native and domesticated species of horticultural crops owing to diverse cultures, tribal and farming communities, climates, eco-geography, agro-ecology, farming systems and cropping situations. India occupies a unique position among major gene rich countries of the world. In fact, the country has been recognized as one of the world's top 12 mega-diversity nations with about 1.15 lakh species of identified and described plants and animals. In flora, the country can boast of 45,000 species which accounts for 15 per cent of the known-world plants. The Western Ghat area in Peninsular India and the North Eastern Himalayas regions are two important treasure houses of species diversity.

Table 3.4: Agro-ecological Regions with Horticultural Crops having Wide Diversity*

Western Himalayas	☆ Pumpkin, cucumber, alliums pp., ginger, brassicae ☆ Pome, stone and nut fruits ☆ Medicinal plants
Indus plains	☆ Okra, cucumis spp. ☆ Jujube, khirni, phalsa
Eastern Himalayas	☆ Pumpkin, cucumber, alliums pp., ginger, chayote, tree tomato, brassicae ☆ Pome and stone fruits
Eastern peninsular region	☆ Taros, yams, elephant foot yam ☆ Banana, mango, lemon/lime, jackfruit, Malabar tamarind ☆ Ginger, turmeric, chilli/capsicum spp., coconut
North Eastern Himalayas	☆ Pumpkin, chayote, cucumber, okra, eggplant, chilli/capsicum spp., pointed gourd, ash gourd ☆ Taros, yams ☆ Citrus spp., banana ☆ Tea, large cardamom, ginger, long pepper
Western peninsular region	☆ Okra, eggplant, cucumber, chilli/capsicum spp. ☆ Taros, yams, elephant foot yam ☆ Jackfruit, banana, lemon/lime, orange, jamun/syzigium spp. ☆ Black pepper, ginger, turmeric, coconut, arecanut
Gangetic plains	☆ Okra, eggplant, bitter gourd, cucumis spp., luffa spp. ☆ Jackfruit, mango, banana, orange, lemon/lime, jujube, Indian goose berry, jamun, melons
Island region	☆ Coconut, chilli, taros, yams

* Mathewkutty (2004)

Threats to Biodiversity

After Aldous Huxley read Rachel Carson's *Silent Spring* (about the loss of songbirds due to DDT), his comment was that "we are losing half the subject-matter of English poetry". Which we may find in real terms, the FAO's estimates that about three-quarters of the genetic diversity of agricultural crops have been lost over last century. And of 6300 animals breeds, 1350 are endangered or already extinct. Clearly, the biodiversity on this planet is under threat. The population exploration, overexploitation of resources and illegal trade are the major threatening factors to biodiversity.

Human Population Growth and its Impact on Biodiversity

The human population is growing at a rate that will lead to doubling the present population on the earth by 2050. Based on the 'Net Primary Productivity' of the earth, it was estimated that the earth's carrying capacity for humans is much less than 15 billion; in fact, probably less than the current population. Rapid population growth has social consequences, leading to environmental destruction. The following estimates indicate the alarming situations on this planet:

- ☆ We have already transformed or degraded 39-50 per cent of the Earth's land surface (agriculture, urban).
- ☆ We use 8 per cent of the primary productivity of the oceans (25 per cent for upwelling areas and 35 per cent for temperate continental shelf areas).

- We have increased atmospheric CO_2 concentration by 30 per cent
- We use more than half of the accessible surface fresh water
- Over 50 per cent of terrestrial nitrogen fixation is caused by human activity (use of nitrogen fertilizer, planting of nitrogen-fixing crops, release of reactive nitrogen from fossil fuels into the atmosphere)
- On many islands, more than half of plant species have been introduced by man; on continental areas the fraction is 20 per cent or more
- About 20 per cent of bird species have become extinct in the past 200 years, almost all of them because of human activity
- 22 per cent of marine fisheries are overexploited or depleted, 44 per cent more are at the limit of exploitation
- 12 per cent decline in per capita irrigated lands
- 21 per cent decline in per capita cropland
- 22 per cent decline in per capita rangeland and pasture
- 30 per cent decline in per capita forestland

The demographic transition pattern and uneven income distribution both have shown deep impact on environment. The widening gap in the distribution of income is a major cause of environmental decline, where both the categories of the population (the rich and the poor) are more likely than those in the middle to do serious ecological damage - the rich because of their high consumption of energy, raw materials, and manufactured goods, and the poor because they are often forced to cut down forest, grow crops and graze cattle in order to subsist on the land. The U.S. (rich nation), with only 4.7 per cent of the world's population, consumes 25 per cent of the world's resources and generates 25 to 30 per cent of the world's waste (pollution).

Compared to an average citizen in India, a typical person in the U.S. uses:

- 50 times more steel
- 56 times more energy
- 170 times more synthetic rubber and newsprint
- 250 times more motor fuel
- 300 times more plastic

Illegal Trade is a Threat to Biodiversity

Many animals and plants are increasingly threatened because of international trade in wildlife and wildlife products. In some cases (pets, rare plants) the living organism is in demand; in other cases, specific body parts like skins, furs, bones, ivory, sex organs, bile, coral skeletons, and claws are highly valued for ornaments, jewellry, or for putative medicinal or aphrodisiac qualities. These are mostly luxury items in demand in industrialized countries, by consumers that have very little knowledge of how endangered some of these plants and animals are. International wildlife trade has grown and become very profitable with increased prosperity, improvements in shipping, and fast air transport. The major exporters are tropical and subtropical countries of Africa, Southeast Asia and South America while the major importers are Europe, the U.S. and Japan. The illegal wildlife imports in the U.S. amounts to be as high as $ 100 - $ 250 million per year.

Biopiracy

Unscrupulous investors have been trying to patent genetic varieties, often when it is clear that they have not "invented" anything, but are simply taking advantage of traditional knowledge. For example, in 1994 the president of the U.S. seed company POD-NERS bought yellow beans in Sonora, Mexico. He brought them home, bred several generations, selecting for yellow seeds. In 1996 the company had produced a "uniform and stable variety with yellow seeds", and applied for a patent. In 1999 a U.S. patent (and Plant Variety Protection Certificate) was granted. In 1999 the company was suing Mexican bean exporters for patent infringement. There are hundreds of other cases of suspected biopiracy which is a specialized form of threat biodiversity.

Biodiversity has also been threatened by urbanization, deforestation, pollution and the conversion of wetland. FAO said "Due to agricultural modernization, changes in diets and population density, humankind increasingly depends on a reduced amount of agricultural biodiversity for its food supply'.

Biodiversity Conservation

Need for Biodiversity Conservation

Reduction of biodiversity entails a reduction of option for ensuring more diverse nutrition, enhancing food production, raising incomes, coping with environmental constraints and managing ecosystem. Recognizing, safeguarding and using the potential and diversity of nature is critical for food security and sustainable agriculture. Global efforts to conserve plants and animals in gene banks, botanical gardens and zoos are vital. But an equally important task is to maintain biodiversity on farms and in nature, where it can evolve and adopt to changing conditions or compete with other species. As custodians of the world's biodiversity, farmers can develop and maintain local plants and trees and reproduce indigenous animals ensuring their survival. The wildlife trade monitoring programme (TRAFFIC) of World Wildlife Fund and World Conservation Union identified 102 medicinal plant species (including the frankincense tree) and 29 medicinal animal species as priorities for early conservation and management action.

Conservation "Hot Spots"

The high levels of endemism in certain areas of the world, coupled with imminent threats from habitat loss in many of these areas, has led to the designation, by World Conservation Union (IUCN) and Conservation International, of "hot spots" for preservation. Coastal California is one of these regions. Other regions are the Atlantic coast of Brazil, Madagascar, and the Indo-west Pacific. Significant Natural Areas are those sites which meet at least one of the following criteria:

1. Areas supporting extremely rare species or natural communities;
2. Areas supporting associations or concentrations of rare species or communities;

3. Areas exhibiting representative examples of common or rare communities;
4. Areas of high species-richness or habitat-richness.

Hi-tech Conservation of Horticultural Germplasm

Maintenance in field gene banks and *in-situ* conservation of germplasm in horticultural crops are very difficult and risky. Successful attempts have been made to develop protocols for *in-vitro* preservation of horticultural crops, especially those propagated asexually. Methodologies for a long-term storage under conditions closer to ambient temperature have also been standardized for a good number of horticultural crops. *In-vitro* storage and cryopreservation are being practiced. Successful cryopreservation of seeds, meristems, somatic or zygotic embryos have been reported in cassava, potato, coconut, citrus, *Prunus* sp., *Allium* sp., taro, yam, black pepper, cardamom and grapes. Freezing of encapsulated meristems is possible in pears, strawberry and potato. *In-vitro* germplasm storage is now routinely followed in horticultural crops.

National and International Efforts to Conserve Biodiversity

The Convention on Biological Diversity

The Convention on Biological Diversity (CBD) was signed by over 150 governments at the 1992 Earth Summit in Rio de Janeiro, and became effective as international law in December 1993. It is the first international agreement committing governments to comprehensive protection of the Earth's biological resources. As of February 2004, 188 countries and the European Union had ratified the agreement; the U.S. was one of the last nations to sign.

The CBD has three overall goals:

1. The conservation of biological diversity,
2. The sustainable use of its components, and
3. The fair and equitable distribution of benefits derived from "genetic resources".

By signing the CBD, participating governments agree to carry out various measures to conserve biodiversity. The measures include (among other things):

1. To create national plans for protection of biodiversity.
2. To identify ecosystems, species and genomes important for conservation and sustainable use of biodiversity
3. To monitor biological diversity and any factors that might have impacts on it.
4. To establish a system of protected areas
5. To manage biological resources to ensure conservation and sustainable use
6. To rehabilitate and restore ecosystems
7. To take measures for ex situ conservation.

The CBD has also published a useful Global Biodiversity Outlook. The CBD also includes agreements for using biological diversity. And governments must provide for "fair sharing" of the benefits derived from genetic resources (*i.e.* compensation for its use or transfer of technology derived from genetic resources).

The Biological Resources Division, USGS

In U.S., the requirements of the CBD are being met by the Biological Resources Division. Geological Survey (formerly the National Biological Service), established in 1993. Its operating mission is "to work with others to provide the scientific understanding and technologies needed to support the sound management and conservation of the nation's biological resources." It sponsors a variety of national scientific research programs as well as a national information infrastructure program. It has the following Strategic Science Plan:

1. Characterize natural processes and identify factors that influence the quality or quantity of the Nation's biological resources at all levels of biological organization.
2. Facilitate sound management of the Nation's biological resources by collaborating with partners in all phases of our work.
3. Provide national and international leadership for the development of a biological information infrastructure to provide access to, disseminate, share, and use biological data, information, and technology.
4. Assess and report the condition of the Nation's biological resources.
5. Integrate biological resources research, inventory, and monitoring efforts with those directed at other natural resources, throughout the USGS.

GAP Analysis

The U.S. Geological Service has developed a method for digitized mapping of many parameters important for conservation, including the distribution of plant communities and of vertebrate animals, and the geographic pattern of land use. This method is called GAP analysis, because it was designed to identify gaps in the system of protected land areas, where insufficient protection was being provided for endangered and other species. However, the technology can also be used for many other purposes where geographic comparisons are important, such as monitoring changes in species distribution caused by climate change; for mapping levels of contaminants so they can be compared with changes in species distribution; for tracking the spread of exotic species and of urbanization and their consequences for species distribution, and so on.

Population Policies- U.N. Conference on Population (Cairo, 1994)

The United Nations has for over forty years been coordinating efforts to bring global population growth under control. At the U.N. Conference on Population in Cairo in 1994, 179 nations endorsed a new "Programme of Action" that called on governments to provide universal access to reproductive health care by 2015 as a global human rights imperative. The goal of the Cairo agreement is to stabilize human population at 7.8 billion by 2050. There are five basic components of which the following is the most important recognition for environmental protection :

- ☆ Recognize that environmental protection and economic development are not necessarily antagonistic, but that economic development is essential for environmental protection. Promote free trade, private investment and development assistance.

Convention on International Trade in Endangered Species of Wild Fauna and Flora (CITES)

International trade in wildlife puts enormous economic pressure on suppliers and threatens the existence of many rare plants and animals. For this reason in 1975 the U.S., along with nine other countries, signed the Convention on International Trade in Endangered Species of Wild Fauna and Flora (CITES). The objective of CITES, which now involves over 150 countries, is to ensure that trade will not cause the extinction of plant or animal species.

WTO Established a Committee on Trade and the Environment

Because of fears that free trade would lead to increased pressure on natural resources, the WTO established a Committee on Trade and the Environment. The WTO does not want to intervene in national or international environmental policies or to set environmental standards. But they believe that free trade is, in general, beneficial to the environment. They argue that "Removal of trade restrictions and distortions, in particular high tariffs, tariff escalation, export restrictions, subsidies and non-tariff barriers, has the potential to yield benefits for both the multilateral trading system and the environment." In general, WTO rulings follow the principle that environmental issues should be dealt with explicitly through multilateral agreements (*e.g.* CITES), not through trade manipulation.

Some WTO rulings:

- ☆ The countries concerned should try to cooperate to prevent environmental damage.
- ☆ The complaining country can act (*e.g.* on imports) to protect its own domestic environment, but it cannot discriminate against other countries. Standards, taxes or other measures applied to imports from the other country must also apply equally to the complaining country's own products.

When the issue is not covered by an environmental agreement (*e.g.* CITES), WTO rules apply, including:

- ☆ Trade restrictions cannot be imposed on a product purely because of the way it has been produced.
- ☆ One country cannot impose its standards on another country.

Traditional Medicine for the Islands (TRAMIL)

In 1982, the TRAMIL research network was launched in the Dominican Republic to provide scientifically proven alternatives to patent drugs, which are becoming scarcer and more expensive due to increasing poverty and dwindling foreign currency reserves. The network – which recognizes that many rural people are more familiar with medicinal plants – aims to ensure the safety, efficacy, and accessibility of natural medicines. In this manner, TRAMIL hopes to preserve the diversity of plant species and indigenous knowledge.

United Nations Convention on Biodiversity

The United Nations Convention on Biodiversity addresses the issues related to biodiversity and has been ratified by 170 countries. Some countries, including the United States, still refuse to sign the agreement, primarily because it contains clauses requiring profits from biodiversity be shared with the species' country of origin.

Patent

In 1998, the Indian Council of Scientific and Industrial Research successfully challenged six patent applications on the medicinal properties of turmeric. The U.S. patent office ruled that turmeric's medicinal properties were part of the traditional Indian knowledge base and were therefore not patentable.

MDG7 - 'Ensuring Environmental Sustainability'

The Millennium Declaration stipulates that prudence must be shown in the management of all living species and natural resources, in accordance with the principles of sustainable development. Since this 2001 consensus, members of the international community have committed themselves to achieving the Millennium Development Goals (MDGs), including MDG7 'Ensuring Environmental Sustainability'. This ambitious goal identifies three targets (9-11) and eight global indicators (25-32) in order to provide countries with an overarching framework to monitor progress towards environmental sustainability. The three targets that fall under MDG7, include: Integrating the principles of sustainable development into country policies and agenda programmes and reverse the losses of environmental resources (9); halving, by 2015, the proportion of people without sustainable access to safe drinking water and sanitation (10).

Info-Bio-Nano Revolution - for Environmental Improvement

"Nanotechnology is not the next industrial revolution, but it will converge with ongoing revolutions in information technology and biotechnology to create it. The environmental community has a chance to guide the coming Info-Bio-Nano Revolution in ways that avoid the mistakes of the first industrial revolution – and harness this one for environmental improvement."

PPVFR Act, 2001 and Biodiversity Act, 2002

Two legislative Acts, one for protection of plant varieties and farmers' rights (PPVFR Act, 2001) and another on sovereign rights over biodiversity (Biodiversity Act, 2002) enacted by the Indian Parliament to safeguard the interest of breeders and farmers have been highlighted to create awareness on benefits sharing as and when their materials (generated/possessed) is commercialized. The protection of Plant Varieties and Farmers' Rights (PPVFR) Act has been brought in place for establishment of effective system for protection of plant varieties, the rights of farmers and breeders to comply with WTO agreement under section 27.3 (b) of TRIPS. Enormous rights to farmers have been protected under the act to safeguard their interests. The researchers have been provided access to protect varieties for bonafide research purposes.

Conclusion

'Biodiversity' is the buzz word of scientific articles now-a-days, but it's a wholesome challenge for present and future of mankind. Both legal and spiritual efforts at national as well as international levels need to gain much momentum to face this challenge, while the participatory approach could be a much fruitful option. We did, and do believe that our nature on this planet has enough potentiality to give a lump sum return tomorrow against a investment today for its conservation.

References

Callicott, J. B., Crowder, L. B. and Mumford, K.(1999). *Conservation Biol.* **13** : 22-35.

Gaston, K. J. (1996). *Biodiversity: a biology of numbers and difference*, Oxford, Blackwell.

Mathewkutty, T. I. (2004). *Intensive Agri.*, **42** (9-10) : 7-13.

Meffe, G.(2000). *Nature and Human Society: The Quest for a Sustainable World,* National Research Council (NRC).

Chapter 4

Post Harvest Management for Generating Livelihood and Income for Farm Women

Ivi Chakraborty

Women community though equal men in population, it is really unfortunate to note that, yet in this modern sophisticated world, many a times she is treated as the second class citizen or subordinate to men. Though various efforts are being made to improve the status of women, they have not taken their maximum use and remain still backward. Women constitute atleast one third of country's economically active population, particularly in the unorganized sector and specifically in the agrarian sector. Low literacy levels (40 per cent), skill based and knowledge grounded technologies alongwith patriarchal social control, lack of access to and control over resources like land, credit, marketing and management; gender biased extension services and limited access to training; lack of support services to play the multiple roles expected of them; gender-biased wage differentials; occupational hazard to health; social restrain which might not enable them to equip themselves to meet the challenges and take advantage of the opportunities provided by the new economic policy (Sobha, 2001). In India only 8 to 10 per cent of the small scale manufacturing units are run exclusively by women entrepreneur that is proportionately very small as compared to other developed or developing countries. In USA alone, about 50 per cent of the business is owned by women (Naik, 2003).

Still it is fact that throughout the world, Women's participation in economic activities is increasing. The role of women entrepreneur is also spreading its dimension. In India too, women are participating in almost all the spheres of economic activities. From village to city, we can see large number of women entrepreneurs contributing towards the national income of the country.

Role of Women in Agriculture

Woman form the backbone of agriculture comprising the majority of agricultural labourers in India. Gender divisions in agriculture are stark, with all activities involving manual labour assigned to women. There is an intense need to free women from gender related shackles, build up their capacity, empower them with information and knowledge and allow them to emerge as leaders with confidence to guide their own destinies. This is relevant all the more in agriculture and allied sectors. According to official estimates, one-third of agriculture labour force and nearly half of self employed farmers are women. Of the total female work force in rural areas 90 per cent are employed in agriculture and allied industrial sectors.

Horticulture plays an important role in the economic development of the country by creating rural employment and on farm income generation. Fruits and vegetables are important ingredients of the human diet as they contain much needed vitamins and micronutrients. Export of horticultural produce and products bring valuable foreign exchange to the country. India with its varied agro-climatic conditions produces a large number of diversified horticultural crops. But unfortunately, a considerable amount of the produce is just wasted as a result of poor post harvest management.

Most of the horticultural produces are highly perishable in nature because of high moisture content (70 to 95 per cent), soft texture and due to their richness in constituents like carbohydrates, fats, proteins and vitamins etc., micro organisms survive on them and may easily extract their requisite nutrition. Most of them are valuable crops having great potential for generating income and employment in rural areas and also high export values provided proper and scientific tools of efficient post harvest managements are taken care of.

Considerable degradation in their quality starts immediately after harvest, which tends to increase cumulatively as the produce moves downwards from production to post harvest. Losses of perishables vary widely from product to product and from place to place and it becomes more complex due to lack of storage facility and marketing problems.

Women's participation in post harvest management is a common practice. It starts from harvesting/picking, threshing, winnowing, curing etc. and continues till its storage and preservation. For post harvest management of floricultural crops also, they play significant roles. Although some of the post harvest operations like harvesting of crop from tall trees, most of the machine operated activities, transportation, marketing etc are generally predominated by men. However, harvest and post harvest operations carried out by women varies from country as well as region. In Bangladesh, women are relatively restricted to do food processing and homeland but hand pounding of paddy was socially accepted form of employment. While harvest in West Bengal and Bangladesh is mainly men's work. In Indonesia, it is traditionally women's work. In Philippines, harvesting is done mostly by women. Threshing varies from region to region in terms of women participation. But winnowing, drying, storing, milling and food preservation is mostly done by woman. Thus in most Asian countries, the rural women are largely responsible for harvesting, on farm handling, storage and processing of food grains.

Women as custodian of food grains and other agricultural produce have following basic roles to play:

1. Minimization of harvest and post harvest losses.
2. Value addition to agricultural produce and by-products for increasing family income.
3. Meeting family requirements of food, feed, fibre and fuel through economical options.
4. Within economic resources provide nutritional security to the family.

Although good amount of work is being carried out on the overall research strategies by various Agricultural Universities in India, relatively little attention has been given to women's post harvest activities. Low cost improved techniques which uses local material, renewable energy sources and easy to construct and maintain structures are needed for on farm crop drying, winnowing, parboiling, husking, storage, handling, processing and preservation of grains, vegetables, fruits and fish.

Post-harvest Technologies Related to Employment and Income Generation

India has a rich biodiversity of horticultural crops growing in its varied agro-climatic regions, fruits, vegetables, flowers, ornamentals and spices are perishable in nature and their shelf life is limited, depending upon environmental and handling conditions. Due to poor Post-Harvest Management Practices and lack on infrastructure facilities, huge losses (20-40 per cent) occur which cause an annual estimated loss of Rs. 23,000 crores and more.

If the rural women folks (according to their literacy level and interest) may be trained with the technical know how regarding the latest Post-Harvest Management/ Processing techniques including scientific handling, grading and storage methods, post harvest disease management, solar/osmotic dehydration techniques, food irradiation, food biotechnology, food safety and quality assurance (HACCP) which will have an impact on reduction of post harvest losses in perishable horticultural produce that will lead to sustainable development of our agrarian economy. The target group of women needs to create capacity that would enable safe handling of produce to process, package and transport to remunerative markets so as to minimize post harvest loss and improve net availability of quality saleable produce to consumers and net return to growers.

Rural and small scale cottage industries have potentialities to absorb women on several jobs. Development in the agricultural and processing sector, especially the growth of agro based industries has tremendous scope for women's participation in economic developments.

Employment Opportunities are there in the Following Field

1. Post harvest management of perishables for extending shelf life.
2. Cut flower preservation and preparing flower decorations, ornaments, and floral arrangements.

3. Food preservation and packaging of prepared products
4. Dehydration of flower and preparation of floral arrangements with dry flower.
5. Utilization of agricultural wastes.

   (*a*) Briquettes; (*b*) Compost; (*c*) Mushroom cultivation; (*d*) Mud cooler'; (*e*) Industrial waste utilization to prepare value added items.
6. Marketing of fresh as well as prepared products.

Food processing and preservation is an area offering immense scope for income generation and prevention of post harvest losses to farm families. The seasonal availability of perishable foods especially fruits and vegetables need to be given utmost attention. During peak harvest season, the farmers sell their perishables at very low cost which are not commensurate with the cost of their inputs and labour. If small scale preservation units for fruits and vegetables could be established on a co-operative basis including villages or blocks, this would go a long way in agricultural marketing and nutritional balance throughout the year. Pickles, juices, jams, jellies, ketchup and sauces can be prepared and sold through the net work of Government stores and Institutions. These net works of preservation projects can easily be managed by farm women provided they are given the technical and managerial expertise.

However, Success of any value adding activity depends upon the capacity building for its appropriate manufacturing and marketing the products at remunerative prices. Since agricultural produce and products are usually perishable and semi-perishable in nature, ability to retain without excessive quantitative and qualitative losses is pre-requisite for successful marketing operations.

Marketing is a competitive job. Laws of demand and supply, compatibility between the product and consumer's preferences, purchasing power, packaging, handling, distribution, motivation and sales promotion etc. govern it. Talents required for marketing are not the same as required for production and processing. Women power sometimes supersedes men successfully in this field.

Factors Facilitating the Transfer of Technology

Farm women must be trained to continue themselves in farm operations using the technologies to their best advantages. Strategies need to be evolved for generating newer appropriate technologies to forecast and realistically assess the implications of the technologies and suitably train the farm women in other alternative pursuits for economic independence.

Farm women are confronted with the triple burden of farm work, household duties and child rearing. Considering this and keeping in mind the potential areas as well as limitations of farm women, the technologies will have to be:

- ✰ Simple without involving highly complicated procedures
- ✰ Introducing the machineries with the ease of handling, operating and maintaining which encourages the users to get experience and realize its utility and potency.
- ✰ Low cost

- ☆ Inland/internally generated technologies have added advantages over those imported
- ☆ User friendly
- ☆ Interesting to the users.

Strategies should be evolved to disseminate to farm women the ideas, information and appropriate technologies developed for reduction of drudgeries on the farm and in the home. A sound feed back mechanism based on systematic monitoring and evaluation of such dissemination and implementation need to be established.

As observed by the International Women's Tribune Centre (1987), efforts to modernize the food industry often ignore the role which women play in this direction. Women do not receive sufficient information, technical advice and credit support for the range of food processing in which they should really be involved. Since farm women are the custodians of what has been produced in the farm both for household use and commercial purposes they need to be trained in all aspects of post harvest technology, which include harvesting, post harvest handling operations, storage and preservation of all food and cash crops.

References

Naik, S. (2003). The need for developing women entrepreneurs. *Yojana* **47** (7) 36-39.

Sobha, I. (2001). Women in agriculture: A case study. *Yojana* **45** (7) 40-43.

Chapter 5

Youth in Agricultural Entrepreneurship: It's Prospects

Chakradhar Satapathy

Youth is a very beautiful and attractive word. The youth is a unique demographic dividend with young person's below the age group of 35. They constitute about 70 per cent of the total population. In India, population within 10-19 years is approximately 225 million, the largest number cohort of young people to make a transition in adulthood. According to National Sample Survey (2004-05), percentage of workers living in BPL households is about 27.6 per cent and youth population represents 23.1 per cent in rural areas. Unemployed youth is about 20.1 per cent of the state and in rural areas it is 17.9 per cent. Likewise, the jobless youth constitute 28.4 per cent and in rural areas it is as high as 27.2 per cent. As per record, in 2007-08, 32.8 per cent of youth (15-24 years) attended educational institutions in India whereas in rural areas the youth of same age group account for 29 per cent attending educational institutions. The youth population in rural areas faces more tension compared to urban areas. The tension among rural youths is due to (i). discrepancy between achievement and aspiration (ii) absence of adequate role models to follow (iii) lack of clear cut ideas about future career (iv) lack of clear cut crystallization about acceptance of traditional or modern values of society (v) unfavourable attitudes of elders in society (vi) absence of interaction between development authorities and youth (vii) urban and rural divide (viii) differential economic back ground such as high, medium and low income groupings (ix) dependence of families on agriculture and non agriculture occupation. About 55 per cent of the young workers were employed in agriculture and allied activities in 2007-08 and the rest 45 per cent were employed in non-agriculture Activities. However, the share of agriculture in youth employment declined faster than adults. It was 54.4 per cent for youth and 57 per cent for adults in 2007-08. It is interesting to note the share of industrial sector increased faster for youth as compared to services. Unlike for adults, the share of industry for youth is higher than services in the year 2004-05. To release the tension and bring rural youths to path of economic prosperity, agri-entrepreneurship or agribusiness appears to be most suitable alternative in rural areas. As our country is of agrarian in origin and living, the agro enterprises provide wider scope for rural youths to accept as career.

Agro-business is ever changing in nature. The youths entering into this field must keep watch on it and prepare accordingly to face the change. Ideas about agro-business are the key to bring success in their entrepreneurship achievement.

Sources of Agro-business

The aspiring youths entering into this line of profession can gather information and get influenced about it from the sources like,

1. Past work experience
2. Interest and hobbies
3. Strength and abilities
4. Friends and family
5. Travel
6. Channels
7. Books and magazines
8. Research
9. Web

Role of Agro-Entrepreneurs

Agriculture is to be viewed as business. The rural youths stepping into this world have to be trained as successful entrepreneurs. Their roles in agro-business have to be clearly defined. Hebert and Link (1989) and Wennekers and Thurik (1999) have specified certain roles of entrepreneurs that are equally applicable in case of agro-enterprisers as following.

1. Association with uncertain risk
2. Availing of financial capital
3. Starting of new ventures
4. Ownership and enterprise
5. Exploring factors of production
6. Taking of decision
7. Possession of leadership quality
8. Knowledge of management
9. Skill of organization and coordination of economic resources
10. Innovative in nature
11. Acting of contractor

Reasons for Growing of Agro-business

At present, rural area and rural products have drawn attention of many business houses. To feed business houses, agro-entrepreneurs have to come forward. This can well be done by providing required trainings to rural youths. The gradual rise of agribusiness in our country can be ascribed to the following reasons.

1. Industry structure
2. New technology
3. Preference for privatization
4. Formation of entrepreneurs group
5. Increase in demand for rural product specifically organic ones
6. Government incentives and subsidies
7. Increasing flow of information to agro-sector
8. Easy access to resources
9. Education in management – establishment of agro-business institutes
10. Regards for self-employment
11. Rising dissatisfaction at job

These independent variables have drawn the attention of policy makers to train rural youths for opting to agro-business. But to start agri-business is not an easy job. Once the youths are trained and motivated, the agri-business will positively provide both employment and income. They are not searching for work because they know or believe that acceptable employment is not available (OHiggins, 2008).

The youths are to be cautioned on certain issues before starting agro-business.

Caution in Agro-business

1. Control over unit
2. Idea generation and scanning
3. Flexibility in approach to employees as well producers
4. Capital arrangement
5. Besides these important cautions, the agro-entrepreneurs also need to know advantages and disadvantages of agri-business

Advantages

1. They are the boss of business and there is no imposition on him
2. All profits go to them
3. Roles are variables and they can take up any required roles
4. Own work can provide them satisfaction
5. Success and profit will bring social recognition

Disadvantages

1. They are alone to look after enterprise
2. They have to take all decisions
3. Losses go to them only
4. Works they do may not be satisfying
5. They have to wait with patience for many activities
6. Loss may bring them bad name
7. Difficult to exist from agri-business
8. May affect family and social life

The youths in agro enterprise ought to know the advantages and disadvantages of it to make up their mind to go with such business

Characteristics and Skill of Agro-Entrepreneur

Timmons (1994) has come out with six important characteristics of entrepreneurs which have applicability in case of agribusiness and agri entrepreneurs.

1. Commitment and determination
2. Leadership
3. Opportunity obsession
4. Tolerance of risk and ambiguity
5. Creativity, self-reliance and ability to adopt
6. Motivation to excel

The above characteristics can help them to know the success of youth in agri-enterprise.

Scanning of Local Resources for Enterprise

Agro-entrepreneurship at present includes something more than just the agriculture. It is simply not restricted and not bearing risk, but also involves generating and implementing ideas. Agro-entrepreneurship is to be taken as dynamic and creative for development of economy. Rapid changes in agriculture have led to change the mind set of people and people in rural areas are opting to make business on agricultural produce.

Motives for Agriculture Entrepreneurship

The motives for agriculture entrepreneurship are depending on several factors. But the youths who are opting for agri-business should have the motives like:

1. To be own boss and manage the business
2. To earn more income and create flow of financial inputs
3. To grow fast and as large as possible

Whatever the motives may be agro-entrepreneur looks for opportunity to revolutionalize the existing business, reshape the market, change the product or bring modification to enhance the business. After all, agro-entrepreneur looks after the opportunities in rural sector ahead of others. The analyzing information available with them, they have to develop their vision and draw support of stake holders to start agri-business.

Scanning Opportunities

Opportunity in business particularly in agro-business is the important aspect for consideration. The chance to do something new with existing situation or outside the social boundary in known as opportunity. The opportunity is associated with threats. The favourable and unfavourable changes act as stimulus in agro-business. Threats and problems are challenges. Agro-entrepreneur looks forward for the solution of problems and to proceed ahead.

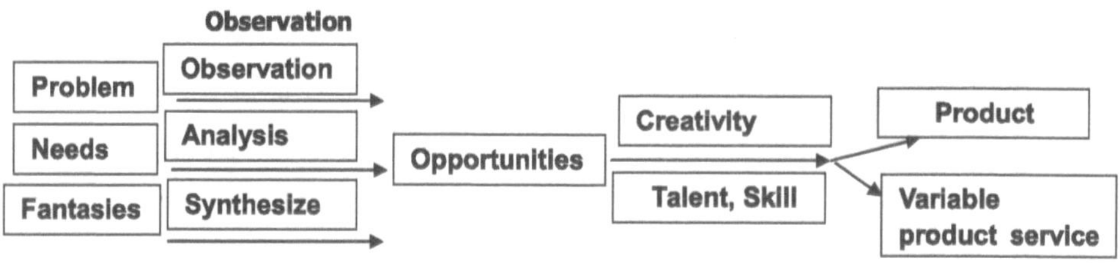

Opportunities

Opportunities are observed and synthesized after need assessment and analysis of problems. To arrive at selective opportunities, agro-entrepreneur has to take serious steps in correct analysis of needs and problems.

Creativity

The creativity of agro-entrepreneur is proportional to their talent and skill that they possess. The studies reveal that more they are exposed to different places and literature; more they are the creative as individuals. Applying this principle, our youth can make proper judgment of the kind of talent and skill that they possess to undertake agro-business. In agro-business youths are to be dreamers with a definite vision. Their own fantasies and dreams about products and service they like to have are also called opportunities. They work on them and develop them into viable product and services.

Product and Services

Agro-entrepreneurs have to develop a viable project based on product and services to run the business. In rural areas the products are numerous like agriculture, animal husbandry, horticulture, fishery, forestry and many others. This creativity based on talent and skill can decide the kind of product and service they can opt for.

For an agro-entrepreneur, the options are found everywhere and may be in the form of need or problem. The solution of these problems may lead to opportunities. In simple word, it is the process of perceiving the needs and problems of people and society and coming at solution through business. The agro-entrepreneur while dealing with opportunities performs to basic roles.

A. An opportunity spotter

B. A profit of champion

Agro-entrepreneur as Opportunity Spotter

(i) Always look for opportunities to earn profit
(ii) Maintain a reservoir of an unexploited opportunities
(iii) They avoid to exhaust their resources on every opportunity
(iv) They execute and act on opportunities that may lead to viable projects
(v) They are fast to gain support and persuade others for help to achieve goal.

Agro-entrepreneur performs two roles to achieve goal which can be presented in words of Bill Botton and John Thompson.

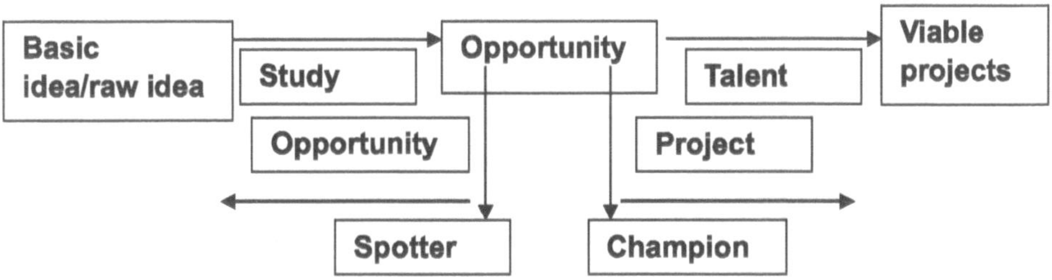

Factors Influencing Scanning of Opportunities

In rural areas not only opportunity but also problems are much more. Agro-entrepreneurs when see an opportunity react to them. They are fast to spot them in terms of solutions and problems. The important factors in sensing the opportunity are:

(i) Ability to perceive and preserve basic ideas: All ideas may not become an opportunity because of problem, difficulties and change in trend, invention and competition. Therefore, the youths who enter into domain of entrepreneurship in agriculture have to carefully decide opportunities for agro-business.

(ii) Ability to gather knowledge from various sources: The opportunity without knowledge is always meaningless. Agro-entrepreneur must have ability to spot opportunity; nurture them and collect information from different sources like books, magazines, net and synthesizing them for identification of right opportunity.

(iii) Vision and creativity: Own creativity and vision give correct shape to opt to business. The studies indicate that out of sample of rural youth, limited individuals have vision and creativity to start agro-business.

Enterprise process diagram

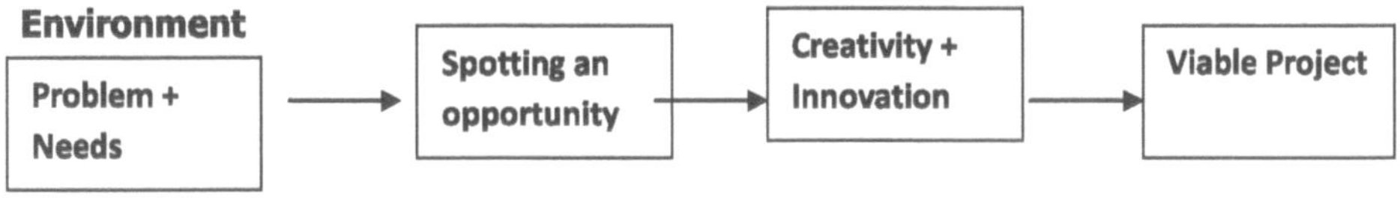

Environment of Agro-business

In agro-business, we refer environment as the sum total of all factors and forces in running agro-business. These are mostly external factors beyond the control of entrepreneur and exert influences in agro-business. An entrepreneur has to work keeping harmony with these factors. These are:

Techno-Economic factors

(i) Social, (ii) political, (iii) economic, (iv) legal and (v) techno factors.

A. Constraint and Opportunity Factors

To summarize socio-cultural factors in agro-business, the following structural framework can be perceived.

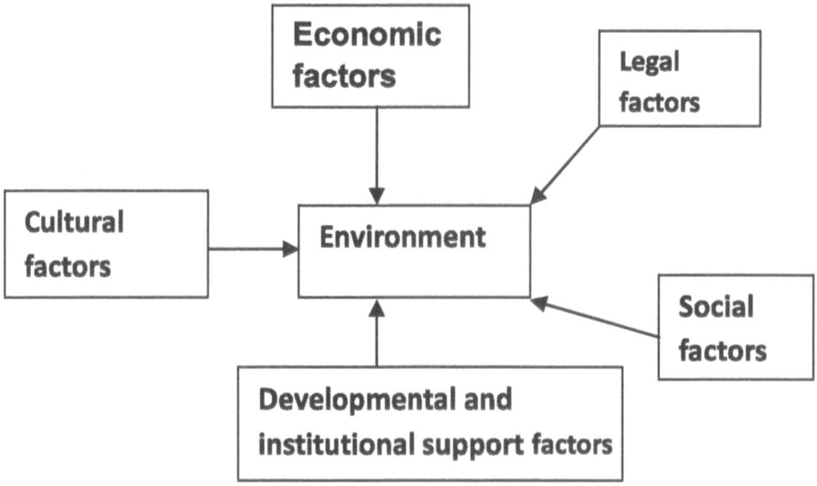

Importance of Environment in Agro-business

Environment affects agro-business. The agro-entrepreneurs are expected to know the following steps.

(i) Scanning the environment
(ii) Generation of ideas
(iii) Identification of products and services

Scanning the Environment

These include checking of ideas whether viable or not.

(a) Spotting opportunity
(b) Deciding of environment
(c) Identification of agro-product
(d) Starting agro-business

Generation of Ideas

Generation of ideas involves creativity and skills for which the common sources are:

(a) Publication
(b) Market research
(c) Brain storming
(d) Attending seminars
(e) Dreaming about idea

The generation of ideas also includes knowledge about existence of produce, natural resources, market, trading and sustainability.

Identification of Products and Services

It includes: the factors like,

(a) Costumers
(b) Preference and choice substitutes
(c) Availability and use level of product
(d) Cost and competition
(e) Turnover and profit
(f) Demand analysis
(g) Supply factor
(h) Cost and price of product
(i) Possible change of products

Avenues for Agro-enterprises

There are a number of avenues to go for agro-enterprises. Some of them are cited below.

1. Vermin composting
2. Marigold cultivation
3. Off-season vegetable cultivation
4. Papaya cultivation
5. Mushroom cultivation
6. Vegetable seed production
7. Raising vegetable seedlings and fruit plants
8. Fish fry production
9. Ornamental fish production
10. Poultry/duck farming
11. Goat/sheep rearing
12. Pig rearing

13. Preservation of local fruits and vegetables
14. Value added agro-products
15. Coir/jute work
16. Leaf plate making
17. Spices powder making
18. Rice processing
    (many more can be added on specific situation and location)

Market Management in Agro Business

Proper market management leads to success. It is the core area of business unit. The young agri- entrepreneurs should have knowledge in marketing management to establish a good unit. The capacity building in marketing should include the following aspects.

1. What is marketing management
2. Constituents of marketing
3. Marketing functions
4. Marketing mix
5. Personal selling
6. Sales promotion

Market and Marketing in Agri-business

Market is a place where people meet to sell and buy. Entrepreneurs find market in their locality, village, daily or weekly market. In local market exchange of goods and services take place. Marketing is the activity of utilities, time and possessions.

Constituents of Market

It involves movement of farm produce, agricultural products, industrial raw materials and finished goods from place of production to place of consumption. Activities like selling, distribution, price determination, intermediary charges, tolls, taxes are very common in marketing. The agri-entrepreneurs should have ideas about these activities to make their unit success.

Importance of Marketing

In each and every activity of marketing, the agri- entrepreneur should possess ideas and knowledge. The important aspects of marketing are (i) producer (ii) consumer (iii) trader and (iv) government The specific objectives of marketing are:

(a) Creation of demand for produce
(b) Managing of sales
(c) Maintaining of sales
(d) Maintaining of profit

Marketing and Selling

Marketing and selling are of closed knits. The need for marketing and selling skill is realized by every business man. Personal selling is gaining importance. Competency in personal selling leads to better prospective in rural areas. Personal selling requires (i) face to face communication (ii) contact with customers may be limited (iii) clarification of doubts if arises (iv) knowledge about psychology of customers and (v) persuasive skill. Agri-business man should see his competency in these regards. Good salesmanship is an indicator of good business. It helps in (i) introducing new product in market (ii) informs producers about market trend (iii) encourage production level according to market position and (iv) helps in decision making process.

Qualities of Good Salesman

1. Knowledge of agro-product
2. Knowledge of production and productivity
3. Knowledge of customers
    - Motive to buy
    - Nature of requirements
    - Viewpoints of customers
    - Capacity to buy

Knowledge about Techniques of Selling

Meeting customers and persuading them to buy the product requires (i) politeness (ii) gaining of confidence (iii) realization of importance of customers in business (iv) sincerity, helpfulness, courteous and clarification of doubts.

Personal Qualities of a Successful Salesman

In business salesman is of great importance. Agri-entrepreneurs must know as whom to select for selling of produce and whether he possess the following traits or not.

(i) Personality
(ii) Mental quality
(iii) Sociability
(iv) Aptitude
(v) Enthusiasm
(vi) Discipline

Marketing Functions

May it be agro product or any other type, but there are certain functions that have to be performed in marketing. These are, selling, buying, demand creation, advertising, price determination, transport, storage, grading, standardizing, quality control, packaging and product planning. Capacity building programme for agri-entrepreneurs should care for these aspects.

Marketing Mix

In management domain "Marketing Mix" is an important issue and concern. It has equal application in agribusiness. These are the marketing tools that agribusiness has to pursue to achieve the objectives. Agri Entrepreneurs should have ideas in marketing to sustain their business. Training programme for capacity building for them should include these aspects.

Product

Agri-products are many and season bound with characteristics of perishabality which requires preservation and storage. Reliance Fresh has entered into this business. Branding, packaging, service and size of products are of great consideration in agribusiness. These are all skill oriented jobs and training programme should be accordingly designed.

Pricing of Product

The most difficult task is to fix price of produce that too in rural areas. This aspect needs careful and critical consideration. Fixing of price of agri-produce depend on factors like, (i) demand in market (ii) cost of production (iii) competition (iv) channel of sale and (v) government policy.

Place

Place in marketing mix refers to the place through which produce pass to consumers. It may be direct sell to consumer in most cases of agr-products but in many cases the products pass through intermediary by transportation. Place of marketing depends on (i) nature of produce (ii) perishability (iii) size of produce (iv) location of market (v) size of customers and (vi) urgency of sale.

Promotion

Promoting of sale is a crucial decision. Agro-entrepreneurs have to decide whether to opt for (i) personal selling (ii) through retailers (iii) wholesalers or by some other means. Advertisement depends on volume of business mostly confined to manufacturing product of business houses.

Anchorage

It is the focal point of the network. Agri entrepreneur is the focal point in agribusiness.

1. **Density:** It is a measure of the extent to which an entrepreneur is connected to network. It is the speed of information flow amongst the entrepreneurs of common business.
2. **Range**: It refers to the number of people in direct contact with agri-entrepreneurs.
3. **Heterogeneity:** It defines the range of net work.
4. **Reachability:** Reachability is how far and how agro-entrepreneurs can contact others in net working.

Entrepreneurial Net Works

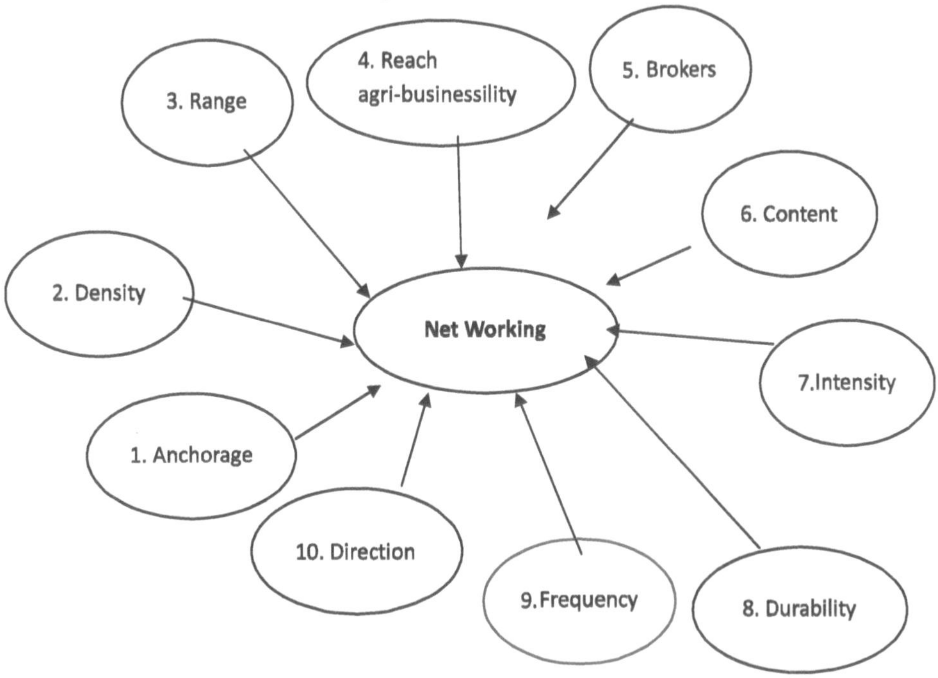

5. **Brokers**: The brokers are the link between agri- entrepreneurs and others in same business.
6. **Content**: Contents refer to the reasons behind the relationship.
7. **Intensity**: It is the strength of bonds between the networked individuals.
8. **Durability**: It is the duration of relationship
9. **Frequency**: It is the amount of time individuals dedicate to the relationship including number of interactions.
10. **Direction**: This is the dimension of power of balance in relationship.

For successful business networking is a must. The networking is an indicator of good business unit. Net working considers the issues of competency, dependability and helpfulness.

Chapter 6

Value Addition, Processing and Marketing of Spices

Jayanta Kumar Das

Agro-processing and value addition are the important aspect for spices. In India, there is only cost addition no value addition in case of agricultural commodities. In developed countries more than 80 per cent of the produce is processed, but in case of India it is only 5 per cent. It is to be noted here that 23.62 per cent of total spices export from India as value added crops. The returns from processing and value addition of species considerably higher as compared with those in the agronomic and horticultural crops. It is not only increase the quality of the produce but also create employment opportunities and enhancement of income. It requires minimum investment at household level. Species are less perishable and less voluminous as compared with other horticultural produce. Value addition and processing may be done under different categories *viz.* solid to solid, solid to dust, green to dry and green to green (in case of ginger). Information or studies regarding this aspect are very scanty, especially in case of spices. Due to the value addition of species price fluctuation could be minimized; excess production could be properly utilized. It also provides the processed/value added product to the consumer at a reasonable rates which would have not been possible in other ways. The taste, preferences, family composition and speed of life are being forced to use the value added or processed product of species.

In the light of this present discussion an attempt has been made to probe deeply into the different intricacies related with the value addition of species. The broad objectives of the study are as follows:

1. To estimate the cost and return of value addition and processing of species at household and commercial level.
2. To estimate the cost and return of value addition (from green to dry) of turmeric at whole seller level.

3. To assess the economics of value addition of ginger (green to green) at commercial level.
4. To identify the different existing marketing channel operating for value added products of species.

The present investigation was carried out in Cooch Behar Sadar and Mathabhanga Sub-division of Cooch Behar District. There are 10-15 entrepreneurs who are engaged in value addition/processing of species at house hold level. Out of that 4 entrepreneurs have been selected randomly for the study. A commercial processing unit of species at Nighiganj and a unit of value addition of ginger at Chakchaka have been studied. Besides, a good number of processors (green to dry) are engaged at wholesaler level, which is operating in the areas of Nishiganj and Goksadanga primary agricultural wholesale market. Out of these 10 processors were selected randomly for this purpose. The data pertaining to the year 2001-2002 were collected with pre-tested interview schedule. Simple statistical tools like, table, percentage etc. were used for analyzing the data.

Table 6.1 represents the value addition at house hold level of different seed species. It appears from the table that in all cases cost of raw materials are the highest like cumin (98.90 per cent), cumin black (97.83 per cent), coriander (97.78 per cent), fennel (97.83 per cent) and ajmund (98.04 per cent), followed by cleaning, washing and drying and packaging charges. Cost of value addition is highest in case of cumin followed by ajmund, fennel and cumin black and coriander. The cost of raw materials is highest in case of cumin than other spices. So far as profit is concerned that coriander occupies the highest rank followed by fennel, ajmund, cumin black and cumin. The profit-cost ration is also following is the same trend.

Value Addition and Processing of spices at house hold level solid to dust is represented in Table 6.2. A perusal of Table shows that the cost incurred by the entrepreneurs for processing of spices is highest in case of raw materials followed by grinding, Packaging and labour charge respectively. The cost of grinding is higher in case of chilli as compare to other two spices viz, turmeric and cumin. One plausible reason is that chilli requires additional manual grinding other than machine. The study also reveals that the weight loss due to conversion from solid to dust is higher for turmeric. So far profit is concerned to the entrepreneur, cumin occupy the first place followed by turmeric and chilli respectively. One interesting feature reveals from this investigation is that though profit per kg. is the highest in case of cumin but profit cost ratio is higher for turmeric, as the volume of business is more than the other spices.

A comparative picture of return per unit of spices and profit-cost ratio of solid and dust at house hold level is delineated in the Table 6.3. In case of solid to solid return per kg is the highest in coriander but profit-cost ratio is higher for ajmund. So entrepreneur may be opted for value addition of ajmund in large quantity besides other spices. The table also reveals that for processing of spices from solid to dust, the return and profit-cost ratio is the highest in cumin.

Table 6.1: Value Addition at House Hold Level (One unit) from Solid to Solid (per annum)

Sl.No.	Particulars	Cumin (A)	Cumin Black (B)	Corriander (C)	Fennel (D)	Ajmund (E)
1.	Raw materials	620kgxRs.90= 55,800 (98.90)	120kgxRs.45= 5400 (97.83)	525kgxRs.30= 15,750 (96.78)	150kgxRs.45 =6750 (97.83)	75kgxRs.50 =3750 (98.04)
2.	Cleaning, Washing and Drying	@Rs.0.75x620kg= 465 (0.82)	@Rs.0.75x120kg= 90 (1.63)	@Rs.0.75x525kg= 393 (2.41)	@Rs.0.75x150kg= 112.50 (1.63)	@Rs.0.75x75kg= 56.25 (1.47)
3.	Packaging charges	@Rs.0.25x620kg= 155 (0.27)	@Rs.0.25x120kg= 30 (0.54)	@Rs.0.25x525kg= 131.25 (0.81)	@Rs.0.25x150kg= 37.50 (0.54)	@Rs.0.25x75kg= 18.75 (0.49)
4.	Total Cost	Rs.=56420(a) (100.00)	Rs.=5520(a) (100.00)	Rs.=16274.25(a) (100.00)	Rs.=6900(a) (100.00)	Rs.=3825(a) (100.00)
5.	Sale Proceeds	620kgxRs.95= Rs.58900(b)	120kgxRs.52= Rs.6240(b)	525kgxRs.40= Rs.21000(b)	150kgxRs.54= Rs.8100(b)	75kgxRs.58= Rs.4350(b)
6.	Profit (b-a)	58900 – 56420= Rs.2480	6240 – 5520= Rs.720	21.00 – 16274= Rs.4726	81000 – 6900= Rs.1200	4350 – 3825= Rs.525
7.	Cost/kg	Rs. 91.00	Rs. 46.00	Rs. 31.00	Rs. 46.00	Rs. 51.00
8.	Sale/kg	Rs. 95.00	Rs. 52.00	Rs. 40.00	Rs. 54.00	Rs. 58.00
9.	Profit/kg	Rs. 4.00	Rs. 6.00	Rs. 9.00	Rs. 8.00	Rs. 7.00
10.	Profit Cost Ratio (b/a)	1.04	1.13	1.29	1.17	1.14
	Total profit per annum (A+B+C+D+E) =Rs. 9651/-					

Figures in the parentheses indicate percentage of respective total.

Value Chain of spices from Solid to Solid.

Producer – Whole sellers – Processor – Retailer – Consumer.

Table 6.2: Value Addition and Processing of Spices at House Hold Level (Solid to Dust/Annum)

Sl.No.	Particulars	Turmeric (A)	Cumin (B)	Chilli (C)
1.	Raw Materials	36Qtl @ Rs. 35/kg = 126000 (92.20)	480 kg @Rs. 100/=48000 (97.04)	720 kg @Rs. 45/=32400 (91.74)
2.	Grinding Charge	36Qtl @ Rs. 2/kg = 7200 (5.27)	480 kg @Rs. 21/= 960 (1.94)	@Rs. 3/-x720 kg = 2160 (6.12)
3.	Packaging	Rs.1125 (0.82)	Rs.192 (0.39)	Rs.288 (0.82)
4.	Labour Charge	@0.65/-x3600kg= Rs. 2304(1.64)	@0.65/-x480kg= Rs.312(0.63)	@0.65/-x720kg= Rs.468(1.32)
5.	Total Cost	Rs.136665(a) (100.00)	Rs.49464(a) (100.00)	Rs.35316(a) (100.00)
6.	Sale proceed	@Rs. 50=X 3600kg = Rs. 180000	@Rs. 130=X 480kg = Rs. 62400	@Rs. 60=X 720kg = Rs. 43200
7.	Wt. Loss due to conversion (100gm/kg)	@Rs.50x360kg= (-)18000=162000 (b)	@Rs.130x36kg = (0.75gm/kg)= (-) 4680=57720 (b)	@Rs.60x54kg = (0.75gm/kg)= (-) 3240=39960 (b)
8.	Profit (b-a)	Rs. 160000 - 136665= 26000	Rs. 57720 - 49464 =8256	Rs. 39960 - 35316 = 4644
9.	Cost per kg.	Rs. 37.95	Rs. 103.05	Rs. 49.05
10.	Sale per kg.	Rs. 45.00	Rs. 120.25	Rs. 55.50
11.	Profit per kg.	Rs. 7.05	Rs. 17.20	Rs. 6.45
12.	Profit – Cost Ratio (b/a)	1.19	1.17	1.13
	Total Profit per annum (A+B+C) = Rs. 38900/-			

Figures in the parentheses indicate percentage of respective total.

Value Chain of Processing of Spices from Solid to Dust.

I) Producer (Solid) – Processor (Dust) – Retailer (Dust) – Consumer (Dust)

II) Producer (Solid) – Whole Seller (Solid) – Processor (Dust) – Retailer (Dust) – Consumer (Dust)

Table 6.3: Comparative Picture of Return per unit (kg) of Spices and Profit-Cost Ratio of Solid and Dust at House Hold Level

Sl.No.	Name of Spices	Solid		Dust	
		Return	Profit-cost ratio	Return	Profit-cost ratio
1.	Cumin	4.00	1.04	17.20	1.19
2.	Cumin Black	6.00	1.13	–	–
3.	Ajmund	7.00	1.29	–	–
4.	Fennel	8.00	1.17	–	–
5.	Corriander	9.00	1.14	–	–
6.	Turmeric	–	–	7.05	1.17
7.	Chilli	–	–	6.45	1.13

The economics of processing unit of spices at commercial level is presented in the Table 6.4. The cost of processing under the present investigation is grouped namely, material and fixed costs. It is evident from the Table that the building, grinding and packet sealing machine and electricity charges. Cost share is 66.67, 28.00, 04 and 4.29 per cent of the total investment cost of a processing unit respectively. The major costs being building and machine which indicates that processing of spices at commercial level is capital intensive and highly labour intensive as well.

Table 6.4: Investment Requirement of Processing Unit of Spices at Commercial Level

Sl.No.	Particulars	Cost(Rs.)	Per cent
*1.	Building	100000	66.67
*2.	Grinding Machine	42000	28.00
*3.	Packet Sealing Machine	1560	1.04
4.	Electricity Charges (Office, unit etc.)	6440	4.29
	Total Cost	150000	100.00

· Depreciation on items No. 1 to 3 has not been calculated, as the unit established very recently.

Labour component of the processing unit.

1. Office Staff – 1
2. Marketing Personnel – 1
3. Labour – (a) Female – 2; (b) Male – 2

Table 6.5 indicates the processing cost, profit and total turnover of different brand of spices is an entrepreneur at commercial level. The cost incurred by the entrepreneur for processing of different spices is highest in case of purchase of raw materials in all items but the cost of raw materials for the items meat masala and herbal tea occupy the highest share of the total cost as because the items are being prepared with the combination of different vegetables and spices. Further more, the table also depicts that the cost of cleaning and washing is highest in sabji masala and

Table 6.5: Average Processing Cost, Profit and Total Turnover of different Brand of Spices at Commercial Level

Sl.No.	Processing Items	Name of the Species							
		Chilli	*Cumin*	*Turmeric*	*Mixed Spices*	*Meat Masala*	*Sabji Masala*	*Fish curry Masala*	*Herbal Tea*
1.	Purchase price/kg	50.00 (72.57)	100.00 (69.17)	35.00 (74.71)	300.00 (82.27)	100.00 (72.95)	100.00 (73.49)	90.00 (71.84)	140.00 (79.10)
2.	Cleaning and washing/Kg	0.50 (0.73)	5.50 (0.35)	–	1.00 (0.27)	1.00 (0.73)	1.00 (3.67)	1.00 (0.80)	1.00 (0.57)
3.	Grinding/Kg	5.00 (7.26)	5.00 (3.46)	2.00 (4.27)	5.00 (13.7)	6.00 (4.38)	5.00 (3.67)	6.00 (4.79)	5.00 (2.82)
4.	Packing cost/kg	1.00 (1.45)	1.00 (0.69)	1.00 (2.13)	1.00 (0.27)	1.00 (0.73)	1.00 (0.73)	1.00 (0.80)	1.00 (0.57)
5.	Cost of packets/Kg	2.40 (3.48)	20.00 (13.83)	2.40 (5.12)	2.40 (0.66)	10.00 (7.30)	10.00 (7.35)	10.00 (7.98)	8.00 (4.53)
6.	Loss due to conversion/kg	5.00 (7.26)	10.00 (6.92)	2.62 (5.59)	30.00 (8.23)	10.00 (7.30)	10.00 (7.35)	9.00 (7.18)	8.00 (4.53)
7.	Interest on fixed capital @ 120 per cent kg	1.00 (1.45)	1.00 (0.69)	1.00 (2.13)	1.00 (0.27)	1.00 (0.73)	1.00 (0.73)	1.00 (0.80)	1.00 (0.57)
8.	Sale Tax @ 8.08 per cent	4.00 (5.81)	8.08 (5.59)	2.83 (6.04)	24.24 (6.65)	8.08 (5.89)	8.08 (5.94)	7.27 (5.80)	13.00 (7.34)
9.	Total cost/kg	68.90 (100.00)	144.58 (100.00)	46.85 (100.00)	364.64 (100.00)	137.08 (100.00)	136.08 (100.00)	125.27 (100.00)	13.00 (100.00)
10.	Sale price/kg	75.00	160.00	50.00	560.00	160.00	140.00	150.00	200.00
11.	Profit/kg (sl. 10-9)	7.10	15.42	3.15	195.36	23.00	3.92	24.73	23.00
12.	Total sale per annum	900 kg	750 kg	6000 kg	150 kg	1200 kg	1800 kg	1050 kg	120 kg
13.	Profit per annum (Rs.)	6390.00	11565.00	18900.00	29304.00	27600.00	7056.00	25967.00	2760.00

Figures in the parentheses indicate percentage to the total cost.

* Interest out fixed capital has been distributed in total turnover (in kg); ** On the basis of profit per kg.

Value Chain of spices at commercial level.

Channel I: Processor – Retailer – Consumer; Channel II: Processor – Wholesaler (Jalpaiguri and Siliguri) – Consumer;
Channel III: Processor – Commission agent – Wholesaler/Retailer – Consumer.

the lowest the mixed masala, for packing cost it is highest in turmeric and lowest in mixed masala, for cost of packet cumin occupies the highest cost and the lowest cost in mixed masala. One of the important features of processing is conversion and it is often found that loss due to conversion is inevitable and in this respect highest loss occurred in mixed masala but lowest is in herbal tea. It is noteworthy to mention that mixed masala is lowest in almost in all the processing cost as because a major portion of the processing are being done manually. A perusal of the table also indicates is the profit per kg. It is highest in mixed spices, followed by fish curry masala, meat masala and herbal tea, cumin, chilli, sabji masala and turmeric respectively.

Table 6.6: Cost of Processing of Turmeric (Green) at Wholesaler-cum-Processor Level (Per MD)

Sl.No.	*Particulars*	*Cost (Rs.)*	*Per cent*
(1)	A Fixed cost (Utensils)	300.00	–
1.	B. Varaible cost Transport	2.00	5.19
2.	Fuel	6.00	15.58
3.	Curing (Bamboo dhara)	1.00	2.60
4.	Labour	20.00	51.94
5.	Polishing (sack)	1.00	2.60
6.	Polishing (female) contractual labour	8.00	20.78
7.	* Depreciation on utensils	0.50	1.30
8.	Total (Sl. No. 2 to 8)	38.50	100.00

Calculated on the basis of total volume of business. Note 1 md. = 40 kgs.

Value Chain of Turmeric from Green to Dry and Dust at Whole Seller-cum-Processor Level (Nishiganj/Gokshadanga).

Channel-I: Producer (Green) – Local Assembler (Green) – Processor (Green) – Whole seller (Dry) – Retailer (Dry) – Consumer (Dry) (Distant place).

Channel-II: Procedure (Green) – Paiker-cum-processor (Green) – Whole seller (Dry) – Retailer (Dry) – Consumer (Dry) (Distant place).

Channel-III: Producer (Green) – Local Assembler (Green) – Whole seller (Green) – Same market)

Producer (Green) – Whole seller (Dry) – Retailer (Dry) Consumer (Dry). (Distant place)

Channel-IV: Producer (Green) – Paiker-cum-whole seller (Dry) – Whole seller cum processor (Dust) – Distributor (Dust) – Retailer (Dust) – Consumer (Dust).

Channel-V: Producer (Green) – Local Assembler (Green) – Whole seller (Green) Processor (Dry) – Whole seller (Dust) – Retailer (Dust) – Consumer (Dust).

Channel-VI: Producer (Green) – Local Assembler-cum-processor (Dry) – Retailer-cum-processor (Dust) – Consumer (Dust).

Table 6.6 presents the cost of processing of turmeric from green to dry at whole

Table 6.7: Value Addition of Turmeric from Green to Dry at Primary Wholesale Agril. Market at Nishiganj/Goksadanga

Variety (Green)	*Price/md. (Dried)*	*Conversion/md. Addition/md. (Rs.)*	*Cost of Value (Dried) (Rs.)*	*Sale Price/kg (Green) (Rs.)*	*Profit/md.*	*Rank*
Ghanakathi	Rs. 160.00	7-8kg	38.50	Rs.216.00 @ Rs. 27 per kg	17.50	VI
Sonamukhi	Rs. 160.00	10kg	"	Rs.275.00 @ Rs. 27.50 per kg	76.50	II
Hemti (local)	Rs. 100.00	5-6kg	"	Rs.159.00 @ Rs. 26.50 per kg	20.50	V
Hemti (Malsira)	Rs. 90.00	5-6kg	"	Rs.159.00 @ Rs. 26.50 per kg	30.50	IV
Bitti (Adaforma)	Rs. 70.00	3-3$^{1}/_{2}$kg	"	Rs.84.00 @ Rs. 24 per kg	(-) 24.50	X
Hemti (Basmati)	Rs. 75.00	4-5kg	"	Rs.125.00 @ Rs. 25 per kg	11.50	VII
Kuchlat	Rs. 150.00	7 kg	""	Rs.192.50 @ Rs. 27.50 per kg	4.00	VIII
Hirat Patnai	Rs. 170.00	9-10kg	"	Rs.256.50 @ Rs. 27.50 per kg	48.00	III
Hemti (Mahisa)	Rs. 120	5-6kg	"	Rs.162 @ Rs. 27.50 per kg	3.50	IX
Aam Haldi	60.00	8-9kg	"	Rs.180.00 @ Rs. 20 per kg	81.50	I

Note: 1 md = 40 kgs.

sale cum processor level. This processing unit is still organized at house hold level. The processing costs have been calculated on the basis of two items of costs viz, fixed and variable costs. The variable cost incurred by the entrepreneurs is the highest for labour (51.94 per cent) and the lowest for the item of depreciation (1.30 per cent). The analysis of the item of the cost in the Table 7.6 also reveals that the processing unit is highly labour intensive but less capital intensive.

The value addition of ginger is done through washing, cleaning, drying grading packaging and lebelling and it is pertinent to note here that all the work mentioned above is performed by the labourers on contractual basis (Rs. 30/- for 60kg) as reported by the entrepreneurs. Most of the labourers are from outside the state.

Table 6.7 depicts the value addition of turmeric from green to dry at house hold level. Several varieties were also identified during investigation, which are mainly used by the whole seller for processing locally. Among the varieties conversion ratio is the highest for sonamukhi and the lowest for Adaforma or bitti. The purchase price/md of the green turmeric varies from Rs. 170 (Hirat or Patnai) to Rs. 60 (Aam Haldi). The Table also indicates that there is significant variation of sale prices of dired turmeric, ranges from Rs. 27.50 per kg (Sonamukhi) to Rs. 20.00 (Aam Haldi). So far as profit per md. Green turmeric is concerned (conversion), the Table also shows that profit is the highest in "Aam, Haldi" and the lowest (negative) in "Bitti" one remarkable feature came out from the study that the area covered by the Aam Haldi is minimum, where as the area covered by the Bitti is moderate as reported by the producers as well as whole seller.

The investment pattern of value addition of ginger of an entrepreneur is presented in Table 7.8. The cost incurred by the entrepreneur is the highest (36.50 per cent) for the items store and office and pump and accessories, followed by processing structure (18.25 per cent) excavation (3.65 per cent, electricity charge (2.92 per cent) utensils (1.46 per cent) and shallow point (0.73 per cent) respectively. The results indicate that the processing unit is capital intensive.

Table 7.8: Investment Requirement of Value Addition Unit of Ginger at Commercial Level

Sl.No.	Particulars	Cost (Rs.)	Per cent
1	Processing structure	25000	18.25
2	Storage and Office etc.	50000	36.50
3	Pumps and Accessories	50000	36.50
4	Shallow point	1000	0.73
5	Excavation, cleaning of pond etc.	5000	3.65
6	Utensils	2000	1.46
7	Electricity	4000	2.92
	Total	137000	100.00

The cost of value addition of ginger at commercial level is being presented in Table 6.9. The important cost items for value addition are labour (570.25 per cent), weight loss (19.08 per cent), Diesel (14.37 per cent). The table also indicates that the processing unit is high labour intensive in nature.

Table 6.9: Cost of Value Addition of Ginger at Commercial Level (per annum)

Sl.No.	Particulars	Cost (Rs.)	Per cent
1.	Running Cost- Diesel	60000.00	14.73
2.	Colour for marking bags	50.00	0.01
3.	Maintenance of pumps	1500.00	0.36
4.	Labour cost (loading, unloading, washing, cleaning, drying and grading) 7970bgs@ Rs. 30/-	239100.00	57.25
5.	Weight loss due to cleaning and washing @ 2 kg/bag = 15940 kg x Rs. 5	79700.00	19.08
6.	Interest on working capital @ 6 per cent	22893.00	5.48
7.	Depreciation on buildings	244.00	0.06
8.	Depreciation on Machines	302.00	0.07
9.	Depreciation on Utensils	120.00	0.03
10.	Interest on fixed capital @ 10 per cent	13700.00	3.28
	Total Cost	417609.00	

Note: Value addition of Ginger cost Rs. 0.873/kg (approx).

Table 6.10: Volume of Business (per annum)

Sl.No.	Months	Quantity (1 bag = 60 kg)	Total (kg)
1.	December	2100bag x 60 kg	126000 (26.35)
2.	January	1800 bag x 60 kg	108000 (22.58)
3.	February	1500 bag x 60 kg	90000 (18.82)
4.	March	1480 bag x 60 kg	88800 (18.57)
5.	April	450 bag x 60 kg	27000 (5.65)
6.	May	250 bag x 60 kg	15000 (3.14)
7.	June	210 bag x 60 kg	12600 (2.63)
8.	July	180 bag x 60 kg	10800 (2.26)
	Total	7970 bag x 60 kg	478200 (100)

Figures in the parentheses indicate percentage to total.

The volume of business of ginger (per annum) as mentioned by the entrepreneurs is being represented in the Table 6.10. The analysis of the Table reveals that the volume of business starts well from the month of Dec. (26.35 per cent) and continued up to March (180.57 per cent) The producers of North-Eastern hill regions sale their produce to this entrepreneur and others, for filling the demand of this processing unit. In the month of Dec. the volume of business in this unit was highest due to

Christmas day, a the producers of those areas required a good amount of money for observe the holy festival. The investigation also reveals that from the month of August to November, the volume of business of this unit is negligible.

Table 6.11: Economics of Value Addition of Ginger at Commercial Level (per annum)

Sl.No.	*Items*	*Unit*	*Value (Rs.)*
1.	Purchase Price	478200kg@ Rs 5/-	2391000.00
2.	Value-Addition	478200kg@ Rs 0.873/- (approx)	417609.00
3.	Sale Price	478200kg@ Rs 9/-	4303800.00
4.	Profit [3 – (1+2)]	Rs. 28,08,609 Rs. 43,03,800	14,95,191.00

Source: Agricultural Situation in India, April, 2005, Publication Division Directorate of Economics and Statistics, Department of Agriculture and Co-operation, Ministry of Agriculture, Government of India.

Chapter 7

Strategy for Agricultural Development in the Context IPR and Farmers' Rights

S.K.T. Nasar

Introduction

Agricultural development and progress of human civilisation are interdependent. Agriculture forms the foundation of Indian socio-culture and economy. The sunlight, air, water, soil, and numerous living organisms were deemed since the birth of organised agriculture as a common heritage freely accessible to all. Similarly, the benefit of knowledge and wisdom, barring exceptions, was a bounty for universal use. Exchange of ideas and experiences in agriculture flowed effortlessly. Plant types, planting materials, seeds and production technology changed hands from one human generation to another and from one geographical location to another without restraint. These gifts of nature gradually attracted human desire for control of, and profitability from these common good.

Intensification of trading and globalisation shaped a paradigm shift over time and several items essential to agriculture were controlled and owned as property. India inherited the colonial Property Rights regime in the form of common laws. The most notable law in the present context was the British Empire's Patent Act of 1856 on Protection of Inventions based on the British Patent Law of 1852. This Act, although hardly used in India till recently, acknowledged for the first time that exploitable intellectual prowess is a property. The colonial Indian Patent laws could not inculcate an aptitude for invention for industrial purposes among common people. The foundation of an intellectual property rights regime was nevertheless laid.

Post-independence India showed a truly remarkable growth of agricultural production, based largely on input intensive agriculture with high yielding and fertiliser responsive varieties in irrigated systems. The integrated intensive farming system inter alia caused increased genetic uniformity of cropped spices and decreased indigenous agricultural biodiversity. New laws, policy shifts, accelerated agri-business and international trade progressively dominated in reshaping Indian agriculture. Technological advancements, climate change, globalisation and profit focusing are among the most prominent factors affecting the production system. Breakthroughs in agriculture related Recombinant DNA (r-DNA) technology and biotechnology, the boom in computer-aided automation and networking, satellite assisted geographical information services, and precision agriculture have together opened extraordinary possibilities. The silver-line among agricultural, nutraceutical and pharmaceutical approaches is obliterating. The climate change is imposing alterations in cropping systems on a regional scale. Global warming is changing the hydrological and precipitation status causing substantial migration of crop associated pest and non-pest organisms. These challenges can be met with commensurate adjustments in farm level operations.

Globalisation, though conceptually an old phenomenon, has gained contemporary significance due to the new international trade order. India is under obligation as a founder member of the World Trade Organisation (WTO) to make far-reaching policy changes within binding legal frames. Most of these, specifically the laws enacted in conformity with Intellectual Property Rights (TRIPs), Agreement on Agriculture (AoA) and Convention on Biological Diversity (CBD) involve agriculture in a big way. New legislations in consequence of international commitments on patenting of microorganism, plant variety, biological diversity, seed, planting material, molecule, gene and gene reconstruct have enormous impacts on agriculture. Agricultural business and trade in domestic and international markets is also noticeably influenced by Sanitary and Phytosanitary (SPS) measures, Geographical Indications (GI), Codex Alimentarius Commission standards and a host of other international instruments.

This presentation briefly discusses the parallelism and convergence of today's agriculture and intellectual property rights regime in the Indian context. It also emphasises that local communities' and farmers' rights must be essentially included in the strategy for sustainable and long-term agricultural developments in India.

Basic Tenets of Agriculture

Organised agriculture fundamentally uses the phenomenon of photosynthesis for the capture and conversion of solar energy into chemical energy utilisable by living organisms. Photosynthesis is the monopoly of cells harbouring a green pigment, chlorophyll, in tiny chloroplasts. Living organisms that lack chloroplasts depend upon photosynthetic autotrophs for energy to sustain life. The flow of solar energy via autotrophic primary producers to the heterotrophic consumers takes the form of a food or energy web. Agriculture depends on characteristic and processes at the levels of biomolecules, cells, tissues, organisms and ecosystems. Man-made alteration at one or more of such levels, even if mediated by natural phenomena, is now patentable.

The patentees hold ownership of such inventions. Agriculture also entails migration, movements and transportation of agricultural commodities, products, knowledge and processes. These components of agriculture are traded in diverse forms and at a variety of levels across boundaries. Unlike in the more visible industry, all pertinent items of sustainable agriculture are within the purview of international patent regime and trading arrangement. Non-tariff barriers are also used for agricultural commodities at a growing frequency. Farmers' interests are, however, missing in the whole scheme of these developments although agriculture is ingrained in the Indian ethos as integral part of tradition, culture, environment and livelihood security of rural masses.

The strategy for agricultural development essentially revolves round land, labour, energy, seed, water, technology, and farmers' skill for production, harvesting and efficiency in disposal of product. Socio-political and economic situations complicate the structuring of strategy for sustainable developments at local and national levels. Farmers have to tend to their individual needs while the government and the market forces have their own agenda. Flood, drought, erratic rainfall, climate change and global warming make the formulation of long-term strategy a daunting task.

Origin and Divergence of Agricultural Development Strategy

Subsequent to the transition from hunting-gathering to organised farming, local to global to the present-day glocal scenario piloted phases of developments in agriculture. Alongside, subsistence agriculture transformed in to surplus agriculture mainly for business and trade at local level expanding finally to the international stage. Infusion of technology-assisted reduction of dependence on seasons and weather, especially rainfall pattern boosted non-farm business and trade. Substitution of the age-old natural agriculture by modern chemical agriculture propelled the whole scenario to a new phase. Localised soil-plant-animal-human continuum with concomitant recycling of water, energy, nutrient of the earlier era gradually led to the cutting off of the elements of production and consumption as separated entities (Figure 7.1).

Most developments in agriculture gained momentum over the last hundred and fifty years peaking in recent decades. There has been a gradual marketisation of agriculture with concomitant increase in competition among communities culminating in competition among countries.

International competitions and desire for profits resulted in the distortion of trade in harvest-level and value added agricultural products. These phenomena led to the formulations of international laws, covenants, treaties and agreements. Such developments focused largely on agriculture related trade and business. Production units run by farmers were not given due attention.

Crisscrossing of Agricultural Knowledge and Technology

Extant cropped varieties with associated traditional knowledge and farming techniques crisscrossed human generations and geographical boundaries with increasing velocity. Communities and nations adapted immigrated species and

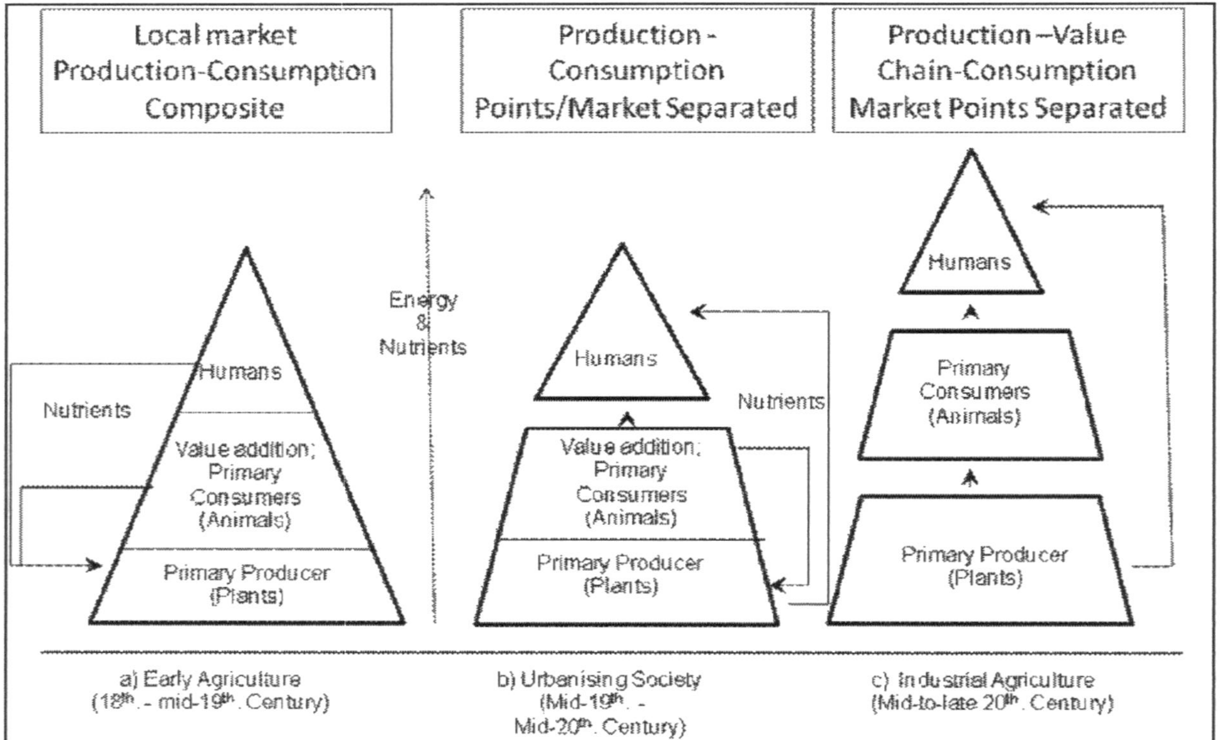

Adapted from John Bellamy Foster and Fred Magdoff 2000 Liebig, Marx and the Depletion of Soil Fertility: Relevance for Today's Agriculture, In Chapter Two, Hungry for Profit- The Agribusiness Threat to Farmers, Food and the Environment; Editors: Fred Magdoff, John Bellamy Foster and Frederick H. Buttel 2000; Monthly Review Press, New York.

Figure 7.1: Phases of Agriculture through Recent Centuries Showing the Separation of Local Producer-Consumer Consortium to Disparate Entities.

varieties of plant, animal and fish with modifications suited to local situations. Exchange, selection and breeding of useful types became routine. Farming communities conserved favoured and marketable types for use and further improvement generation after generation. Later, organised breeding programmes executed by trained scientists yielded improved varieties comparatively more quickly.

In More Recent Years

Recent years have seen extraordinary growth of farming-for-profit, corporatisation of agriculture, value-addition chains for agriproduce, agriculture for industrial and pharmaceutical inputs, international trade in agricultural inputs and outputs, applications of biotechnology to agriculture, molecular farming etc. The age old farming based on location specific natural resources gave way to chemical farming with a quantum jump for external inputs of seed, water, energy, technology, mechanisation, finance, and marketing channels. Seasonality of agriculture began breaking down due to quick transportation of agricultural commodities far away from production points to consumption ends. Consequently, modern agriculture was faced with the challenge of growing demand for food, feed, nutrition, medicinal ingredients, and agricultural inputs for industry to satisfy the needs of modern societies yet keeping production profitable to the farming communities on a sustainable basis. In the meantime, effects of global warming and climate change, pollution in and by agriculture came to the fore as added challenges. Ecological concerns about the depletion of natural soil fertility through loss of soil nutrients and about the loss of forest cover to accommodate agriculture were raised during the middle of the eighteenth century in both Europe and North America. Such concerns remain on the agenda till date.

Another important result of unprecedented change in agriculture was the shifting of importance of farmers' field production factors to non-farm production-relatoed factors (Figure 7.2).

Farmers till land, invest in production, manage field operations, spread seeds from previous harvest, irrigate fields with water as a free natural resource, use compost or farmyard manure, and, finally consume bulk of the produce with as much to spare as possible. The surplus is marketted in local mandis. With the growing population, and concomitant increase in demands for agricultural producuce, the government joined hands with market leaders in restructing production systems with increasing amounts of inputs referred to as non-production system factors in Figure 7.2. External factors influenced agriculture in new ways. This phenomenon led to a spate of supply, manufacture and marketing of agricultural inputs such as seeds, planting materials, fertilisers, agrochemicals, farm machinery and, of course new technologies. The government supplied irrigation water, enrgy and technical inputs free of cost or at subsidised rates to farmers. Indian farms got cut off from traditional systems. However, the Green Revolution touched India in patches such as the states of Punjab, Haryana, Tamil Nadu, Maharashtra etc. While some states excelled in the the green revolution production systems with reference to selected crops, the North-Eastern states by and large lagged behind. Till a few years back the agricultural development in the states was measured more by fertiliser consumption index rather than long term

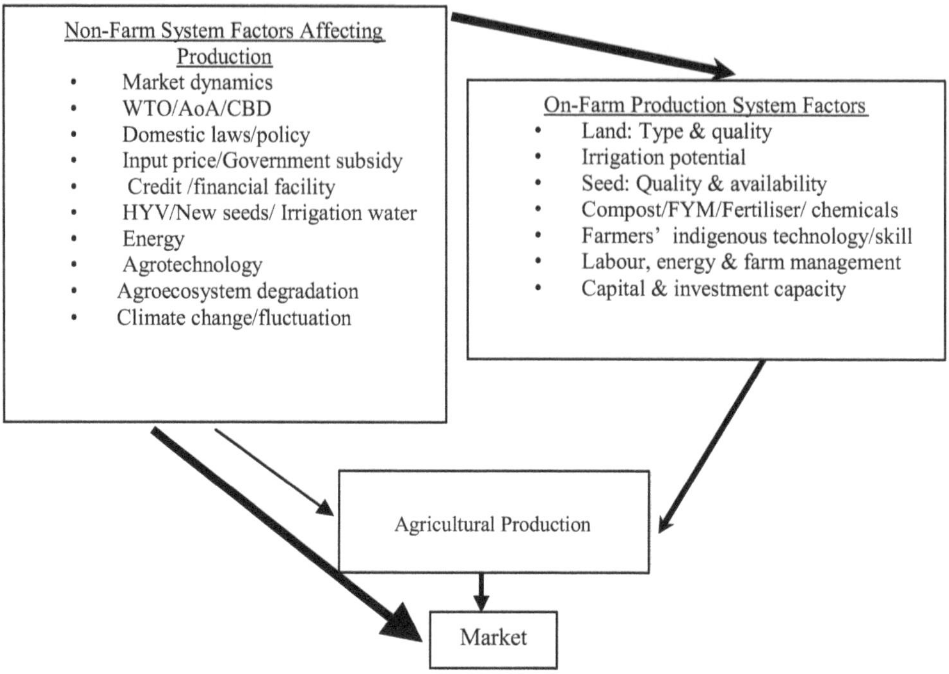

Figure 7.2: The Shift of Farm Production System Factor to Non-Farm Production System Factors Due Mainly to Marketisation of Agriculture.

sustainability of agroecosystems. States began competing with each other in agricultural production for earning profits and petro-dollars through exports. Maketisation of agriculture brought in trade related operation in abig way and drew in patenting systems along with it. Increasingly efficient information dissemination systems and experiences of farming communities highlighted issues relating to loss of soil factor productivity, rising input and production costs, and above all their own profitability. The perception of both the farming and the consumer communities that intensified agriculture caused pollution of agricultural commodities, groundwater and environment made them distrustful of the advertised benefits of new technologies.

Organic Farming Movement

Organic farming is indeed the original form of agriculture in practice for millennia. Chemical agriculture based on the use of synthetic fertilisers originated early in the 20th century supplemented with synthetic pesticides in the second half of the last century. India launched the first Green Revolution on a massive scale in the 1960s with policy supports and subsidies for inputs. This strategy for intensive and extensive agriculture with HYVs, chemical fertilisers, synthetic pesticides, irrigation systems and technology transfer programmes paid dividends initially but gradually inviting serious questions. Downsides of Green Revolution have led to the need of a second Green Revolution based on ecofriendly, sustainable, and rebiologised agriculture with equitable access to livelihood security. Strategies for introducing reinvigorated and high performance organic agriculture are under consideration of many Indian states.

The concept of modern organic farming is attributed to Sir Albert Howard. Sir Albert, an agricultural advisor in India from 1905 to 1924, conducted most of his field experiments at the then Imperial Agricultural Research Institute (IARI) at Pusa in Bihar. In his book 'An Agricultural Testament' (1943), Sir Albert propounded "the Law of the Return" that defines the fundamental tenets of organic farming to date. Chemical agriculture, on the other hand, was launched in Germany during the mid-nineteenth century with Justus von Liebig concretising and popularising his "Law of the Minimum" (1862) first advanced by Carl Sprengel (1828). This principle determined the application of synthetic fertilisers and other agrochemicals in modern agriculture.

India adopted chemical based agricultural technology on a massive scale during the mid-sixties of the last century to commence the first Green Revolution (GR) that paid dividends at a crucial phase of food security pitfall. Several facts emerged as GR took roots and expanded. GR policies largely cut Indian agriculture off the age-old organic cultivation system. Irrigation facilities in the form of huge dams, canals and shallow to deep tube wells were created along with subsidised electricity supplies. Agricultural technologies, HYV seeds and agrochemicals were imported to augment the inadequate indigenous industrial productions. Huge subsidies are still required for inputs, especially for petroleum-based N fertilisers. Pollution in and pollution by agriculture is a formidable challenge. Soil health is a matter of great concern. Partial factor productivity of inputs has dipped alarmingly. Water disputes at local and national levels are at an all time high. GR has created a situation of rising agricultural cost to India and diminishing livelihood prospects for the tiller-farmers. With the four-fold increase in food grain production during the period 1950-1998 there was a concomitant 138-fold increase in fertiliser application. The second generation problems of GR1 intensive exploitative agriculture, such as rapid decline in soil fertility, resurgence of pests and diseases and environmental pollution started showing up. Intensive agriculture has created environmental concerns, which are (1) excessive consumption of irrigation water resulted in over drawl of groundwater, (2) replacement of a rich diversity of traditional crop varieties with few high yielding varieties, and (3) loss of natural soil fertility and destruction of soil biological nutrient cycling system, in addition to water pollution.

Intellectual Property Rights

Arable land, water bodies, marine ecosystems are created by nobody, yet the system of its ownership was established millennia ago followed by the ownership right over such property. Of late, the outputs of human ingenuity suitable for profit, business and trade were granted time bound rights in the form of intellectual property rights (IPR). Patent in the form of right over novel and distinctive varieties of crop plants was extended to the breeders as individuals or as corporate bodies. Novel forms of patenting of living organisms, biomolecules or part thereof are a latest phenomenon. In consequence, the result of human ingenuity in the form of novel, distinct, uniform and stable microbes, crop plants and, in some cases animal and fish are now extended *sui generis* patents. Likewise, novel genomes, genes, biomolecules are included into the patent regime. Unprecedented and quick expansion of intellectual property rights also invited criticisms. Intellectual property rights over living organisms especially crop plants are considered by many as being of benefit mainly to the holders of patent rights while it is restrictive for the producing communities. The clause of value for cultivation and use was also added for cropped plant and animal species by some nation-states.

A Major Criticism of Treaties such as WTO and CBD is that Indian Famers are given only Limited Rights.

The World Intellectual Property Organisation (WIPO) of UNO is the international disputes settlement and negotiation agency. However, the International Court of Justice as WIPO's enforcement arm never heard an IPR dispute case.

Such generally concurrent but often inconsistent developments in different parts of the world at different times culminated in the formation of international consensus leading to the establishment of international bodies of which two are extremely important for this discussion. The World Trade Organisation (WTO) was established in 1995 under the aegis of United Nations Organisation (UNO), based on General Agreement on Trade and Tariff (GATT; 1948 to 1993) (see World Trade Organization 1994). This global development much away from local farmers and their cultivated fields finally dictated the implementation of novel policies and operational strategies by the governments of countries of the world, India included. The second important international instrument is the United Nations Convention on Biological Diversity (CBD) and its Cartagena Protocol on Biosafety. India signed CBD on 5 June 1992 at Rio de Janeiro and Cartagena Biosafety Protocol on 29 September 2000 at Montreal. The emerging global circumstances and commitments required matching laws and regulatory systems in India. Other laws relevant to Indian agriculture and farmers' rights framed under international instruments exist in addition WTO and CBD.

International Treaties and Laws in India

Article 253 of the Constitution of India provides for "Legislation for giving effect to international agreements - Notwithstanding anything in the foregoing provisions of this Chapter, Parliament has power to make any law for the whole or any part of the territory of India for implementing any treaty, agreement or convention with any other country or countries or any decision made at any international conference, association or other body"

International treaties are either binding or non-binding in consequence of India signing the agreements on completion of negotiations. Simple bilateral agreements between India and another country are binding. Other types of international agreements involve ratification and/or legislation by the Parliament. The latter is exemplified by multilateral treaties such as the Final Act of the Uruguay Round of Trade negotiations under General Agreement on Trade and Tariff (GATT).

International forums are also referred to as international instruments. Many a time, legislations are prompted by international commitments made by India in different fora notwithstanding anything contrary to Constitutional provisions. Legislatures enact laws for the welfare of the people. Laws in India are made within the Constitutional framework by the legislature – Parliament and State Assemblies. Such laws are known as Acts divided into sections. Every Act authorises the government to frame rules consistent with connected laws. Acts and rules are announced through Gazette Notifications to be in effect. Authorised executives issue orders.

Agriculture-related International Institutional and Legal Framework

Agriculture-related International Legal and Institutional Framework in addition to WTO, TRIPS and AoA are: Consultative Group on International Agricultural Research (CGIAR; 1971), International Undertaking for Plant Genetic Resources (IUPGRFA; 1983), International Treaty on Plant Genetic Resources for Food and Agriculture (ITPGRFA; 2001), International Convention for the Protection of New Varieties of Plants (UPOV Convention; adopted 1961 and revised 1972, 1978 and 1991); Convention on Biological Diversity (CBD; 1992), UN Covenant on Economic, Social and Cultural Rights (ESCR Covenant; adopted 1966; entry into force 1976). However, laws in India made under obligation of WTO-TRIPS and CBD are most relevant for agriculture and its developmental strategies. India enacted Plant Varieties Protection and Farmers' Rights (PVP and FR or PVPFR) law in consonance with section 27.3 (0) of TRIPS. Almost simultaneously, India embraced decisions of Convention on Biological Diversity (CBD) and framed laws under The Biological Diversity Act. Strategy for agricultural development must take into consideration the realities mentioned.

Agreement on Agriculture (AoA)

WTO member governments including India consider AoA as a significant move towards fairer global competition with the least distortions in agriculture-related trade, improved market access and reduced trade-distorting subsidies in agriculture. The purported objective of AoA is to reform trade in agriculture by making corresponding market-friendly policies to improve predictability and security for importing and exporting countries. These relate to enhanced two-way market access, reduced domestic support as subsidies and programmes that promise raised farmgate prices and farmers' incomes, and decreased export subsidies and other methods used to make exports artificially competitive. Governments began to plug loopholes in AoA by fixing and reducing tariffs, freeing agriculture of bans or restrictions on import, and phased subsidy reduction, both in domestic markets and on exports. Marketisation of Indian agriculture has been given greater importance than farmers' traditional benefits.

Protection of Geographical Indications

Geographical indications identify a good as originating in the territory of a country, or a region or locality in that territory, where a given quality, reputation or other characteristic of the good is essentially attributable to its geographical origin. The Protection of Geographical Indications has been considered under Section 3 of Article 22 of WTO-TRIPS.

Sanitary and Phytosanitary Measures (SPS)

Part VIII: Article 14 of AoA deals with Sanitary and Phytosanitary Measures. This agreement concerns food safety and animal and plant health regulations. The agreement recognises that governments have the right to take sanitary and phytosanitary measures but that they should be applied only to the extent necessary to protect human, animal or plant life or health and should not arbitrarily or unjustifiably discriminate between Members where identical or similar conditions prevail.

Ratification and Enactment of New Laws in India

The ratification by India of TRIPs changed the situation. Consequently Protection of Plant Variety and Farmers' Rights Act (2001), 'new' Patent Act (Amendment Ordinance; 2004), Biological Diversity Act (2002), Geographical Indication Act (2001) and amendments to Seed Act (2004) and other laws and instruments relevant to agriculture were promulgated in quick successions. Genetic Use Restriction Technologies (GURTs) and V-GURTs, plant varieties with 'terminator' genes that mimicked self-perpetuating IPR system are spared from patentability. World Trade Organization (WTO), World Intellectual Property Organization (WIPO), Union for the Protection of Plant Varieties (UPOV), International Plant Genetics Resources Institute (IPGRI), Convention on Biological Diversity (CBD), Food and Agriculture Organization (FAO), and World Health Organization (WHO) are among selected international agencies specifically dealing with intellectual property rights (IPRs) concerning agriculture. There exist a host of multinational commissions and committees involving agriculture. Important international undertakings in agriculture broadly are UPOV (1961, 1972, 1978, 1991, 1999), CBD (1992), WTO-TRIPS (1995), International Treaty on Plant Genetic Resources for Food and Agriculture (IT PGRFA 2001) and Codex Alimentarius Commission along with their latest versions. Other international instruments are also operational.

Core items of agriculture and agricultural biotechnology included in IPR system in India are: extant varieties, farmers' varieties, land races, geographical Indicator varieties, new varieties, microorganisms, biochemicals, microbiological processes and products, processes involved in value addition, bacterial transformation, waste management, water purification and imparting nutritive value and disease-, pest- and weed-resistance, genes, gene sequences, transgenesis used for genetic improvement, new farm machinery and implements and indigenous knowledge on material and processes. India is under obligation to implement almost all commitments on or before 1 January 2005. Some laws have been enacted hurriedly as a consequence. The national legislations governing IPRs Indian agriculture are Patents (Ordinance 2004), Geographical Indicators of Goods (Registration and Protection) Act (1999, 2001), PPV and FR Act (2001), Biological Diversity Act (2002) and Seed Act (1966; Amendment 2004). New Regulations and rules are in operation in many cases.

The Government of India in fulfilling the commitments to the Budapest Treaty on the International Recognition of the Deposit of Micro-organisms for the Purposes of Patent Procedure at Budapest on 28 April 1977 promulgated the Patents (Amendment) Ordinance, 2004 on 27 December 2004 amending for the third time the 1970 provision.

The Ordinance introduces product patent protection for drugs, food and chemicals. The new Ordinance extends the product patent regime to agro-chemicals, food and biotechnology products, apart from drugs and pharmaceuticals. The new amendment has not categorically excluded seeds developed by novel means. The Act of 1970 had excluded life forms from patentability and specifically precluded the patentability of methods of agriculture and horticulture. Product patent was also excluded. Seeds of transgenic variety, microorganisms and gene reconstruct are patentable under the new law. This has generated heated debates.

The Protection of Plant Varieties and Farmers' Rights Act (2001)

Article 27.3 (b) of TRIPS under WTO requires member countries to protect plant varieties either by patents, or by an effective sui generis system of protection. Indian Parliament elected to comply using the sui generis option and passed the Protection of Plant Varieties and Farmer's Rights Act (PPVFR) on 9 August 2001. The purpose of PPVFR is "to provide for the establishment of an effective system for protection of plant varieties, the rights of farmers and breeders, [and] to encourage the development of new varieties of plants....". The key element of PPVFR is the protection of intellectual property rights for plant varieties by a registration process. A variety is a plant grouping, except for microorganisms within a single botanical taxon of the lowest known rank, which can be defined by certain characteristics. Five types of varieties such as new varieties, extant varieties, essentially derived varieties, transgenic varieties and farmers' varieties can be registered by the Plant Varieties and Farmer's Rights Protection Authority under the Act. For this purpose, the variety must conform to the criteria of novelty or newness, distinctiveness or distinguishable by at least one essential characteristic, uniformity or constancy and integrity of the distinguishing characteristic or characteristics, and stability or non-changeability across crop generation and ecological situations. DUS or NDUS is an essentiality that must be certified by authorised agencies and institutions. The variety should be a notified variety under Section 5 of the Seed Act. Transgenic varieties cannot be registered under this Act but can be released by the Genetic Engineering Approval Committee (GEAC) under the Environment protection Act and in the Ministry of Environment and Forests. The Act treats the researchers and the farmers almost equally. The scientist-breeders and farmers have been provided their rights over the variety they have developed, maintained and perpetuated in their own ways. The farmers' rights of the Act define the privilege of farmers and their right to protect varieties developed or conserved by them. Farmers can save, use, sow, resow, exchange, share and sell farm produce of a protected variety except sell under a commercial marketing arrangement (branded seeds) [Section 39 (1), (i)–(iv)]. The certificate of registration for a variety issued under this Act shall confer an exclusive right on the breeder or his successor or his agent or licensee, to produce, sell, market, distribute, import or export of the variety [Section 28 (1)]. The researchers may have access to protected varieties for bona fide research purposes [Section 30]. 'Nothing contained in this Act shall prevent (a) the use of any variety registered under this Act by any person using such variety for conducting experiments or research; and (b) the use of a variety by any person as an initial source of a variety for the purpose of creating other varieties provided that the authorisation of the breeder of a registered variety is required where the repeated use of such variety as a parental line is necessary for commercial production of such other newly developed variety'. A community as a legal person can have protection under the Community rights. Creation of a National Gene Fund and establishment of Plant Variety Protection Appellate Tribunal are integral to the law. PPVFR Rules 2003 are in place. The full implementation of the Act in all its components can be achieved only after removal of technical, legal and social constraints.

The duration of protection of registered varieties is different for different crops: trees and vines - 18 years, other crops - 15 years and extant varieties - 15 years from the date of notification of that variety. Under the Plant Act, the researcher has the liberty to conduct experiment with a registered variety, and the farmer has the right to save, use, sow, re-sow, exchange, share or sell his farm produce including seed or a variety protected under the Plant Act. The farmer does not accrue the right to sell seed of a protected variety under a brand name. On the other hand, the breeder holding the certificate of registration for a variety issued under the Act or his successor, agent or licensee may produce, sell, market, distribute, import or export the variety. Any person, who produces, sells, imports or exports any variety or uses denominations to confuse the public without the permission of the rights-owner, infringes his lawful rights. Infringement of any right under PVP and FR Act attracts action under Civil and Criminal laws, the latter involving sentence of up to two years RI and fine of INR 50,000 or its dollar equivalent.

A Farmers' Variety includes all plant varieties, except modern improved varieties, that have been selected, improved and conserved by farmers. Depending on the level of improvement and use in cultivation, they are called traditional varieties, folk varieties and land races. All of them are essentially farmers' varieties. The knowledge associated with each farmer's variety is the decisive factor for assessing the importance of these varieties to agriculture, in the present and in the future. Such indigenous technology knowledge does not find a place in PPVFR laws but forms an important feature of BD Rules.

The Biological Diversity Act (2002)

The Biological Diversity Act 2002, in concurrence of India's commitment to Convention on Biological Diversity, provides for the establishment of a three-tier biodiversity control and management system with the National Biodiversity Authority at the Centre (Chennai), a State Biodiversity Board for each State and a Biodiversity Management Committee for each Local Self Government such as the Panchayat or other local body. The Act is designed to promote conservation; sustainable use and equitable sharing of benefits accrued from the commercialisation of products developed from biodiversity. This provision concerns all living forms and the ecologies of their niches. The Biological Diversity Act, per se, is not a patenting system but provides ownership of life forms for the purpose of benefit sharing among retainer-users of biodiversity. This Act does, however, restrain unscrupulous use of the Patent and PPVFR Acts in equal measure. Each organism indigenous to India is a national property to be controlled, managed and profitably used at local, state and national levels. The Community Biodiversity Registers are legal documents that come in handy in case of national and international disputes. Biological Diversity Rules 2003 have been notified. Again, these rules will see profitable use only with the involvement of all concerned as in the case of PPVFR Rules 2003.

The Geographical Indications of Goods (Registration and Protection) Act (1999)

The act comes into force on 15 September 2003 while the Geographical Indications (Registrations and Protections) rules were framed by the Government of India in 2002.

Referred casually as Geographical Indications (GI) Act covers all goods, such as agricultural goods, manufactured goods and processed goods where a given quality, reputation or other characteristics of such goods is essentially attributable to its geographical origin. This Act has profound impact on trade of especially value added agricultural commodities like Guntur Sannam chilli, Bhalia wheat, Gir Kesar mango, Laxman Bhog mango, Basmati rice, Darjeeling tea, Murshidabad silk, Joynagar Moya, Fazli mango and such plant, animal and fish cultivars, agricultural products, and such other commodities may now be registered for protection to farmers' and traders' advantage.

SPS Measures and Codex Alimentarius

Sanitary and Phytosanitary Measures and Codex Alimentarius Commission standards are not the elements of patent systems. Physical, chemical and biological contaminations considered injurious to life forms above the standards set by WTO agencies, FAO and WHO are unacceptable. These measures and standards, from the manner by which they are applied in a growing number of instances with subjectivity in the definition of quality, are the most important non-tariff barriers allowed under WTO. Strict adherence to the Codex standards is mandatory under the system. Member-states are free to have stricter national standards of their own. The quality standards of the commodity and the increasing standards of quality in international market, especially of phytosanitary nature, such as pesticide residue, heavy metal contaminants, propagules of harmful organisms, mycotoxins, bacterial and similar contamination deter or block trade. These measures can, and indeed do negate advantages, if any, gained by the protection allowed under the instruments of intellectual property rights.

Indian agriculture is a gigantic producing-consuming market. It is faced with unprecedented challenges. The global developments in cutting edge technologies in conjunction with open market economy encourage the national potentials to gain a comparative advantage. The IPR regime encompassing Indian agriculture calls for common action to national benefit. Agriculture has grown larger than mere cropping of organisms. India needs manpower to steer the agriculture sector through international disputes, trade intrigues and local-level litigations. Science and technology involving agriculture should be developed. People's Biodiversity Registers and documentation of innovations of all kinds and at all levels should become indispensable through mass awareness and action. We suggested that economic, legal and developmental aspects of the new agriculture should be seriously considered for inclusion in curricula and co-curricula.

The Emergence of Farmers' Rights in India

WTO and Farmers' Rights

WTO, after a long drawn and tortuous GATT (1986–1994; and also General Agreement on Trade in Services; GATS) finally agreed to allow countries under Trade Related Intellectual Property (TRIPS) to formulate their own sui generis (*i.e.* 'of its own type') system.

PPVFR and Farmers' Rights

Protection of Plant Varieties and Farmers Rights Act (PPVFR), 2001 is the main law in India dealing with Farmers' Rights. The government of India drafted of a bill to grant Plant Breeders' Rights (PBR) in 1993-1994. This was based on provision of *Union Internationale pour la Protection des Obtentions Végétales* / International Union for the Protection of New Varieties of Plants (UPOV), an intergovernmental organization in Geneva, was established by the International Convention for the Protection of New Varieties of Plants. The Convention was adopted in Paris in 1961 and it was revised in 1972, 1978 and 1991. UPOV does not contain farmers' rights. The Ministry of Agriculture prepared a second draft in 1996 and a third one in 1997. The third draft added the words 'Farmers' Rights' in the title and was labelled the Plant Variety Protection and Farmers' Rights Act. This Act is loaded in favour of breeders. However, some specific rights granted to farmers by PPVFR Act are Right to Register Varieties; Right to Reward and Recognition; Right to Information and Compensation for Crop Failure; Right to Compensation for Undisclosed use of Traditional Varieties; Protection against Innocent Infringement and Right to Free Services. Under BD Act on major right is Benefit Sharing. The GI Act does not mention farmer *per se* but indirectly provides for the rights of the farming community. All these rights have been achieved by organised farmers' assertions through decades. An example of farmers being recognised as breeders is the release of rice variety Gontra Bidhan-1 (IET 17430) [Maturity: 110-118 days/Grain type: MS/Ecosystem: Irrigated/Resistant to pests/diseases: Tolerant to Sheath Blight and Sheath Rot and moderately Resistant to BPH; MR-BPH/Recommended for state/region: Irrigated areas of Punjab, West Bengal; Notification Number S. O. 2458 (E) 16.10.2008 (CR); S.O. 2456 (E)] notified under the Seed Act that was developed by Gontra Samabaya Krishi Unnayan Samiti Ltd., Ghetugachi, Nadia, West Bengal in collaboration with Bidhan Chandra Krishi Viswavidyalaya. This co-operative society also maintains substantial agrobiodiversity with focus on rice varieties. This exemplifies that farmers' groups are not only recognised as legal persons under PPVFR and other relevant acts, but may also influence agricultural policy matters with right lawful approaches.

CBD and Farmers' Rights

CBD has recognised, albeit indirectly, several farmers' rights. However individuals or even farmers' communities are not given rights; the entire biodiversity belongs to the government and the local-self governments. The enormous contribution that the local and indigenous communities and farmers of all regions of the world particularly those in the centres of origin and crop diversity have made and will continue to make for the conservation and development of plant genetic resources which constitute the basis of food and agriculture production throughout the world has been duly recognised. Farmers' rights include protection of traditional knowledge relevant to plant genetic resources for food and agriculture, the right to equitably participate in sharing benefits arising from the utilisation of plant genetic resources for food and agriculture and the right to participate in making decisions, at the national level, on matters related to the conservation and sustainable use of plant genetic resources for food and agriculture. All of these except the right of 'making decisions' have been conferred to farmers' communities via the government under BD Act.

BD Act and Farmers' Rights

The Parliament enacted the Biological Diversity Act 2002 (Act No. 18 of 2003) on 5 February 2003.

Traditional Knowledge (TK) Defined

Government of India, on the recommendations of the National Biodiversity Authority framed in 2009 the 'Rules for Protection, Conservation and Effective Management of Traditional Knowledge Relating to Biological Diversity'.

Traditional Knowledge (TK) has been defined under these rules as:

"the collective knowledge of a traditional community including of a group of families, on a particular subject or a skill and passed down from generation to generation, either orally or in written form, relating to properties, uses and characteristics of plant and animal genetic resources; agricultural and healthcare practices, food preservation and processing techniques and devices developed from traditional materials; cultural expressions, products and practices such as weaving patterns, colors, dyes, pottery, painting, poetry, folklore, dance and music; and all other products or processes discovered through a community process including by a member of the community individually but for the common use of the community".

TK associated with biological resource has been used since time immemorial for community services and for commercial benefits. It has provided for development of useful products, processes and technologies. The modern industry has taken advantage of valuable leads provided by TK to save time, money and investment into any research and product development. Hence, a share of benefits must accrue to creators and holders of TK. The international community recognises that for the traditional knowledge to be maintained the social and economic context in which it developed has to be maintained. It is necessary to recognise and respect the rights of holders of TK. TK must be protected against misappropriation and biopiracy.

Strategy for Agricultural Development vis-a-vis IPR and Farmers' Rights

Agriculture has always been dynamic. Strategies for agricultural developments should take into account several factors. Table 7.1 enumerates some selected phase wise strategies adopted by India. The days ahead are likely to throw up more complexity of factors. These include the flux in international trade systems, climate change, and increasing pollution in and by agriculture. Decline in productivity of agriculture related factors, raising cost of production factors and above all the growing disenchantment of the farming community with agriculture. New laws relating to agriculture and related biotechnology are adding to farmers' concerns. Age old rights of farmers are perceived to have been scuttled.

Table 7.1: Phases of Agricultural Development Strategy in India Since 1947

Strategy Phase	*Selected Action/Plan*
1947 to mid-1960s	Agrarian/land reforms; India's membership of the failed ITO; Development of major irrigation projects; Initial debate over organic vs. chemical agriculture; Cooperative credit institution; Community Development and Intensive Area Development Programmes; Agricultural universities/research systems; Agricultural Price Commission was set up (1964)
Mid-1960s to 1980	New agricultural strategy – HYVs, multiple/mixed cropping, modern farm practices, spread of irrigation facilities; achievement of self-sufficiency; Research, extension, input supply, credit, marketing, price support and spread of new agro-technology; Spread of agrochemical manufacture/import
1980 to 1991	Intensification/diversification of production systems; Growth in non-foodgrain output like pharmaceuticals, milk, fishery, poultry, vegetables, fruits etc.; Increased subsidies and support to agriculture sector; Globalisation, liberalisation and marketisation of agriculture
1991-1992 till date	Economic reforms - deregulation, emphasis on privatisation/PPP model; New laws based on WTO, TRIPS, UPOV, AoA, CBD etc.; Release of national agriculture policy;

Adapted from Tripathi and Prasad (2009) and modified.

Domestic and Inter-territorial Conflicts and Disputes will Rise

Biotechnology is labelled as 'the Sunrise Technology' by the government of India. This has lead to a spate of development of genetically modified (GM) crops in India such as Bt-cotton and Bt-brinjal. The latter is facing a moratorium. There are two cases against the developers of Bt-brinjal: one of biopiracy under BD Act pending action by National Biodiversity Authority, and another one under Common Law with focus on PPVFR and GE Approval rule pending decision by higher courts.

Farmers and other consumers are more aware now about environment protection laws and would not accept infringement of their rightful heritage and future.

Strategies for agricultural development must take into due consideration people's aspirations, laws related to agriculture and the dynamics of domestic needs and international desires.

References

Howard, Sir Albert 1943. An Agricultural Testament, Oxford, UK: Oxford University Press. [First published in UK 1940 and in USA 1945; Accessed at @ http://www.zetatalk3.com/docs/Agriculture/An_Agricultural_Testament_1943.pdf], retrieved 9 June 2011 pdf per Special Rodale Press Edition, 1976.

Lee Peoples 2004. Features - International Trade in Agricultural Products: A Research Guide, LLRX.com; http://www.llrx.com/node/867/print.

Liebig Justus von 1862. Die Chemie in ihrer Anwendung auf Agricultur und Physiologie. 7th edition. 2 vols. Brunswick, Germany: Friedrich Vieweg and Sohn.

Magdoff F, Foster JB and Buttel FH (Editors) 2000. Hungry for profit: The agribusiness threat to farmers, food and the environment, Monthly Review Press, New York

Ravi S Bala 2004. Manual on Farmer's Rights; M. S. Swaminathan Research Foundation, pp. 82. Available at www.mssrf.org.

Sanyal SK and Nasar SKT 2002. Arsenic contamination of groundwater in West Bengal (India): Build-up in soil-crop systems. In Analysis and Practice in Water Resource Engineering for Disaster Mitigation, New Age (P) Publishers, New Delhi, pp. 216-222.

Tripathi A and Prasad AR 2009. Agricultural Development in India since Independence: A study on progress, performance and determinants; Journal of Emerging Knowledge on Emerging Markets, Volume 1 (1), pp. 66.

WTO 1994. World Trade Organization: WTO legal texts; General Agreement on Tariffs and Trade. Retrieved from http://www.wto.org/english/docs_e/legal_e/06-gatt_e.htm.

Search in the internet for WTO, UPOV, WIPO, FAO. The Constitution of India and relevant laws.

Chapter 8

Perception of Disaster and Risk in Enterprise Management

Manas Mohan Adhikary and Sankar Kumar Acharya

What is Disaster?

Disaster is some rapid, instantaneous or profound impact of the Natural environment upon the Socio-economic System (Alexander, 1993).

Disaster is an extreme event, concentrated in time and space, which threatens a society (Turner, 1976).

Major planks of Disaster are:

- ☆ Rapid
- ☆ Extreme event
- ☆ Profound Social and Ecological Impact.

In a general sense, disaster can be defined as sudden, calamitous event bringing great damage, loss or destruction.

Disaster and Hazards

Hazards are referred to any potentially damaging phenomenon whereas, it is called a disaster when it occurs in a density-populated area, and results in a large destruction.

Disaster and Development

Although the most important issues of recent era is whether the development is doing any harm to the environment resulting in disastrous consequences in the long run, but most interestingly an inverse relationship between the level of development and loss of human lives in the case of disaster has been found.

As much as 95 per cent of the deaths occur in developing countries, where more than 42.00 million people live in comparison to the developed countries. While the great floods in Bagladesh in 1988 claimed 1410 lives incurring a loss of US $1.1 billion, in contrast, the loss of Mississipi flood of 1993 with toll of 30 people and estimated loss of US $15.8 billion.

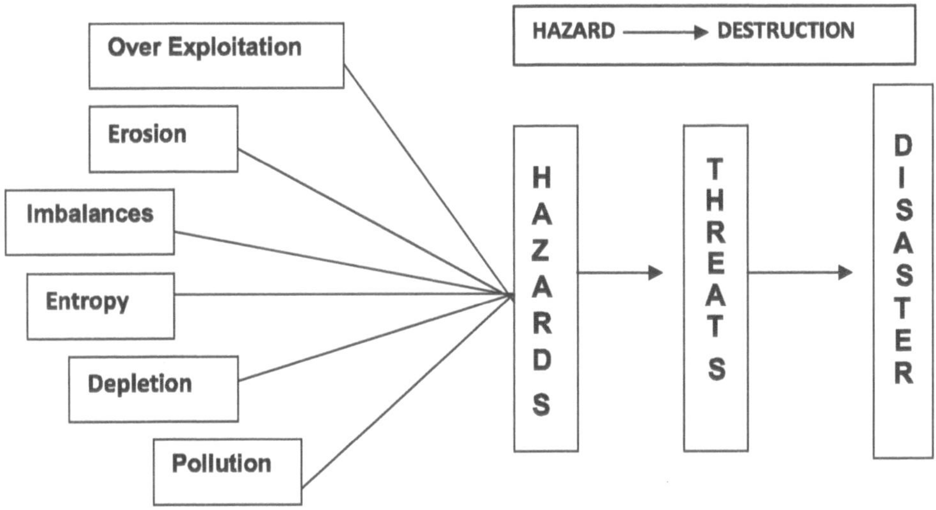

Disaster Management and Disaster Preparedness

Disaster Management is basically a management intervention for reducing the negative impacts of disaster on life, property and ecosystem, building preparedness for accepting and coping up challenges and risks generated from disaster.

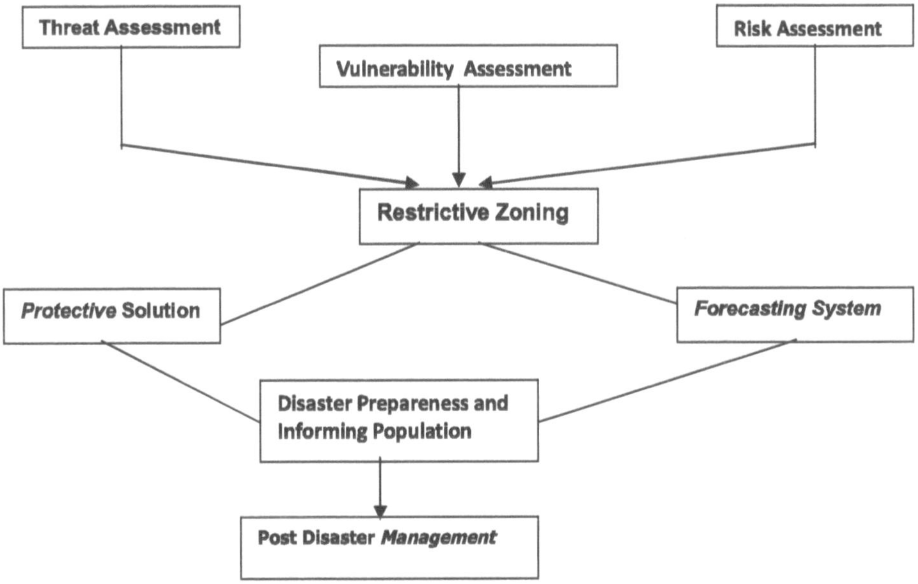

Flow Chart for Disaster Management

Disaster Preparedness

The term "disaster preparedness" reforms to a specialized type of management strategy, for unavoidable circumstances, such as the natural disaster, where, the communities or systems are made preferred not to succumb at the time of disastrous consequences.

Component of Disaster Preparedness

- ☆ Assessment of hazards and risks;
- ☆ Preparation of disaster plan;
- ☆ Anticipation of damage to critical facilities;
- ☆ Disaster training exercise;
- ☆ Building warning information and forecasting system;
- ☆ Organizing monitoring and control systems;
- ☆ Evaluation plan;
- ☆ Designing damage inspection, repair and recovery procedures

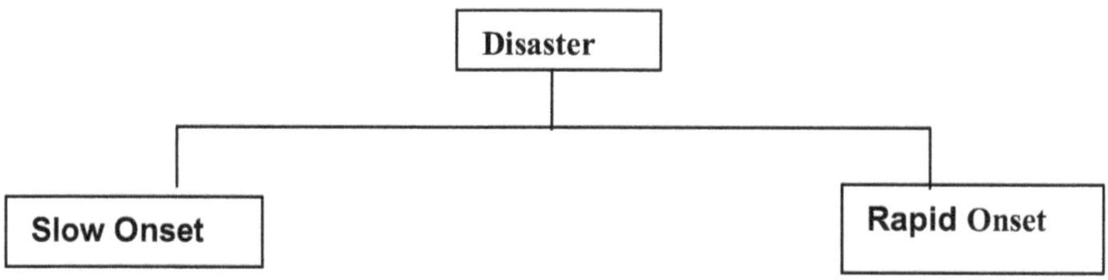

Slow Onset

- ☆ Erosion
- ☆ Depletion
- ☆ Fragile Gene base
- ☆ Global Warming
- ☆ Endemic nature
- ☆ Entropy

Slow Onset

- ☆ The rate of relative semi-level rise is presently approaching 3.14 mm/yr (Sagar Island)
- ☆ Sundarbans will lose 14 per cent of its land area by 2020 and making more than 50,000 people homeless.
- ☆ Several mouzas are engulfed by sea at Digha, East Midnapur, in the last 30 years.
- ☆ In India 60 per cent of the total area is earthquake prone;
- ☆ 15 per cent area is susceptible to annual landslide;
- ☆ 40 million hectares of land are prone to annual flood;
- ☆ 8 per cent of area is visited by cyclones almost once in two years

Rapid Onset Disaster (Lonergan, 1994)

These are considered instantaneous extreme events with profound socio-economic-ecological impacts.

- ☆ Flood
- ☆ Earthquake
- ☆ Tsunami
- ☆ Cyclone
- ☆ Katrina
- ☆ Avalanche
- ☆ Landslide

When	*Where*	*What (Lose of lives)*
1755	Lisbon	60,000
1883	Western Java	36,000
1896	Japan	27,000
1946	Alaska	159
1958	Alaska	200
1960	Chile	1,00
1964	Alaska	120
1976	Philippines	8,000
1998	Papua New Gunieu	2,200
2004	India, ES Asia, Africa	2,00,000+

Earthquake in India

When	*Where*	*What (Lose of lives)*
1737	Calcutta	3,00,000
1793	Mathura	Not known
1819	Kutch	200
1828	Kashmir	1,000
1833	Bihar	1,000
1886	Dacca	2,30,000
1887	Shillong	1,500
1905	Kangra	20,000
1934	Bihar, Nepal	10,700
1950	Assam	1,500
1967	Satara	1,000
1988	Bihar	850

When	Where	What (Lose of lives)
1991	Uttar Pradesh	2,500
1993	Maharashtra	11,000
1997	M.P	50
1997	Latur	20,000
2001	Bhuj	10,000

Classification of Disaster

Methodological

- ☆ Drought
- ☆ Dust Storm
- ☆ Flood
- ☆ Lighting
- ☆ Cyclone
- ☆ Katrina

Geomorphological

- ☆ Earthquake
- ☆ Tsunami
- ☆ Mitch
- ☆ Avalanche
- ☆ Subsidence
- ☆ Ground Water Pollution
- ☆ Coastal Erosion
- ☆ Glacial Lake Outburst
- ☆ Coal fires
- ☆ Volcanic eruption
- ☆ Landslide

Ecological

- ☆ Crop disease
- ☆ Insect infestation
- ☆ Forest fire
- ☆ Mangrove decline
- ☆ Coral reef decline
- ☆ Biodiversity erosion

Socio-Technological

- ☆ Armed conflict
- ☆ Landmines
- ☆ Air – Land – Sea Traffic Accidents
- ☆ Oil Spill
- ☆ Nuclear Disaster
- ☆ Electrical Power Breakdown
- ☆ Collapses of dams
- ☆ Terrorist attacks
- ☆ Pesticides pollution
- ☆ Gas leakage

Global Environmental

- ☆ Acid rain
- ☆ Atmospheric pollution
- ☆ Green house effect
- ☆ Sea Level rise
- ☆ El-Nino and La-Nino
- ☆ Ozone depletion
- ☆ Entropy

Extra Terrestrial

- ☆ Asteroid impact
- ☆ Meteors
- ☆ UFO

Flood: The Indian Scenario

Floral Affected Area (km^2)	*Year of Flood*	*Total Number of Years*
Below 500	1985,1989,1992,1994,1997	5
500–2000	1962,1963,1964,1965, 1972,1975,1996,1966	8
2000–5000	1960,1961,1967,1969,1970, 1974, 1976, 1980,1981, 1982	10
5000–10,000	1973,1977,1993,1995,1998	5
10,000–15,000	1968, 1979, 1983, 1990, 1999	4
15,000– 20,000	1971, 1986, 1987, 1988	4
Above 20,000	1978,1984,1991, 2000	4

DISASTER: Flood

Benefits of Flood Forecasting and Warning

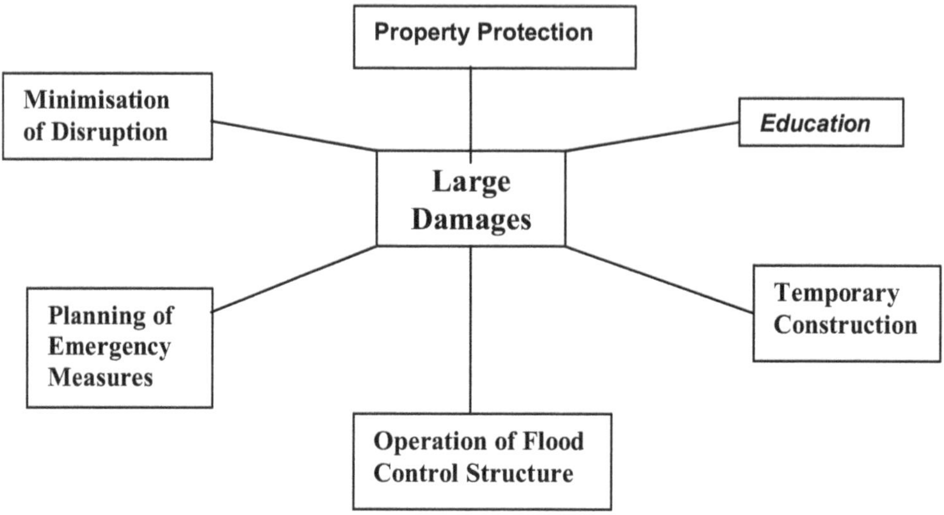

Extension Strategy for Disaster Management

Approaches

- ☆ Identification of high risk hazards
- ☆ Delineation of areas to be affected
- ☆ Assessment of risks
- ☆ Constant monitoring and surveillance
- ☆ Vulnerability of the infrastructure
- ☆ Retrofitting of weak structure
- ☆ Use of space technology like GIS
- ☆ Creating Disaster Information System
- ☆ Multi hazards approach

Cyclic Approaches

- ☆ Prevention
- ☆ Mitigation
- ☆ Preparedness
- ☆ Recovery
- ☆ Rehabilitation
- ☆ Reconstruction

Socio Natural Disaster

Items	*Social Impact*	*Natural Impact*
Poverty	Poor Access to the means of safe and decent life, to mere existence even	✰ Pressure on CPR ✰ Extraction and Depletion of forest, river, lands ✰ Killing of wild lives ✰ Forest fire ✰ Grazing ✰ Selective gene erosion and so many
School drop-out	✰ Social erosion disenchant-ment child labour ✰ Social entropy ✰ Juvenile delinquency	✰ Collection of "mins" and destruction of icthyofaunal diversity ✰ Extraction of fire woods, leaves, wild fruits ✰ Billing of birds, wild lives.
Terrorism and Intimidation	✰ Spurt of social and political violence ✰ Alienation ✰ Extremism ✰ Altruism ✰ Hegemony Genocide	✰ Illicit extraction of natural resources to purchase arms and arsenals ✰ Fruit fire ✰ Landmines to destroy topography ✰ Oil spills
Ethnic clashes and Communal Violence and Wars	✰ Mass billing ✰ Migration ✰ Fractured Nationalism ✰ Identity Crises ✰ Destroyed economy and other Social Institution	✰ Destruction of natural habitats ✰ Pollutions even through radio-active elements ✰ Impotency of soil, genetic materials, livestock e.g. Vietnam, Iraq and ✰ Afganisthan ✰ Earthquake ✰ Killing of sea birds as in gulf war

Hurricane Horror Foretold

.........In January this year (2005) it has been discussed ad nauseam. For years, the federal Government has placed a hurricane-hitting New Orleans Square as one of the most dire threats to the US....

Yet, it seems few listened and fewer cared....

And 10,000 gone with water

Source National Geographic Article

Times of India, September 2005

Bring the unwilling spring
down to earth,
Let there be pristine smile
of
angels and children
let there be
endless showering
of
dreams and piece
of love and bliss

Chapter 9

Water Potentiality in Agriculture vis-à-vis Value Addition

Anandamoy Puste

Introduction

The decline of agricultural productivity has become an intrinsic character across the world. In India the situation is grave. The holding size is just about 0.8 ha to support a family, ecologically and economically. In a great stride to explore untapped arable land, wetland presents a stupendous opportunity to exploit this natural resource to generate both income and livelihood within ecological ambiance. Each of the wetlands in India and anywhere in the world has got geogenic as well as anthropogenic history. While transforming this apparently stale water mass into a vibrant aquatic economy, the participation of the local people with their skill, knowledge and praxis is an essential ingredient to this entire process.

The sociology of water management is earning exponential importance in terms of people participation in the entire process application of traditional knowledge towards maintaining existence and efficacy of natural water bodies and transforming these natural resources in to social capital. In getting all these done, we need to go for social engineering of motivation management and leadership creation. The wetlands projects have gone deeper in ushering a perceptual change over the water resources and intuitioning it as economic source towards generating livelihood and social security too. The present paper is an outcome of social participation in wetland management through creating and effective mechanism for people's participation.

Wetlands: The Concept

Wetlands are the places where the two great natural components that cover our earth - land and water - wet and mingle to support life forms that are often different from those that dwell only on land or only in water". Or in otherwords, "wetland is an area of land where the water level remains near or above of the surface of the ground for most of the year. Simply, "The wetlands are land transitional between terrestrial and aquatic system where the water table usually at or near the surface or the land is covered with shallow water". Moreover, this definition includes several attributes, which are:

1. At least periodically the land must support predominantly hydrophytes,
2. The substrate is predominantly hydric soil and

The substrate is non-soil and is saturated with water or covered by shallow water sometimes during the growing season of each year.

Wetlands: The Role and Contribution

1. Survival of human civilization is inextricably linked with wetlands,
2. Wetlands sustain economic stability to hundred millions of people,
3. On a short time scale, wetland is useful as sources, sinks and transformers of a multitude of chemical, biological and genetic materials,
4. Wetlands clean the polluted water, prevent flood and recharge ground water aquifer,
5. It provides a unique habitat for a wide variety of flora and fauna,
6. The swampy environment of the carboniferous period produced and preserved many of the fossil fuels on which we depend now,
7. James (1995) has rightly termed wetland as *'Natures kidney'* because of the natural functions they performed,
8. Rice, fish, other food and non-food crop production is derived from wetlands,
9. The development and management of wetlands are an important part of the integrated watershed management plans,
10. Wetlands are immensely valuable in terms of economics and quality of life, and
11. A vast area of coastal wetland is very significant today for conservation of mangrove forests as well as wildlife when environmental pollution is a burning question of the world.

Wetlands: Its Proper Utilization

Practically, farmers' are using these vast wetlands as per their traditional need, which are not always agro-ecologically suitable, subject to vacant oftenly without any proper use.

Not systematically, but in scattered way these swampy, fertile, productive wetlands are utilized occasionally by the associated farmers' for the production of nutritious and popular aquatic food crops, economically viable non-food crops, medicinal, aquatic fodder-crops and other important aquatic plants including bio-fertilizers. However, over all production of the system and the economic outturn is not at all encouraging due to technical back up and financial stringency. In fact, this fertile and productive soils of wetland can effectively be utilized following proper agro-techniques for better production, which may turn economically viable for the rural sector of the people (Puste, 2004), *viz.*

The Crops in Wetland Ecology

Aquatic Food Crops

1. Deep water rice (*Oryza sativa* L.)
2. Deep water rice cum fish culture
3. Water chestnut or *singhara* (*Trapa bispinosa* Roxb.)
4. *Makhana* or gorgon nut or fox nut (*Euryale ferox* Salisb.)
5. Swamp taro or *pani kachu* or *jal kachu* (*Cyrtosperma* sp.)

Aquatic Food-cum-Ornamental Plants

6. Lotus or padma (*Nelumbium speciosum*)
7. Water lily or saluk (*Nymphaea* spp.)
8. Royal water lily (*Victoria amazonica*)

Aquatic Non-Food Economic Crops

9. Matreed (*Cyperus* spp.)
10. Shital pati (*Clinogyne dichotoma*)
11. Hogla (*Typha elephantina*)
12. Rattan palms or cane (*Calamus* spp.)
13. Shola (*Aeschynomene aspera*)

Aquatic Fodder Crops

14. Para grass or water grass (*Brachiaria mutica*)
15. *Coix* or gurgur grass (*Coix* spp.)

Aquatic Medicinal Plants

16. Brahmi (*Bacopa monnieri*)
17. Kesuti or keshori (*Eclipta alba*)
18. *Ipomea* or swamp cabbage (*Ipomea aquatica*)
19. *Talmakhana* or *kule khara* (*Hygrophila auriculata*)
20. *Enhydra* or helencha (*Enhydra fluctuans*)
21. Indian pennywort or thankuni (*Hydroctyle asiatica*)
22. *Marsilea* or susni (*Marsilea quadrifolia*)
23. Water cress or swamp forest (*Nasturtium officinale*)

Aquatic Aromatic Plant

24. Bach (*Acorus calamus*)

Aquatic Weed-cum-Plants of Organic and Biofertilizers

25. Water hyacinth (*Eichhornia crassipes*)
26. Duck weed or taka pana (*Pistia stratiotes*)
27. Jal jhangi or Hydrilla (*Hydrilla verticillata*)
28. Azolla or pana (*Azolla pinata*)
29. Algae

Extent and Distribution

Wetlands possess a significant position in respect their distribution and coverage in the world. Wetlands are found not just in remote parts of the world, they are everywhere around us and are most productive ecosystem on earth. Wetlands occupy approximately 6.4 per cent of the world total area (Maltby and Turner, 1983 and Anonymous, 1986). The approximately area covered by wetland in the world, India and West Bengal is presented in Tables 9.1 and 9.2.

Table 9.1: Coverage of Wetlands at a Glance

Place	*Area*
World	855.8 million ha
India	329.0 million ha
West Bengal	5.25 lakh ha

Source: Maltby and Turner, 1983 and Anonymous, 1986.

Table 9.2: Wetlands in West Bengal

Sl.No.	*Types*	*Areas ('000 ha)*
1.	*'Beel'* or almost permanent waterbodies	31
2.	Inundated areas	44
3.	Marshy areas	43
4.	Saline affected areas of Sundarban	400
5.	Wetland areas near the reservoir or wetland beside the dam	6.5

Methodological Approaches and Socialization of Technology

Study Zones

West Bengal is a part of Eastern India located between 20°31′ and 29°14′N latitude and 85°59′E longitude. This state is bounded at south by the Bay of Bengal, at north by Sikim and Bhutan, at east by Assam and Bangladesh border and at the west by Bihar and Orissa. It has six agro-climatic zones (New Alluvial and Old alluvial, Coastal, Red and Laterite, Terai and Hill zone) based on vivid pattern of rainfall, temperature, soil type and topographical situation of land, varied from highly fertile to degraded land situation, humid to semi-arid situation, soil class and weather variability. Almost in all the zones, there is a reasonable area under wetland situation, however, predominantly prevails in New and Old Alluvial, Coastal zone of the state of West Bengal. It is to be marked that it is not so being utilized properly, nor taking so initiatives from any part, except at the farmers' level, which is not sufficient for making them to productive in nature.

Out of six agro-climatic zones of West Bengal, New Alluvial, Old Alluvial and Coastal zones are important and most potential zones of the state. Besides, in all the zones, there is a reasonable area is under waterlogged or low-lying areas, predominantly occurs during rainy months. With this, this areas also posses a demarcated area that are exposed to waterlogging (in different nature as per depth and durability of the condition) during the peak rainy months, almost in every year, suffering much in respect to production system, their livelihood and social scenario to a great extent, may be reached in extreme situation, even in unprecedented form.

Indeed, during the peak rainy months, a sizeable area (about 25-30 per cent area, which covers about 23.5 million ha in almost every year and even, sometimes it may goes up to >40.00 million ha) in the coast and north-eastern plains of India is exposed to waterlogging and the duration of such submergence depends upon the topographical situation of lands and textural class of the soil. The depth of water varies from 1.0 to sometimes >5.0 m, depending upon the situation of lands. These zones particularly, comprises of flat alluvial plains intersecting with the main river system (The Ganges, Mahanada, Rupnarayan, Ajoy, Damodar, Kangsabati, Keleghai, Haldi etc.) and its so many tributaries, canals, low-lying areas, nayanjali subject to frequent waterlogging and flood making them swampy and almost turns to out of normal cultivation during the peak *kharif* or rainy months of the year, resulting unproductive, suffers a lot. This has, however, been aggravated with less storing or intaking capacity proportionately in fewer numbers of natural depressed watershed ponding system of the projected area, as it flooded in a very non-systematic manner over the entire area within a short period.

Moreover, this area aqua-terrestrial ecosystem subjected to face day by day (i) rapid urbanization, (ii) Industrialization, (iii) prolong siltation or sedimentation (iv) delinked with the main river system through tributaries and canals (v) dereliction of productive areas (vi) non-systematic using pattern of the system (vii) lack of proper technology for systematic use of the area (viii) socio-economic condition, particularly for low and poor marginal farmers and (ix) absence of proper training how best to use the situation in terms of making production and economically viable.

Problem Perception and Appraisal

With the effective foot-steps of the project activities at the different centres of the project, it was tried from the project's sources to develop all sorts of possible ways to overcome these problems in respect to need-based cleaning, excavation, renovation for reshaping it further fruitful utilization for the benefits of farming community. The following methodological approaches were undertaken for study as follows : The PRA exercise for :

1. Problem prioritization,
2. Estimation of social osmosis,

3. Constraint analysis,
4. Crop productivity and
5. Benefit-cost ratio.

It is undoubtedly an effective foot-step to organize and involvement of small and marginal group of farming community including well-shaping state Panchayat system in the areas for performing a series of working activities related with the development of aqua-terrestrial ecosystem (survey, selection, demarcation, awareness campaigning, training programme, need-based developmental works and finally technological adaptation) in the targeted rural villages. It undoubtedly was accelerated by putting and involvement of NGOs (i. Taldi Netaji Sangha, Taldi, Canning, ii. WATER, Nabagram, Burdwan and iii. Berhampore SARVODAYA, Berhampore, Murshidabad), SHGs etc. of the areas.

By the adoption of different improved package of practices on the aquatic food and non-food crops along with valuable fish genotypes, in respect to :

1. Selection of location specific cultivars,
2. Integrated and balanced nutrient and
3. Water management practices, which is able to produce more with quality under aqua-terrestrial ecosystem, so as to achieve sustained and high productivity per unit area per unit time as well.

Comparatively less quantity of inorganic fertilizers combined with organic sources was brought the soil health as well as water holding capacity (WHC) of the soil without further degradation of causative factors.

Problem that Could be Solved as a Result of Project Study

The following problems of the study zones were solved through the integrated development approach in problematic waterlogged aqua-terrestrial ecosystem on the basis analyses of these waste wetland areas, as because these areas are:

1. Waste and unused wetlands, which suffers a lot in the area
2. Underutilized crops (aquatic food and non-food)
3. Unbalanced and fragile ecosystem
4. Runoff and soil erosion
5. Unutilized surplus water
6. Less groundwater contribution/recharging of the area
7. Less per plant/crop availability of soil moisture and
8. Over all, technically backward community

Project's Motto/Development through Implementation

1. To conserve rainwater for multiple use including maximization of ground water recharging,

2. Demonstration and dissemination of low-cost farming technology for productive utilization of wastelands without further degradation,
3. To adopt improved agro-techniques for enhancing production system of aquatic food crops, fruit trees and others variables in waterlogged wastelands areas of the areas,
4. To sustainable increase in land and water productivity, profitability of farming and economic stability of inhabitants in degraded problematic lowland ecosystem and
5. To develop a productive technically sound farming community, especially with fishery-based farming system under aqua-terrestrial ecosystem.

Justifications

1. Increased ground water recharging, resulting in more availability of conserved rain in the WHT along with ground water contribution of the area,
2. Conversely, there is less chances of occasional flash-floods in the area, as intake/storage capacity of WHT and dykes with more conserving of runoff water simultaneously increases productivity through the watershed approach,
3. Run off as well as soil erosion drastically reduces due to higher rate of conserved water through watershed hydrological systems,
4. More available *in situ* water vis-à-vis recycling of water as live-saving irrigation during subsequent dry seasons for the crops,
5. More productivity and profitability through multistoried aquatic crops-fish culture cum tree - vegetables under aqua-terrestrial ecosystem,
6. For the participating farmers it will have far-reaching productive impact in waste low land management, the benefit will increase gradually over the years after the completion of project in the form of increased sustainable productivity,
7. Because of the low-cost technology used in the project, the expected confidence and trust of the farmers on the reliability of the scientific technology packages being propagated would be higher and
8. It is thus imperative to utilize and development, the vast lowland wastelands for the shake of development of diversified agriculture, as the system is socially acceptable and environmentally sound and non-polluting.

To realize the facts, with great significance some specific and precise update agro-techniques in unutilized wetlands ecosystems were undertaken in location-specific at the different project sites of West Bengal. Subsequent improvised agro-techniques (Puste *et al.*, 2004) on the variables following the system approach of integration was conducted at the locations during the subsequent pre to post-rainy months, respectively for proper implementation of the project, emphasizing with following objectives of the project –

1. To identify, designate the zone-specific aqua-terrestrial ecosystem
2. To train up and increasing awareness about the system approach and feed back the technical back-up to the contract/participatory farmers at the different sites of the project,
3. To study the improvised agro-techniques on crop cultivars for its productivity and quality produce,
4. To evaluate the integration of variables (aquatic food, non-food crops including fishes) for total productivity of the area,
5. To utilize water resources efficiently through rainwater conservation and recycling it as live-saving irrigation, particularly during the drier months of the year for its optimum yield level of the area and
6. To increase more return over cost of production ha^{-1} in respect to integration of all variables and to generate more employment and avenues.

So, encouragement of scientific culture of integration of crops - cum - fish farming in wetland areas is, therefore, most desirable due to the following sharing of advantages, was possible to utilize at its maximum levels,

1. The synergistic effect of fish on aquatic crops (food, non-food etc.).
2. Control of aquatic weeds and associated insects by fish,
3. Increased efficiency of resource utilization, reduced investment risk through crop diversification and additional sources of food and income,
4. More frequent visits to the field particularly for fish genotypes by the farmers, resulting in better crop management.
5. Low risk for poor crop-cum-fish farmers and modest capital investment,
6. Year round employment opportunity for the farm family and
7. Consequently, improvement of farm family income and nutrition level.

The main aim and objectives of integrated farming system (IFS) are :

- ☆ Proper utilization of arable and/or aqua-terrestrial or aquatic ecosystem,
- ☆ For enhancing productivity per unit area per unit time and
- ☆ For generating income and livelihood development that are sustainable in nature, particularly for small and marginal group of farming community.

Emphasized on the resource utilization of integration of aquatic crops (deep-water rice, water chestnut/*makhana*, lotus/water lily, *Colocasia* etc.) integrated with fish variables for total productivity of the area as well as economic sustainability, was undertaken precisely in the wetland ecosystem of the areas, are as follows :

In Comparative Deep to Semi-deep Water System

1. Deep-water rice - cum - fish culture (as food crop)
2. Water chestnut - cum - fish culture (as food crop)
3. *Makhana* - cum - fish culture (as food crop)
4. Lotus/water lily - cum - fish culture (as food - cum - ornamental crop)

In Comparative Semi-Deep Water System

1. *Lati kachu* - cum - fish culture (as food crop)
2. Shola/hogla - com - fish culture etc. (as non-food crop)

Moreover, utilizing the adjacent terrestrial ecosystem, which was developed through need-based cleaning, excavation and renovation works of the project sources, were utilized through the adaptation of seasonal vegetables and plantation of tree spices on it suitably. Due to high availability of subsoil water, the tree spices and seasonal vegetable were grown very fast in *Gangetic* West Bengal, but it needs sustained scientific input with improved agro-techniques as developed by the different project sources. The system approaches were as like :

In Comparatively Shallow-Submergence Situation

Aquatic or swampy medicinal plants (as food crops) - cum - horticultural plantation (vegetable and other plantation) under aqua-terrestrial ecosystem.

Techno-Social and Economic Outcome of the Study

Analysis of Socio-Economic Status of Wetland Based Agrarian Community

In these areas, the average land holding size were varied from <1.0 ha of 29.24 per cent, 1-2 ha of 54.71 per cent and >2 ha of 16.05 per cent, respectively, dominated mostly poor and marginal farmers of both the communities of Hindu and Muslims. Sometimes, they are mostly belongs to below or equally poverty lines as the areas are mostly faces by flash-flood, could be in unprecedented form. In the same area, annual gross income of a farm family varied accordingly <Rs.10,000/- of 25.0 per cent, Rs.10,000-15,000/- of 41.67 per cent and >Rs.15,000/- of 33.33 per cent, respectively with their family size 5-7 of different age group, which faces the former two groups an acute stringency of money and bound to maintain their minimum livelihood. Prior to technological intervention through the project sources in these areas, subtracting the cost of cultivation (with the maximum utilization of household labours), the production of crop variables was practically less remunerative one, mostly existed in these sub-zones. Indeed, it was only Rs.15,000/- to 20,000/- per ha/year with the utilization of 5-7 farm family members and thus, it was more pronounced when these are raised as mono-crop, traditionally without implication of any suitable agro-techniques on it. This was, however, further aggravated with the socio-economic condition of the area as well as soil fertility and degraded land situation in these areas.

However, it has been tried successfully from the project's end to improve all the condition in respect to: (i). proper utilization of wetlands (waste/unused in nature) including adjacent terrestrial land system, (ii). for making them productive in nature through intensive cropping programmes including culture of fishes in integrated system in aquatic ecosystem and cultivation of seasonal vegetables and plantation of valuable tree spices etc. (iii). economic sustainability of the areas including the improvement of farming condition of the farming community for making improvement of their livelihood pattern, and (iv). in addition, women community as well as tried to promoting SHGs of the targeted projected areas of the respective zones.

Nature of Problem Identified and its Characterization

These zones (New Alluvial, Old Alluvial and Coastal zone of West Bengal) under aquatic ecosystem comprises of flat alluvial plains intersecting with the main river systems - The Ganges, Padma, Mahanada, Mahanadi mostly under Indo-Gangetic basins, and its so many tributaries, canals, *jheels*, *beels*, *nayanjali*, boars, low-lying areas subject to frequent waterlogging and flood making them swampy, which almost deteriorates the normal cultivation during the peak *kharif*/rainy season. It is one of the richest zones effective for production system due to inherent capacity of the soil profile, but with some problem that faces occasionally, especially lying with the wetland ecosystem mainly during the peak rainy months of the year. However, the problems of the zones related to wetland ecosystem that has been identified in the projected areas (Table 9.3) has been summarized as follows:

1. Topographical situation : unusual saucer-shaped depressed land area bounded by terrestrial land, especially mid portion of surrounding villages without drainage scope causing waterlogging mostly,
2. More susceptible to soil erosion, conversely siltation and hence, dereliction of lowland productive areas day by day,
3. Waterlogging causing flash-floods occasionally in the adjoining areas, erosion of top fertile soil with runoff water,
4. Decreased ground water recharging due to sub-soil compaction,
5. Low survival of plantation due to water stagnation and unplanned land shaping and irregular topography,
6. Severity of unwanted aquatic weeds, sometimes it is in dense, severe form,
7. Development of fragile ecosystem in the area causing low production and productivity per unit area and time

Problem Prioritization and Perceptual Evaluation through PRA Exercise

Through PRA exercise the related problems delineating of wetland ecosystem at the different projected areas have been identified for taking further actions (Table 9.3), particularly for development in respect to :

1. Nature of wetlands,
2. Their cropping pattern,
3. Adoption of improvised farm technology for aquatic crops and
4. Integrated system of crop + fish culture etc.

The identified problems may be summarized as follows:

1. Low-lying areas subject to waterlogging, soil erosion, dereliction of productive areas
2. Shriken low-lying areas and siltation, uncontrolled runoff water,

Table 9.3: Perception and Quantification on Site-Specific Problem through PRA Exercise

Sl.No.	*Problems Identified*	*Site-Wise Extent of Problems*				
		Taldi, 24 Parganas (S)	*Kalinara-yanpur, Nadia*	*Nabagram Burdwan*	*Ahiron, Murshi-dabad*	*Mohanpur Nadia*
1.	Mostly unused or waste in nature	++	++	++	+++	++
2.	Low-lying areas subject to waterlogging	+++	+++	++	+++	++
3.	Uncontrolled runoff sometimes flush-flood occur	+++	++++	++	++++	++
4.	Soil erosion	+++	++++	++	++++	+
5.	Siltation or sedimentation and barrier for moving runoff water during peak rainy months	++	++++	+	++++	+

+: Mild; ++: Moderate; +++: Severe; ++++: Very severe.

3. Uneven low-lying and uncontrolled runoff system,
4. Unusual natural aqua-terrestrial ecosystem and flash-flood subject to water-logging,
5. Low-lying depressed areas along with uncontrolled runoff water, ground water depletion,
6. Low-lying areas subject to flush-flood and uncontrolled runoff water and sedimentation, ground water depletion,
7. Low-lying depressed areas subject to flush-flood and uncontrolled runoff water, ground water depletion and
8. Unusual natural aqua-terrestrial ecosystem and flash-flood subject to water-logging, erodability in nature, ground water depletion etc.

Wetland Crops and its Production Potential

The production yield and its potentiality of valuable aquatic crops is so important (Table 9.4) and also are equally important to the common markets and locals *bazaar*, mostly preferred by the rural and urban people. All the aquatic crops (food, non-food commercial) are covering a reasonable area in the different agro-zones of this sub-continent and gaining importance day by day. But, in the country very limited works on improvised agro-techniques have been so far done on the utilization of these underutilized crops, although these would have immense importance for developing in water harvested basins as well as to make it more productive and sustainable in nature. These have great importance, particularly in the north-eastern part of the country due to prevailing of humid to sub-humid accomplished with sub-tropics in nature, which is really congenial for growth and development of these aquatic crops.

Table 9.4: Yield Potentiality of different Wetland Crops at the Different Agro-zones

Sl.No.	*Description*	*Average Production Potentiality (ha^{-1})*
A.	**Food Crops**	
	Deep-water paddy (*Oryza sativa* L.)	2.5 – 3.5 t ha^{-1}
	Water chestnut (*Trapa bispinosa* Roxb.)	7.5 - 10.0 t ha^{-1}
	Makhana (*Euryle ferox* Salisb.)	2.5 - 3.0 t ha^{-1}
	Lati kachu (*Cyrtosperma* sp.)	40 t $ha^{-1}yr^{-1}$ (as *lati*); 18,750 pieces t $ha^{-1}yr^{-1}$ (corm + cormels)
	Cocoyams (*Xanthosoma sagittifolium*)	30-45 t ha^{-1}
	Water-lily and Lotus (*Nelumbo* sp.)	40,000 – 50,000 numbers ha^{-1}
B.	**Non-food Commercial Crops**	
	Matreed (*Cyperus tegetum*)	2,250 bundles of mattable sticks ha^{-1} (1st cutting in Sept-Oct); 1,500 bundles ha^{-1} (2nd cutting in Feb-Mar); 2,250 bundles ha^{-1} (3rd cutting in Jun-July), respectively.
	Hogla (*Typha elephantina*)	500 bundles of mat ha^{-1}
	Sitalpati (*Clinogyne dichotoma*)	750 mats ha^{-1} in 1st cutting; 1125 mats ha^{-1} in 2nd cutting and 3000 mat ha^{-1} in 3rd cutting, respectively.
	Shola (*Aeschynomene aspera*)	It is available in the hat or in market in bundle basis (2.0-2.5 m circumference of 0.75 - 1.5 m length). Single harvesting : 175-225 bundles and multiple harvesting 225-275 bundles ha^{-1}

Zone-Specific Adoption of Improvised Cropping System

As per the situation specific, choice of crops, integration of crops and fish culture under different wetlands (permanent, semi-permanent and temporary in nature) would be greater importance for proper utilization of aqua-terrestrial ecosystem at its maximum level as well as obtaining higher productivity, which may fetch more monetary return per unit area per unit time in a confined area (Table 9.5).

Table 9.5: Utilization of Aqua-Terrestrial Ecosystem Based on Multiple 3-tier Watershed System

Situation	*Depth and Durability of Wetlands*	*Adaptation of Cropping System Utilizing Aqua-Terrestrial Ecosystem*
Situation I	Very shallow-submergence	Mat-making sedges, *Colocasia*, aquatic medicinal plants etc.
Situation II	Shallow-submergence	*Colocasia* + fishes, low-lying paddy + fishes, *makhana* + fishes, aquatic medicinal plants etc.
Situation III	Medium-submergence	*Colocasia* + fishes, deep-water paddy + fishes, water-lily or lotus or *makhana* + fishes, water chestnut + fishes etc.
Situation IV	Deep submergence	Water-lily or lotus or water chestnut + fishes etc.
Situation V	Adjacent terrestrial lands	Field and horticultural crops, vegetables, tree plantations etc.

Undoubtedly, this integrated system as a whole would able to be benefited >2-3 folds (about Rs.60,000 to 80,000/- per ha per annum subtracting the cost of cultivation with a B-C ratio >2.0), even sometimes it may go up to 3.0 than that of obtained previous one (d"1.0). This practice is not only recast the socio-economic condition of the resource poor farmers' in these particular sub-zones but also improve the fertility status of the soil system as a whole.

Integrated Farming System (IFS) Approach

It is realized that number of advantages lying with the integrated farming system than that of monoculture. In fact, improvised demonstrations, particularly on integration of aquatic crops like, deep-water rice (Ghosh and Saha, 1980, Jhingran, 1983, Dutta *et at.*, 1984, Sinhababu *et al.*, 1998, Puste and Bandyopadhyay, 2000 and Bandyopadhyay and Puste, 2001), water chestnut, *makhana* and water-lily, padma etc. (Puste and Das, 2001, Puste *et al.*, 2004, Puste *et al.*, 2007, Puste *et al.*, 2010 and Puste *et al.*, 2011) in different wet situation with different variables along with fish variables in mixed farming system, especially with energy-rich live-fishes are very meager but can sustain with more productivity and outturn of the system in this sub-tropics. For the feasibility of mixed farming rather monoculture may bring about the yield improvements, thereby increasing the net profit of the system as a whole.

Wetland Ecology and Community Mobilization

The protection and conservation of wetland ecosystem has been sought here through community mobilization by the following components in sequence :

- ☆ **Component 1:** General awareness creation amongst the stakeholders of the wetlands, by nature, who are core, intervening and peripheral,
- ☆ **Component 2:** Micro-planning process has been conducted by involving the stakeholders for identifying the elements and areas of interventions,
- ☆ **Component 3:** Generation and cataloguing of traditional knowledge on indigenous aquatic crops grown in the wetlands and their food, economic and ecological values,
- ☆ **Component 4:** PLA exercises through a multi-stakeholders approach,
- ☆ **Component 5:** Integrations between exotic and indigenous knowledge for exploring the possibility of economic production of aquatic enterprises and crops and these have included water chestnut (*Trapa bispinosa* Roxb.), *makhana* (*Euryale ferox* Salisb.), water-lily (*Nymphaea* spp.), lotus (*Nelumbium speciosum*), *Lati kachu* or swamp taro (*Cyrtosperma* sp.) etc.
- ☆ **Component 6:** Standardization and social customization of ecological farms,
- ☆ **Component 7:** The social osmosis between wetland ecosystem band social ecosystems has been studied to identify the point of convergence between water ecology and social ecology.

Point of Convergence

Intervention and Osmosis: Social and Wetland Ecology

Table 9.6: A Representation of the Intervention and Osmosis in Respect to Social and Wetland Ecology

Intervention	*Wetland ecology*	*Osmosis*	*Social ecology*
Economic production system	Suitable for various aquatic crops including fish culture	Income changes from Rs. 500/- to Rs. 1500/- per head	Need for generation for income
Water conservation measures, creating watershed basins, seepage tank etc. and terrestrial or dykes developed.	Water retentivity was there but not enough for the purpose	Cropping intensity and entrepreneurships developed	Need for irrigation to summer rice and live-saving irrigation to various arable crops
Makhana (*Euryale ferox* Salisb.) was included for economic production system	Different types of traditional aquatic crops but was neglected with poor agro-techniques	Social market for value-added mobility (popped-form) as an new income generation	New food for starch, nutritious
Water chestnut (*Trapa bispinosa* Roxb.) was included for more production system	Water chestnut with the system with poor management	Market developed as fresh fruit in absence of other fruits during the time and ground form as *'atta'*	New food item with enriched starch, carbohydrates, protein, minerals etc.
Pani-kachu or *jal kachu* (*Cyrtosperma* sp.)	It was with their practice but with agro-technques, low inputs	Market developed as corm, stolons as vegetables	New food item with enriched starch, carbohydrates, protein, minerals etc.
Intervention of energy-rich live-fishes integrated with aquatic crops as IFS models	System was in the wetland ecology but with poor agro-techniques.	Total productivity per unit area per unit time developed along with market accessibility	Foods from aquatic crops as well as fishes with the farming system enhanced, nutritious to the farm family also
Seasonal vegetables and other valuable tree plantation developed on the dykes	It was with the system but not in a systematic manner	More income generation from IFS in short as well as long term basis.	Developed IFS based on aqua-terrestrial ecosystem, enhanced total production system round the year

Impact of Technological Intervention and Livelihood Development

Through studies on the beneficiary group of farming community in the areas were done systematically for knowing the impact/outcome of technological intervention and livelihood development in respect to before technological interventions.

Condition before Intervention

Cropping intensity (134.8 per cent), ground water recharge (5.76 ft) and income was low, uncontrolled soil loss and runoff and life-style was not so good.

Condition after Intervention

1. Crop Productivity : has been increased to the tune of 2.14 t ha^{-1} than previous situation as like 1.46 t ha^{-1},
2. Cropping Intensity: Increased 68 per cent (227.1 per cent) due to increased no. of crops and duration of working season
3. Ground water recharge: Increased 31 per cent (raised 1.78 ft) over previous situation due to controlling runoff
4. Erosion control: Cleaning and excavation works increased storing of runoff water and controlled runoff
5. Income: Monthly income raised at about Rs.3,800/- than that of previous Rs.2,500/- system,
6. B-C ratio: B-C ratio has been changed remarkably from previous system (1.18) raised to 2.35 – 2.64 adopting improvised agro-techniques on IFS on "Rice-cum-fish culture-cum-vegetables" at farmer's level and
7. Livelihood change : Owned a grocery shop for his unemployed son adjacent with his house due to more production as well as more out turn of the system.

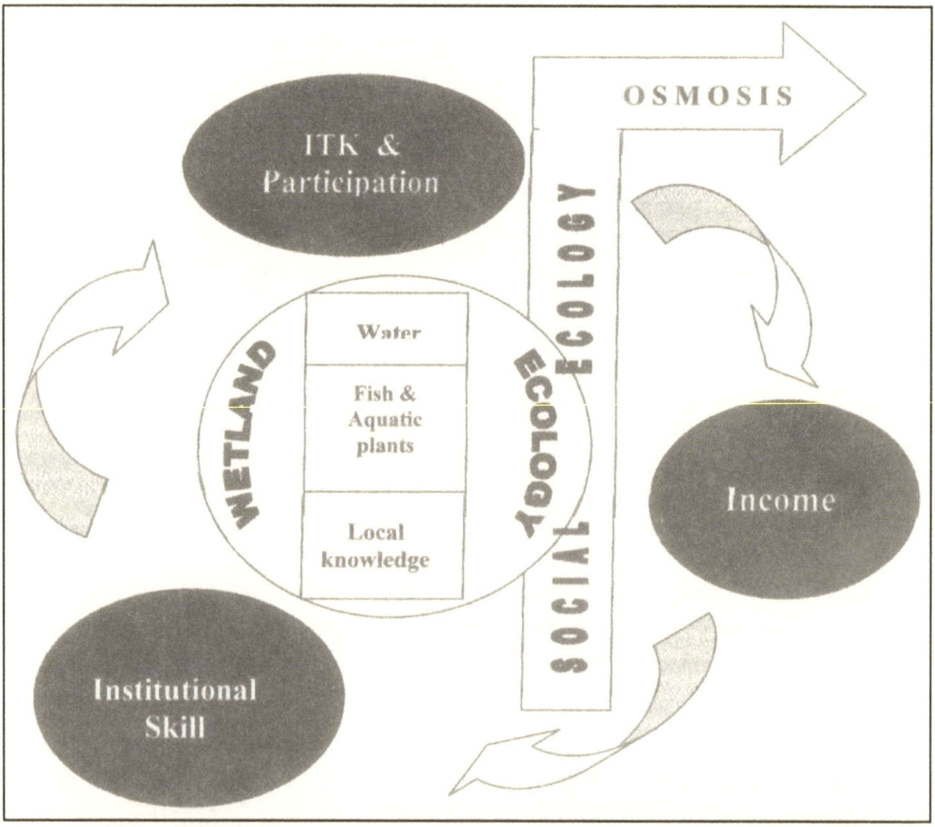

Figure 9.1: The Economy of IFS in Wetland Ecology: An Empirical Study.

Success Stories with Interpretations

Water Potentiality of Humid to Very Wet Regions

These regions broadly included the regions of humid, wet and very wet, where rainfall potentiality is high enough. It ranged from 1000 to >2,000 mm, but somewhere it is >10,000 mm per annum. Due to prevailing of soil characteristics, geomorphic situations and climatic condition there is a greater chance to accumulate more water in depressed low-lying situations, invariably after satisfying *in situ* demand of the soil. Moreover, it accelerates due to prevailing of more number of depressed saucer-shaped areas like oxbow lakes (*mauns, chaurs, jheels, beels, baors, nayanjali* etc.) predominant in Indian subtropics, which has been intersected by the main river system and it's so many tributaries (The Ganges, Brahmaputa, Padma, Mahanadi, Mahananda, Ichhamati, Churni, Jalangi, Rupnarayan etc.), mostly in Indo-Bangladesh regions.

Like arable or terrestrial land system, wetlands are also equally important for (i) conserving runoff water, (ii) cultivation of aquatic crops, (iii) fish culture, (iv) integration of fish-cum-aquatic crops, (v) recharging of ground water, moreover (vi) it maintains and sustained for ecological balance in the area. So, wetland has a greater role for all concerning including its sustainability in the areas.

Technological Adaptation and Outcome of the System

The inland fresh-water fish culture in the rice field played a significant role in the economic utilization of waste wetlands in food production as duel purpose. It will however, contribute substantially to increase the production of fish as well as grain production per hectare from inland water. Besides food production, bottom soil of the pond also improved considerably in consequence due to integration of both culture. Keeping in view a details field experiment was carried out to visualize the impact of fish-feed and type of different carps on the yield performance of improvement of soil status as a whole (Puste and Bandyopadhyay, 2000 and Bandyopadhyay and Puste, 2001).

Application of fish-feed (powdered mustard oilcake + rice bran in 1:1 ration @ 6 times body weight of fish applied at weekly interval) increased the grain and straw yield of rice [Sabita (NC 492), a tall *Indica*] by 19.0 and 13.0 per cent, respectively. The addition of fish-feed also helped to increase the fish yield by 7.0 per cent and different types of carps influenced the grain and straw yield of rice and fish yield significantly (Table 9.7). Higher grain (2.433 t ha^{-1}) and straw yield of rice (4.301 t ha^{-1}) was achieved by the plots, where Katla (*Catla catla* Ham.) was allowed. Among the carps, Katla produced highest amount of fish yield (1.344 t ha^{-1}) and the performance of Rohu (*Labeo rohita* Ham.) and Mrigal (*Crrihinus mrigala* Ham.) was statistically at par. Introduction of Katla, Rohu and Mrigal in rice field might increase the CO_2 concentration by respiration in crop canopy level and accumulated higher percentage of nitrogen and other elements in soil through their faeces and thereby, increased the grain and straw yield to the tune of 24-58 and 15-53 per cent, respectively. On the other hand, an additional higher yield of fish was also observed (1.082 to 1.344 t ha^{-1}) under rice cum fish culture ecosystem, which provided several natural planktons needful for rapid growth of fish. Application of fish-feed not only increase the fish yield by providing better feeding material but also helped to increase the productivity of rice due to its manurial value.

Table 9.7: Effect of Fish Feed and Carp on Grain, Straw Yield of Rice and Fish Yield

Treatments	*Pooled Yield (t ha⁻¹) of Ywo Years*		
	Rice Grain	*Rice Straw*	*Fish*
Fish-feed			
Without fish feed	1.869	3.303	0.886
With fish feed	2.225	3.734	1.513
S.Em ±	0.006	0.098	0.048
C D (p=0.01)	N.S.	N.S.	0.396
C D (p=0.05)	0.208	0.440	0.215
Types of carps	Rice grain	Rice straw	Fish
No fish	1.541	2.802	–
Rohu	2.299	3.738	1.172
Katla	2.433	4.301	1.344
Mrigal	1.915	2.233	1.082
S.Em ±	0.078	0.103	0.036
C D (*p*=0.01)	0.317	0.419	0.155
C D (*p*=0.05)	0.231	0.307	0.111

Fish-cum-Aquatic Food Crops

An intensive case study was carried out in RBD replicated four, where, individual aquatic food crops and fish genotypes [T_1 - LF (Singi - *Clarias batrachus*) and T_2 - LF (Magur - *Heteropneustes fossilis*); T_3 - SW fish (Rohu - *Labeo rohita*) and T_4 - SW fish (Katla - *Catla catla*) and T_5 - Water chestnut (*Trapa bispinosa* Roxb.) and T_6 - *Makhana* (*Euryale ferox* Salisb.)] were considered as six sole treatments along with four mixed treatments combined of crops with fish variables (T_7 - Water chestnut + LF (both in 1:1 ratio); T_8 - Water chestnut + SW fish (both in 1:1 ratio); T_9 - *Makhana* + LF (both in 1:1 ratio); T_{10} - *Makhana* + SW fish (both in 1:1 ratio) altogether ten treatments. In the sub-plot treatments, four combinations of both fish groups were allowed for their competitive effect of two variables (Puste *et al.*, 2004). Results from the study, fish-crop diversity was much important and substantial enough for more productivity per unit area per unit time. For easy comparison among variables, yield of fishes and crops were converted to MEY (*Makhana* equivalent yield) due to heterogeneous characteristics. Sole yield of all fishes (0.59, 0.54, 1.18 and 1.32 t ha^{-1}) and water chestnut and *makhana* (8.39 and 2.36 t ha^{-1}) dominated individually in respect to their associated yield. However in MEY study, among the combinations, superior results obtained with L-fishes (1.90 and 1.35 t ha^{-1}) + *makhana* (1.72) in respects to their cumulative yield (4.97 t ha^{-1}) followed by L-fishes (2.05 + 1.49 t ha^{-1}) + water chestnut (1.28 t ha^{-1}) due to their compatibility with the system. Regarding economic outcome, GMR, NP and B-C ratio (US$ 3,244 and 2,557 ha^{-1} and 3.71) were also very promising

Table 9.8: Yield and *Makhana* Yield Equivalent (t ha^{-1}) of Water Chestnut, Makhana and Fishes

Treatments	*Individual Yield (t ha^{-1})*						*Makhana Yield Equivalent (t ha^{-1})*						*Total*
	M	*S*	*R*	*K*	*WC*	*Mak*	*M*	*S*	*R*	*K*	*WC*	*Mak*	
T_1 - Sole M	0.59	—	—	—	—	—	2.95	—	—	—	—	—	2.95
T_2 - Sole S	—	0.54	—	—	—	—	—	2.52	—	—	—	—	2.52
T_3 - Sole R	—	—	1.18	—	—	—	—	—	2.36	—	—	—	2.36
T_4 - Sole K	—	—	—	1.32	—	—	—	—	—	2.42	—	—	2.9
T_5 - Sole WC	—	—	—	—	8.39	—	—	—	—	—	1.68	—	1.68
T_6 - Sole Mak	—	—	—	—	—	2.36	—	—	—	—	—	2.36	2.36
T_7 - WC + LF	0.41	0.32	—	—	6.38	—	2.05	1.49	—	—	1.28	—	4.82
T_8 - WC + RK	—	—	0.86	0.98	5.82	—	—	—	1.72	1.8	1.16	—	4.68
T_9 - Mak + LF	0.38	0.29	—	—	—	1.72	1.9	1.35	—	—	—	1.72	4.97
T_{10} - Mak + RK	—	—	0.76	0.78	—	1.48	—	—	1.52	1.43	—	1.48	4.43
CD (*P=0.05*)	0.08	0.07	0.19	0.21	0.89	0.25	—	—	—	—	—	—	—

with the system, which economically viable particularly for rural farm families. Among the combined system, lowest values obtained with SW-fishes + *makhana*, due to their non-compatibility characteristics (Tables 9.8 and 9.9).

Table 9.9: Gross Monetary Return and Benefit-Cost Ratio of the System

Treatment	*Net Profit (Rs.'000 ha^{-1})*	*Benefit-Cost Ratio*
T_1 - Sole M	66.97 (1456)	3.11
T_2 - Sole S	55.39 (1204)	2.74
T_3 - Sole R	51.24 (1114)	2.62
T_4 - Sole K	52.98 (1152)	2.7
T_5 - Sole WC	31.86 (693)	1.72
T_6 - Sole Mak	52.54 (1142)	2.88
T_7 - WC + LF	112.19 (2439)	3.46
T_8 - WC + RK	107.69 (2341)	3.29
T_9 - Mak + LF	117.6 (2557)	3.71
T_{10} - Mak + RK	100.86 (2193)	3.15

Production Potential of Fish-Crops Ecosystem

Likewise, fisheries sector plays a pivotal role in the natural economy in view of its contribution to the food basket. The total fish production during last few years was around 6.15-6.50 million tones, with marine and inland fisheries contributing 46.2 and 53.8 per cent, respectively to the total. Fish is a source of cheap animal protein and the current per capita consumption of fish in India is around 9.0 kg per annum as compared to 11.0 kg recommended by WHO. Fortunately, presently the contribution of fisheries sector to the GDP and agriculture GDP has been estimated to 1.2 and 4.2 per cent, respectively (Ayyappan and Biradar, 2004).

Intensification of practices, improved and economic fish feed as organic sources, comprehensive environmental management, systematic measures against fish diseases, community aquaculture and proper guidance are the prominent strategies to increase fish production, not only in the country but also equally important to other fish growing countries in the world.

Step II: System Adopted on Community Basis Individual Fish Yield

In the field study, these were implemented on community basis in each zone covering 3-4 villages in each cluster. In the demonstration cum improved trials, all the fish variables were cultured individually but for convenience for presentation individual fishes were grouped into two; *e.g.* FG_1, composed of Magur, Singi, Shoil and Gajer; and FG_2, which composed of Rohu, Katla, Mrigal and Silver carp. These were reared with normal but improved nutritional status (powdered mustard/ groundnut oilcake + rice bran in 1:1 ration @ 6 times body weight of fish at weekly interval) along with proper management practices. Fingerlings, fish-feed and other necessary inputs were supplied to the farmers timely from the institutional project's sources. Individual fish yield in cluster I and II in both the groups comparatively least performed than that of cluster III and IV, may be due water quality and depth of submergence of the respective ponds. It is more contrasting and comparable enough with the local practice. The magnitude of yield increased in all the individuals were to the tune of 46.0 to 78.0 per cent, more prominent in Magur, Singi, Mrigal and Silver carp (Table 9.10). A significant price differences were noted in between local market or *hat* and zonal trade centres of the zones. In most of the cases, fish farmers are sale out their output directly to the middlemen in the trade centres, with reasonable profit margins. Fewer cases they sale out their fishes locally with small harvests.

Table 9.10: Individual Fish Yield and their Price Study in Local Zonal Trade in the Study Areas

Variables	*Av. Yield (t ha⁻¹)*						*Av. Price of Fishes (Rs. t⁻¹)*	
	Cl. I	*Cl. II*	*Cl. III*	*Cl. IV*	*Av. Yield*	*Av. Local Yield*	*Local Market*	*Zonal Trade*
FG_1								
Magur	0.57	0.59	0.68	0.64	0.62	0.38	1,30,000 (2,826.1)	1,50,000 (3,260.9)
Singi	0.52	0.56	0.62	0.58	0.57	0.32	1,20,000 (2,608.7)	1,40,000 (3,043.5)
Shoil	1.32	1.38	1.46	1.52	1.42	0.96	1,10,000 (2,391.3)	1,30,000 (2,826.1)
Gajer	1.38	1.44	1.5	1.52	1.46	0.92	1,10,000 (2,391.3)	1,30,000 (2,826.1)
FG_2								
Rohu	1.14	1.19	1.27	1.24	1.21	0.83	50,000 (1,086.9)	60,000 (1,304.3)
Katla	1.28	1.35	1.43	1.38	1.36	0.86	40,000 (1,086.9)	55,000 (1,195.6)
Mrigal	1.09	1.15	1.26	1.22	1.18	0.68	40,000 (1,086.9)	50,000 (1,086.9)
S. carp	1.18	1.26	1.39	1.33	1.29	0.74	40,000 (1,086.9)	45,000 (978.3)

Av: Average; Cl: Cluster; Parenthesis indicates US$ (1US $ = INR 46.00).

Group Fish Yield

Application of eco-friendly and judicious fish-feed (powdered mustard/ groundnut oilcake + rice bran in 1:1 ration @ 6 times body weight of fish applied at weekly interval), applicable on fish variables in grouped basis were remarkably influenced the fish yield, practiced in farmers' fish ponds on community basis of different villages of the zones. Almost all the feed items were more or less equally effective for such increment of fish yield and it significantly differed from control pond, where fishes were nourished without any fish-feed (Table 9.11). The increment was to the tune of 82.2 to 116.4 per cent in FG_1 and 98.5 to 131.0 per cent, respectively in FG_2. However, among the 3 fish-feed materials highest results obtained with F_1 (powered mustard oilcake + rice husk in 1:1 ratio @ 6 times body weight of fish at weekly interval), although it quite statistical at par with other treatments in the field studies. The practice of proper fish-feed was quite effective to rearing and to increase fish yield irrespective of fish variables in the different 4 clusters of the zones (Puste and Basu, 2004).

Table 9.11: Fish Yield in Groups as Influenced by Fish-Feeds Used by the Farming Community

Variables	Cl. I		Cl. II		Cl. III		Cl. IV		Av. of Clusters	
	FG_1	FG_2	FG_1	FG_2	FG_1	FG_2	FG_1	FG_2	FG_1	FG_2
Fish feed										
F_1	1.02	1.28	1.19	1.34	1.21	1.33	1.18	1.32	1.15	1.32
F_2	1	1.22	1.12	1.18	1.18	1.26	1.15	1.21	1.11	1.22
F_3	1.04	1.18	1.13	1.22	1.12	1.24	1.09	1.22	1.09	1.21
F_4	0.56	0.59	0.55	0.58	0.64	0.67	0.61	0.64	0.59	0.62

Practical orientation and Analysis of -Integrated Farming System Model (IFS) on "Rice-cum-fish culture-cum-vegetables" for productivity and livelihood development utilizing aqua-terrestrial ecosystem.

Objectives

1. Proper utilization of aqua-terrestrial ecosystem, particularly which are waste or unused in nature,
2. For enhancing productivity per unit area per unit time
3. For generating income and livelihood development that are sustainable in nature, particularly for small and marginal group of farming community.

System Approach

Implementation steps at sites :

- ☆ Survey and Layout of the area for implement
- ☆ Need-based manipulations of areas (earth works etc.), *viz.*
  - Development of existing watershed ponds for rainwater conservation and recycling,
  - Layout and shaping of wetlands as per situation (shallow, medium and deep),

- Development of aqua-terrestrial ecosystem (IFS models),
- Development and Installation of watering system etc.

☆ Adaptation of field crops including fishes

- Adaptation of improvised agro-techniques in respective areas,
- Utilization of almost all the essential inputs in time and quality materials (seeds and fingerlings, plant foods and feed materials and need-based bio-degradable protection measures) and
- Time to time monitoring, analysis, harvesting and marketability are to be emphasized for better return and economics.

Working Plan

An income generated IFS (Integrated Farming System) model on **"Rice-cum-fish culture-cum-vegetables"** utilizing 1.0 ha unit area of aqua-terrestrial (land) may be developed in which rice (and/or other aquatic food, non-food commercial, food-cum-ornamental plants may be considered equally), fish and prawn may be grown during wet season in the low-lying field, and *kharif* and winter vegetables, fruits like papaya and coconut may be developed on field bunds or adjacent land system. Watermelon or other trailing type crop may be considered after harvest of rice. Such a system not only gives good productivity and cash return to the farmer but also provides employment opportunity to the farm family round the year (Table 9.10).

Due care, however, must be taken not to use the chemical plant protection measures as far as possible but may be used bio-degradable and/or organic pesticides. As a matter of fact in lowland rice system, insecticides hardly work because of standing water remaining up to heading stage of rice, which greatly dilutes the concentration of pesticides. However, a pest tolerant rice variety along with the common and major carps like Catla, Rohu, Mrigal, besides freshwater to giant prawn species (*M. rosanbrgi* and *N. malcolmsonil)* are compatible with lowland rice.

Approximate Specific Area for the system

1. Aquatic or wetland system (60 per cent) *i.e.* 0.60 ha and
2. Terrestrial or land system (bund and adjacent land) at about 40 per cent *i.e.* 0.40 ha.

An income generated IFS (Integrated Farming System) model on **"Deep-water rice-cum-fish culture-cum-vegetables"** utilizing 1.0 ha unit area of aqua-terrestrial (land) may be developed in which rice (and/or other aquatic food, non-food commercial, food-cum-ornamental plants may be considered equally), fish and prawn may be grown during wet season in the low-lying field, and *kharif* and winter vegetables, fruits like papaya and coconut may be developed on field bunds or adjacent land system. Watermelon or other trailing type crop will be considered after harvest of rice. Such a system not only gives good productivity and cash return to the farmer but also provides employment opportunity to the farm family round the year (Table 9.12). Approximate specific area may be provided for cited this example, like – (i) for aquatic or lowland or wetland system (65 per cent) *i.e.* 0.65 ha and (ii) for terrestrial or land system (bund and adjacent land) at about 40 per cent *i.e.* 0.35 ha may be provided, but in individual farmer's level will depend on availability of land.

Table 9.12: Utilization of Aqua-Terrestrial Ecosystem vis-à-vis Production Potential and Economics of Rice-Based IFS System (1 ha unit)

	Item	*Actual Cost*	*Annual Depreciation @ 5 per cent on Actual Cost*	*Annul interest @ 12.0 per cent on Actual Cost (Rs.)*
I.	**Fixed Capital Cost (Rs.)**			
	Earth Work	30,000.00	1,500.00	3,600.00
	Sub-Grand Total			**35,100.00**
II.	**Variable Operational Cost (VOC)**			
1.	Rice			10,000.00
2.	Crop after rice (watermelon)			5,000.00
3.	Fish and prawn			15,000.00
4.	Vegetable on pond bund or adjacent areas			
	a) *Kharif* vegetables			5,000.00
	b) Winter vegetables			6,000.00
5.	Horticultural plantation crops/trees :			
	a) Papaya (150 female plants)			3,000.00
	b) Banana 100 plants (both green and ripening type)			5,000.00
	c) Coconut 50 plants (production starts from 5th year) for plantation and subsequent maintenance			**6,000.00**
	From 5th year maintenance for 50 coconut plants (Rs.200/- per plant)			**10,00,.00**
	TOTAL			**55,000.00**
	Interest on VOC @ 12 per cent per annum for half of the period			6,600.00
	Sub-Grand Total			**61,600.00**
	GRAND TOTAL			**96,700.00**
	GRAND TOTAL including coconut maintenance			**1,06,700.00**

So, total annual investment: (Fixed capital cost + VOC + annual interest on VOC) = Rs.96,700.00 and Rs.1,06,700.00.

III. Returns

	Crops	*Area (sq. m)*	*Produce (kg)*	*Av. rate (Rs. kg $^{-1}$)*	*Total Value (Rs.)*
Rice					
a)	Grain	6,500	2,750	14.00	38,500.00
b)	Straw	6,500	5,350	2.00	10,700.00
	Fish	1,300	1,019	80.00	81,520.00
	Prawn	8,000	180	300.00	54,000.00
Crops after rice (watermelon)		6,500	8,000	5.00	40,000.00

Contd...

Table 9.12–*Contd...*

Crops	Area (sq. m)	Produce (kg)	Av. rate (Rs. kg $^{-1}$)	Total Value (Rs.)
Vegetables on bund or adjacent lands				
a) *Kharif*	500	950	12.00	11,400.00
b) Winter	600	1,800	10.00	18,000.00
Horticultural crops				
a) Papaya (150 female plants)	300	3,000	10.00	30,000.00
b) Banana plantation 100 plants (both green and ripening type)	500	100 per pl	8.00	40,000.00
c) Coconut 50 plants (from 5th year)	300	150 per plant	10.00 per coconut	75,000.00
Gross annual income (from 1st to 4th year)				**3,23,850.00**
Gross annual income (from 5th year onwards)				**3,98,850.00**

B-C Ratio Over the System or Specific System

1st Year to 4th Year

Cost of cultivation : Rs.96,700.00

Gross monetary return (GMR) : Rs.3,23,850.00

So, net profit (NP) of the system : Rs.3,23,850.00 – Rs.96,70.00 = Rs.2,27,150.00

$$\text{So, B-C ratio of the system} = \frac{2,27,150.00}{96,700.00} = 2.349 \text{ or } 2.35$$

5th Year and Onwards

GMR : Rs.3,23,850.00 + Rs.75,000.00 = Rs.3,98,850.00

For coconut maintenance on and from 5th year onwards Rs.10,000.00

So, Cost for the produce : Rs.96,700.00 + Rs.10,000.00 = Rs.1,06,700.00

NP of the system : Rs.3,98,850.00 – Rs.1,06,700.00 = Rs.2,92,150.00

$$\text{So, B-C ratio of the system} = \frac{2,92,150.00}{1,06,700.00} = 2.738 \text{ or } 2.74$$

N. B.:

1. Need-based excavation, renovation and cleaning of waste wetland is require (permanent cost),
2. Plants (especially cocoanut on bunds or boarder areas), which will produce approximately from 5th year or onwards,
3. Other tree (timber) plantation may be done, especially on the bund or boarder areas, which will be valuable approximately after 25-30 years and
4. Over all utilization of water in watershed may be available as irrigation to arable crops (as life-saving irrigation).

Some Useful Produces and its Value-Added Products of Wetland Ecosystem

Grains of Deep-Water Rice

Different type of deep-water rice cultivars (available in CRRS, Chinsurah), which are normally grown under varying water depth, starting from very low-lying/ waterlogged to deep-water situations, usually submerged by the monsoon water remains of months during rainy and post-rainy months. Categories, it is a special type of paddy called 'deepwater' or 'floating' paddy variety, which is capable of raising its leaf tips above the water surface with the increase in water level and serves well even under adverse situation. There are so many varieties, which developed and selected fitted with the wetland ecosystem. There are mostly photo period sensitive with yielding ability to 3.5 to 5.0 t ha^{-1}. Some useful varieties of low-lying ecosystem are Sabita, Hanseswari, Ambika, Suresh, Jogini, Nalini, Jalaplaban etc.

Plants and Water Chestnut

Water chestnut is a floating (rooted) aquatic annual of perennial plant, found commonly floating on water surface of lakes, tanks and pools adhering with their root system in muddy soils. It is an aquatic food crop grown almost through out India and other similar countries. The common cultivated species is *Trapa bispinosa* Roxb., although other two species (*T. quadrispinosa* and *T. natans*) are found somewhere in the family Trapaceae under the natural order Onagraceae. The origin of the crop is in India due to prevalent in the country since remote past. The common name in India is '*Singhara*' and 'Paniphal' in Bengali.

Fruits of Water Chestnut

The fruits of water chestnut or *singhara* are grown in India mainly for human consumption either in the form of fresh half-matured fruits, vegetables, flour or in the shape of sweet dishes. The kernels are sweet in taste, rich in carbohydrates (23-24 per cent), also contains protein (4.7-5.0) and mineral matter (1.1 per cent) with negligible amount of fat (0.3 per cent). Generally, two types of plants and fruits to grow *viz.* greenish (more weighty and big but comparatively less sweet in taste) and reddish (comparatively small in size but more sweet in taste) in colour. Generally, picking of fruits is start from September and continues up to November and sometimes in December. The yielding ability of the crop (fresh fruits) is 8.5 – 10.0 t ha^{-1}.

Water Chestnut Flour or *Atta*

The ground form of or flour or *atta* produced from matured seed kernels or fruits of water chestnut. So, it is a value-added product of water chestnut. Its flour is considered superior to rice in its food value. It compared well with other foods and is a good source of minerals. It abounds in starch, resembles the chestnut in flavour and is eaten either raw or cooked or in the preparation of *roti* or *chapattis*. It is an easy digestible article due to presence of negligible fat percentage and helpful for dieting patients. Now-a-days, some of the organizations are showing their interests for manufacturing this ground flour from water chestnut seeds kernels.

Makhana Seeds

Matured seed kernels of *makhana* (*Euryle ferox* Salisb. under the family Nymphaeaceae), is another most important aquatic food crop next to water chestnut in India and probably in other similar countries. The seed contains ample carbohydrates (76-77 per cent), good amount of protein (9.7 per cent) and other mineral matters. The seeds may be eaten directly through roasting or as fresh starchy edible materials after husking the hard seed coats. Sometimes starches are used in textile industries. *Makhana* flour (in ground form) is also available in the ability of crop (seed kernels) is 2.5-3.0 t ha^{-1}. *Makhana* is commercially cultivated in North-eastern part of Bihar and some areas Maldah district of West Bengal.

Flour or Popped-Form of *Makhana* (*khoi*)

The ground-form or the popped-form of *makhana*, are prepared from *makhana* seed kernels, are important value-added products of wetlands. Both are available in the market and is used for farinaceous food, for making bread, pudding, halwa etc. *Makhana* flour or *atta* is also used for milk based food preparation like *kheer*, curry due to rich nutritive food value. But the most convenient and useful of *makhana* is as a roasted fried-form, which is widely used especially in the vegetarian belts, even now exports to other foreign countries (Rs.250.00 - 260.00 per kg in packet form). These are used for the preparation of curry, moa, halwa, pudding etc. and prefers the vegetarian people due to nutritive value of these products.

Lati or *Pani Kachu* or *Cyrtosperma* sp. Plants or Seedlings

It is a perennial tuberous plant, called *Jal* or *Pani kanchu* (*Cyrtosperma* sp.) under the family Araceae, grown mostly marshy or shallow wet situation, prefers humid climate. All parts of the plant are edible (corms, leaves, stalks or petioles, lati etc.), mainly by cooking as curry. Locally common name – *pani* or *jal kanchu*, *ghat kanchu*, *sola kanchu* etc. It is vegetatively propagated by using side suckers or tillers, corms etc. Form the corms, numbers of products are being produced and marketed in the package form. Presence of carbohydrates, protein, fat, mineral matters all are vary from parts to parts of the plants. The production in terms of numbers it may be 18,000 – 20,000 corms or cormels ha^{-1} with proper management practices in the field.

Stolons or *Lati* of *Cyrtosperma* and its Palatable Products

The stolons or *lati* of *Cyrtosperma* sp. are produced from the basal or from root-stock or from rhizome of *jal* or *pani kachu* or *lati kachu*. It is favourable and palatable food item for the preparation of curry. The crop is generally cultivated in wetland puddle situation, where proliferation and vigour both are developed properly. During the rainy season it is common in the market along with other items of taro. Besides production of corms or cormels, production of *lati* or stolons may further be added extra, may be of 15-20 t ha^{-1} y^{-1}r. As a whole, total production of the crop is more productive and remunerative, if all the follow up actions are taken in the field and it goes in favour of the cultivators. Now-a-days, so many palatable products of cormels or corms are produced and available in the market and oftenly, it is commonly seen with the hawkers in different common places.

Lotus or Padma (*Nelumbium* spp.)

Lotus or padma (*Nelumbium* spp.) is an aquatic perennial herb with rhizomatous stem has ornamental as well as food value, very familiar to Indian context. Seeds are ground to flour and bread is made. Seeds are also eaten as raw, cooked, candied and canned. Rhizomes are marketed as fresh or dried and it also produces a kind of arrowroot, used in dysentery or diarrhoea.

Water Lily (*Nymphaea* spp.)

Water lily (*Nymphaea* spp.) is an aquatic perennial herb under the family of Nymphaeceae. The seeds are roasted and eaten. Tubers and seeds are eaten and the long peduncles are eaten as vegetables, familiar to the rural people. The decoction of flowers is prescribed for palpitation of heart, flowers removes impurities from blood. The powered rootstock is given in dysentery, diarrhoea and piles.

Madur-kati and its Products

The culms of mat-making sedges (*Cyperus* sp.) are generally used for making matreed, is s post-harvest technological approach as like cottage industries, performs mostly aged men and women members of the family. The culm or main shoot of the plant is used for the preparation of matreed of different sizes. These are dependent mainly on the technical expertise with experience, particularly of aged person of a family. Besides, the marvelously fine-splitted forms are being used for the preparation of number of costly, decorative colourful appliances (ladies bag, mobile cage, door and window-screen, chappals, table-mat, files, baskets etc.), even now is being exported to foreign countries. These are the value-added products of *madur-kati* or mat-making sedges, are available in the market and different organizations, NGOs, SHGs are engaged with this job, sometimes on community basis

Shola Products

From shola or shola-pith (*Aeschynomene aspera*), number of value-added products are being made utilizing with specialized skilled manpower. After harvesting, dry shoal-piths are mainly used for preparation of beautiful, decorative as well as costly handicrafts of different kinds through marvelously fine splitted forms of different shapes and sizes. Besides, it is also used for surface dress, fish-net for floats, fish-baskets etc. generally, some pieces of shoal-piths may be of Rs.25-30/- but with making of such materials in to a beautiful, decorative product, the added cost may goes up to Rs.800-1,000/- or even more.

Bet or *Calamus*

Bet or *Calamus* (*Calamus* sp. Linn. under the family Palmae) is an important commercial, perennial marshy crop. These are mostly inhabitants of tropical to sub-tropical countries, grown in moist places along the edges of freshwater swamps or banks of rivers throughout the evergreen and semi-evergreen forests of Assam, West Bengal, Tripura, Kerala, Karnataka and Maharastra in the country. The long slender somewhat climbing type of stems or sticks is used for the preparation of valuable, decorative, long-lasting different home appliances, which are used in most of the family.

Products of bet or *Calamus*

So many products are being made using the stems or sticks of Calamus or bet, an important, economically valuable wetland crop (*Calamus* spp., Linn. under the family Palmae). These are mostly inhabitants of tropical to sub-tropical countries, grown in moist places along the edges of freshwater swamps or banks of rivers throughout the evergreen and semi-evergreen forests of Assam, West Bengal, Tripura, Kerala, Karnataka and Maharastra in the country. The long slender somewhat climbing type of stems or sticks is used for the preparation of valuable, decorative, long-lasting different home appliances, which are used in most of the family.

Hogla

Hogla or elephant grass or Indian reed-mace or cats-tail is an aquatic or palustrial herbs, distributed in the temperature as well as tropical regions. The fibrous stems and leaves are used for many purposes throughout the country and other similar countries in the world, where it grows abundantly. In West Bengal, it is mainly grown in Howrah, Midnapore (East), 24-Parganas (both) and partly Nadia, Murshidabad and Hooghly districts. The crop is mainly cultivated for its leaves, which are used much to make temporary shades, roofs, tents, thatched house etc. The dry leaves are also used to prepare 'gola', kind of bins for storing food grains. Besides, this aquatic crop has the significant importance in various industrial uses.

Para Grass or Water Grass

Para grass or water grass [*Brachiara mutica* (Forsk.) Stapf under the family Poaceae (Grammineae)] is an aquatic annual or perennial grass. The grass is suited to warm and humid climate mostly prefers the moist, waterlogged or low-lying situation. Animals are generally prefers to feed this succulent grass and it is generally fed as a green, fresh valuable fodder (organic matter – 88.82, crude protein – 5.95, ether extract – 2.31 per cent and reasonable amount of crude fiber) to animals. After planting in the field, several cuttings are possible from the same plants. Sometimes, it is also used as hay.

Coix or Gurgur Grass

Coix or gurgur grass [*Coix aquatica* Roxb. Under the family Grammineae (Poaceae)] popularly known as gurgur grass or 'Jobs Tears'. Categorically, it is an emergent amphibious and marginal plants, prefers mostly swampy, marshy wet environment found throughout the plains of India. This crop is considered for cultivation as fodder crop due to its nutritive value in marshy wetlands during the rainy season of the year in the topics. Besides, the hard-coated dry, spherical seeds are also used for ornamental purposes.

Aquatic Medicinal Plants

So many aquatic valuable medicinal plants [Brahmi (*Bacopa monnieri*), Kesuti or Keshori (*Eclipta alba*), *Ipomea* or swamp cabbage (*Ipomea aquatica*), Talmakhana or kule khara (*Hygrophila auriculata*), Enhydra or helencha (*Enhydra fluctuans*), Indian pennywort or thankuni (*Hydroctyle officinale*)] are inhabitants in marshy, swampy and waterlogged areas. All these are small perennial aquatic herbs, commonly grown in moist, wet, marshy places throughout the plains of India, even extended to an altitude of 1,300 m. It is valuable due to its medicinal property (alkaloids along with other valuable ingredients) and its wide use in locally, ayurvedic industries in nationally and internationally. All the plant parts (leaves, stems, roots, flowers etc.) are used extensively as succulent foliage as leafy-vegetables in almost daily diets and for medicinal purposes to all age groups.

Aquatic Weeds-cum-Plants of Biomass

This group of aquatic plants is considered as weeds cum plants of biomass (water-hyacinth, Pistia or takapana, *Hydrilla* or *Jal jhanjhi*, *Azolla*, Algae etc.). Almost all are these are aquatic free-floating, stemless aquatic herb (water- hyacinth, *Pistia* or takapana, *Hydrilla* or *Jal jhanjhi*, *Azolla*) found almost in all water bodies to float on the surface of water. However, some one is rooted submerged leafy water plants found almost all over the world (*Hydrilla* spp. or commonly known as *Jal jhanjhi*). Although all there are considered as the aquatic weeds, however, now it is being used extensively for various purposes (production of biomass, industrial use for making food and other items) due to its medicinal properties.

Conclusion

As water is the key and integral elements of natural resources, it would obviously be utilized in precise and wise way for benefits of mankind's, even in sub-humid to very wet areas and sustainability in the world, either may be ground, surface and/or wetland sources. In the country, in addition with this, vast wetland ecosystem may effectively be utilized through the cultivation of so many aquatic crops, fish variables which not only valued to human beings but also imperative for the upliftment of resource poor rural economy as well, who are inextricably linked with the system. Besides, wetlands are continuously enriched by adding large quantities of biomass, enriched in sequence, which has possible to utilize this enriched biomass effectively for production of adjacent arable crops as well as improvement soil nutrient status in consequence in the zones.

Wetlands are highly exploited. Sufficient commercially exploitable species and aquaculture make the wetlands more valuable. Rice, fish and other underutilized crops, food and non-food crop and their products, icthyofaunal diversity are derived from wetlands. Substantial quantities of fuel and materials for weaving and thatching are also obtained from wetlands. Wetlands provide water for domestic and industrial consumption and in arid regions provide water for irrigation. The development and management of wetlands should form an important part of the integrated watershed management plans. A vast area of coastal wetlands is very significant today for conservation of mangrove forests as well as wild life when environmental pollution is a burning question around the world. Yet wetlands are disappearing at an alarming rate with the expansion of civilization.

Wetlands provide water for domestic and industrial consumption and in other regions provide water for irrigation. The development and management of wetlands should form an important part of the integrated watershed management plans. A vast area of coastal wetlands is very significant today for conservation of mangrove forests as well as wild life preservation, when environmental pollution is a burning question around the world. Yet wetlands are truly disappearing at an alarming rate with the expansion of civilization, - needs study, developing awareness for its sustainability.

Acknowledgement

The authors are grateful to the Department of Agriculture, Govt. of West Bengal, I.C.A.R. (Indian Council of Agricultural Research), NWDPRA and TDET of DoLR, Ministry of Rural Development, Govt. of India, New Delhi for providing their financial assistance, the Institute, Bidhan Chandra Krishi Viswavidyalaya and the local farming community for providing incessant facilities in conducting such field trials in the respective zones.

References

Anonymous (1986). *In*: Indian Agriculture in brief. *Geographical Magazine* 55 (21st ed.), pp. 12-17.

Ayyappan, S and Biradar, R. S. (2004). Enhancing global competition, In, *The Hindu, Survey of Indian Agriculture*, pp. 97-99.

Bandyopadhyay, S. and Puste, A. M. (2001) Effect of carp and fish feed on yield and soil nutrient availability under integrated rice-fish culture. *Asian Fisheries Science* 14: 437-441.

Dutta, S. K.; Konar, D.; De, S. K and Mukhopadhyay, P. K. (1984). Paddy and air-breathing-fish culture: effects of supplemental feed on the growth and yield of rice and fish. *International Rice Research Newsletter*, 9 (2): 23-24.

Ghosh, A. and Saha, S. K. (1980). Scope for paddy-cum-fish culture in India. *Proceedings of 5th International symposium. Tropical Ecology*, 16-21 April, 1979, Kualalampur, Malaysia, pp: 1009-1015.

James, E. J. (1995). Managing the wetlands and their watersheds. *Yojana*. 39 (182), pp. 43-50.

Jhingram V. G. (1983) In: *Fish and Fisheries of India*, Hindustan Publishing Corporation, Delhi India, (Second edition), pp: 475-47.

Maltby, E. and Turner, R. E. (1983). Wetlands of the World. *Geographical Magazine*, 55: 12-17.

Puste, A M and Das, D. K. (2001). Optimization of aquatic-terrestrial ecosystem in relation to soil nitrogen status for the cultivation of fish and aquatic food crops of the Indian sub-tropics. *TheScientificWorld*, USA, 1(S2) : 130-134.

Puste, A M, Ray Pramanick, B, Jana, K, Roy, A. Dasgupta, M, Maiti, A K and Basu, D. (2007). Utilization of wetland ecosystem through fish-crop diversification for enhanced productivity and economic stability for the fish farm community of the Indian sub-continent. Proceedings of the 'Int. Conf. on Community Based approaches to Fisheries Management (CBFM-2)' held in Radisson Hotel, Dhaka, Bangladesh during March 05 - 08, 2007 organized by WorldFish Center, Penang,

Malaysia jointly with South-Asia Office WorldFish Center of Bangladesh : 20-21 and 102-110.

Puste, A M, Ray Pramanik, B, Dasgupta, M, Maity, T K and Maiti, A K. (2011). Integrated wetland management - a holistic approach for development of diversified farming system for production of aquatic crops and fishes for rural sustainability in Indian sub-continent, published in the Proceedings of the International Conference on "Tropical Island Ecosystems : Issues Related to Livelihood, Sustainable Development and Climate Change" held at CARI (ICAR), Port-Blair, Andaman and Nicobar Islands, India during March 23-26, 2011, pp: 90-91.

Puste, A. M. (2004). In: *Agronomic Management of Wetland Crops*, published by Kalyani Publishers, Ludhiana, India during 2004.

Puste, A. M. and Bandyopadhyay, S. (2000) Effect of different types of carps and fish feed on the growth and yield from paddy cum fish culture. *Environment and Agriculture: At the Crossroad of the New Millennium* (Nepal), pp. 527-529.

Puste, A. M. and Basu, D. (2004) Information and communication technologies (ICTs) imperative for updating agricultural production system and its economy, presented and published as a full paper in the *Proceedings of the 'Joint Conference of The 4th International Conference of the Asian Federation Technology in Agriculture and the 2nd World Congress of Computers in Agriculture and Natural Sciences (AFITA/ WCCA2004)'* held at the Four Seasons Hotel, Bangkok, Thailand, August 9-12, 2004, 805-814.

Puste, A. M. and Das, D. K. (2001). Optimization of aquatic-terrestrial ecosystem in relation to soil nitrogen status for the cultivation of fish and aquatic food crops of the Indian subtropics, *TheScientificWorld*, 2001, 1(S2) : 130-134.

Puste, A. M., Ray Pramanik, B., Dasgupta, M. and Maity, T. K. (2010). Integrated management of aqua-terrestrial ecosystem for enhancing productivity and promoting rural sustainability in agro-climatic zones of Indian sub-continent, published in the Proceedings of the National Seminar on "Integrated Rural Development and Management: Issues, Strategies and Policy Options" held at the Ramakrishna Mission Vivekananda University, Kolkata during December 17-18, 2010, pp : 13-14.

Puste, A. M., Sarkar, P. K. and Das, D. K. (2004) Wetland ecosystem - a case study on fish-aquatic food crops diversity for enrichment productivity and economic stability for fish-farm families in Indian subtropics. *Proc. of the 'Twelfth Biennial Conference of the International Institute of Fisheries Economics and Trade ('IIFET 2004 JAPAN')* held in TUMSAT, Tokyo, Japan, July 20-30, 2004 and published by IIFET, Oregon State University, USA: Theme: Aquaculture Economics and Management, Group 3: 1-8.

Sinhababu, D. P.; Jha, K. P.; K. C and Ayappan, S. (1998). Integrated Farming System: Rice-fish for lowland ecologies. *Proceedings of the International Symposium on Rainfed Rice for Sustainable Food Security*, September 23-25, 1996, Cuttack, Orissa, pp: 203-222.

Chapter 10
Strategies for Entrepreneurship Development and Management

Puranjan Das

Entrepreneur

- ☆ The word "Entrepreneur" derived from the French verb 'entrepredre' meaning 'to undertake'
- ☆ An entrepreneur is an individual who introduces something new in the economy
  - A method of production not yet tested by experience in the branch of manufacture concerned
  - A product with which consumers are not yet familiar
  - A new source of raw material or of new market and the like
  - Entrepreneurship is the process of identifying opportunities in the market place, arranging the resources required to pursue these opportunities and investing the resources to exploit the opportunities for long term gains

Types of Entrepreneurs

On the Basis of Ownership

- ☆ **Private-** motivated by profit and would not enter those sectors of the economy in which prospects of monetary rewards are not very bright
- ☆ **Public-** government takes the initiative to share enterprises.

On the Basis of Scale of Enterprise

- **Small Scale-** do not posses the necessary talents and resources to initiate large scale production and introduce revolutionary technological changes.
- **Large Scale-** posses the financial and necessary enterprise to initiate and introduce new technical changes

Working Models for Entrepreneurship Worldwide

- **The free market model: *e.g.*: United States and Canada-** Emphasizes minimal government intervention in the economy and deregulation
- **Guided individualism model: *e.g.*: Singapore and Taiwan-** Encourages individual enterprises but emphasizes the use of public policy and government to influence the direction of entrepreneurial activity
- **India- the mixed economy model**
  - An economic system in which both the state and private sector direct the economy, reflecting characteristics of both market economies and planned economies. While both public and private sectors coexist, a central role is assigned to the state's planning machinery for resource allocation across sectors.
  - The stated primary objectives of the planning process have been economic growth, social justice and self-reliance.
  - The basic idea of the mixed economy is that the means of production are mainly under private ownership; that markets remain the dominant form of economic coordination; and that profit-seeking enterprises and the accumulation of capital remain the fundamental driving force behind economic activity.
  - However, unlike a free-market economy, the government would wield considerable indirect influence over the economy through fiscal and monetary policies.

Agriculture Production Scenario in India

	Agricultural Commodity	*Production (million tonne)*
☆	Foodgrain production (2010-11 crop year)	241.0
☆	Milk production (2011)	121.7
☆	Fish Production	7.8
☆	Vegetable Production	141.3
☆	Fruit production	75.7

Major Issues to be Addressed

- Lower growth in farmers' income(1.5 per cent) and higher growth in expenditure (4-5 per cent)

- Only few secondary agriculture industries exist and most of the agro-industries unorganized
- Low Consumption of value added fruits and vegetables compared to fresh fruits and vegetables and the inclination towards processed foods is mostly visible in urban centers
- Value addition only in 8 per cent of total food production
- Only 2.2 per cent of the horticulture products processed, while upto 35 per cent fruits and vegetables spoiled

Post Harvest Loss

Sl.No.	Type of food Commodity	Present Level of Production (million tonne)	Per cent	Quantity (million tonne)	Monetary Value (Crore)
1	Durables	230	10	23	23,000
2	Semi-perishables	40	15	6	1,800
3	Perishables	210	20	42	63,000
4	Total/average	480	14.8	71	87,800

Source: Publication by Dr. R. P. Kachru, ADG, ICAR.

Levels of Food Processing Across Segments in India

Segment	Level of Processing (per cent)
Fruits and Vegetables	2.2
Fisheries	26
Poultry	6
Buffalo meat	20
Milk	35

Source: MOFPI Annual Report 2007-08

Possible Areas of Entrepreneurship in Agriculture

- **Agro produce processing units:** These units do not manufacture any new product. They merely process the agriculture produce

  e.g., Rice mills, Dal mills, Decorticating mills
- **Agro produce manufacturing units:** These units produce entirely new products based on the agricultural produce as the main raw material

  e.g., Sugar factories, Bakery, Strawboard units
- **Agro-inputs manufacturing units:** These units produce goods either for mechanization of agriculture or for increasing productivity

  e.g., Fertilizer manufacturing plants, insecticides production units, food processing units, agricultural implements etc.

- **Agro-service centers:** These include the workshops and service centers for repairing and serving the agricultural implements used in agriculture.
- **Agribusiness**
  - In the developed countries, agribusiness is defined as the total output arising from farm production and product processing at both pre- and post farm gate levels.
  - In developing countries like India, the agribusiness sector encompasses four distinct sub-sectors, *viz.* agricultural inputs, agricultural production, agro-processing, and marketing and trade
  - All these add value or utility to the goods
  - Agribusiness is emerging as a specialized branch of knowledge in the field of management sciences. In this context, agribusiness can be defined as science and practice of activities, with backward and forward linkages, related to production, processing, marketing, trade, and distribution of raw and processed food, feed and fiber, including supply of inputs and services for these activities.

Need for Agripreneurship

- **Competitive advantages for many primary production activities in agriculture**: Rain-fed farming, tropical fruits and vegetables, livestock, animal husbandry, aquaculture, wild craft, etc. are produced through real low cost production methods.
- **Private sector is willing to enter in to agribusiness at all levels of operation:** Changing consumer demand and retail revolution have opened the doors for investments by private sector in agribusinesses like Reliance, Bharati, Pantaloon, Carrefour, etc.

Agribusiness Sector in India (2004-05)

- Agro-industries include both organized and unorganized sector units.
- There are 17.0 million units in the unorganized sector, of which 13.91 million are agro-based out of which:
  - 12.32 million are own account manufacturing establishments (OAME)
  - 1.2 million are non-directory manufacturing establishments (NDME) and
  - 0.39 million are directory manufacturing establishments (DME)
- The units in the organized sector are few in number but account for a bulk of the total gross value addition. There are:
  - 35,000 modern rice mills
  - 20,000 pulse mills
  - 5198 fruit and vegetable processing units
  - 400 fish processing units

- 426 sugar mills, and
- 3619 grinning and pressing units

☆ The primary producing units are around 121 million operational holdings, of which 63 per cent (76 million) are of less than one hectare in operating size, on an average, 0.4 hectare of land.

☆ There are around 5 million wholesale traders and 11.2 million retailers of agricultural and other commodities out of which 3.7 million are estimated to be food retail outlets.

☆ The value of agro-processed products during the year 2004-05 was Rs 4169 billion, in which Rs 2960 billion was from registered manufacturing units and Rs 1209 billion from unregistered manufacturing establishments.

☆ During 2004-05, value addition from trade and hotels and restaurant activities (GDP) has been estimated as Rs 4178 billion and Rs 416 billion, respectively.

☆ Most of the food processing units are in the unorganized sector and Indian food market continues to be dominated by the fresh food segment. According to one estimate, Indian food market comprises:
  - 10 per cent processed segment
  - 15 per cent semi-processed segment and
  - 75 per cent as fresh food segment

☆ Processing is reported to be around 2 per cent in fruits and vegetables, 37 per cent in milk, 21 per cent in meat, 6 per cent in poultry and per cent in marine fish.

☆ Fruit and vegetable processing is projected to go up to 10 per cent by 2010, taking the food processing segment to 32 per cent of the total food market.

☆ The overall value addition in food products, which is currently 8 per cent, is likely to increase to 35 per cent by 2025.

Emerging Areas of Agribusiness

Agribusiness Opportunities:

Production

☆ Production of high-yielding seeds

☆ Production of high-quality planting material, including use of tissue culture methods of micro-propagation

☆ Nurseries, including hardening nurseries

☆ Organic farming

☆ Production of microbial cultures and vermicompost, and

☆ Floriculture

Processing

- ☆ Fruit and vegetable processing, including dehydration, canning, aseptic packaging, processing of underutilized fruits, and processing for other products like grape raisin, osmo air-dried fruits, fruit toffee, bleached dry ginger and spices powders
- ☆ Processing of maize for starch and feed through improved mini/small mills and dry milling plants
- ☆ Processing of millets for various purposes, including malt from finger millets and RTE (Ready-to-Eat) products
- ☆ Processing of sugarcane for various jaggery products like spiced jaggery, powdered jaggery, and jaggery cubes
- ☆ Processing of herbal and medicinal plants
- ☆ Processing of dairy products
- ☆ Processing for poultry products, including poultry dressing, and
- ☆ Processing of livestock products and livestock wastes

Infrastructure

- ☆ Cool chain infrastructure, including cold stores
- ☆ Storage and warehousing
- ☆ Specialized transport services
- ☆ Packaging infrastructure, including pack houses, and
- ☆ Agri-clinics and service centers

Trade and Others

- ☆ Procurement through contract arrangements, including contract farming
- ☆ Retailing
- ☆ Supply chain management, and
- ☆ Capacity building, including human resource development in agribusiness

Emerging Areas of Agribusiness

- ☆ Increasing demand of organic and quality food both in India and abroad (Figure 10.1).

Supply Chain

Supply chain is a coordinated system of organizations, people, activities, information, and resources involved in moving a product or service in physical or virtual form from supplier to the customer.

- ☆ Supply chain activities include transformation of raw material and components into finished products that are then delivered to the end customers

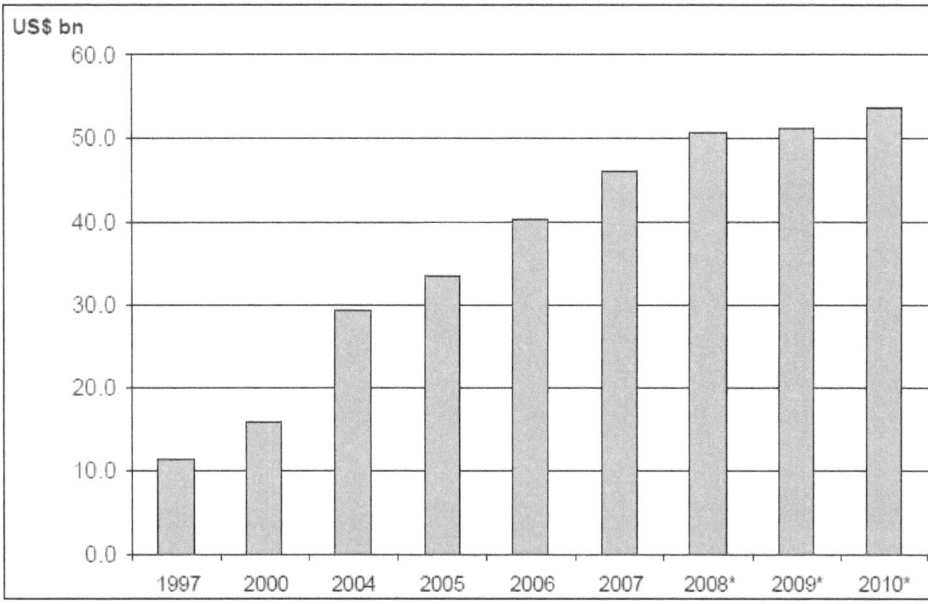

Figure 10.1

*** Figures for 2008-10 are Projections.**

Food Processing Value Chain in India

Farm - Inputs-Farming - Marketing/Aggregator - Processing - Logistics - Retail/Food Services - Consumer

Financial and Business Services: Transport Services/Infrastructure, Quality Control: Market Intelligence, Product Design: Distribution: Marketing

India's Retail Sector

- Indian retail sector is estimated to be worth US$ 330 billion or Rs 13.2 trillion. It is projected to increase to US$ 427 billion by 2010 and to US$ 637 billion by 2015. Given this rate of growth, it is increasingly attracting the attention of corporate, both domestic and foreign.
- In India, food is the largest segment of retail industry. There are around 3.7 million food retail outlets with an estimated turnover of Rs 7400 billion (Chengappa et al., 2007),), employing about 21 million people.
- Food retailing in India is, by and large, unorganized, highly fragmented and predominantly small, family owned businesses (Singh, 2007)
- Unorganized food retail segment consists of kirana shops, selling dry food products, and fruit/vegetable shops and hawkers (pushcart walas) selling

wet food products. As the unorganized retail outlets are under-capitalized, these are not able to cater to the consumer demand for value-added services, thus providing the edge to the organized retail sector.

- Super markets use various channels to manage supplies (Chengappa *et al.*, 2007; Singh, 2007). Some of these are:
- Direct, uncontracted purchases from farmers at individual super markets
- Purchases from wholesalers, who either work directly with farmers or through wholesale markets
- Purchases through independent procurement companies (dedicated suppliers) who often work with farmers approved by super market chains
- Purchases through government-sponsored distribution centres
- Purchases through informal farmers' groups, farmers' associations or cooperatives

Agro-Processing

- Agro processing is defined as set of techno-economic activities, applied for conservation and handling of agricultural produce, originating from agricultural farm, livestock, aquacultural sources and forests for their conservation, handling and value-addition to make them usable as food, feed, fibre, fuel or industrial raw materials
- Some estimates suggest that in developed countries, up to 14 per cent of the total work force is engaged in agro-processing sector directly or indirectly. However, in India, only about 3 per cent of the work force finds employment in this sector revealing its underdeveloped state and vast untapped potential for employment.

Issues in Agro-Processing Sector

- India's share in international food trade is negligible less than 1.5 per cent share of global agri -trade.
- Poor Linkages between Public and Private sector
- Policy Issues in Food Processing (APMC ACT, Commodities Act, Taxes, Credit) Traditional and Subsistence Farming, Market Linkages and Infrastructure. Unorganized Agricultural Sector small farms (in all 100 million, with farm size of less than 1.2 hectares divided into 3-10 separate plots), which constrains productivity levels or yield per hectare
- Value of wastages is at over Rs 50,000 crore
- Long value chain leads to Cost differential between the *Mand*i and the market price to 60-100 per cent

Cost Increase in Various Stags of Supply Chain of Agricultural Produce

Marketing Chain	*Cost Increase (per cent)*
Small Farmer > Village Consolidator	10
Village Consolidator >Commission Agent	8
Commission Agent>Wholesaler	10
Wholesaler > Sub-wholesaler	6
Sub-wholesaler > Retailer	25
Retailer > Consumer	10-11
Total	**70**

Increase in Trade Value at Various Stages in Supply Chain

Increase in Trade Value at Various Stages	*Cost Increase (per cent)*
Farmers to wholesalers	30-35 per cent
Wastage	15-25 per cent
Transportation	10-12 per cent
Total	**60-75 per cent**

Supporting Institutions

- ☆ 1986-Agriculture and Processed Food Products Export Development Authority (APEDA) initiated to develop agricultural commodities and processed foods and their export.
- ☆ Agri-Clinics and Agri Business Centers in association with NABARD launched programmes to enhance the reach of modern farming technologies to farmers across country.
- ☆ National Institute of Food Technology Entrepreneurship (NIFTEM)- Objective is development of world class food technologists, creation of Entrepreneurs and Managers in the food Industry.
- ☆ 1994-Small Farmers Agri Business Consortiums (SFAC)- Objective to set up an autonomous body for the production and development of small farmers agri business activities.

National Institute of Food Technology Entrepreneurship (NIFTEM)

Objective is development of world class food technologists, creation of Entrepreneurs and Managers in the food Industry

- ☆ The Ministry of Food Processing Industries (MFPI) has formulated a Vision 2015 with an objective of trebling the size of the industry,
  - increasing the level of processing from 6 per cent to 20 per cent,
  - value addition from 20 per cent to 35 per cent and
  - share in global food trade from 1.5 per cent to 3 per cent by 2015.

- MFPI has identified priority areas for implementation of the strategic goals, prominent among them- SME Capacity Building, Human Resource Development.

Agri-Clinics

- **Agri-Clinics** are envisaged to provide expert advice and services to farmers on various technologies including soil health, cropping practices, plant protection, crop insurance, post harvest technology and clinical services for animals, feed and fodder management, prices of various crops in the market etc. which would enhance productivity of crops/animals and ensure increased income to farmers.
- **Agri-Business Centers** are commercial units of agri ventures established by trained agriculture professionals. Such ventures may include maintenance and custom hiring of farm equipment, sale of inputs and other services in agriculture and allied areas, including post harvest management and market linkages for income generation and entrepreneurship development.

Inefficiencies in Agriculture and Agri-Business

Lack of Infrastructure

- Post Harvest: Cold storage, Warehouse
- Collection centers
- Processing units
- Distribution centers
- Perishable cargo centers
- Private markets
- Refrigerated vans

Lack of Marketing Reforms

- APMC Act
- Allowing of FDI in retail sector/Infrastructure
- Amendment in PPP sector

Lack of Research and Extension

- New crop production technologies such as organic farming
- Research on high yielding seeds
- Dissemination of information on crop certification
- Handling and storage of produce

Levels of Value Addition

- **Level-1:** Post Harvest level/Primary Processing: Proper Cleaning, grading and Packaging particularly of fruits and vegetables

- ☆ **Level-2:** Secondary processing (basic processing, packaging and branding) of post harvest produces particularly of foodgrains (*i.e.* Atta Suji, Besan, etc.) and milk produce
- ☆ **Level-3:** High- End Processing (Complex processing technologies, equipment finance and global in nature)

Approaches to Promoting Entrepreneurship

There could be three major stages of training intervention for promoting entrepreneurship *viz.*

- ☆ pre-start-up stage
- ☆ start-up stage; and
- ☆ survival and growth stage

Accordingly, all the EDPs could be classified into the following three categories:

- ☆ Entrepreneurship Orientation and Awareness Programme
- ☆ New Enterprise Creation Programme
- ☆ Existing Entrepreneurs' Programme for Small Business Survival and Growth

Entrepreneurship Orientation and Awareness Programmes

Entrepreneurship Orientation Programmes (EOPs) mostly focus on general information about entrepreneurship so that the students may start thinking in terms of taking up entrepreneurship as a career.

- ☆ Entrepreneurship Awareness Programmes (EAPs) normally refer to short-term courses conducted for arousing interest in entrepreneurship as a career among participants who, in most of the cases, come from outside the education system but are willing to do something which may lead them to become their own masters, rather than serving somebody else.

New Enterprise Creation (NEC) Programmes

- ☆ Defined as a comprehensive training package evolved to develop competencies in trainees/participants, which would lead to self-employment of the promoter, setting up of an enterprise creating further employment.
- ☆ The programme aims at developing knowledge, skills and attitudes in a potential entrepreneur to make him/her an actual owner-manager of an enterprise.
- ☆ It is offered on a regular basis and includes an adequate planned post-training follow-up support.

Existing Entrepreneurs' Programme for Small Business Survival and Growth

Programmes meant for assisting entrepreneurs to cross the 'survival stage' of enterprise establishment

Performance Improvement Programmes

- focus on streamlining business processes to manage an enterprise effectively
- targeted at those entrepreneurs who have been in business for a couple of years or so and who did not have the benefit of formal training in small enterprise management prior to launching their ventures
- such enterprises are found to be operating at or just above the break-even level and are yet to be stabilized

Growth Programmes

- developed by EDI to help entrepreneurs pursue a planned and sustainable approach to growth
- The training-cum-counselling intervention targeted at well-established entrepreneurs is spread over the three phases

Private Sector Agencies Engaged in Entrepreneurship Development

- Confederation of Indian Industries (CII)
- Federation of Indian Chambers of Commerce and Industry (FICCI)
- Federation of Association of Small Industries of India (FASII)
- Federation of Indian Micro and Small and Medium Enterprises (FISME)
- Associated Chamber of Commerce and Industry (ASSOCHAM)
- All India Manufacturers ' Organization (AIMO)
- World Association for Small and Medium Enterprises (WASME)
- Asian Centre for Entrepreneurial Initiatives (ACENT)
- Association of Women Entrepreneurs of Karnataka (AWAKE)
- Consortium of Women Entrepreneurs of India (CWEI)
- Association of Lady Entrepreneurs of Andhra Pradesh (ALEAP)

Agri-Export Zones (AEZ)

- The concept of the Agri Export Zone attempts to take a comprehensive look at a particular produce/products located in a contiguous area for the purpose of developing and sourcing the raw materials, their processing and packaging, finally exporting them. It is a part of agripreneurship development.

Key components of AEZ include:

- Cluster approach of identifying the potential products and geographical region in which these products are grown
- Adopting an end-to-end approach of integrating the entire process, right from the stage of production till it reaches the consumption stage
- Integration of the activities of various agencies connected with the development of the product.

"To produce what could be sold rather sell what is produced"

Concerns and Challenges of Entrepreneurship Education in India

More specific concerns include:

- Entrepreneurship education is prevalent only at higher levels of education
- Entrepreneurship education is not sufficiently differentiated from business management education in many institutions, and is not holistic enough
- No degree-awarding programs at academia. Limited focus on research and publications
- Difficulty with finding talented faculty. Issues with faculty awareness, opportunities, and interest
- Linkages of university and R and D centers with entrepreneurs is still weak
- Policies need to be more responsive to emerging trends, both local and international
- Absence of a mentor pool for entrepreneurs at all stages
- Shift in societal attitude towards failure is required

Food Processing Sector in India

- The food processing industry ranks fifth in size in the country and employs 19 per cent of the industrial labour force.
- 14 per cent of the industrial output is contributed by the food processing industries, which is 5.5 per cent of the GDP of India.
- The estimated turnover of the food processing industry is Rs.1,440 billion of which about 75 per cent is estimated to be in the unorganized sector and that is in the rural areas.
- A study conducted by Mckinsey and Company has visualized the year 2000 and beyond as a period of immense growth and development for the food processing industry in India. Not only the food processing industry is estimated to reach a gross value of Rs.5 trillion but value-added foods are expected to grow at a much faster rate, *i.e.*, to expand from Rs.800 billion to Rs.2,250 billion by 2005 AD. The study predicts that by 2005 around 200 million people will move from subsistence foods, like cereals and pulses, to basic processed foods.

Major Constraints Faced in the Development of Food Processing Industries

- The value addition of food fortification in India is only 7 per cent in comparison to other countries like China (23 per cent); Philippines (45 per cent); and U.K. (88 per cent).
- Although India is one of the largest producers of raw material for the food processing industry in the world, the food industry is still at a nascent

stage. Less than 2 per cent of fruit and vegetable production is processed compared Thailand (30 per cent), Brazil (70 per cent), Philippines (78 per cent) and Malaysia (80 per cent).

- Agriculture has largely been for subsistence and has not been market-driven. This has not yielded adequate surpluses for processing.
- The small and unorganized sector accounts for 75 per cent of the total food processing industry. There are only a few organized large industrial houses who make their presence felt in the food-processing arena in India.
- Multiple and complicated tax regimes have made the food processing industry uncompetitive.

Value Addition in Agriculture

- **Value added** is the net output of a sector after adding up all outputs and subtracting intermediate inputs. It is calculated without making deductions for depreciation of fabricated assets or depletion and degradation of natural resources. The origin of value added is determined by the International Standard Industrial Classification (ISIC), revision 3.
- It is a production/marketing strategy driven by customer needs and preferences
- "Valueadded" is used to characterize food products converted from raw materials through processes that give the resulting product an "incremental value" in the market place either through higher price or expanded market.

Agribusiness Sector in Bangladesh

- While agricultural production grew by about 37 per cent during the 1990s, agribusiness grew by about 66 per cent –or roughly 5.2 per cent annually over a decade.
- The growth of agribusiness, though faster than the 3.2 per cent annual expansion in agriculture in that period, still was insufficient to generate the employment and income growth required to reduce poverty significantly and accelerate the overall growth of the rural economy.
- Demand-driven agribusiness activities are critically important to Bangladesh economy.
- Agribusiness accounts for about 10 per cent of the country's gross domestic product (GDP), according to estimates. Because the market demand is income-elastic the
- Growth of agribusiness is much higher than the growth of agricultural production.
- With proper support, however, agribusiness growth could accelerate to an estimated 7 per cent a year or more, which would speed progress toward poverty alleviation.

- In Bangladesh, agribusinesses comprise about 3 million enterprises, of which 90 per cent have less than 10 workers or employees and 36 per cent have only one worker.
- Nonetheless, they are estimated to account for half of household income for about 15 million people.
- The agri-businesses in Bangladesh may be divided into three layers:

Small and Landless Farmers

Bangladesh has many household level, cottage or micro-industries concerned with processing agricultural and related raw materials, including non-food items, involving jute, cotton, bamboo, or timber.

- Non-farm households (*i.e.*, the landless) mange about half of these, while mostly small farmers (owning less than 1.0 hectare) manage the rest.
- Often, non-government organization (NGO) microfinance funds them. Investments in this category are typically below $1,000 per household. These enterprises usually serve a localized market around the village or sub-district headquarters.

Rural Enterprises

The next layer of agribusiness is made up of rural based enterprises. Bangladesh has about 3 million such enterprises, which make up the bulk of the country's agribusinesses.

- These typically include small poultry, dairy, and fisheries enterprises, as well as rural produce and input traders and transporters.
- This segment of rural business is fairly dynamic and market driven. Investments are $1,000–$10,000 per enterprise.

Urban Enterprises

Medium-sized, generally urban-based, agroprocessing enterprises constitute another layer.

- These include commercial juice, jelly, jam, pickles, and snack food preparation; potato cold stores; fish freezing plants; and farmers' service cooperatives.
- Most of them have investments well below $100,000.

Malaysia (S.E Asia)

Ninth Plan

Emphasis on agriculture- During the Ninth Plan period, the agriculture sector will be revitalized to become the third engine of growth.

The emphasis will be on **New Agriculture which will involve:**

- Large scale commercial farming

- ✰ Production of high quality and value-added products
- ✰ The participation of entrepreneurial farmers and skilled workforce
- ✰ Wider application of modern technology
- ✰ Unlocking the potential in biotechnology
- ✰ Increased convergence with information and communications technology (ICT)
- ✰ One of the most important mechanisms is through the extension and technical advisory services of the Department of Agriculture (DOA) and supported by Malaysian Agricultural Research Institute (MARDI), Farmers Development Board, Rubber Industry Smallholders Development Authority (RISDA) and Federal Agriculture Marketing Authority(FAMA).
- ✰ The government of Malaysia has established few agencies within the Ministry of Agriculture and Agro-based Industry (MOA) such as Farmers' Organization Authority (FOA) to provide extension on technical and motivational training to individual farmers to become agri-entrepreneurs

Trends in Brazilian Agribusiness

Agribusiness GDP represented 22.3 per cent of Brazilian GDP in 2010, a total of R$ 821 billion. Agribusiness is the main responsible for the Brazilian trade balance surplus, representing 37 per cent of total exports.

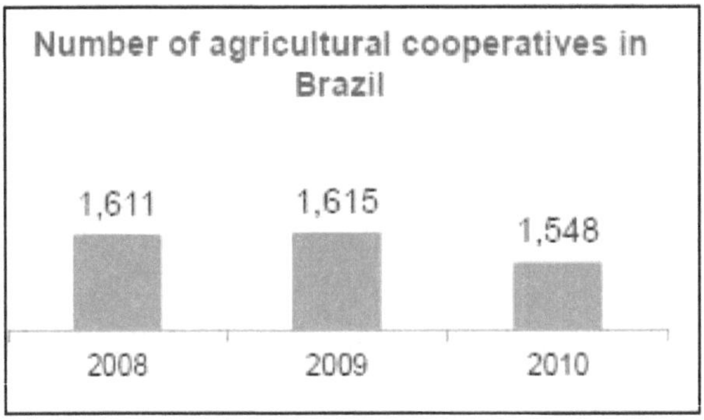

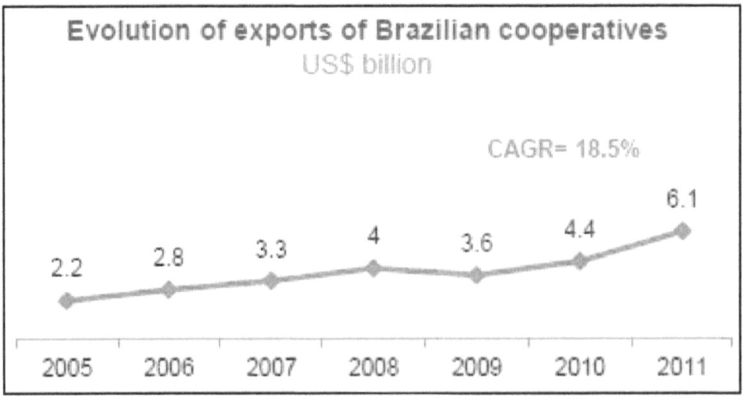

Today, more than 50 per cent of all agricultural production in the country pass, directly or indirectly, by a cooperative.

Agribusiness in Japan

- ☆ All farmers in Japan are members of multi-purposes agricultural cooperatives.
- ☆ In Japan, Cooperative farm corporations (CFC) have been traditionally distinguished from Joint stock companies (JSC) and limited responsibility companies, while the farm operated under either of these management forms has been called "organized farm".
- ☆ CFCs are a form of management, which is operated either by a single farm manager or a group of farmers, and are controlled by the Agricultural Cooperatives Act. These are independent organizations dedicated to earning private profits.
- ☆ CFC can be characterized as follows:
  - (a) It is a form of management organization evolved from the conventional family-based farms since farmers and their family are only eligible to be members, and
  - (b) There is always agricultural production division as a principal part of business activities which makes use of farmlands owned by the corporation or its members.
- ☆ According to the Agricultural Census for 1995, there were about 6,500 organized farm business management bodies whose business was to sell agricultural products (the custom farm work services organizations were excluded). These organizations were:
- ☆ Joint-stock companies (JSCs): 1,000 (500)
- ☆ Limited responsibility companies: 2,000 (800)
- ☆ Cooperative farm corporation (CFCs): 1,500 (400)
- ☆ Agricultural cooperatives or other agricultural associations: 500 (100)
- ☆ Non-registered corporate organizations: 1,500 (30)

(The figure in parentheses shows the numbers of organization with annual sales of farm products amounting to 100 million/or more. 1 US$=120/-)

- ☆ Of the total farmland acreage in Japan (4.8 million ha), only about 1 per cent is occupied by these organized farm business management bodies. In recent years, however, the trend has moved toward organized form of farm business management bodies, and the number of such bodies has been increasing gradually.
- ☆ Over the past 20 years, the entry of enterprises into agriculture-related businesses has been increasing. These enterprises were engaged in the contract-based purchase and processing and/or distribution of agricultural products with the sales of packaged materials and technical instructions.

- ☆ The main activities of these enterprise farms are:
  - (a) organic products (or products produced with essentially less chemical fertilizers and pesticides),
  - (b) vegetables and flowers cultivated in controlled facilities or by using hydroponics,
  - (c) cut and ready vegetables, tomatoes for processing, herbs, etc., and
  - (d) poultry, egg production and raising swine.
    - ☆ In recent years, enterprises have begun to actively enter into the business of importing land cultivation based vegetables and fruits, which are produced in developing countries at their own initiative

China

Agricultural Enterprise Development in China

Enterprise and Organization	*1996*	*1998*	*2000*
Total number	11824	30344	66000
Farm families involved(number/Percentage)	1995.1/10	4923.3/15	5900/25
Leading enterprises and organizations (number/Percentage)	5381/45.51	15088/49.93	27000/41
Intermediary organizations (number/Percentage)	3384/28.62	8024/26.44	22000/33
Specialty organizations (number/Percentage)	1450/12.26	4848/15.98	7600/12
Other organizations (number/Percentage)	1600/13.61	2384/7.85	9600/14

Africa

Agri-Enterprise Development for Stimulating Rural Economies in Africa (UNDP)

The proposed regional programme framework is a joint initiative between FAO, IFAD, ILO, UNDP and UNIDO, working in partnership with Songhai Centre, an international NGO based in Benin.

Programme operations will initially focus on a number of selected countries in West, Eastern and Southern Africa, *i.e.*, Benin, Burkina Faso, Cote d' Ivoire, Gabon, Ghana, Guinea, Kenya, Liberia, Sierra Leone, Malawi and Togo.

Programme Objective

To promote access of youth, men and women to appropriate agri-entrepreneurial, leadership and management skills required for their effective participation in the creation of and investments in commercially viable agro-enterprises in selected African countries.

- ☆ The programme will have five interrelated components aimed at:
- ☆ Component 1: Facilitating and supporting the establishment of a Regional Centre of Excellence for Agribusiness and Entrepreneurship Development in Africa.

- Component 2: Reinforcing the capacity of relevant national institutions to establish National Centres for Agri-Enterprise Development (NCAED) in participating countries
- Component 3: Developing agri-entrepreneurial skills and capabilities of youths, women and men, particularly those from rural areas

The comprehensive integrated training program will focus on the following:

- Primary sector (Livestock, aquaculture, crop production)
- Secondary sector (post-harvest, food processing and appropriate technologies)
- Tertiary sector (marketing, quality assurance, agribusiness and services)
- Management (leadership, community development)
- Mechanization and rural energy generation and use
- Natural resource management and bio-dynamics
- Irrigation and water management
- Component 4: Creating platforms to facilitate effective linkages between agribusinesses to credit, market and business support service providers
- Component 5: Improving institutional and business environment for small- and medium scale agri-business development

The programme will be implemented in two phases.

The first phase will consist of:

1. Country institutional and sector scoping, developing country specific plans, organizing institutional framework for the implementation of country specific programmes, and
2. Supporting the establishment of a Regional Center of Excellence for Agri-business development embedded within the Songhai Centre.

The second phase will largely focus on country level activities. Implementation of activities will be carefully sequenced taking into consideration resource availability and the level of preparedness of each country participating in the programme.

- There are three integrated components to the **Songhai development model**:
    1. Agricultural centers involved in financially and environmentally sustainable production of primary and value added products.
    2. Training of farmers as entrepreneurs and socio-economic leaders emphasizing zero waste and sustainable agricultural production, including, pre and post harvesting techniques.
    3. Outreach to a community of trained farmer graduates who create clusters of agri-enterprises and manage their own farms, through a network that provides services and a route to markets.

Programme Outputs

- ✰ A Regional Centre of Excellence for Agribusiness and Entrepreneurship Development established.
- ✰ National Centres for Agri-Enterprise Development (NCAED) countries established in participating countries
- ✰ Viable agri-businesses established in participating countries
- ✰ Enhanced linkages between agri-business to credit, market and business support service providers through networking leading to improved productivity and profitability
- ✰ Improved business environment for SMEs facilitating agri-businesses registration and employment generating approaches

Chapter 11

Market Led Strategy for Accelerated Agri Development: Role of Private (Agriculture) Entrepreneurs

Pradip Kumar Mazumdar

Vision Needs a Paradigm Shift

- A shift from "yield increase approach" to " income driven approach" - from "technology driven" to "market driven".
- *Technology would still be needed to make the vocation viable.*

Road Map to Achieve the Vision

- Growth rate in agricultural and Horticultural crops for next 8 years to be fixed at > 8 per cent.
- Growth rate for food grains to be fixed at 4 per cent, but for oilseed it is to be at higher level and for horticultural crops it be targeted at 10-12 per cent.
- Higher growth be possible by planned development in Poultry, Goatary and Fishery
- All these need to be viewed through 1 Common State Agricultural Plan (SAP), a pre-requisite for development.

Strategy to Reach the Goal

- For a market led Strategy to succeed
- Farmers Field Schools (FFS) are essential

☆ FFS would encourage:

- Formation of Farmers Interest Groups (FIG)s and Producer Companies.
- Thus Organized, Farmers would lead to aggregation of produce, linking to the market through end to end projects, and PPP farming with actively involving Pvt Enterprise etc.

Forward and Backward Linkage to the Market

1. Set focus on production of Pulses and Oilseeds, Corn, Local Aromatic Rice and horticultural crops which are in demand in the market.
2. Set higher production target for Dairy, Livestock and Fishery,
3. Integrate "Jal Dharo Jal Bharo" and similar other Water Resources Programme with the Production Strategy.

Stepping Stones to move forward

1. Train FIGs to aggregate their produce,
2. Pvt Enterprise or a State Marketing Corporation may act as the Pivot regarding i) imparting training, ii) buying the produce and iii) arrange to aggregate and supply to Corporate/processing Industry and Marketers, thus effectively reducing the number of Intermediaries.

Stepping Stones to Address Gray Areas

1. From Programmes to Projects

Integration of all the desired interventions in a given area under Govt. Programme, is missing; need to treat as end to end Projects involving all the related sectors like agriculture, Horticulture, ARD, Water Resources, Fishery etc.

2. End to End Projects in PPP Mode

Govt. Intervention may also be through one Agency/Private player to groups of farmers provided all managerial inputs and marketing arrangements are made by that Player.

3. From Divergence to Convergence

☆ Agriculture has been sub-divided into large no. of Departments and to achieve higher growth rate immediate convergence of actions is the pre-requisite.

☆ Formation of Farmers Groups, Federations and Co-operatives are the need of the day as negotiating power of individual small and marginal farmers is highly limited.

☆ FFSs are the building blocks of this farmers organizations and successful implementation of multi-sector end to end projects with room for Pvt Enterprise too.

Integrating Dev of Horticulture Crops

☆ Efficient Resource Utilisation by integration and convergence of schemes, integration of multiple organizations, re-defining roles of key-functionaries.

- ☆ Project Mode of operation for pre-determined outcome, providing complete technology solution to hike productivity, higher per cent of quality produce, plant protection, PHM, storage, transport and marketing.
- ☆ Ensure availability of quality seeds and planting materials in sync with area expansion programmes
- ☆ Prioritize on the basis of Focus Crop and Focus Production Cluster/ Production Area
- ☆ Promote capital investment for development of critical infrastructure for horticulture development

4. Integrating and Developing Water Resources of the State

- ☆ Set goal of increased development of Watershed by next 8 years in the back drop of 40 per cent of Cultivable Area under Actual Irrigation.
- ☆ Convergence of resources under MGNREGA, other Schemes of P and RD, WRIDD including: "Jal Dharo Jal Bharo".
- ☆ District wise AIA vis-à-vis UIP/CCA under each season to be identified and "land Shaping" and other models need to be applied.

Development of Seed Village

- ☆ Seed a critical input, could remain in the hands of Private Players, more if it is novel technology intensive.
- ☆ Seed Village could provide capability to produce quality seeds, even hybrids, to farmers through FFS.
- ☆ Seed Certification Dept. needs to be revamped to provide necessary certification for FIGs.
- ☆ End to end Projects could be formed with this concept.

Encourage Agro-based micro-enterprises (MSME)

- ☆ Dal Mills, Grading and Packing Units, at least one for each producer Groups/Company to be established.
- ☆ Where time is required to form producer Groups/Company, existing FPOs, PACS be targeted.
- ☆ Units producing bio-fertilizers, bio-pesticides, small storage structures, grading and packing units, oil crushers and even combined harvesters and trans planters can be made locally available through MSME, run by educated SHGs/FPOs/select PACS.

Create Storage Structures

- ☆ To avoid distress sale, decentralized storage units are needed;
- ☆ One per *Mandi*/block has already been planned in WB
- ☆ Under Warehousing (Dev. and Reg.) Act, 2007, warehouses can be accredited, qualifying for interest subvention on Bank Credits. This would be the beginning of farmers themselves managing their Produce without intermediaries

- ☆ Private entrepreneur can play a pivotal role to mitigate farmers' distress and yet make reasonable margin.

Strengthen Cooperative Institutions

- ☆ Agri-based Cooperative institutions had been formed [but many are destroyed due to lack of vision] to be linked with value chain.
- ☆ PPP mode should ve new definition where private entrepreneur [PE] join hands with cooperatives and Govt. plays the role of facilitator.
- ☆ PACS/PE could assume the role of input suppliers, bio-fertilizer producer, warehouse keeper, agri-implement hirer besides credit disbursement agency.

Fixing the Policy Priority

1. To achieve a quantum jump in growth rate with all encompassing increase in farmers' income, the following Priorities are identified:
2. Moving away from schemes and programmes to end projects.
3. Incentivizing PPP projects, where backward integration in value chain is State's responsibility and Forward linkages are of PBAMC and Private Entrepreneurs.

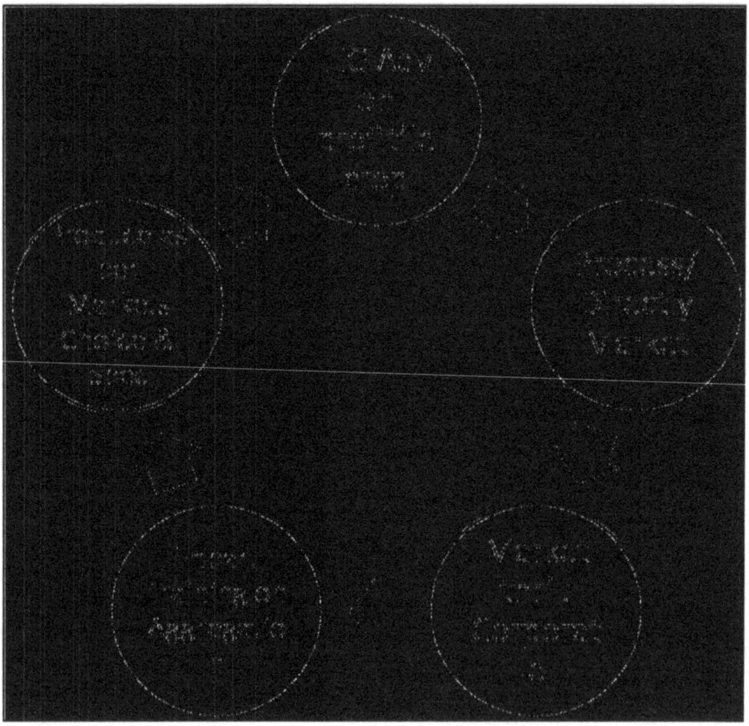

Forward Linkages thru PBAMC

– Part II –

Entrepreneurship and Organization

Chapter 12

Transforming Organizations: Concept of Co-Intrapreneurship

Manas Mohan Adhikary and Sankar Kumar Acharya

Epilogue

Seeking alternative to command and control, new post-bureaucratic models of organization, based on co-ordination, emphasized the need for decentralizing authority and development of team oriented designs. Decision making process and entrepreneurial activities need to be conveyed to the level of each individual in the organization in order to top their specific competencies and knowledge for future success. The modern context of a post industrial economy, the workforce can no longer be understood as a factor of production, but has become a strategic determinant of success (*e.g.* Blyton and Turnball, 1992; Jackson and Schuler, 1999).

Entrepreneurship and Co-Entrepreneurship

In the modern Centex of a post industrial economy, the workforce can no longer be understood only as a factor of production, but has become a strategic determinant of success.

[Jackson and Schuler, 1999]

Human Resources in particular are not primarily to be seen or treated as 'Cost component', but as a potential source for added value and Long term Investment.

[Ulrich et al., 1999]

Entrepreneurship can be defined as the process of eliciting or evoking value from unique and new combination of resources in an uncertain environment, consequently, Entrepreneurship can be understood as co-operating organization members, who innovate, identify and create business opportunities, assemble ad coordinate new combination or arrangement of resources so as to yield or enhance value. Co-intrepreneurial development can be understood as socially effective processes that support all members of an organisation and their co-operating interaction.

Co-Intrepreneurship and Acculturation

Acculturation is generally defined as "Changes induced in (two cultural) systems as a result of the diffusion of cultural elements in both directions"

[Berry, 1980]

Model: Acculturation of Mergers and Acquisitions

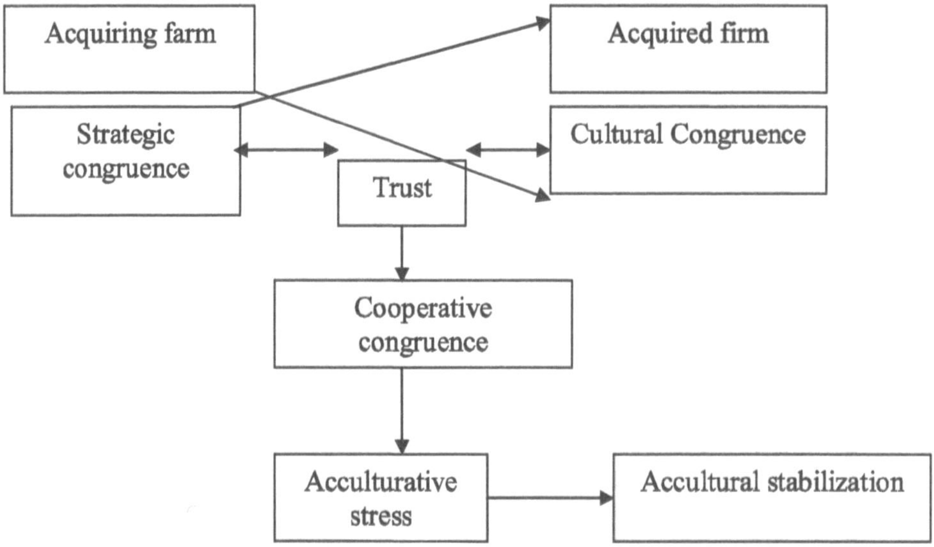

Changing Scenario: Entrepreneurship: Bureaucracy Entrepreneurship

Definition of task: A task is any job well defined in terms of role, responsibility, time, rewards and accomplishment leading to a goal achieving satisfaction. In modern day entrepreneurship and for any innovative organization there are two set of task.

a) Flexible task: A task is having role flexibility, time and responsibility flexibility is called flexible task. Flexible task leaves the scope for innovation.

b) Stereo type task: A task which is hard structure leaving no scope for time resource and cost flexibility call stereo type task. Stereo type task leaves little or no scope for alternative thinking or innovation.

The Intrapreneur's Ten Commandments

- ☆ Come to work each day, willing to be fired.
- ☆ Circumvent any orders aimed at stopping your dream.

- ☆ Do any job needed to make your project work, regardless of your job description.
- ☆ Find people to help you.
- ☆ Follow your Institution about the people you chose, and work only with the best.
- ☆ Work underground as long as you can, publicity triggers the corporate immune mechanism
- ☆ Never bet on a race unless you are running in it.
- ☆ Remember it is easier to ask for forgiveness than for permission.
- ☆ Be true to your goals, but be realistic about the ways to achieve them.
- ☆ Honour your sponsors

The Key Elements of Transformation Process

Goal

Context

Potential

Coordination

Policy

To Safeguard and Increase the Company Value

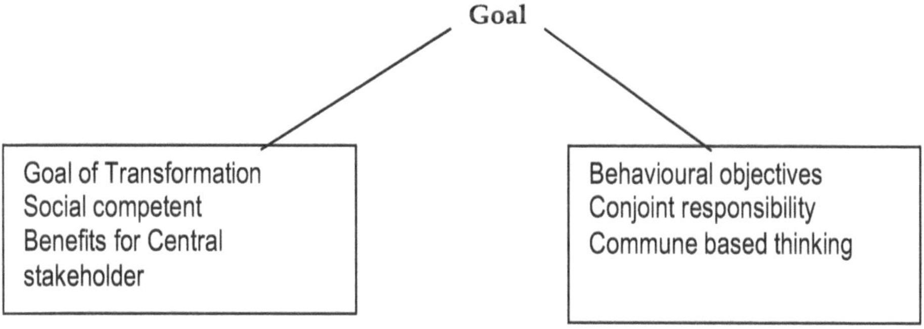

Context

The macro and micro context, as well as the resources, are the basic conditions for the co-intrapreneurial transformation process.

Potential

Selection and Advancement concentrates on a few key co-intrepreneurial

Competencies rather than a catalogue of general, technical or job specific characteristics

(Ulrich, 1997)

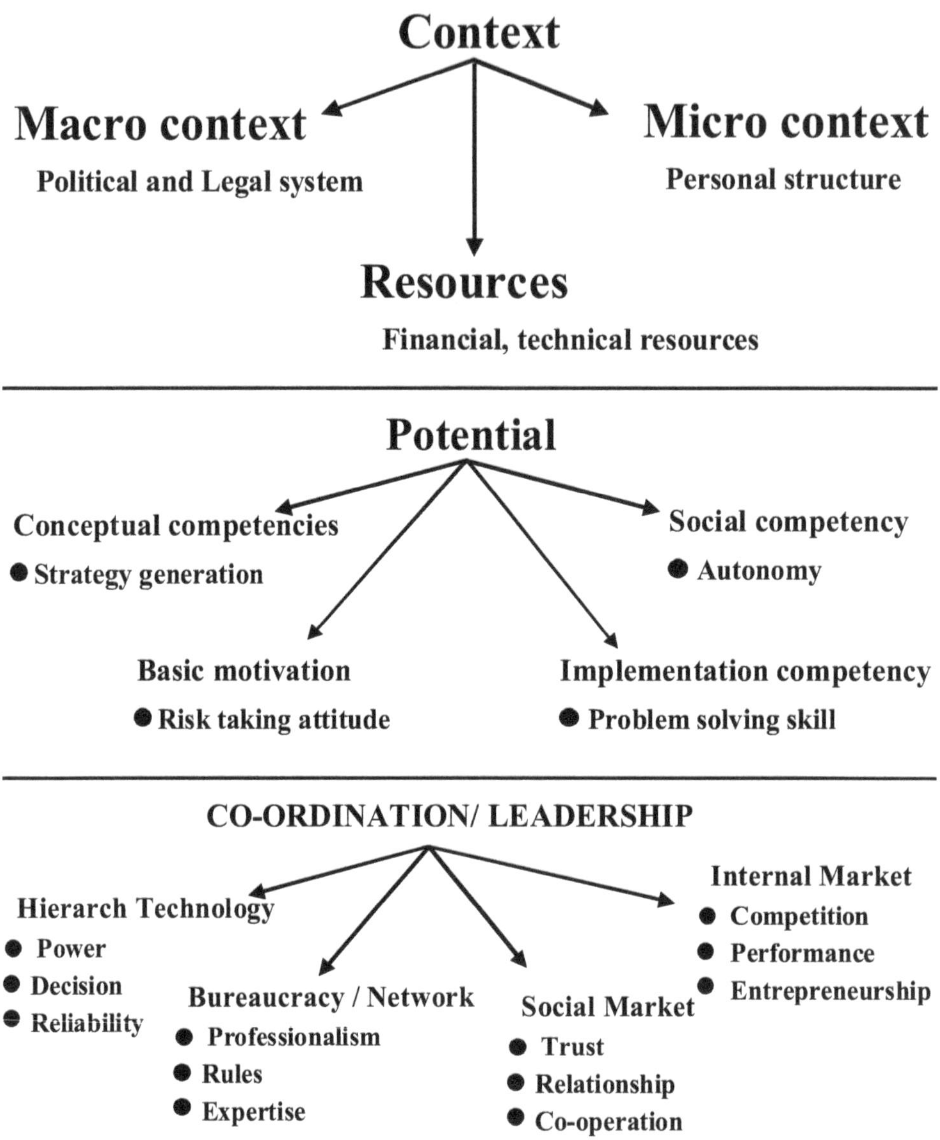

Co-ordination/Leadership

A configuration of market and social network as systems of co-ordination modify and substitute the classical leadership model of hierarchy and bureaucracy

(Axelord, 1984)

Policies

Differentiated selection, placement and development in accordance with a specific typology approach prevent demands for exclusive or utopian collective behaviour.

(Wanderer, 2001)

Acculturation Stress

Three Vicious circles vicious Circle (1) : Global vs Local Activities

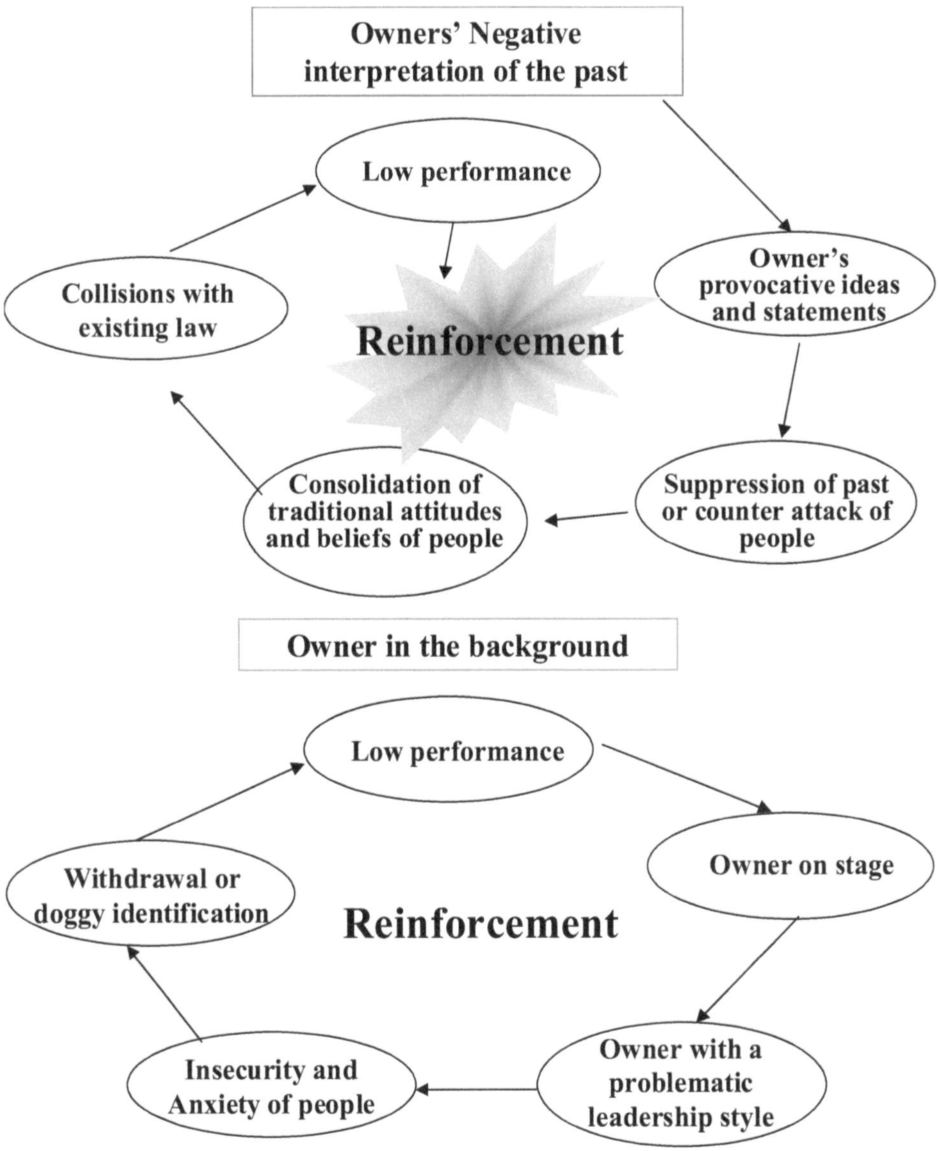

Various circles and Structure

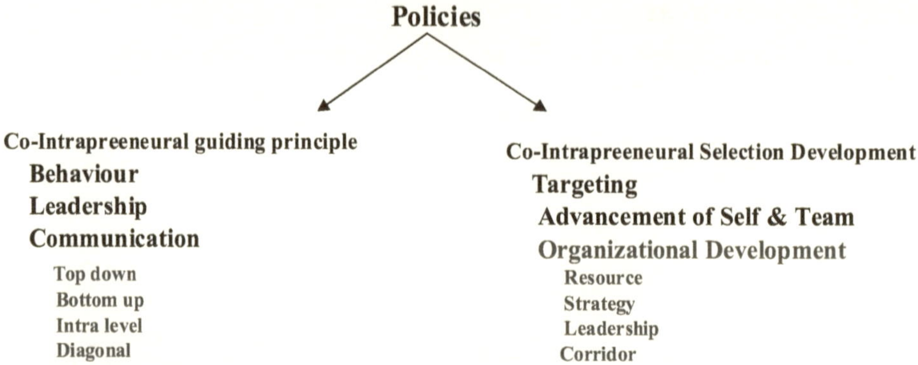

Vicious Circle (2) : Different Interpretation of the past

Ref : Acculturation of Cross-border Acquisition : Cob ways in Germany by Thomas Steger

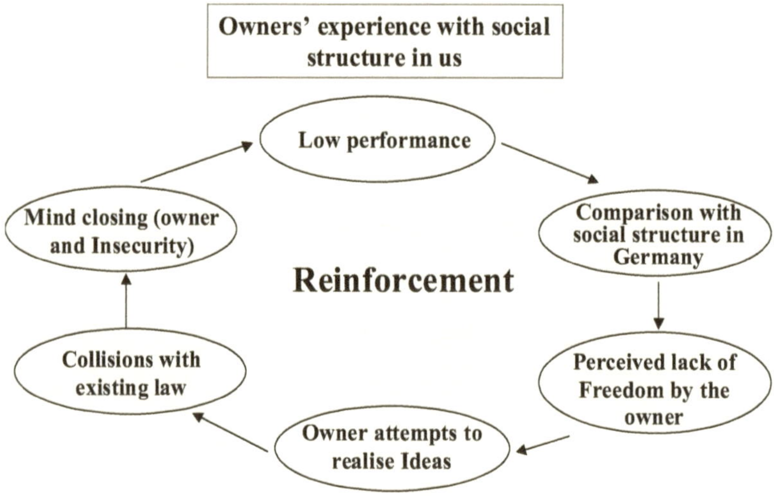

Actoral Leaver: Sense Making in Change Environment

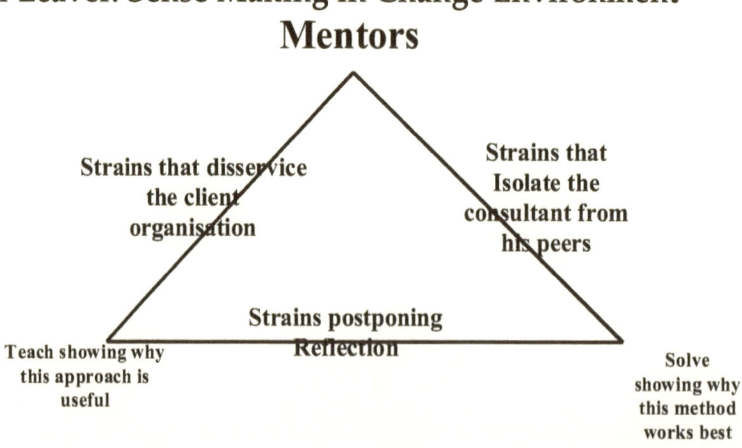

Research: Sense making in change intervention

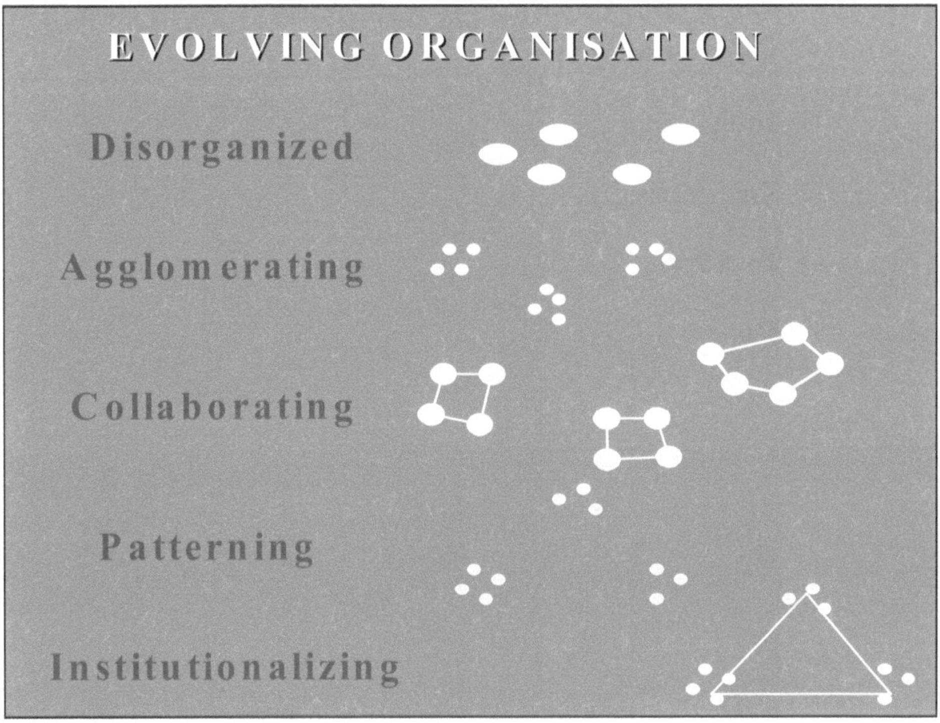

Partnership : Homogeneity vs Heterogeneity

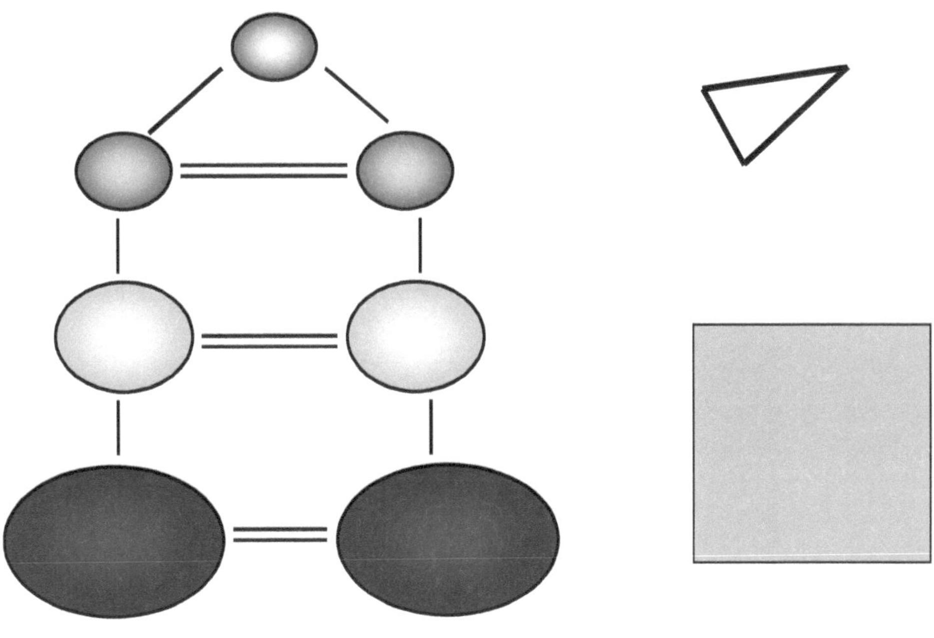

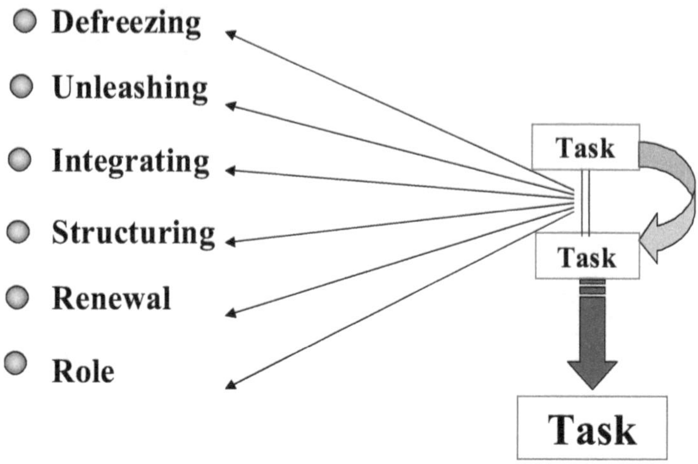

J. A. Mc. Caffery, 1986

Chapter 13

Conflict Management in Enterprise Development: Concept, Factors and Interventions

Sankar Kumar Acharya and Manas Mohan Adhikary

Whenever and wherever people work collectively, conflict is likely to occur. Conflict is daily reality for everyone, it is inevitable.

Conflict and stress are common when complex entities such as individuals and small groups are brought together.

The Natural Resource Management scenario in India, the other way round does present the depiction of conflicts at different levels, in various dimensions and to diverse destination as well.

What is Conflict?

Conflict may be viewed behaviourally as a form of opposition.

It is opponent concern, based on incompatibility of goals, aims or values of opposing parties

Conflict is a process in which one party perceives that its interests are being opposed or negatively affected by a party.

The natures of conflict are as follows:

- ☆ Policy conflicts
- ☆ Ownership Conflicts
- ☆ Gender Conflicts
- ☆ Resource Conflict

- ☆ Role Conflict
- ☆ Stake Holder Conflict
- ☆ Cultural Conflict

Conflict	*Stake Holder*	*Impact*
Policy conflicts	Political Leaders, Government, Organizations, National & International Agencies, Parliaments, Legal systems, Research Organizations etc.	New policy emerges, Conventional policies refined and ratified, New roles emerges, empowerment results, Confusion simmers etc.
Ownership Conflicts	Forest dwellers, Forest officials, Peripheral villagers, Panchayat leaders, Timber merchants etc.	Illicit felling, Unabated erosion, Forest fire, Poaching, Resolution committee, Creation of FPC & EDC, New monitoring mechanism etc
Gender Conflicts	Tribal women, Traditional Women folk, Male folk, Old men and women, Women belonging to culture group etc	Gender sensitive forest management policy, Disintegration of families, Alcoholism etc
Resource Conflict	Forest dwellers, Forest officials, Peripheral villagers, Panchayat leaders, Timber merchants, Political leaders, Ecologist, Stake Holders of Green Movement, Multinational Company etc.	Illicit felling, Unabated erosion, Forest fire, Poaching, Resolution committee, Creation of FPC & EDC, New monitoring mechanism, Creation of Eco tourism, Resource conservation techniques etc.
Role Conflict	Forest dwellers, Forest officials, Forest planners, Peripheral villagers, Panchayat leaders, Timber merchants, Political leaders, Ecologist, Stake Holders of Green Movement, Multinational Company etc.	Illicit felling, Unabated erosion, Forest fire, Poaching, Resolution committee, Creation of FPC & EDC, New monitoring mechanism, Creation of Eco tourism, Resource conservation techniques, Emergence of Joint Forest Management Concept etc.
Stake Holder Conflict	–do–	–do–
Cultural Conflict	Culture groups, Political leaders, Local players, Media personnel, People from the walks of art and culture, environmentlogist etc.	Disintegrations, Communal clashes, Social disaster etc.
Territorial Conflicts	Different state leaderships, Political groups, Commission on rivers, forest etc.	Truce, Arbitration, Negotiation, MOU, Promulgamation of Laws, Amendments etc.
Natural Conflicts	Wild Lives, Tribals and forest dwellers etc.	Conservation policy, Research and eco-systems etc.
Political conflicts	Political Leaders, Administrators, NGO's, Institutions etc.	Disintegrations, Intimidation, Coercion, Emergence of new organizations etc.
Operational Conflicts	All stake holders mentioned in the column.	All impact mentioned in the column.

- Territorial Conflicts
- Natural Conflicts
- Political conflicts
- Operational Conflicts

What is Conflict?

The conflict may be viewed behaviourally as a form of opposition, which is an opponent concerned, based on incompatibility of goals, aims or values of opposing parties.

Types of Conflict

- Interpersonal
- Intrapersonal
- Intergroup
- Intragroup

Assumption of Conflict

- Conflict is an organisational reality
- Conflict typically involves a relationship
- No single way is there to manage conflict
- Conflicting participants are emotionally and attitudinally diverse

Potential Sources of Conflict

- Aggressive Nature of People
- Competition For Limited Resources
- Interest and Value Clashes
- Role Ambiguity
- Power seeking and self-interest
- Poorly defined responsibilities
- Lack of compatibility of change
- Intidimidation and Hegemony

Impact of Conflict

Impact of conflict is an effectual disposition of conflict process that can be catalogued as positive and negative impacts.

Positive

- Constructive Change
- Motivation and energy
- Creativity
- Productivity through Competition

Negative

- ☆ Stress
- ☆ Waste
- ☆ Non-Performing
- ☆ Distortion of Goals
- ☆ Entropy

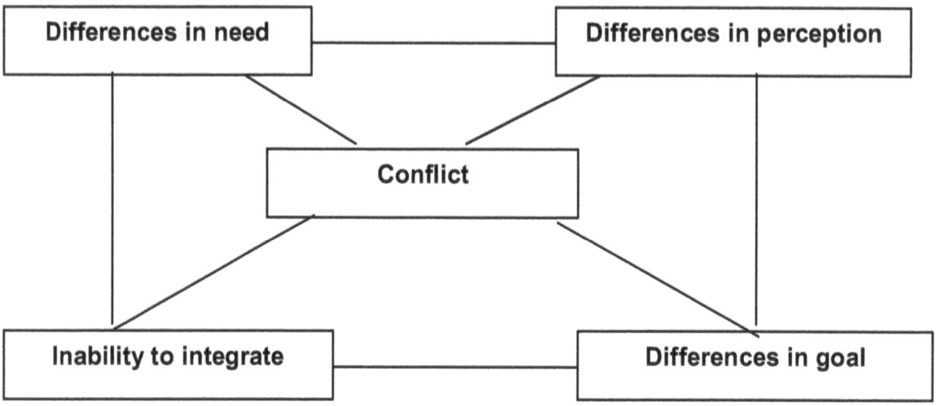

The Johari Window: this window generated in USA in Michigan State University is extensively used for personality analysis.

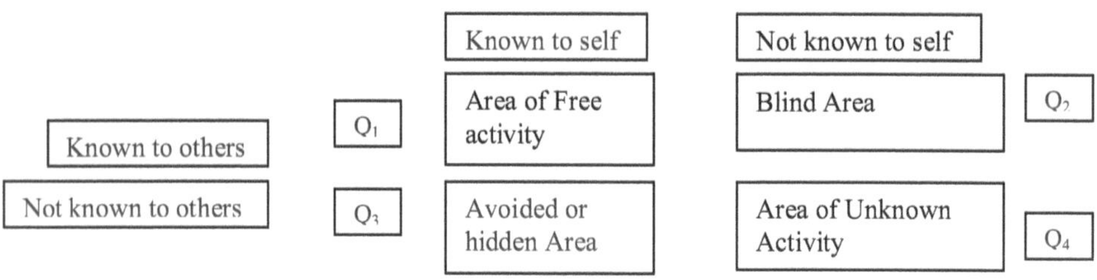

The Johari Window

Reactions to Conflict

Every conflict is wielding a response as well as reactions to conflicts. These reactions are having an expense of very modest to extreme disposition of psychic interactions.

Avoidance ⇨ Withdrawal

Aggression ⇨ Overt attacks

Manipulation ⇨ Covert attacks

Avoidance ⇨ One giving in

Theorizing Conflict

The conflict process can be put to a theoretical paradigm by accommodating and retrofitting some of the elements as follows:

- Assumption
- Reflection
- Behaviour
- Feedback
- Reinforcement – competition and collaboration

Constructs of conflict

⇩ ⇩

Conflict situation conflict episode ⟹ Behaviour

The conflict has four components

1. Parties involved
2. Field of conflict
3. Dynamics of participation
4. Management of conflict

Conflict Resolution Strategy

- Avoidance
- Diffusion
- Confrontation- power and negotiation

Motivational outcome – I WIN, YOU LOSE
YOU WIN, I LOSE
I LOOSE, YOU LOSE
WE WIN

Where the mind is without fear and the head is held high; where knowledge is free, where the world has not been broken up into fragments by narrow domestic wall where word come out from the depth of truth, where tireless striving stretches its arms towards perfection.

Chapter 14

Issues and Perspective of Drudgery in Gender Studies

Manas Mohan Adhikary and Sankar Kumar Acharya

You can tell the condition of a nation by looking at the status of its women.

Prime Minister Jawaharlal Nehru

However, much a mother may love her children, it is all but impossible for her to provide high-quality child care if she herself is poor and oppressed, illiterate and uninformed, anaemic and unhealthy, has five or six other children, lives in a slum or shanty, has neither clean water nor safe sanitation, and if she is without the necessary support either from health services, or from her society, or from the father of her children.

Vulimiri Ramalingaswami, "The Asian Enigma"

The women who participate in and lead ecology movements in countries like India are not speaking merely as victims. Their voices are the voices of liberation and transformation... The women's and ecology movements are therefore one, and are primarily counter-trends to a patriarchal mal development.

Vandana Shiva

***Amartya Sen - the Unheeded Conscience:** we will lionise him, but will we ever listen to what he's saying?*

Sen points out that when he took up issues of women's welfare, he was accused in India of voicing "foreign concerns." "I was told Indian women don't think like that about equality. But I would like to argue that if they don't think like that they should be given a real opportunity to think like that."

Parmita Shastri, Outlook India, 1998

Looking through the lens of hunger and poverty, there are seven major areas of discrimination against women in India:

- ✰ ***Malnutrition:*** India has exceptionally high rates of child malnutrition, because tradition in India requires that women eat last and least throughout their lives, even when pregnant and lactating. Malnourished women give birth to malnourished children, perpetuating the cycle.
- ✰ ***Poor Health:*** Females receive less health care than males. Many women die in childbirth of easily prevented complications. Working conditions and environmental pollution further impairs women's health.
- ✰ ***Lack of education:*** Families are far less likely to educate girls than boys, and far more likely to pull them out of school, either to help out at home or from fear of violence.
- ✰ ***Overwork:*** Women work longer hours and their work is more arduous than men's, yet their work is unrecognized. Men report that "women, like children, eat and do nothing." Technological progress in agriculture has had a negative impact on women.
- ✰ ***Unskilled:*** In women's primary employment sector - agriculture - extension services overlook women.
- ✰ ***Mistreatment:*** In recent years, there has been an alarming rise in atrocities against women in India, in terms of rapes, assaults and dowry-related murders. Fear of violence suppresses the aspirations of all women. Female infanticide and sex-selective abortions are additional forms of violence that reflect the devaluing of females in Indian society.
- ✰ ***Powerlessness:*** While women are guaranteed equality under the constitution, legal protection has little effect in the face of prevailing patriarchal traditions. Women lack power to decide who they will marry, and are often married off as children. Legal loopholes are used to deny women inheritance rights.

India has a long history of activism for women's welfare and rights, which has increasingly focused on women's economic rights. A range of government programmes have been launched to increase economic opportunity for women, although there appear to be no existing programs to address the cultural and traditional discrimination against women that leads to her abject conditions.

The Inextricable Link

The greatest tragedy facing humanity today is the persistence of chronic hunger – an intolerable phenomenon that takes the lives of 24,000 of us every day. For fully one-fifth of humanity, life is a daily struggle to survive in conditions of relentless poverty. Day after day, the lives of one billion individuals are cut short or terribly diminished by chronic, persistent hunger. Day after day, one billion people are denied the opportunities they need to lead healthy and productive lives.

People living with chronic hunger exist in conditions of severe poverty. What they lack is the chance to change their situation, to develop their own self-sufficiency.

The most potent confirmation of this fact can be seen in the lives of women. They, along with their children, are the main victims of hunger, and they are also most lacking in opportunities to end their own and their families' hunger.

The Hunger Project has come to the recognition that the persistence of hunger in India – and elsewhere in the world where hunger is still an overriding social issue – is, to a large degree, due to the subjugation, marginalization and disempowerment of women. Furthermore, women's suppression is rooted in the very fabric of Indian society – in traditions, in religious doctrine and practices, within the educational and legal systems, and within families.

Ironically, much of the essential work of ending hunger rests in women's hands.

Traditionally, women bear primary responsibility for the well being of their families. Yet they are systematically denied access to the resources they need to fulfill their responsibility, which includes education, health care services, job training, and access and freedom to use family planning services.

In order to gain a shared understanding of the condition of the status of women in India and its impact on the persistence of hunger, this document surveys papers done by leading scholars in Indian development issues. It is organized in a framework of seven issues that characterize the plight of resource-poor women, with a focus on rural women, in India: malnutrition, poor health, lack of education, overwork, lack of skills, mistreatment and powerlessness.

The link between these issues and the persistence of hunger in India was underscored in a 1996 study: *The Asian Enigma,* by Vulimiri Ramalingaswami:

In short, the poor care that is afforded to girls and women by their husbands and by elders is the first major reason for levels of child malnutrition that are markedly higher in South Asia than anywhere else in the world.

India – An Overview

India, with a population of 989 million, is the world's second most populous country. Of that number, 120 million are women who live in poverty. India has 16 per cent of the world's population, but only 2.4 per cent of its land, resulting in great pressures on its natural resources.

Over 70 per cent of India's population currently derive their livelihood from land resources, which includes 84 per cent of the economically-active women. India is one of the few countries where males significantly outnumber females, and this imbalance has increased over time. India's maternal mortality rates in rural areas are among the world's highest. From a global perspective, Indian accounts for 19 per cent of all lives births and 27 per cent of all maternal deaths.

"There seems to be a consensus that higher female mortality between ages one and five and high maternal mortality rates result in a deficit of females in the population. Chatterjee (1990) estimates that deaths of young girls in India exceed those of young boys by over 300,000 each year, and every sixth infant death is specifically due to gender discrimination." Of the 15 million baby girls born in India each year, nearly 25 per cent will not live to see their 15th birthday."Although India was the first country to announce an official family planning programme in 1952, its population grew from 361 million in 1951 to 844 million in 1991. India's total fertility rate of 3.8 births per woman can be considered moderate by world standards, but the sheer magnitude of population increase has resulted in such a feeling of urgency that containment of population growth is listed as one of the six most important objectives in the Eighth Five-Year Plan."

Since 1970, the use of modern contraceptive methods has risen from 10 per cent to 40 per cent, with great variance between northern and southern India. The most striking aspect of contraceptive use in India is the predominance of sterilization, which accounts for more than 85 per cent of total modern contraception use, with female sterilization accounting for 90 per cent of all sterilizations.

The Indian constitution grants women equal rights with men, but strong patriarchal traditions persist, with women's lives shaped by customs that are centuries old. In most Indian families, a daughter is viewed as a liability, and she is conditioned to believe that she is inferior and subordinate to men. Sons are idolized and celebrated.

May you be the Mother of a Hundred Sons is a common hindu wedding blessing

The origin of the Indian idea of appropriate female behaviour can be traced to the rules laid down by Manu in 200 B.C.: "by a young girl, by a young woman, or even by an aged one, nothing must be done independently, even in her own house". "In childhood a female must be subject to her father, in youth to her husband, when her lord is dead to her sons; a woman must never be independent."

A study of women in the Swayam Shikshan Prayog (SSP), based in 20 villages in four districts in Maharashtra state was introduced in this way:

The primary issue for which all women in the SSP were struggling with was that of everyday survival. Insufficient incomes and the lack of employment were reported to be their most pressing concerns. Survival is a constant preoccupation and at its most basic, survival means food (Chambers 1983). *The most common problems were the lack of basic amenities such as food, water, fuel, fodder and health facilities. In addition, the deterioration of the natural environment and the fact that many of their traditional occupations were no longer viable were conditions that were making it increasingly hard for women to continue sustaining their families, as they had done in the past.* SSP is a loose, informal network of women's collectives, voluntary organizations, action groups and unions.

Women are Malnourished

The exceptionally high rates of malnutrition in South Asia are rooted deeply in the soil of inequality between men and women.

"...the poor care that is afforded to girls and women by their husbands and by elders is the first major reason for levels of child malnutrition that are markedly higher in South Asia than anywhere else in the world."

This point is made in the article, *The Asian Enigma*, published by Unicef in the 1996 Progress of Nations, in which the rates of childhood malnutrition in South Asia are compared with those in Africa. We learn that malnutrition is far worse in South Asia, directly due to the fact that women in South Asia have less voice and freedom of movement than in Africa. "Judgement and self-expression and independence largely denied, millions of women in South Asia have neither the knowledge nor the means nor the freedom to act in their own and their children's best interests."

"Gender disparities in nutrition are evident from infancy to adulthood. In fact, gender has been the most statistically significant determinant of malnutrition among young children and malnutrition is a frequent direct or underlying cause of death among girls below age 5. Girls are breast-fed less frequently and for shorter durations in infancy; in childhood and adulthood, males are fed first and better. Adult women consume approximately 1,000 fewer calories per day than men according to one estimate from Punjab. Comparison of household dietary intake studies in different parts of the country shows that nutritional equity between males and females is lower in northern than in southern states." Nutritional deprivation has two major consequences for women: they never reach their full growth potential and anaemia. Both are risk factors in pregnancy, with anaemia ranging from 40-50 per cent in urban areas to 50-70 per cent in rural areas. This condition complicates childbearing and result in maternal and infant deaths, and low birth weight infants.

One study found anaemia in over 95 per cent of girls ages 6-14 in Calcutta, around 67 per cent in the Hyderabad area, 73 per cent in the New Delhi area, and about 18 per cent in the Madras area. This study states, "The prevalence of anaemia among women ages 15-24 and 25-44 years follows similar patterns and levels. Besides posing risks during pregnancy, anaemia increases women's susceptibility to diseases such as tuberculosis and reduces the energy women have available for daily activities such as household chores, childcare, and agricultural labor. Any severely anaemic individual is taxed by most physical activities, including walking at an ordinary pace."

Women are in Poor Health

Surviving through a normal life cycle is a resource-poor woman's greatest challenge.

"The practice of breast-feeding female children for shorter periods of time reflects the strong desire for sons. If women are particularly anxious to have a male child, they may deliberately try to become pregnant again as soon as possible after a female is born. Conversely, women may consciously seek to avoid another pregnancy after the birth of a male child in order to give maximum attention to the new son."

A primary way that parents discriminate against their girl children is through neglect during illness. When sick, little girls are not taken to the doctor as frequently as are their brothers. A study in Punjab shows that medical expenditures for boys are 2.3 times higher than for girls. As adults, women get less health care than men. They tend to be less likely to admit that they are sick and they'll wait until their sickness has progressed before they seek help or help is sought for them. Studies on attendance at rural primary health centers reveal that more males than females are treated in almost all parts of the country, with differences greater in northern hospitals than southern ones, pointing to regional differences in the value placed on women. Women's socialization to tolerate suffering and their reluctance to be examined by male personnel are additional constraints in their getting adequate health care.

Maternal Mortality

India's maternal mortality rates in rural areas are among the highest in the world.

A factor that contributes to India's high maternal mortality rate is the reluctance to seek medical care for pregnancy – it is viewed as a temporary condition that will disappear. The estimates nationwide are that only 40-50 per cent of women receive any antenatal care. Evidence from the states of Bihar, Rajasthan, Orissa, Uttar Pradesh, Maharashtra and Gujarat find registration for maternal and child health services to be as low as 5-22 per cent in rural areas and 21-51 per cent in urban areas.

Even a woman who has had difficulties with previous pregnancies is usually treated with home remedies only for three reasons: the decision that a pregnant woman seek help rests with the mother-in-law and husband; financial considerations; and fear that the treatment may be more harmful than the malady.

It is estimated that pregnancy-related deaths account for one-quarter of all fatalities among women aged 15 to 29, with well over two-thirds of them considered preventable. For every maternal death in India, an estimated 20 more women suffer from impaired health. One village-level study of rural women in Maharashtra determined on the basis of physical examinations that some 92 per cent suffered from one or more gynecological disorder.

Contraception Use

Women's health is harmed by lack of access to and the poor quality of reproductive services.

"About 24.6 million couples, representing roughly 18 per cent of all married women, want no more children but are not using contraception. (Operations Research Group, 1990). The causes of this unmet need remain poorly understood, but a qualitative study in Tamil Nadu suggests that women's lack of decision-making power in the family, opportunity costs involved in seeking contraception, fear of child death, and poor quality of contraceptive service all play an important role." (Ravindran 1993).

Some estimates suggest that some 5 million abortions are performed annually in India, with the large majority being illegal. As a result, abortion-related mortality is high. Although abortion has been legal since 1972 in India, "studies suggest that although official policy seeks to make pregnancy-termination services widely available, in practice guidelines on abortion limit access to services, particularly in rural areas. In 1981, of the 6,200 physicians trained to perform abortions, only 1,600 were working in rural areas."

Job impact on Maternal Health

Working conditions result in premature and stillbirths.

The tasks performed by women are usually those that require them to be in one position for long periods of time, which can adversely affect their reproductive health.

A study in a rice-growing belt of coastal Maharashtra found that 40 per cent of all infant deaths occurred in the months of July to October. The study also found that a majority of births were either premature or stillbirths. The study attributed this to the squatting position that had to be assumed during July and August, the rice transplanting months.

Impact of Pollution on Women

Women's health is further harmed by air and water pollution and lack of sanitation.

The impact of pollution and industrial wastes on health is considerable. In *Environment, Development and the Gender Gap*, Sandhya Venkateswaran asserts that "the high incidence of malnutrition present amongst women and their low metabolism and other health problems affect their capacity to deal with chemical stress. The smoke from household biomass (made up of wood, dung and crop residues) stoves within a three-hour period is equivalent to smoking 20 packs of cigarettes. For women who spend at least three hours per day cooking, often in a poorly ventilated area, the impact includes eye problems, respiratory problems, chronic bronchitis and lung cancer. One study quoted by WHO in 1991 found that pregnant women cooking over open biomass stoves had almost a 50 per cent higher chance of stillbirth.

Anaemia makes a person more susceptible to carbon monoxide toxicity, which is one of the main pollutants in the biomass smoke. Given the number of Indian women who are anaemic – 25 to 30 per cent in the reproductive age group and almost 50 per cent in the third trimester – this adds to their vulnerability to carbon monoxide toxicity.

Additionally, with an increasing population, diseases caused by waste disposal, such as hookworm, are rampant. People who work barefooted are particularly susceptible, and it has been found that hookworm is directly responsible for the high percentage of anaemia among rural women.

Women are "Uneducated"

Women and girls receive far less education than men, due both to social norms and fears of violence.

India has the largest population of non-school-going working girls.

India's constitution guarantees free primary school education for both boys and girls up to age 14. This goal has been repeatedly reconfirmed, but primary education in India is not universal. Overall, the literacy rate for women is 39 per cent versus 64 per cent for men. The rate for women in the four large northern states – Bihar, Uttar Pradesh, Rajasthan and Madhya Pradesh – is lower than the national average: it was 25 per cent in 1991. Attendance rates from the 1981 census suggest that no more than 1/3 of all girls (and a lower proportion of rural girls) aged 5-14 are attending school. Although substantial progress has been achieved since India won its independence in 1947, when less than 8 per cent of females were literate, the gains have not been rapid enough to keep pace with population growth: there were 16 million more illiterate females in 1991 than in 1981. Sonalde Desai in *Gender Inequalities and*

Demographic Behaviour asserts that "parents' reluctance to educate daughters has its roots in the situation of women. Parents have several incentives for not educating their daughters. Foremost is the view that education of girls brings no returns to parents and that their future roles, being mainly reproductive and perhaps including agricultural labour, require no formal education. As more and more boys are engaged in education, there is a growing reliance on the labour of girls. Girls are increasingly replacing their brothers on the farm while carrying on their usual responsibilities in housework. A large proportion of the roughly 40 million "nonworking" girls who are not in school are kept at home because of responsibilities in housework."

The role of parents is to deliver a chaste daughter to her husband's family. Sonalde Desai goes on to point out that "another disincentive for sending daughters to school is a concern for the protection of their virginity. When schools are located at a distance, when teachers are male, and when girls are expected to study along with boys, parents are often unwilling to expose their daughters to the potential assault on their virginity."

There is little response to counter these obstacles: school hours remain inflexible to the labour demands of girls; many villages do not have a school; and fewer than 1/3 of India's primary and middle-school teachers are women.

According to *Mapping Progress*, "educational funds were cut by 801.3 million rupees in the 1991-92 budget. Funds for the mass literacy movement, in which women participate enthusiastically, have been reduced by 5 per cent from the previous year. Budgetary provisions for non-formal education have been cut by 17 per cent, leading to closure of many night schools and adult education programs in which working-class women participate. Reduction in government expenditures on higher education and encouragement to private colleges will reduce women's opportunities for higher education since privatization in education promotes only male-dominated professional and technical courses, as they are lucrative."

Women are Overworked

Women work longer hours and their work is more arduous than men's. Still, men report that "women, like children, eat and do nothing."

Hours Worked

Women work roughly twice as many hours as men.

Women's contribution to agriculture – whether it be subsistence farming or commercial agriculture – when measured in terms of the number of tasks performed and time spent, is greater than men. "The extent of women's contribution is aptly highlighted by a micro study conducted in the Indian Himalayas which found that on a one-hectare farm, a pair of bullocks works 1,064 hours, a man 1,212 hours and a woman 3,485 hours in a year."

In Andhra Pradesh, (Mies 1986) found that the workday of a woman agricultural labourer during the agricultural season lasts for 15 hours, from 4 am to 8 pm, with an hour's rest in between. Her male counterpart works for seven to eight hours, from 5 am to 10 am or 11 am and from 3 pm to 5 pm.

Another study on time and energy spent by men and women on agricultural work (Batliwala 1982) found that 53 per cent of the total human hours per household are contributed by women as compared to 31 per cent by men. The remaining contribution comes from children.

The linking of agricultural activities to male dominance is described by Roy Burman (in Menon 1991):

The anxiety of man to monopolize his skill in plough culture is reflected in the taboo that is observed almost all over India, against the women's handling the plough. In many societies, she is not even allowed to touch it. Mies further observed that "whereas operations performed by men were those that entailed the use of machinery and draught animals, thereby using animal, hydraulic, mechanical or electrical energy, women almost always relied on manual labour, using only their own energy." Rice transplantations, the most arduous and labour intensive task in rice cultivation, are carried out entirely by women without the help of any tools.

"Girls learn to assist their mothers in almost all tasks, and from the age of 10 years participate fully in the agricultural work done by women. Mies cites the case of Laxmi, a three-year-old infant who, along with her mother, pulled seedlings for transplanting. Boys on the other hand were seldom seen transplanting or weeding though they did help out in ploughing or watering the fields."

"Not only do women perform more tasks, their work is also more arduous than that undertaken by men. Both transplantation and weeding require women to spend the whole day and work in muddy soil with their hands. Moreover, they work the entire day under the intensely hot sun while men's work, such as ploughing and watering the fields, is invariably carried out early in the morning before the sun gets too hot. Mies argues that because women's work, unlike men's, does not involve implements and is based largely on human energy, it is considered unskilled and hence less productive. On this basis, women are invariably paid lower wages, despite the fact that they work harder and for longer hours than do men."

In contrast, a study in Uttar Pradesh reports that men "only reluctantly conceded that their womenfolk really work. The researchers in this area were repeatedly told that women, like children, simply eat food and do nothing."

The Invisibility of Women's Work

Women's work is rarely recognized.

Many maintain that women's economic dependence on men impacts their power within the family. With increased participation in income-earning activities, not only will there be more income for the family, but gender inequality should be reduced. This issue is particularly salient in India because studies show a very low level of female participation in the labour force. This under-reporting is attributed to the frequently held view that women's work is not economically productive.

In a report of the National Commission on Self-Employed Women and Women in the Informal Sector, the director of social welfare in one state said, "There are no women in any unorganized sector in our state." When the Commission probed and asked, "Are there any women who go to the forest to collect firewood? Do any of the women in rural areas have cattle?" the director responded with, "Of course, there are many women doing that type of work." Working women are invisible to most of the population.

If all activities – including maintenance of kitchen gardens and poultry, grinding food grains, collecting water and firewood, etc. – are taken into account, then 88 per cent of rural housewives and 66 per cent of urban housewives can be considered as economically productive.

Women's employment in family farms or businesses is rarely recognized as economically productive, either by men or women. And, any income generated from this work is generally controlled by the men. Such work is unlikely to increase women's participation in allocating family finances. In a 1992 study of family-based textile workers, male children who helped in a home-based handloom mill were given pocket money, but the adult women and girls were not.

The Impact of Technology on Women

The shift from subsistence to a market economy has a dramatic negative impact on women.

According to Sandhya Venkateswaran, citing Shiva, the Green Revolution, which focused on increasing yields of rice and wheat, entailed a shift in inputs from human to technical. Women's participation, knowledge and inputs were marginalized, and their role shift from being "primary producers to subsidiary workers." Where technology has been introduced in areas where women worked, women labourers have often been displaced by men. Threshing of grain was almost exclusively a female task, and with the introduction of automatic grain threshers – which are only operated by men – women have lost an important source of income. Combine harvesters leave virtually no residue. This means that this source of fodder is no longer available to women, which has a dramatic impact on women's workload. So too, as cattle dung is being used as fertilizer, there is less available for fuel for cooking. "Commercialization and the consequent focus on cash crops has led to a situation where food is lifted straight from the farm to the market. The income accrued is controlled by men. Earlier, most of the produce was brought home and stored, and the women exchanged it for other commodities. Such a system vested more control with the women."

Women are Unskilled

Women have unequal access to resources.

Extension services tend to reach only men, which perpetuate the existing division of labour in the agricultural sector, with women continuing to perform unskilled tasks. A World Bank study in 1991 reveals that the assumption made by extension workers is that information within a family will be transmitted to the women by the men, which in actual practice seldom happens. "The male dominated extension system tends to overlook women's role in agriculture and proves ineffective in providing technical information to women farmers."

Mapping Progress, states, "in the farm sector, the process of mechanization of agricultural activities has brought in tendencies for gender discrimination by replacing men for a number of activities performed by women and also by displacing the labour of women from subsistence and marginal households. Women are employed only when there is absolute shortage of labour and for specific operations like cotton-picking.

"To supply food-processing industries being set up with foreign collaboration, there has already been a major shift from subsistence farming method of rice, millet, corn and wheat to cash-crop production of fruit, mushrooms, flowers and vegetables. This shift has led to women being the first to lose jobs."

A number of factors perpetuate women's limited job skills: if training women for economic activities requires them to leave their village, this is usually a problem for them. Unequal access to education restricts women's abilities to learn skills that require even functional levels of literacy. In terms of skill development, women are impeded by their lack of mobility, low literacy levels and prejudiced attitude towards women. When women negotiate with banks and government officials, they are often ostracized by other men and women in their community for being 'too forward.' Government and bank officials have preconceived ideas of what women are capable of, and stereotypes of what is considered women's work.

Women are Mistreated

Violence against women and girls is the most pervasive human rights violation in the world today.

Opening the door on the subject of violence against the world's females is like standing at the threshold of an immense dark chamber vibrating with collective anguish, but with the sounds of protest throttled back to a murmur. Where there should be outrage aimed at an intolerable status quo there is denial, and the largely passive acceptance of 'the way things are.'

Male violence against women is a worldwide phenomenon. Although not every woman has experienced it, and many expect not to, fear of violence is an important factor in the lives of most women. It determines what they do, when they do it, where they do it, and with whom. Fear of violence is a cause of women's lack of participation in activities beyond the home, as well as inside it. Within the home, women and girls may be subjected to physical and sexual abuse as punishment or as culturally justified assaults. These acts shape their attitude to life, and their expectations of themselves. *The insecurity outside the household is today the greatest obstacle in the path of women. Conscious that, compared to the atrocities outside the house, atrocities within the house are endurable, women not only continued to accept their inferiority in the house and society, but even called it sweet.*

In recent years, there has been an alarming rise in atrocities against women in India. Every 26 minutes a woman is molested. Every 34 minutes a rape takes place. Every 42 minutes a sexual harassment incident occurs. Every 43 minutes a woman is kidnapped. And every 93 minutes a woman is burnt to death over dowry.

One-quarter of the reported rapes involve girls under the age of 16 but the vast majority is never reported. Although the penalty is severe, convictions are rare.

Selective Abortions

The most extreme expression of the preference for sons is female infanticide and sex-selective abortion.

A study of amniocentesis in a Bombay hospital found that 96 per cent of female fetuses were aborted compared with only a small percentage of male fetuses.

"Government officials event suspect that the disproportionate abortion of female fetuses may be a major underlying cause of the recent decline in the nation's sex ratio. In 1971 there were 930 females for every 1,000 males. A decade later this figure had increased to 934, but by 1991, instead of continuing to rise, the ratio dropped to 927, lower than the 1971 figure. This sex ratio is one of the lowest in the world."

Sonalda Desai reports that there are posters in Bombay advertising sex-determination tests that read, "It is better to pay 500 Rs. now than 50,000 Rs. (in dowry) later."

Government has passed legislation to curb the misuse of amniocentesis for sex selection and abortion of female fetuses. Women activists have been critical of this act because of its provision that calls for punishing the women who seek the procedure. These women may be under pressure to bear a male child.

Statistical Summary

Social Indicator	*India*	*World*	*Developing World*	*Kerala*	*Sri Lanka*
Infant Mortality Rate, per 1000 live births	73	60	66	16	17
Maternal Mortality Rate, per 100,000 live births	570	430	470		140
Female Literacy, per cent	58	77.6	70.4	86.93	90.2
Female School Enrollment	47	62	57	63	67
Earned Income by females, per cent	26	58.0	53.0	49.8	35.5
Underweight Children, per cent	53	30	30		38
Total Fertility Rate	3.2	2.9	3.2	1.9	2.1
Women in Government, per cent	6	7	5		9
Contraception usage, per cent	44	56	54	56	66
Low birth weight babies, per cent	33	17	7		25

Kerala and Sri Lanka are added for comparison as they are recognized as having made exceptional efforts in the field of female education.

Drudge

- ✰ A person who does mental, distasteful, dull or hard work
- ✰ *A person who works in a routine, unimaginative way to perform distasteful, dull, hard work under compulsion*

What is Drudgery?

Mental, distasteful, dull, hard, repetitive work

Drudgery and Poor home

The extreme drudgery experienced by the women through harsh and sultry household chores has amounted to the creation of poor home.

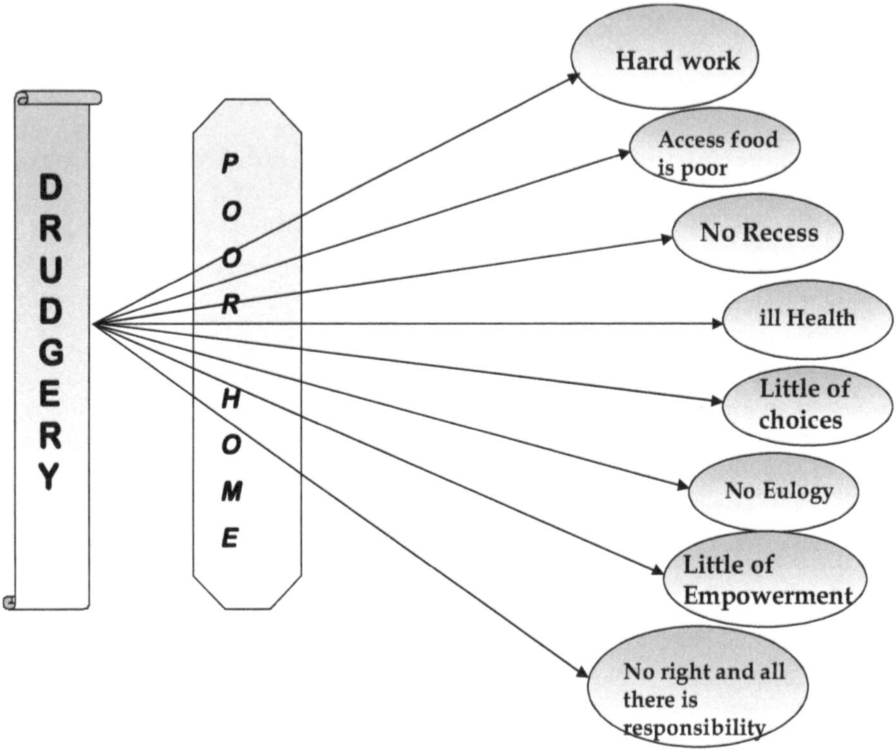

Fractured Motivation

A poor home thus generated by extreme drudgery acted upon women folk is supposed to generate fractured motivation damping the prospect of good entrepreneurship.

Poor Farm and Drudgery

A poor home and a fractured motivation have gone together for creating a poor farm sustaining decelerating productivity, stale crop rotation, fragile enterprise plan and non viable farm operations.

Changing of Classical crop rotation, insipid attitudinal disposition, non-supportive family, subsistence farm economy, crystal and prescriptive management orientation are all helpful to build up creative entrepreneurship.

So, there is a spiral mode of management rotation, which would translate promises into policy, policy into execution and so on. This follows an evolutionary pathways and the pace and kind of transformation of this gender empowering process would be governed by certain objective and conditions.

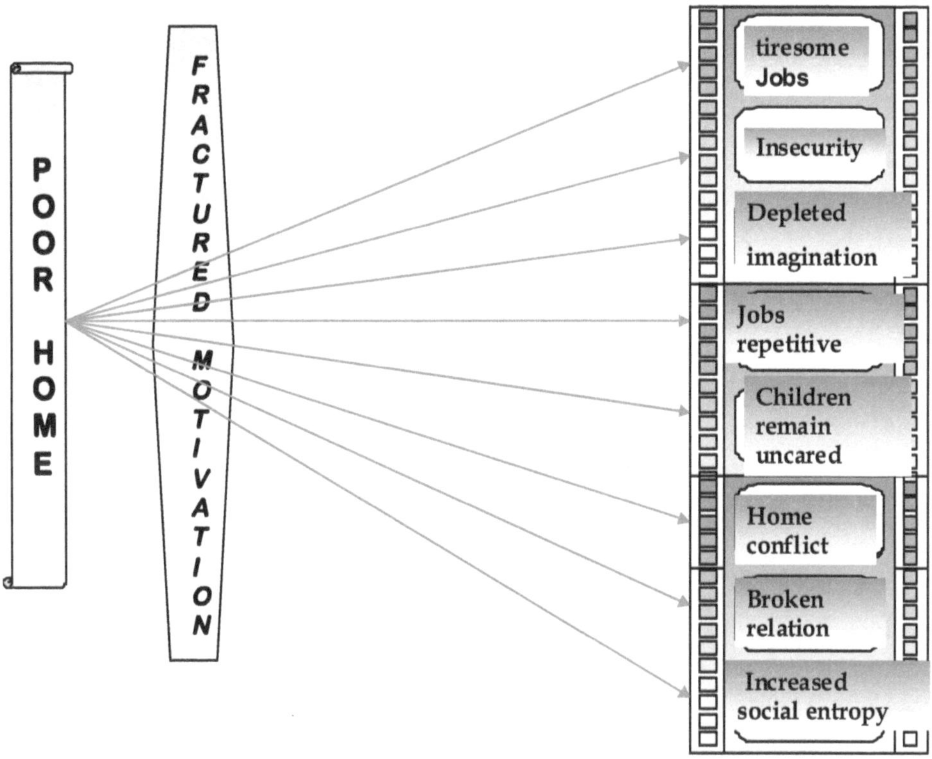
POOR HOME
FRACTURED MOTIVATION
tiresome Jobs
Insecurity
Depleted imagination
Jobs repetitive
Children remain uncared
Home conflict
Broken relation
Increased social entropy

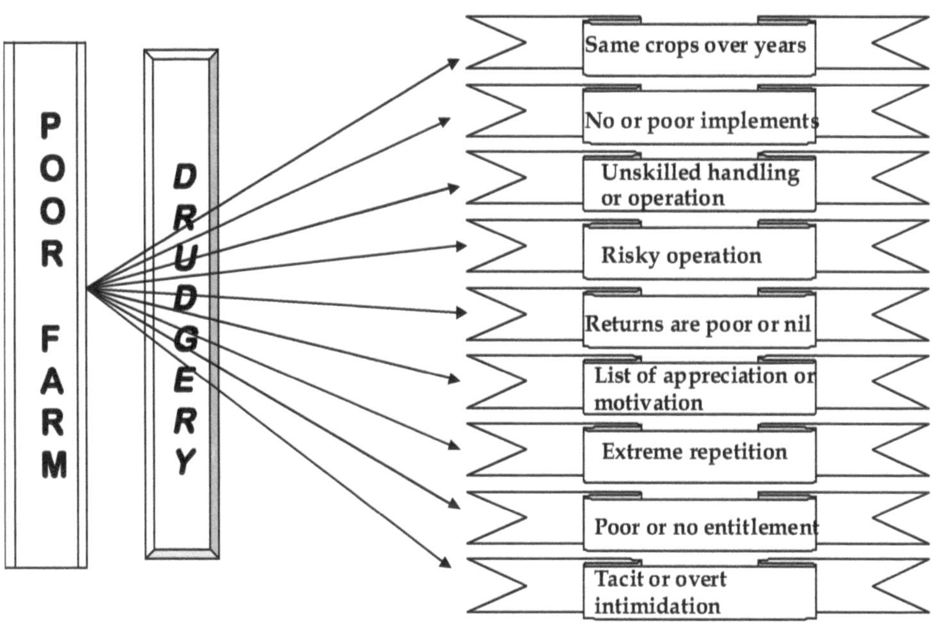
POOR FARM
DRUDGERY
Same crops over years
No or poor implements
Unskilled handling or operation
Risky operation
Returns are poor or nil
List of appreciation or motivation
Extreme repetition
Poor or no entitlement
Tacit or overt intimidation

So, the outcomes are

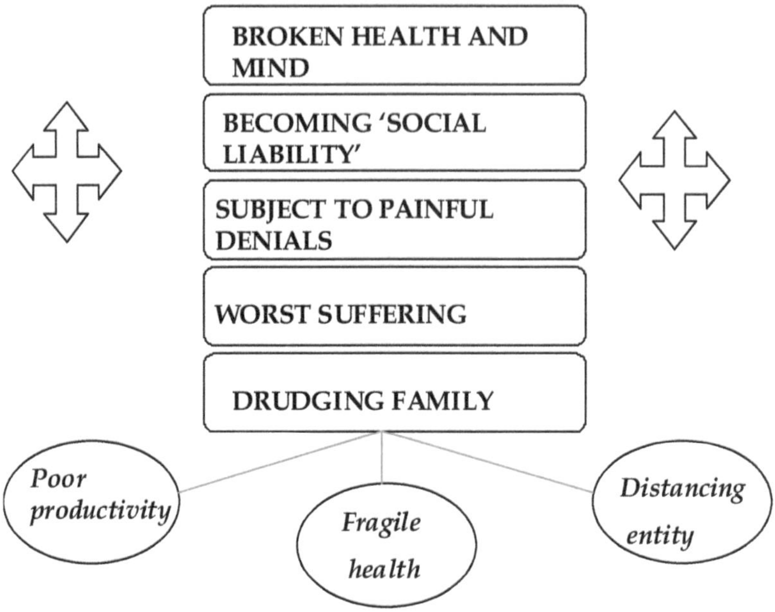

Change the go of the day....

How?

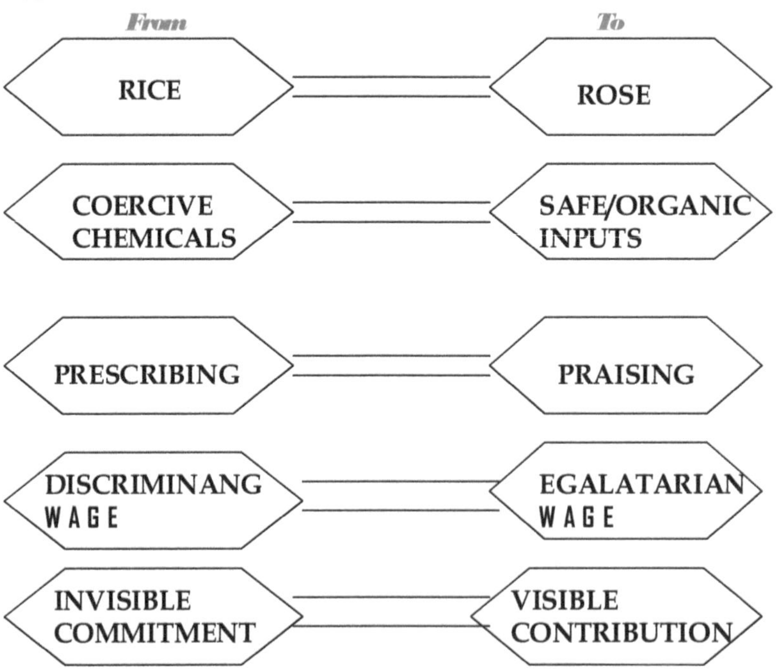

Hence, the spiral mode of evolution is....

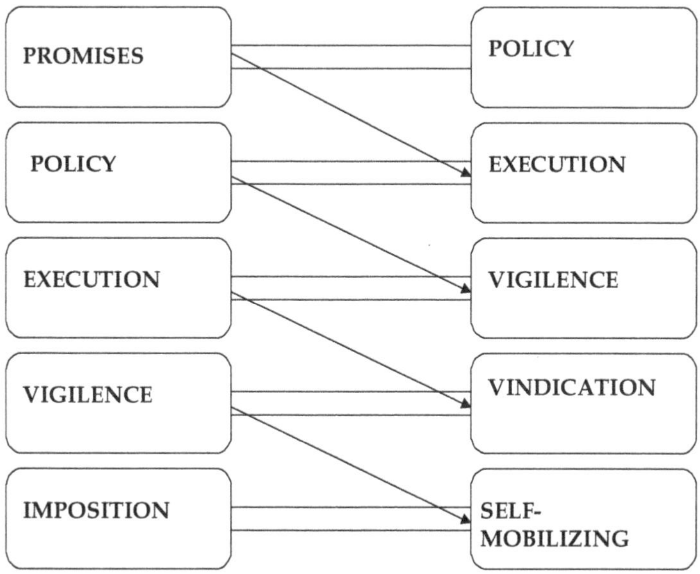

Extreme Drudgery and its Exposition

Extreme drudgery kept happening to the women, be it in agriculture, industry, service sector or household chores, it would generate serious and sometimes lethal consequences including psychosomatic and psychosocial resultants as follows

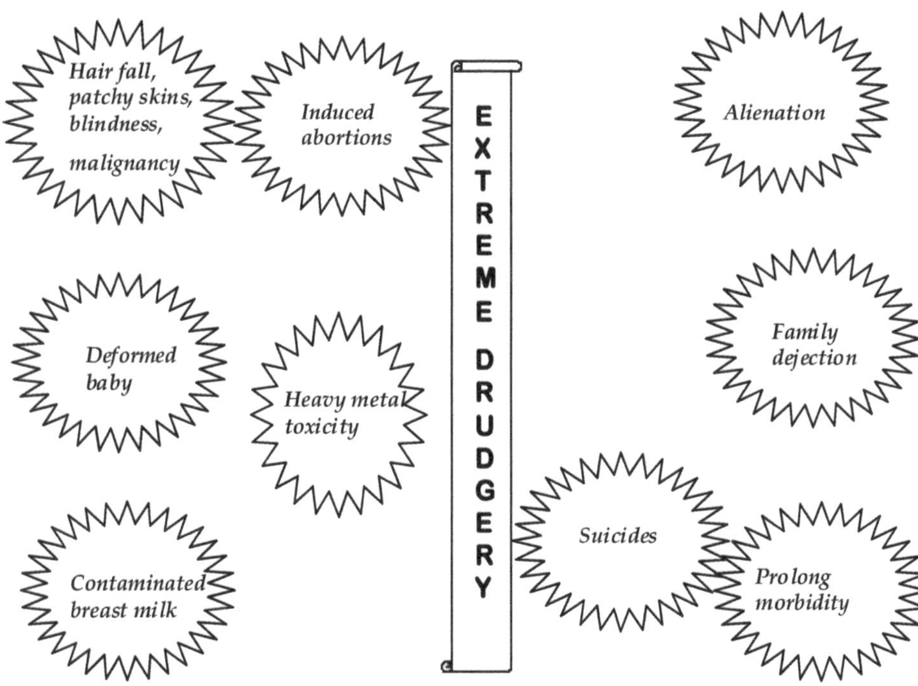

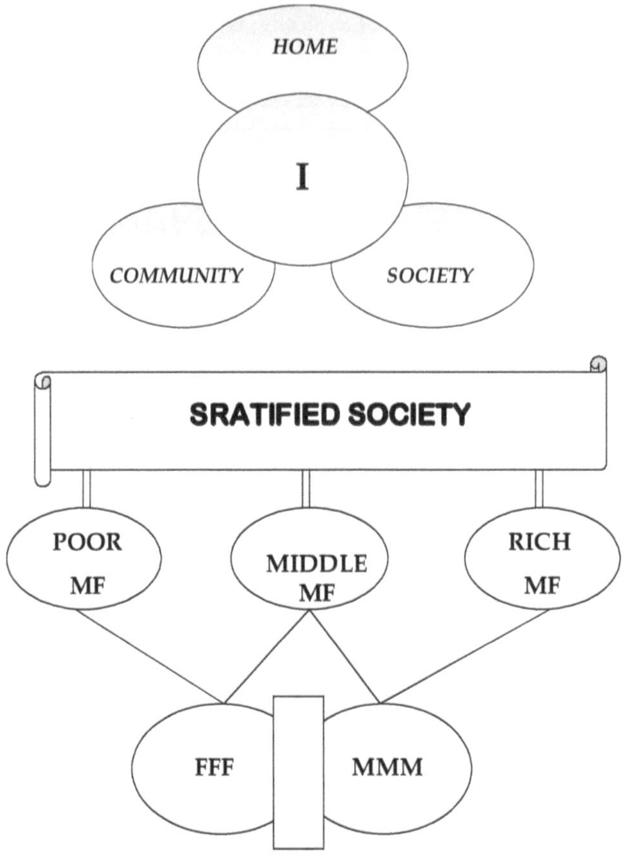
HOME
I
COMMUNITY
SOCIETY
SRATIFIED SOCIETY
POOR MF
MIDDLE MF
RICH MF
FFF
MMM

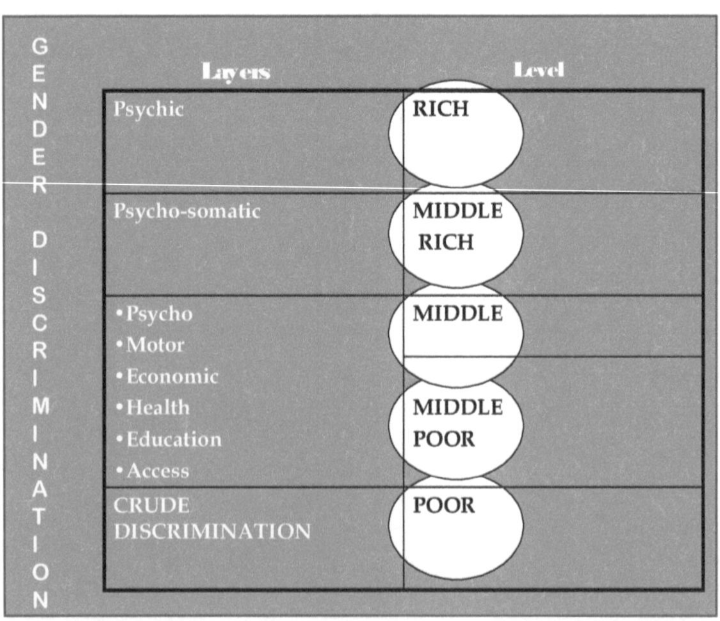
GENDER DISCRIMINATION
Layers
Level
Psychic
RICH
Psycho-somatic
MIDDLE RICH
•Psycho
•Motor
•Economic
•Health
•Education
•Access
MIDDLE
MIDDLE POOR
CRUDE DISCRIMINATION
POOR

The stratified society supports and disposes hierarchical exposition of biophysical and psychosocial resultants. While the society, as a whole, has been dichotomized into male and female genders, it has got class and culture based disposition as well. The discriminations, by nature and element has become gross and stubborn for the women of the poorest families, the elite women on the contrary have become the subject to subtle and finer deprivation being basically psychic in nature. In most cases, however, the 'I' of any women has been extremely diluted being ruthlessly engulfed by the greater echelons of family, community and society.

Conclusion

As UN Secretary General Kofi Annan has stated, "Gender equality is more than a goal in itself. It is a precondition for meeting the challenge of reducing poverty, promoting sustainable development and building good governance."

This recognition is currently missing in India. Transforming the prevailing social discrimination against women must become *the* top priority, and must happen concurrently with increased direct action to rapidly improve the social and economic status of women. In this way, a synergy of progress can be achieved.

As women receive greater education and training, they will earn more money. As women earn more money - as has been repeatedly shown - they spend it in the further education and health of their children, as opposed to men, who as women rise in economic status, they will gain greater social standing in the household and the village, and will have greater voice. As women gain influence and consciousness, they will make stronger claims to their entitlements - gaining further training, better access to credit and higher incomes - and command attention of police and courts when attacked.

- ☆ As women's economic power grows, it will be easier to overcome the tradition of "son preference" and thus put an end to the evil of dowry.
- ☆ As son preference declines and acceptance of violence declines, families will be more likely to educate their daughters, and age of marriage will rise.
- ☆ For every year beyond 4th grade that girls go to school, family size shrinks 20 per cent, child deaths drop 10 per cent and wages rise 20 per cent.
- ☆ As women are better nourished and marry later, they will be healthier, more productive, and will give birth to healthier babies.

Only through action to remedy discrimination against women can the vision of India's independence – an India where all people have the chance to live health and productive lives – be realized.

References

Abzug, Bella., and Davis, Susan. 1998. "India." *Mapping Progress: Assessing Implementation of the Beijing Platform.*

Bunch, Charlotte. "The Intolerable Status Quo: Violence Against Women and Girls." *The Progress of Nations* 1997 New York: UNICEF

Burns, John F. "Though Illegal, Child Marriage is Popular in Part of India." *The New York Times.* May 11, 1998.

Carr, Marilyn., and Chen, Martha., and Jhabvala, Renana. 1996. *Speaking Out: Women's Economic Empowerment in South Asia.* Southampton Row, London: Intermediate Technology Publications LTD.

Desai, Sonalde. 1994. *Gender Inequalities and Demographic Behavior: India.* New York: The Population Council, Inc.

Carol S. Coonrod, "Chronic Hunger and the Status of Women in India" *June 1998*

Neft, Naomi., and Levine, Ann D. 1997. *Where Women Stand: An International Report on the Status of Women in 140 Countries.* New York: Random House.

Omvedt, Gail. 1990. "Violence Against Women: New Movements and New Theories in India." *Kali Primaries.*

Purushothaman, Sangeetha. 1998. *The Empowerment of Women in India: Grassroots Women's Networks and the State.* New Delhi: Sage Publications.

Rajan. "Will India's Ban on Prenatal Sex Determination Slow Abortion of Girls?" Internet.

Ramalingaswami, Vulimiri., and Jonsson, Urban., and Rohde, Jon. "The Asian Enigma." *The Progress of Nations.* 1996 New York: UNICEF

Reardon, Geraldine. 1995. *Power and Process.* Oxford: Oxfam

Times of India. "Protest Against Atrocities in Women." Internet. p.3

Tinker, Anne. 1996. *Improving Women's Health in India.* Development in Practice Series. The World Bank

Venkateswaran, Sandhya. 1995. *Environment, Development and the Gender Gap.* New Delhi: Sage Publications.

Vinayak, Ramesh "Victims of Sudden Affluence." *India Today.* December 15, 1997

Chapter 15

Behavioural Skill in Extension Management

Siddhartha Deb Mukhopadhyay

Extension Management System essentially includes Extension teaching, Extension research, extension activities/work and extension administration. Extension programmes typically target and measure only cognitive behaviours – knowledge and information. Hence, behavioural skill is very important for an extension professional as well as the extension management system to become successful and effective. The basic requirement for proper behavioural skill among extension professionals is to bring about desirable behavioural changes among the members of the client system.

I. Behavioural Objectives

Extension programs are increasingly being challenged to produce measurable outcomes and behaviour change in learners or clients. Best practice or evidence-based programmes are those that have been shown through rigorous evaluation to reliably result in desired behaviour change. Behaviours can be written for one of three "domains of learning." The cognitive domain deals with the acquisition of facts, knowledge, information, or concepts. Psychomotor behaviours use the mind in combination with motor skills (physical activities). Affective behaviours have to do with changes in attitudes, values, aesthetics, and appreciation.

Behavioural objectives written in the "cognitive" domain can be further divided into six levels of Bloom's Taxonomy (Bloom, Engelhart, Furst, Hill, and Krathwohl, 1956) that includes:

- ☆ **Knowledge:** remembering of previously learned material; recall (facts or whole theories); bringing to mind.

- ☆ **Comprehension:** grasping the meaning of material; interpreting (explaining or summarizing); predicting outcome and effects (estimating future trends).
- ☆ **Application:** ability to use learned material in a new situation; apply rules, laws, methods, theories
- ☆ **Analysis:** breaking down into parts; understanding organization, clarifying, concluding
- ☆ **Synthesis:** ability to put parts together to form a new whole; unique communication; set of abstract relations
- ☆ **Evaluation:** ability to judge value for purpose; base on criteria; support judgment with reason.

In today's world of shrinking budgets, evaluation and accountability are necessary to justify the existence of Extension programmes and faculty. The key to solving the accountability issue is to develop clear, concise behavioural objectives as a part of each educational activity. Behavioural objectives are a vital part of the planning process of all educational activities. Clearly defined objectives provide a sound basis for the selection or design of instructional materials, content, and/or teaching methods.

In addition, stating objectives sharply helps determine if the objective has, in fact, been accomplished and an impact has occurred. Having clear behavioural objectives that state specifically what you want the leaner to be able to do after attending your programme is vital to the programme as well as the evaluation of its impact. Extension educators struggle to develop evaluation instruments for their programs. Clearly stated objectives are vital to evaluating the outcome(s) of instruction.

Clearly stated objectives have four characteristics.

1. First, the instructional objective must state the audience for the educational activity.
2. Second, the observable behaviour(s) expected of the audience must be identified.
3. Third, the conditions under which the behaviour is to be accomplished must be included.
4. Finally, the degree to which the behaviour is to be completed must be specified. Together these four components make up a measurable behavioural objective that will provide the cornerstone to planning, delivery, and evaluation efforts.

II. Principles of Behaviour Change

The principles of behaviour change can be summarised as follows : Extension programmes typically being working on cognitive behaviours – knowledge and information, the first step toward more powerful behaviour change education is to acknowledge the importance of social, psychomotor, and affective/emotional dimensions of change. The second step is designing educational programs based on solid principles of behaviour change. Psychological and educational research has shown that behavioural change occurs and is maintained when interventions incorporate particular principles of behaviour change. Although some of these principles are frequent elements of Extension education programmes, others will be newer ideas for Extension educators.

III. Domain of Behavioural Skill for Extension Management

A. Technical Competence

Extension Management is a practical subject which aims to have an impact on the real world, so it requires functional knowledge and skill in applying a body of professional practice. The knowledge of product technology, marketing techniques, engineering, accountancy, knowledge of relevant legislation, knowledge of basic management principles and theories which purport to help the practitioner with 'best practice' in planning, organising and controlling (Pedler *et al.*, 2001). Although all levels of management require a technical base, one of the unique aspects of management is that the further one ascends the hierarchy of management, the less direct use is made of techniques and the more emphasis is placed on the social, cognitive and political skills of managing extension programme. A major problem for developers is that managers often get promoted to a high level of authority because of their mastery of technical skills. Although they may be highly professional specialists, they may not have had much opportunity on the way up to develop their people skills or to gain a broad appreciation of the organisation as a whole. Once installed in a generalist role, they often find that performance requires strategic and political skills for which their former experience has not adequately prepared them (Garratt 1994).

Typically, in extension management system, the professionals find that their technical skills are not much use but they are not sure how they should behave. This role ambiguity can cause not only great anxiety but also behaviour which may be dysfunctional for the organisation as a whole. It is tempting for them to emphasise what they know best and act as higher level technicians with a narrow problem-solving view of their job, interfering in issues which should be left to those lower down. The beginnings of wisdom in management is often knowing when you are no longer paid to perform a professional-technical role and the job now requires a strategic appreciation and the use of social-political skills to harmonise the parts in the service of a greater whole (Garratt 1994).

B. Self-awareness

At the higher levels of management, technical skills are assumed and effectiveness in management seems very much linked to social and interpersonal skills. Researchers such as Mangham and Pye (1991) and Watson (2001) have tried to give a detailed picture of the micro-processes of management. The ability to do executive work, Mangham concludes, requires an awareness of self interacting with others. What extension executives do is very much shaped by their own perceptions of their role, their goals, their values, their feelings, their assessment of personal strengths and weaknesses. To act effectively they need a well-grounded awareness of the self; the skills of introspection. This makes sense. How can anyone manage others unless they have first mastered themselves? The knowledge we have acquired about ourselves, which makes up our self-concept, is central to improving our management skills. Knowing ourselves helps us understand our own taken-for-granted assumptions, our categories for defining situations and people, our sensitivities, strengths and weaknesses. This knowledge is self-empowering. It allows us to capitalise on our talents. It is the first step in making any changes to ourselves which are needed to develop the skills we think we need.

Self-reflection is also important because it helps us make our interactions with others more effective and insightful. By understanding ourselves we become more sensitive to the differences and also the similarities between people; it makes us more empathic and more skilful in our repertoire of behaviours. From this 'self/other' awareness, the extension professionals are supposed to be in a position to look in on his/her own behaviour from the perspective of others. This is what the professionals mean by 'managing the self so that we can manage others'. By developing a sense of self in the world which is consistent with how we are seen, we have the self-knowledge to present ourselves well – to use the words and take the actions which will influence others through quiet skills of persuasion and coordination.

C. Interpersonal and Social Skills

These skills are hard to define but they largely mean working with and through other people and using careful judgement. They imply sensitivity to people and situations and skill in persuading people to achieve a common goal. A lot of management is watching, sensing, doing readings of others' behaviour. In a sense, extension management are doing what we all do in social situations, but they are doing it with purpose and through complex webs of relationships within very diverse role sets. Mangham (1986) talks of this as 'the theatre of the skull'. Managers engage in a form of 'internal dialogue' to decide on the best line of action. This is really a form of 'role taking': that is, it requires social empathy in which they imaginatively and sympathetically put themselves in the minds of others, anticipating their reactions to the flow of events. By accurately modelling others' behaviour and correctly attributing intention we are well placed to act thoughtfully and appropriately. Ethnographers of management (Watson, Hales, etc.) have found that as we interact with others in management we are judging others in terms of their significance for our plans. What are their assumptions? What values guide their management? What do they hope to achieve here? Where do they stand on various issues? Is there a difference between what they claim and what they really want? Through these 'readings', managers develop a sense of the strategy best suited to the circumstances and most likely to allow their agenda to be implemented. The same empathy, feeling and judging are involved in other aspects of the social process.. That means having a 'feel' for how things are interconnected that is, how a micro-situation has implications for the strategy as a whole; how changing the technical system will have a 'knock on' effect for the social system, for example, motivation, morale, culture and the informal relations between groups. Social skills are needed for defining the situation in a way which encourages people to consider an alternative categorisation of experience and to persuade them to have a sense of ownership of it. This means working with groups, explaining what the abstractions of the plan imply for practical action, encouraging people to reflect on their habitual scripts' recognise their limitations and disengage from them. It also means acting as a catalyst, helping people to find the confidence to look at things anew and experiment with new ways of behaving. The social skills involved here are many and need to be employed with great subtlety (Strauss 1978) around new behaviours. Ultimately this involves the extension professionals defining the situation in terms which commands the broadest possible consensus.

IV. Specific Behavioural Skill Required for Extension Management

Specific behavioural skills in relation with extension management is summarised below.

A. Target and Build Specific Knowledge and Specific Skills

New knowledge creates the foundation for new behaviours. To increase the likelihood of actual incorporation of knowledge into participants' lives, however, it is important to identify and support the specific knowledge and skills needed to support new behaviours.

B. Reduce the Environmental Conditions that Support Negative Behaviours, and Increase the Conditions that Support Positive or Desired Behaviours

It is well established that physical and social environments affect learning and behaviour (Gordon, 2002). In fact, the basic premise of behaviour modification is operant conditioning, which maintains that behaviour can be changed by altering both the antecedents (cues) that precede behaviour and the consequences that follow a behaviour (Sarafino, 1996).

Extension educators may apply this principle through activities that encourage participants to identify the specific environmental situations or cues that lead them to behave.

C. Modify Behaviour Sequentially and Reward Progress

Effective behaviour change must be realistic; educators should help participants set small very specific targets that can be achieved over a realistic time frame

D. Train in Naturalistic Conditions and Settings

When an individual learns new behaviours outside of his or her normal milieu, it can be difficult to transfer those new behaviours to everyday life. That is why best practice programs emphasize application in naturalistic settings.

E. Teach Specific Cognitive Rules or Principles that can Guide Behaviour in New Settings

The idea that cognitions guide behaviour is the basis of many counselling and therapy approaches, including cognitive restructuring and cognitive-behavioural therapy (Sarafino, 1996). Applied to behaviour change, simple cognitive rules can help learners remember and apply important concepts.

F. Model and Offer many Opportunities for Participants to Practice the Desired Behaviour

Instructors who are excellent role models are inspirational. Moreover, providing many opportunities to observe and practice new behaviours within the educational setting increases the likelihood of behaviour change.

G. Facilitate Social and Other Supports Over Time to Support and Sustain Behaviour Change

Social support improves the behaviours of both the receiver and giver of support (Cameron and Pierce, 2002; Donatelle, Hudson, Dobie, Goodall, Hunsberger, and Oswald, 2004; Jensen, 1989).

H. Emphasize the Individual as the Key to Behaviour Change

Three self-perceptions are critical to behaviour change. First, does the person perceive that a specific change is personally important? Second, is the person "ready" to change a specific behaviour? Third, does the person have a "sense of self-efficacy" or the personal belief that he or she can actually make a specific change (Rollnick, Mason, and Butler, 2005)? In other words, for behaviour change to occur, a programme participant must recognize that a particular change is essential to his or her

V. Behavioural Deficit

In spite of all the efforts of the overall extension education system to bring about desirable behavioural skill among the extension professionals as well as the client system, still there exists a deficit of that. This section will delineate the differences between skill-based and performance-based social skills deficits and present intervention approaches in each area.

A. Skill-based Deficits

A skill-based deficit exists when a person has not learned how to perform a given behaviour.. Unfortunately, however, we sometimes become angry with those persons when they don't demonstrate the social skill we desire them to display. A critical issue is whether those persons actually possess the desired skill.

We may determine if a person has a skill deficit by observing whether the desired skill has ever been performed. If not, one may hypothesize that the skill is not in the person's repertoire. This may be tested further by providing strong incentives to perform the desired behaviour. If the person fails to perform under these conditions, it is likely that the problem stems from a skill deficiency.

Intervention to Skill-based Deficit: Teaching Social Skills

Generally, a skill-based deficit is due to lack of opportunity to learn or limited models of appropriate behavior (Gresham and Elliott, 1989). In these instances, direct instruction, or skill training, is necessary. The same principles apply to teaching social skills as to academic skills: provide ample demonstration/modelling, guided practice with feedback, and independent practice. Hazel, Schumaker, Sherman, and

Sheldon Wildgen (1981) listed eight fundamental social skills which can be taught through direct instruction:

1. Giving positive feedback (*e.g.*, thanking and giving compliments),
2. Giving negative feedback (*e.g.*, giving criticism or correction),
3. Accepting negative feedback without hostility or inappropriate reactions,
4. Resisting peer pressure to participate in delinquent behaviour,
5. Solving personal problems,
6. Negotiating mutually acceptable solutions to problems,
7. Following instructions, and
8. Initiating and maintaining a conversation.

They recommended teaching these skills by providing definitions, illustrations with examples, modelling, verbal rehearsal, behavioural rehearsal, and additional practice. Similarly, Walker, Colvin, and Ramsey (1995) recommended a nine step direct instructional procedure, the ACCEPTS instructional sequence. The steps include:

1. Definition of the skill with guided discussion of examples,
2. Modelling or video presentation of the skill being correctly applied,
3. Modelling or video presentation of incorrect application (non example),
4. Review,
5. Modelling or video presentation of a second example with debriefing,
6. Modelling a range of examples, coupled with hypothetical practice situati ons,
7. Modelling or video presentation of another positive example if needed,
8. Role playing, and
9. Informal commitment from student to try the skill in a natural setting.

B. Performance-based Deficits

A performance-based deficit exists when a person possesses a skill but doesn't perform it under the desired circumstances. This may occur if there is a problem with either motivation or with ability to discriminate as to when to exhibit the appropriate behaviour.

Motivational Deficit

When a motivational deficit exists, the person possesses the appropriate skill, but doesn't desire to perform it. A motivational deficit may be hypothesized if observations reveal that the person has acquired the desired skill, but motivational conditions are not sufficiently strong to elicit it. The hypothesis may be confirmed if the person performs the behaviour following introduction of a motivational strategy. In situations such as this, behavioural interventions are effective.

Intervention to Motivational Deficit : Motivational strategies

Motivational deficits can manipulate contingencies that will encourage performance of pro-social behaviours by using the principles of Applied Behaviour Analysis (ABA). The steps include defining the target behaviour operationally, identifying antecedents and consequences related to the behaviour, and finally developing and carrying out a plan to alter the antecedents and consequences so that the desired behaviour will occur. To summarize, once identified, motivational deficits can be remediated using behavior management techniques, either by the adult in the situation, or by the student in question.

C. Discrimination Deficit

A person with a discrimination deficit has the desired skill in his or her repertoire, is motivated to behave properly, but can't discriminate, (*i.e.*, doesn't know when to exhibit the desired behaviour). A discrimination deficit may be confirmed if the person frequently performs the desired behaviour, but fails to perform it under specific conditions. This may be due to an inability to glean relevant information from social situations. When a discrimination deficit exists, the person possesses the desired behaviour but may not be sure as to when, where, and how much to engage in that behaviour.

Intervention to Discrimination Deficit: The Hidden Curriculum

Lavoie (1994) suggested assessment of the student's knowledge of the hidden curriculum as a step in teaching the student to discriminate the appropriate behaviour for a given situation. The first step is to determine the hidden curriculum, or culture. Once the hidden curriculum is identified, the next step is to assess the person's knowledge in key areas. In short, the hidden curriculum must first be identified and then the person's level of understanding of it must be assessed.

VI. Important Behavioural Skill Required for an Extension Professional

We must agree to the point that extension basically falls within the gamut of human resource development and management. Apart from the Job performing skills, like an HR Professional, success of extension professionals lies in their behavioural skills. Even if the extension professional is strong enough in technical knowledge required to perform a task, but not having a required attitude and behavioural skills, he/she will fail; on the other hand if the professional has excellent behavioural skills and attitude, they can perform any task easily.

Commonly there are nine important behavioural skills are required for an extension professional which has been discussed in details below :

1) Clarity of Thoughts

An extension person often faces the situation where they have to choose one way "Either do it or don't ". Sometimes situation becomes so confusing that even we don't want to do it, but the job demands the decision and role becomes confusing. For an extension professional clarity of thoughts has to be there. Until he/she (who is suppose to be a deciding factor) is not clear in thoughts, the decision would not turn out to be accurate and fruitful.

2) Efficiency in Time Management

Extension professionals are expected to be efficient in time management. Most of the time, it has been seen that they don't give any time limit for completion of an activity. You approach them for any work and they respond by saying, " OK, it will be done". But when? All goals, all dreams, all activities, all commitments looks good and are achievable, if there is a closing date for that. Uncertainty, Ambiguity, Open-Ended commitment, Confusion, these are things, that are not expected from any extension professional.

3) Compare Situations and Performances

This is general human tendency to compare two situations or two people, which often creates confusion, wrong decisions or injustice towards one. Hence for an extension professional it is very much required that they have to come out of this general tendency and do not compare two people or two situations as in order to reach an authentic and fruitful decision which is required for smooth functioning of an organization as well as human resources.

4) Knowledge about the Work and Situation

Many a times it is been observed that extension people are not aware of actual enterprise, strength and weaknesses they are working with, which again lead them to wrong path, misleading decisions and poor performance harming the organizations interests and goals.

5) Vision and Goal

An extension professionals should be clear with the goals to be achieved. They must have a positive and futuristic vision about the growth and development of the organisation they are working with as well as the the outcome of the activities they are undertaking for the clients.

6) Zeal for Data/Calculations/Analysis/Projections

Reason why most of extension professionals despite of having excellent ideas and strategies fails in performing is that they do not support their ideas and strategies with data, analysis and calculations. Behind this fact is that most of people join this profession considering it very easy task and a job where calculations and mathematics is not involved.

7) Enthusiasm for the Professional Development

We all know that "Knowledge is a wealth; the more you distribute/share the more you gain". In this profession; we are privileged and are in a very unique position where we can develop people. We can develop their career. We can develop their life. We can help them in changing their perception and behaviour. We can groom them, so that they can be successful in their life. Extension professionals should have sufficient enthusiasm for development of others

8) Self Discipline

It is universal truth before you guide else, you should implement on self. For an extension professional this quality is must. If there job is to set standards and ethics for others, they should set standards, ethics and rules for self first so to motivate others and be an example before they guide them. Discipline in life is a must to grow, prosper and being successful.

9) Trust Worthy

Most required quality to succeed in any profession is been trust worthy. Specially for a critical profession like extension. Extension professionals need to win trust of the clients and then maintain and keep that trust.

If an extension professional follows above mentioned qualities religiously and implement in themselves naturally, nobody can stop him/her to succeed in the profession and marking a difference in the extension management system.

References

Bloom, B. S., Engelhart, M. D., Furst, E. J., Hill, W. H., and Krathwohl, D. R. (1956). *Taxonomy of educational objectives book 1: Cognitive domain.* New York: David McKay Company, Inc.

Cameron, J., and Pierce, W. D. (2002). *Rewards and intrinsic motivation: Resolving the controversy.* Westport, CT: Bergin and Garvey.

Donatelle, R. J., Hudson, D., Dobie, S. Goodall, A., Hunsberger, M., and Oswald, K. (2004). Incentives in smoking cessation: Status of the field and implications for research and practice with pregnant smokers. *Nicotine and Tobacco Research,* 6(2), S163 - S179.

Gordon, J. A. (2002). Beyond knowledge: Guidelines for effective health promotion messages. *Journal of Extension* [On-line], 40(6). Available at: http://www.joe.org/joe/2002december/a7.shtml

Harris, M.B.(1970). Reciprocity and generosity: Some determinants of sharing in children. *Child Development*, 41, 313–328.

Jensen, M. E. (1989). Motivating clients to change: The bottom line. *Journal of Extension* [On-line], 27(2). Available at: http://www.joe.org/joe/1989summer/a1.html

Rollnick, S., Mason, P., and Butler, C. (2005). *Health behavior change: A guide for practitioners*. London: Elsevier.

Sarafino, E. (1996.) *Principles of behavior change*. New York: Wiley.

Walker, H. (1990). *The Acting Out Child*. Sorporis West.

— Part III —

Managing Project for Enterprise Generation

Chapter 16

Micro Planning and Project Analysis in Natural Resource Management

Manas Mohan Adhikary and Sankar Kumar Acharya

The 'myths' and 'realities' of green revolution, as experienced by the soil and society of India, opened a new chapter of discourses and some which could be termed as 'grand discourses'. The so-called scale neutrality of technology intervention had come out to be grim reality for making "rich, richer and driving the poor, poorer". The decade of eighty's went on enthralling the philanthropic intellects all over the world on devising means and methodologies of technology generation vis-à-vis rural development process purported to the cause and confidence of the poorest of the poor. Robert Chambers (*Putting the last first*, 1983) came down heavily on the pernicious flamboyancy of rural development tourism, which to his mind had and had been developed and directed self-deceptive approaches in the conventional rural development programmes.

In search of a 'socialized methodology', imbued with humane techniques, social scientists endeavored time and again to redefine and refine rural peoples pristine commend on folk knowledge, wisdom, experience, intrinsic ability of estimation properties of process and matter in a given rural system. This had been only a beginning, across the globe on the quest of exploring and esteeming time tested methodology, have been nurturing, evolving within the people themselves.

Redefined Participation in Planning Process

The concept of participation's had a long way to evolve and mutate over the myriad variation of social context. People's participation in decision making process and their empowerment in the form of creating their own plans had been conceptualized in terms of availing *social choices, entitlement to resources, capability building and realizing empowerment.*

The conventional idea of participation may pay off little and sometimes go coercive instead of creating real benefit to the people. Prof. A. Sen (*Poverty and Famine,* 1984) while defining "real democracy", he stressed on people's participation resulting in building of capability among poorest of the poor to access social privileges. In his concept of techniques of social choices, he basically implied on generation of skills and capability with a view to attaining and upgrading quality of life.

The ultimate goal of providing people's participation in the programme is imparting the community with shills, strategies, opportunities and support to enhance its own functioning towards self-sufficiency and self-development. In concrete terms, the goal of people's participation could be helping people to build their own organization and institution equipped with a capable acceleration their own development. With this backdrop, the importance and implication of micro-planning can be perceivable in terms of its methodology, techniques, objectives and goals and the following will keep on delineating some of the alternatives of micro planning towards people's participation in the planning process.

Micro-planning

Micro-planning is basically a participatory planning process for a relatively smaller geographical distribution in order to accommodate and address highest possible micro-variation.

The earlier plans (1950 onwards), as developed and implemented, had a clear dent for blanket coverage (CDP, IADP, HYVP), no matter whether they had been destined for a comprehensive rural development or agricultural revolution. The history and evolution of planning process have got both the socio-political and techno-economic contexts. The limitations of conventional planning approach are:

1. Top-down approach (carried with the idea of trickle down effect)
2. Blanket coverage –smaller, tiny and trivial variations were purposefully ignored;
3. Subject to certain biases on: Resource, project, people, season;
4. Excessive reliability on secondary sources of information;
5. No room for objective evaluation – multiple realities were not esteemed.
6. Domination of bureaucracy and hence, every likelihood of red-tapism;
7. Planning has been a long and protractive process;
8. No such scope for making people empowered.

Advantages of Micro-Planning Process

Micro-planning is a historical development in the entire process of planning. The success of micro-planning depends on the following factors:

1. Organizing group,
2. Building goal clarity,

3. Experience profile of the participants
4. Climate setting
5. Group dynamics and group communication.
6. Planning skill
7. Resolving ability and leadership
8. Motivation.
9. Team building
10. Mutual faith and confidence.

Steps in Micro-planning

1. Unfreezing and generation of information;
2. Assigning roles,
3. Analysis of problem and deriving its nature, distribution and impact;
4. Prioritization of problems;
5. Formulation of objectives based on needs of the people;
6. Designing intervention plans;
7. Facilitating the implementation;
8. Participatory evaluation;
9. Micro-level workshops and resetting objectives.

Some Tools and Techniques of Micro Planning

The following are some of the tools and techniques for evaluating designing and implementing micro-plans.

- ☆ LFA (Logical Framework Analysis)
- ☆ ASA (A Systematic Approach)
- ☆ Net work analysis
- ☆ Stake holder analysis

The Logical Framework Approach: A Tool for Planning and Evaluation

The logical framework is a powerful tool for project design and management, which distinguishes between, lower and higher-level project objectives, showing how these objectives are linked. The logical framework approach is appropriate for any level of planning (except strategic planning) and is used for designing and managing broad, long-term programmes to individual research projects. It is frequently used as a framework for brainstorming and planning by small groups of managers.

The Logical Frame Approach (LFA) was initially developed for the US Agency for International Development (USAID) in 1969. It was originally intended as a project evaluation tool to help USAID increase its accountability to congress. The LFA is an aid to logical thinking and a means by which a project may be structured and described for analytical purposes. LFA is descried as "A set of interlocking concepts, which must be used together in a dynamic fashion to permit the elaboration of a well-designed, objectively described and valuable project".

However, LFA is not an integrated set of procedure and nor is it a set of guidelines for the evaluation of a particular type of project.

If used correctly, the log frames can provide a sound mechanism for developing a project concept into a comprehensive project design document.

In fact this logical framework approach provides a set of designing tools that, when used creatively, can be used for planning, designing implementing and evaluating the effectiveness, efficacy and relevance of programmes and projects.

Purpose of using logical framework approach

The Purposes of the LFA are:

1. To assist in the planning, management and evaluation of development activities and research projects.
2. To facilitate the systematic identification and specification of intermediate and final objectives of the programme and project concerned by the planners and managers.
3. Plan for required inputs and desired outputs.
4. Give sound guidance on the structuring of a monitoring and evaluation system by defining the indicators for monitoring and evaluation performance.
5. To provide a comprehensive, practical flexible and effective means to strengthen international research project management.
6. To provide a structure which will allow project planners and evaluators to specify the components of their activities and identify the logical linkages between a set of means and a set of ends.

Basic Structures of Logical Framework Approach

The structure of the Log Frame is deceptively simple: it consists of a 4 x 4 matrix in which the rows represent the level of project objectives, including the means required to achieve them (the vertical logic), while the columns indicate how the achievement of these objectives can be verified (the horizontal logic).

Vertical Logic

The left-hand column in the table is a "narrative summary" of the four levels of objectives of a research project or programme. These are represented by rows of the table and are referred to as the "Vertical Logic" in LFA. This vertical logic identifies what the project intends to do, clarifies the relationship between project means and ends, and specifies the uncertainties concerning both the project itself and the social/physical/political environment within which the project is located.

Narrative Summary (NS)	Objectively Verifiable Indicators (OVI)	Means of Verification (MOV)	Important Assumptions (External Factor)
Goal	Measures of goal achievement	Source of information	Assumptions for achieving goal
Purpose	End of project status (conditions that will indicate purpose has been achieved)	Sources of information, methods used	Assumptions for achieving purpose
Output	Magnitude of outputs planned, completion date	Source of information methods used	Assumptions for generating outputs
Inputs/ Activities	Nature and level of resources necessary. Cost of input planned. Starting date.	Source of Information	Assumptions for providing inputs, initial assumptions about the project.

- ☆ Goal – This is the ultimate objective to which the research project or programme contributes. It is the reason for undertaking the project.
- ☆ Purpose – This is what it is expected to achieve upon completion within the allocated time. It is the motivation behind the production of outputs.
- ☆ Output – These are the specific results to be produced by the management of inputs. The desired results of the research projects or programme derived directly from management of inputs.
- ☆ Inputs/Activities – These are the activities to be undertaken and the human, physical and financial resources available and required to produce the desired outputs. The qualify and quantity of inputs should be specified.

Causes-Effect Linkages

The vertical logic is based on the principle of causality in a means end relationship. There is a sequence of three causal relationships from input to goal. The project is characterized as a set of linked hypothesis of IF-THEN relationship.

If inputs are provided

THEN outputs will be produced;

IF outputs are produced

THEN purpose will be achieved;

IF purpose is achieved

THEN the goal may be realized

Important Assumptions

Log Frame specifies what assumption related to external factors is critical to project success. Important assumptions are the key threats to the project that exist in the external environment. They are necessary conditions linking each level of objectives in the project plan. The important assumptions concern conditions which could affect the progress or success of the project but over which the project manager has not control and which may arise from many sources. The assumption column helps to keep the research manager realistic in their expectations.

The Horizontal Logic

The vertical logic does not provide sufficient detail of a project for implementation or evaluation purposes. It needs to be more precisely defined. The horizontal logic of the LFA indicates how the achievement of the objective can be determined.

The objective of the horizontal logic is the measurement of the resources and results of a project through identification objectively verifiable indicators and means of verification for these indicators.

Verifiable Indicators

The LFA emphasizes the need for information to measure project progress. The objectively verifiable indicator represents a set of criteria, which indicate that expected result, have been achieved, in concrete terms. The verifiable indicators should demonstrate that the desired results are being accomplished and specify the type of evidence needed. For each level of the vertical logic there will be a set of verifiable indicators that are appropriate to the objectives at the level and which constitute proof of achievement at that level.

Means of Verification

In the LFA the means of verification ensure that the previously defined verifiable indicators can be measured effectively. They specify where that evidence can be found and how it can be measured. They provide information on how data will be collected about the indicators. The means of verification are the sources of data that enable verification the status of each indicator.

Thus the 'indicators' and 'means of verification' lay the basis for project monitoring and evaluation plans and systems.

Advantages of LFA

- Provides a comprehensive, practical, flexible and effective means to strengthen international research project management.
- Foster reflection within the project implementing institution.
- Increases accountability.
- Helps to ensure that project managers and their superiors develop realistic understanding of what they and their staff can and cannot realistically hope to achieve, as well as offering them a useful reminder of their key managerial functions.
- Generate early warnings before things begging to go wrong and allow a corrective action.
- It is an aid to project management.

- The way in which objectives, outputs and inputs are set out in such a manner that enables the links between them to be readily seen and hence the logic of the project be exposed to scrutiny.
- LFA specifies in advance what the project is attempting to achieve by indicating the means by which this achievement may be measured.
- It is simple to understand.
- Very useful for monitoring and evaluation. It is also useful for impact evaluation.
- LFA can be used as a delegating tool with targets group over relatively short periods (monthly or quarterly) to be reviewed at the end of the period and updated for the forthcoming period according to learning processes.
- LFA offers programme and project managers a way to introduce vigour into the planning process so that implementation does not just happen without any though of the relevant means and relationships.
- By specifying general consideration and a set of important relationships the approach concentrates attention on the operationalization of the project – Who is doing, What, to Whom. When. Why and How.

Limitations of LFA

- It beings with identifying problems which often produces poor results because of the initial negative focus that ends limited vision of potential solutions. It is also suitable for situations where thee is a great deal of uncertainty.
- At times the rigidity of the approach stifles creative thinking and adaptive management.
- Log frames are at times seen and used as a requirement of funding agencies only and not used as a key planning and evaluation tool and is often developed after the project has been designed.
- At times the project cannot be put into a log frame.
- Sometimes the framework does not properly describe the project especially when a single project is standardized to fit different situation.
- It does not considered potential alternative actions.
- The 'IF-THEN' relationships assumed, among the various project components and elements in the environment are an over-simplification. It does not take into account the uncertainties.
- LFA requires explicit indicator identification to measure different objective and result achievements. It also demands that MOU are identified and the assumptions stated. Frequently this task is not completely understood or it is done quickly and inadequately simply in order to respond to the requirements of funding agencies.

Thus, the Logical Framework Approach helps give structure to the process of planning and communicating essential information about projects. It can be used at all stages of project preparation, execution and evaluation. For best results it should be developed as a participatory team effort. Once established it should be modified and improved at several stages during project preparation, executing and evaluation. The output of planning and evaluation is not a final result, but rather starting point or another cycle of analysis, reflection and action to go forward towards a more sustainable approach to research and development.

A Systematic Approach (ASA)

A systematic Approach is based on the way people think when they are getting things done in a purposeful and efficient way. Naturally, therefore, many people find the approach familiar. It may akin to other pattern of logical though they know of which may have been developed for other uses such as problem solving or systems analysis according to their purpose such system may omit certain stages of Systematic Approach and expand others.

A systematic Approach is directed at getting things done, and importantly, at drawing lessons from experience. These features are especially significant faces us with situations we are not able to predict. Unable to prepare for every particular situation, it is left to us cultivate the thinking skills that will cope with the general case. We can develop confidence, not through knowing the answers, but through having the means findings them. The sequence of steps, which make up a Systematic Approach, can be entered at any point. It is not unusual, for example, to begin by reviewing some significant event in the past and then move into the information stage relating to a future action. Likewise, there are times when the most important thing is to act, now and fast without formal planning.

Each stage of Systematic Approach may need handling differently for different applications. The information stage may call for logical research and analysis on one occasion, and for organized brainstorming on another. Statement of what has to be done may take the form of a simple job list, or a complex critical path network. No system is a substitute for judgment, and Systematic Approach must remain the servant and become the master. Skill and discernment sit use come from practice.

Use of a systematic approach:

- ✰ Getting things done purposefully and efficiently
- ✰ Logical stages of thoughts.
- ✰ Common language – cooperation.
- ✰ Particularly helpful with tackling new and open produces.
- ✰ Confidence of not having to know all the answers and having the means of finding them.

Systematic Approach

In participatory project building, stakeholder analysis (SA) is a simple and important tool to identify the stakeholders related with the project building, implementation and its carrying over from short term tenure to a protracted period. A stakeholder is any person/s or institutions/s who has/have a stake (interest) in any process or programme. The stakeholders are of primary, secondary and Tertiary in nature and approach. The respective roles, contribution and impacts of stakeholders can be assessed through a structured format. But, one facilitator should go careful and meticulous in exercising this format to identify different stakeholders from a target group or community. Triggering of quality group dynamics as well as group communication would certainly breed on fostering the process.

ASA : A Schematic Description

Work Concern/Problem/Objective/Job to be done

Write a concise

Task Instruction

PURPOSE: Why is the task to be

	Undertaken to what purposes to what benefits, who is it for ?
Clarify *the* AIMS	END RESULT: Are these definable, What do I want to achieve to End up with?
	SUCCESS CRITERIA/STANDARDS : What Yardsticks will be used to measure Progress or achievement?
Examine relevant INFORMATION	KNOWN : Background, facts, past experience, resources, (time, money, materials, People, skills), ideas options, risks
	UNKNOWN AND NEEDED
State WHAT HAS TO BE DONE (WHTBD)	

WHIBD By me	WHTBD by somebody else By me to get their Cooperation	WHTBD by me to get it WHTBD

PLAN	PLAN	PLAN
To do it	to do it	to do it

NOTE - Systematic Approach is cyclic in its application Plans need to be followed by Action and then by review, so leading into the next cycle of PREPARATION – ACTION – REVIEW

Stake Holder Analysis (ASA)

In participatory project building, stakeholder analysis (SA) is a simple and important tool to identify the stakeholders related with the project building, implementation and its carrying over from a short term tenure to a protracted period. A stakeholder is any person/s or institution/s who has/have a stake (interest) in any process or programme. The stakeholders are of primary, secondary and Tertiary in nature and approach. The respective roles, contribution and impacts of stakeholders can be assessed through a structured format. But, one facilitator should go careful and meticulous in exercising this format to identify different stakeholders from a target group or community. Triggering of quality group dynamics as well as group communication would certainly breed on fostering the process.

SA Format

Name of programme

Objectives

Stakeholder	*Levels*	*Activity/ Role*	*Value (V)*	*Power (P)*	*Impact (V x P)*	*Rank*
	Primary :					
	Secondary :					
	Tertiary :					

Observations

Flow Diagrams

They show the cause-effect relationship amongst different factors and can be effective for understanding the inter-linkages amongst different aspects of life. The villagers can be asked to identify different causes of a problem, which they are facing and then asked to interconnect the different aspects so as to have a causal analysis of the problem at hand.

Venn Diagram (Chappati Diagrams)

They are visual methods to reprint. The role of individuals/institutions and the degree of their importance in decision-making. In such diagrams circles of different sizes represent an individual or an institution. Whose size shows its degree of importance in decision making in a village and the overlapping of circles indicate their webbing in decision making.

Matrix Scoring

In matrix scoring exercise different attributes and criteria are listed. Which are ranked either on the basis of fixed scoring (any scoring out of few) or free scoring (as many/less) according to their relative importance.

Wealth Ranking and Grouping

Wealth ranking or grouping is a method of ranking/grouping of households on the basis of income, wealth and other local measures of well being. Different criteria can be used for wealth ranking or grouping based on rural perceptions.

Livelihood Analysis

Livelihood analysis is an analytical tool by which peoples income and expenditure capability can be identified indicating their socio-economic status. As per wealth ranking category of the households derived from the key informants, one representative member has to be selected from each category and to be interviewed. This tool is the main indicator of the people's social behaviour like purchasing ability, living standard decision-making ability and coping strategy.

Micro-farming Situation Mapping

A micro-farming situation map describe. The smaller systems operative within a larger agro-ecosystem. It includes soil types, water regisms crop enterprise in terms of mix and sequencing, constraints and scope and such many other varieties characterizing the functioning of agro-ecosystem. A meticulous micro situation analysis provides the basis for micro planning for effective land use, including new enterprises, networking different enterprises etc.

Land Use Mapping

It's a participatory mapping on the land resources using local perceptions of fertility status, topography etc. It includes crop mix and sequences in differential topography, social and cultural dimension of land use.

Crop Profile

Crop profile is a schematic and non-descriptive study of crops according to their production practices. It shows a sequential flow of different agril operations. A crop profiles also depicts cost based studies, spill over effect from previous crop to follow up crop. The important components of crop profiles are :

1 Operational stages of a crop enterprise
2. Component wist cost break-up
3. Identification of problem areas affecting production performance.
4. Participatory estimation of cost, profit and return and hence, delineation of a strategic intervention for improving any enterprise.

Activity Planning and Network Analysis

Every rural development project has to be implemented with the help of a time schedule and managed effectively to minimize cost escalation and avoid time overruns.

The planning of these activities in the project life-cycle is know as implementation planning. In the process of management it is generally served by functions of organizing, staffing and directing. A proper implementation planning exercise helps developing a time table for efficient management of the project.

Activity Planning

The first participatory management function is planning. Therefore, once a project has been accepted for implementation, it becomes imperative to plan for each activity of the project. The planning is carried out in a sequential manner. The steps are given below :

Step 1 : Prepare a list of all activities that need to be executed under the project. Sometimes a project may require the help of a number of departments/sectors/ agencies for its execution. It is most desirable to list the activities according to agencies or departments or sectors. This list will provide the Work Breakdown Structure.

Step 2 : Specify a logical sequence in which the activities need to be executed. This will indicate (1) the order in which activities will follow; and (2) activities, which can be undertaken simultaneously.

Step 3 : Estimate the specific duration of time required for each activity in the project. If you have executed similar projects in the past under identical conditions, you can specify the duration based on past experience. Otherwise, estimate the duration, using your judgment and statistical measures.

Step 4 : Assemble the activities in the form of a flow diagram. This is known as Network Plan.

Step 5 : Analyze the flow diagram. This is called Network Analysis.

Let us take an example to illustrate the above steps for preparing a project implementation plan.

Suppose that we have to establish a milk chilling plant. The following are the assumption underlying the plan :

- ☆ The feasibility and viability of having a milk chilling plant in the area has been examined and the results are favourable. The area has a number of dairy farmers and on an average 4,000 litres of milk can be procured per day.
- ☆ The plan for setting up a milk chilling plant has been drawn and it requires the approval of the district administration, finance department and the dairy development corporation.
- ☆ There is a building, which can be used for housing the milk chilling plant.
- ☆ Staff will be recruited for operating the milk chilling plant. They will be trained in another plant of similar type and capacity, while our plant is under construction.
- ☆ The electricity line for the milk chilling plant will have to be obtained from the Electricity Board, following which the electrical wiring will be carried out.

- The milk chilling plant will be purchased and installed. After the installation of machinery, training of the staff and installation of electrical wiring, the milk chilling plant will be operated on a trial run over a period of one month before it is handed over to the manager of the milk chilling plant.

The first step is to list the various activities as given below in terms of water shed management:

- Project plan approval
- Recruitment of staff
- Purchase of inputs
- Selection of water shed leaders
- Training of leaders
- Land shaping
- Plantation Programmes
- Impact Analysis

In the following Table we furnish the list of activities with their codes for our project Establishment of Milk Chilling Plant.

Table 16.1: Coding of Activities

Activity Codes	*Activity Description*
A	Project Plan Approval
B	Recruitment of staff
C	Purchase of inputs
D	Selection of water shed leaders
E	Training of leaders
F	Land shaping
G	Plantation Programmes
H	Impact Analysis

The next step is to specify a logical sequence in which the activities need to be undertaken. For this we need to consider each activity separately and examine whether it has to wait for any other activity in the package. Then we will indicate the immediate predecessor or previous activity in a separate column as given in Table 16.2.

For example, the project plan approval need not wait for any other activity in our package but, on the contrary, other activities will have to wait for project plant approval. In a similar way the immediate predecessor activity/activities are determined for each activity presented below :

Table 16.2: Activity Sequencing

Activity Code	*Activity Description*	*Immediate Predecessor*
A	Project Plan Approval	None
B	Recruitment of staff	A
C	Purchase of inputs	A
D	Selection of water shed leaders	A
E	Training of leaders	B
F	Land shaping	C
G	Plantation Programmes	D
H	Impact Analysis	E, F, G

Let us see the method of estimating the duration of each activity in the package. Collect data on each activity duration from a number of Project Managers (a fairly large number is suggested to get a more realistic picture) who have undertaken a similar project under more or less identical conditions.

First of all, let us look at the activity: project plant approval. The data collected from say 32 Project Managers, presented below, indicates the number of weeks taken by each to obtain the project plant approval in the process of establishing milk chilling plants under similar conditions.

(1)	(2)	(3)	(4)	(5)	(6)	(7)	(8)	(9)	(10)	(11)	(12)	(13)	(14)	(15)	(16)
14	21	**6**	**6**	5	11	18	**6**	17	8	19	**6**	12	2	**6**	14
(17)	(18)	(19)	(20)	(21)	(22)	(23)	(24)	(25)	(28)	(27)	(28)	(29)	(30)	(31)	(32)
6	2	3	**6**	5	15	**6**	7	20	**6**	22	**6**	10	16	**6**	**6**

Note: Figure in parenthesis gives the serial number. The figure below each parenthesis gives the number of weeks taken by Project Manager.

In the above sample, two Project managers have taken two weeks each; none have taken less than that duration to get their project plan approved. One of them has taken 22 weeks; no one has taken more than that duration. Majority of the project managers in the sample (*i.e.* twelve) have taken six weeks. The maximum (longest duration) in the sample (22 week) is known as the pessimistic time; minimum in the sample (two weeks) is called the optimistic time; and the majority in the sample (six weeks) is known as the most likely time. Using the three numbers from the sample, we follow the formula given below to arrive at the time estimate :

$$\text{Time Estimate: } t = \frac{t_{op} + 4t_{ml} + t_{pc}}{6}$$

where,

t_e: Time estimate

t_{op}: Most optimistic time

t_{ml}: Most likely time

t_{pe}: Most pessimistic time

In the above example, the time estimate comes to :

$$t_e = \frac{2+4\times 6+22}{6} = \frac{48}{6} = 8 \text{ weeks}$$

Similarly, for each activity in the package we can estimate the time duration. The time duration estimated must be expressed in the same unit : days, weeks, or months for each activity. In the case of a project, which takes, say, five years, the duration of each activity may be expressed in terms of months. For a project of short duration, say, nine months, the activity duration may be expressed in terms of weeks. A project which takes, say, three months, the duration may be expressed in terms of the days. The estimated time duration of the activities in our illustrative project is given below (Table 16.3).

Table 16.3: Time Estimates

Activity	*Activity Description*	*Immediate Predecessor*	*Activity Duration (Months)*
A	Project Plan Approval	None	2
B	Recruitment of staff	A	3
C	Purchase of inputs	A	5
D	Selection of water shed leader	A	2
E	Training of leaders	B	2
F	Land shaping	C	4
G	Plantation Programmes	D	2
H	Impact Analysis	E, P, G	1

Network Analysis

Step numbers 4 and 5, as given in the preceding section, constitute network plan and network analysis. Therefore, it is proper that we start from the basics. Network analysis uses symbols, notations, forward and backward movements, etc. Let us get familiarized with these terms and their definitions.

Every activity in a project has to begin with an event and end with another event. An event occurs at a point of time, whereas an activity takes place over a period of time. The following are examples of activities and events:

Activity	*Event*
☆ Conducting a survey to identify beneficiaries	Survey conducted
☆ Selection of sites for sinking bore wells	Sites identified
☆ Appointment of staff	Staff appointed

An event in the flow diagrams is represented by a circle and an activity is represented as an arrow joining two circles. The event with which an activity begins is known as the predecessor event of the activity and the event with which it ends is it successor event. An example follows:

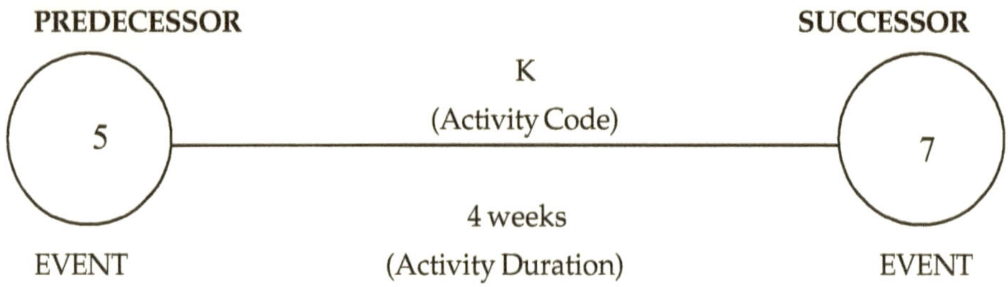

Figure 16.1

An event may give rise to more than one activity on its occurrence. For example, as soon as the plan is approved (which is an event), it will give rise to activities such as land acquisition, appointment of staff, obtaining power supply, purchasing machinery, purchasing machinery purchasing vehicles and obtaining telephone connection, etc. In such cases all the activities cannot be shown as a straight line from the event 'Plant Approved', but can be shown as bent lines.

An event, which gives rise to more than one activity, is called a burst event. An event is said to occur only when two or more activities are accomplished. For example, a poultry complex can be handed over to the beneficiary only after construction of poultry complex, procurement of poultry birds and chicks, providing veterinary services and creating marketing facilities and so on. Such an event, providing veterinary services and creating marketing facilities and so on. Such an event is called a merge event, meaning that several activities merge at that event.

Two sets of activities may be undertaken simultaneously, like selection of site (followed by selection of beneficiaries) and the other stream of activities consisting of procurement of building materials (followed by constructing the poultry sheds) under IRDP. If the logic is to construct the complexes after the survey as well as procurement of materials, then the linkages between activities need to be shown by introducing a dummy activity (dotted line with arrow).

Dummy activity has zero duration. It does not consume resources. Dummy activities are used only to depict the required logic on the flow diagram. Using the above symbols, notations and terms let us develop a network plan for the project we have chosen.

Chapter 17

Logical Framework Analysis: An Innovate Tool for Project Monitoring and Evaluation

Sankar Kumar Acharya and Manas Mohan Adhikary

What is Logical Framework Analysis (LFA)?

Logical Framework Analysis

The Logical Framework Analysis is a powerful tool for project design and management, which distinguishes between, lower and higher-level project objectives, showing how these objectives are linked. The logical framework approach is appropriate for any level of planning (except strategic planning) and used for designing and managing broad, long-term programmes to individual research projects. It is frequently used as a framework for brain storming and planning by small groups of managers.

Purpose of LFA

- ☆ To assist in the planning, management and evaluation of development activities and research projects.
- ☆ To facilitate the systematic identification and specification of intermediates and final objectives of the programme and project concerned by the planners and managers
- ☆ Plan for required inputs and desired out puts
- ☆ Give sound guidance on the structuring of a monitoring and evaluation system by defining the indicators for monitoring and evaluation performance

- To assist in the planning, management and evaluation of development activities and research projects.
- To facilitate the systematic identification and specification of intermediates and final objectives of the programme and project concerned by the planners and managers
- Plan for required inputs and desired out puts
- Give sound guidance on the structuring of a monitoring and evaluation system by defining the indicators for monitoring and evaluation performance

Vertical Logic

The components of LFA format:

- **Goal**: This is the ultimate objective to which the research project contributes.
- **Purpose**: This is what it is expected to achieve upon completion within the allocated time. It is the motivation behind the production of out put.
- **Output**: These are the specific results to be produced by the management of inputs. The desire results of the research projects derived directly from management of inputs.
- Activities: These are the activities to be undertaken and the human, physical and financial resources available and required to produce the desired outputs

Horizontal Logic

The Horizontal logic of the LFA indicates how the achievement of the objective can be determined.

Objectively Verifiable Indicators (OVI)

The objectively verifiable indicator represents a set of criteria, which indicate that expected result; have been achieved, in concrete terms.

Means of Verification (MOV)

MOV specifies where that evidence of OVI can be found and how it can be measured.

Assumption or Exogenous Factors Impacting on the Project Function

Key threats to the project that exist in the external environment Cause Effect Linkage:

Vertical logic is based on the principle of causality in a means, ends and relationship.

If-Then Relationship

If inputs are provided **then** outputs will be produced.

If outputs are produced **then** purpose will be achieved

If purpose is achieved **then** the goal may be realized

Advantage of LFA

- Provides a comprehensive, practical, flexible and effective means to strengthen international research project management
- Foster reflection within the project implanting institution.
- Increase accountability.
- Helps to ensure the project managers and their superiors develop realistic understanding of what they and their staff can and cannot realistically hope to achieve, as well as offering them a useful reminder of their key managerial functions
- Generate early warnings before things begging to go wrong and allows a corrective action
- It is an aid to project management.
- The way in which objectives, outputs and inputs are set out in such a manner that enables the links between them to be readily seen and hence the logic of the project be exposed to scrutiny.
- LFA specifies in advance what the project is attempting to achieve by indicating the means by which the achievements may be measured
- It is simple to understand
- Very useful for monitoring and evaluation.
- LFA can be used as a delegating tool with targets group over relatively short period to be reviewed at the end of the period.
- LFA offers programme and project managers a way to introduce vigour into the planning process so that implementation does not just happen without any though of the relevant means and relationship.
- By specifying general consideration and a set of important relationship the approach concentrates attention on the operationalization of the project- Who is doing, What, to Whom. Why and How.

Why Use LFA?

Because Most Donors Prefer it?

LFA can be a useful tool, both in the planning, monitoring and evaluation management of development projects. It is not the only planning tool, and should not be considered an end in itself, **but using it encourages the discipline of clear and specific thinking** about what the project aims to do and how, and highlighting those aspects upon which success depends.

LFA also provides a **handy summary to inform project staff, donors, beneficiaries and other stakeholders**, which can be referred to throughout the lifecycle of the project. LFA should not be set in concrete. As the project circumstances change it will probably need to reflect these changes but everyone involved will have to be kept informed.

What is so Intimidating about using LFA?

Logical Framework Analysis: An alternative vision

The LFA is a way of describing a project in a logical way so that it is:

- ✰ Well designed.
- ✰ Described objectively.
- ✰ Can be evaluated.
- ✰ Clearly structured.

TERM **MEANING** **EXAMPLE**

Narrative Summary

The goal, purpose, outputs and activities of the project as described in the left-hand column of the logical framework. (the Objectives column)

See below.

Goal

The ultimate result to which your project is contributing - the impact of the project.

Jamaica's dominance of competitive bobsledding.

Purpose

The change that occurs if the project outputs are achieved - the effect of the for bobsledding project.

Jamaica wins the gold medal at the 2002 Winter Olympics.

Outputs

The specifically intended results of the Team members selected by project activities - used as milestones of what has been accomplished at various stages during the life of the project. (Date).

Team at full fitness by (date) etc.

Activities

The actual tasks required producing the desired outputs.

Develop training schedule.

Find practice venue.

Publicity campaign to recruit team member, etc.

What Does the Jargon Mean?

Some of the terminology involved in LFA may seem rather intimidating. Do not be put off by the language.

Remember that the goal, purpose, outputs and activities are all objectives but at different levels of the project hierarchy. Different donors use slightly different terminology, but the logical frameworks are all the same in principle. You will come across the following terms:

Indicators

Also referred to as measurable or measuring progress and whether project and agreed by x, etc.

1. Team members capable of objectively verifiable indicators (OVI) running x metres in x seconds by quantitative and qualitative ways of x time.
2. 4 year training schedule, outputs; purpose and goal have been budget and outcomes developed achieved.

Means of Verification

M.O.V is the information or data

1. Fitness report from team required to assess progress against indicators and their sources.
2. Schedule written and agreed (signed) by coach, team members and team doctor.

Assumptions

Factors external to the project, which is likely to influence the work of the project management has little control, and which need to exist to permit progress to the next level in the LFA.

The Jamaican team qualities for the Games. It snows enough for the Games to be held, etc.

Super Goal

The long-term results of continued achievement of the goal of the project.

Hot countries seen as serious competitors in winter sports.

Inputs

What materials, equipment, financial and human resources are needed to carry out the activities of the project?

Funding, Coach, Bobsled, Snow, Medical Advisor, etc.

BOND Guidance Notes Series 2

The reality of funding proposals and completing log frames usually means a desk officer in the UK trying to summaries a project outline for a funding application.

However, if used correctly as a planning tool, LFA ought to be developed first by, or working closely with the person most closely involved in project implementation who will most likely be your project co-ordinator or partner organisation overseas. The project detail can be more easily developed from the log frame than the other way round. Writing log frames in the UK is not participatory, which has led to criticism of the log frame as a planning tool. Ideally it should be produced 'in country' so that during the planning stage participatory approaches can be used to feed into the log frame as it is developed. Whilst project beneficiaries may not identify easily with the concept of LFA, they may be able to identify the factors that are critical to project success, as well as the most appropriate indicators of progress. In this way, participatory techniques can be used to inform LFA. The log frame can also provide a guide as to what information needs to be gathered through participatory processes, and can be enhanced by combining the outcomes of other planning tools, such as social mapping, wealth ranking, and problem and objective trees.

If the log frame has to be written, adapted or changed in the UK, it is important to feed the detail back to fieldworkers, partners and other relevant stakeholders overseas.

The input to a log frame should be a team effort, as much as possible.

If you are not the person closest to the project, it is important to engage that person in developing the log frame. Always consider:

- What impact the objectives and indicators will have on their work.
- What is realistically achievable?
- Will they have enough time to collect the information you are asking for?
- Are the assumptions are you making realistic to them? Are you fully aware of their working conditions?

Bond

Who Should be Involved?

Many people find it useful to start by developing a Problem Tree. Try to identify what is the real problem the project is to tackle and write it in the middle of a sheet of paper. Then consider the direct causes of the problem and write them in a horizontal line below the problem. Next, repeat the process for each of those on another horizontal line and continue to repeat as necessary. All the statements must be written in negative terms. This will give you a problem tree in which a cause and effect relationship operates from the bottom to the top. You may also find it useful to work upwards from the original problem, identifying its effect etc. Then changing the wording of each item into positive terms will change the Problem Tree into an Objective Tree. Next, decide whether the original problem is still

Now, Where do I Need to Start?

What do I Need to Produce a Logical Framework?

- Supply of large sheets of paper, (preferably flip chart sheets).
- Pencil, eraser and 'Post-it' notes or cards, so you can adjust and amend as you go along.

- Somewhere to work without distractions.
- Ideally, someone to discuss and 'bounce' ideas around with.
- As much information about the planned project as possible - preferably do it 'on site'.

Objectives

- Measurable indicators
- Means of verification
- Important assumptions

Goal

Wider problem the project will help to resolve Quantitative ways of measuring or qualitative ways of judging timed achievement of goal Cost-effective methods and sources to quantify or assess indicators

(*Goal to super goal*)

External factors necessary to sustain objectives in the long run

Purpose

The immediate impact on the project area or target group i.e. the change or benefit to be achieved by the Project Quantitative ways of measuring or qualitative ways of judging timed achievement of purpose

Cost-effective methods and sources to quantify or assess indicators

(*Purpose to Goal*)

External conditions necessary if achieved project purpose is to contribute to reaching project goal

Outputs

These are the specifically deliverable results expected from the project to attain the purpose Quantitative ways of measuring or qualitative ways of judging timed production of outputs Cost-effective methods and sources to quantify or assess indicators

(*Outputs to purpose*)

Factors out of project control which, if present, could restrict progress from outputs to achieving project purpose

Activities

These are the tasks to be done to produce the outputs

Inputs

This is a summary of the project budget Financial out-turn report as agreed in grant agreement **(*Activity to output*)** *Factors out of project control which, if present, could restrict progress from activities to achieving outputs DFID describes the Logical Framework as "a*

tool to help designers of projects think logically about what the project is trying to achieve ***(the purpose)****, what things the project needs to do to bring that about* ***(the outputs)*** *and what needs to be done to produce these outputs* ***(the activities)****. The purpose of the project from the DFID viewpoint is to serve our higher level objectives* ***(the goal)****".*

NOTE: The two boxes in the centre of the "Activities" row are not used for Measurable Indicators and Means Of Verification as the progress and success of the Activities are measured at the Outputs level. Remember, the Activities are carried out to achieve the Outputs. These "spare" boxes can therefore be used to provide any useful additional information such as Inputs and Budgeting requirements.

Taken from DFID's "Guidelines on Humanitarian Assistance", May 1997

Going to be the main focus (or goal) of the project. This may now be higher or lower on the Objective Tree.

Having made that decision, the purpose, outputs and activities should be present in the next three horizontal lines of the Objective Tree. It will probably be necessary to decide what to include at this stage, as the project has to be of a manageable size. (*See Figure 1 below: A Logical Framework -DFID model*)

The key to completing log frames is to use the information generated in the Objective Tree and:

Start at the Top and Work Down

The objectives column - what is the project going to achieve?

Then Think Laterally

How can the progress of the project be measured against its objectives?

Then Reflect Back Up

What assumptions are to be included and what are their implications?

As you work down each step of your objectives, think:

- ☆ How the outputs and activities can be measured.
- ☆ What indicators can be used to measure achievement against?
- ☆ What information will be needed, and how it can be gathered?
- ☆ What problems, obstacles or barriers might arise to prevent the project from progressing as planned?
- ☆ How can their impact be minimised?

Stage Two: Work Across (Measurable Indicators and Means of Verification)

1. Goal

Starting at the top and using the information from the Objective Tree consider the overall **goal** of the project. What issue or problem is the project trying to address?

The goal may be beyond the reach of this project on its own. What ultimate objective is the project contributing to? This should be a brief statement or summary.

Example

To increase literacy among young people in the sub- Sahel region.

2. Purpose

What final result are you trying to achieve? This is the **purpose** of the project. This should be clear and brief.

Example

School attendance and literary skills of 6-14 year olds in (named region) of the sub-Sahel is increased.

3. Outputs

What are the particular **outputs** needed to achieve the Purpose of the project? There may be several outputs.

Example

'Five new language teachers recruited and trained. ''Classes running in all schools at times when children are not required to complete family duties. 'Information sessions for families with school age children held in each village demonstrating benefits of literacy.'

4. Activities

List the **activities**, which are needed to achieve these outputs. There may be several for each output. Statements should be brief and with an emphasis on action words.

Example

'Hold publicity campaign in (named region) to recruit language teachers by (date)'. 'Agree and arrange selection process and schedule for teacher recruitment involving existing teaching staff'. 'Hold training sessions for new teachers covering aims, approach, customs, potential problems holding information sessions, record-keeping, expenses'. 'Arrange appropriate accommodation for teachers in each village at least before arrival'. 'Plan and hold meetings in each village for families with school age children to demonstrate the benefits of literacy' etc.

5. Inputs

When required to do so provide additional information, such as the **inputs** which are needed to carry out these activities. Again, there may be several for each activity and it will help to run through each individually, listing required inputs (resources, equipment, tools, people). Group the inputs and list each once rather than repeatedly. This may include a summary of the project budget.

Example

Budget, Training space, Accommodation, Support for existing teaching staff, Teaching materials, Transport to village, Project Co-ordinator/Fieldworker etc.

6. Indicators

Starting either from the top or the bottom of your hierarchy of objectives, begin to work across the log frame, identifying the indicators for measuring your progress. Indicators need to define 'QQT', Quality, Quantity and Timing:

7. Information Sources

Next, try and work out your means of verification for each indicator. **What information will you need, and how and from where can it be gathered?**

Will project staff or others need to keep records, or can they get the information from somewhere else?

Consider the cost implications, if any, and build this into the project budget.

Don't exclude anecdotal evidence (e.g., views expressed by project beneficiaries, etc) if this is the most appropriate source of information, but remember that donors can be wary of this evidence, and it may later be necessary to demonstrate your claims!

Note

If you are confused about indicators - don't panic! *Identifying indicators requires some practice.* ***Try to find a mentor with more*** *experience* ***who can advise you.*** *Or move on to look at means of verification - thinking about what information you need, and how to get it, may help to define what the indicator should be. If you come up with a long list of possible indicators try to narrow it down to the essential ones.*

Stage One: Top Down (Objectives)

Quality - The kind (or nature) of the change. Quantity - The scope (extent) of the change ie. by how much, how many.

Timing - By when the change should have taken place.

Three Stage to Success

- ☆ Stage One: Top Down (Objectives)
- ☆ Stage Two: Work Across (Measurable Indicators and Means of Verification)
- ☆ Stage Three: Bottom Up (Assumptions)

8. External Factors

What external factors (outside your control) could affect the success of your project or prevent work from progressing? These may be climatic, political, economic, etc. but should be real (possible) risks rather than a list of everything that could go wrong.

Reflecting up from the bottom of your log frame, consider how, if each assumption holds, it will be possible to move to the next stage of the project.

Example

There are sufficient rains to ensure that children are not required to replant crops and therefore unable to attend school. And/or, Sufficient teachers with knowledge of local dialect are recruited. And/or, Conflict from neighboring region doesn't spread into local area so that emergency needs take precedence over education.

9. Double Check

Following completion of your log frame, go over it, from bottom to top, to check the logic of it:

- Will the inputs and activities clearly lead to the outputs required to achieve the purpose and contribute to the goal?
- Will the indicators and means of verification effectively measure the progress of the project?
- Are the assumptions reasonable or do they indicate a level of risk, which suggests that the project is unlikely to get off the ground or be completed? (The killer assumption)
- Is the project staff committed to the objectives and indicators identified and see them as realistic and achievable?
- Are there any changes, which could be made which will make the project it more practical and workable?

Stage Three: Bottom Up (Assumptions)

There are two kinds of indicators you will need to use: **Process indicators**

Which measure the extent to which you have achieved your stated objectives.

Example

How many children attending school by (specified) time.

Impact indicators

Which help to monitor the achievement and the impact of your work.

Example

How many children pass (specified) literacy test at (specified) time. In addition, indicators can also be:

Direct Example

The Number of children attending school.

Indirect Example (also known as soft or proxy).

More books borrowed from the school library -suggesting more children reading, therefore wider literacy.

Examples Using Indicators at the Purpose level:

School attendance of 6-14 year olds is increased by 200 per cent within 4 years. 90 per cent 6-14 years olds (in named region) of the sub-Sahel to have gained (particular level) of literacy skills within 4 years.

Using Indicators at the Outputs Level

Five language teachers recruited and trained by (date). At least two classes running in all schools at times when children are not required to complete family duties by (date). At least three information sessions for families with school-age children held in each village demonstrating benefits of literacy by (date).

10. Write it Up

When the log frame has been checked (and rechecked) and it is truly logical, and representative of the project, type (or write) it up onto A4 sheets. At this point **all the relevant stakeholders should have had a chance to contribute to (and agree) the completed log frame.**

11. Staying Informed

And still you haven't finished; remember that LFA is a flexible tool for planning, managing, monitoring and reporting your project. **As the project progresses and situations change, return to the log frame and revise it accordingly.** Agree these changes with the donor and other stakeholders and **keep everyone informed so that they are able to keep up to date with current progress of the project and its future direction.**

Top Tips

Don't leave it to the last minute. Start working on your log frame when you begin planning the project. You may need to gather information as you go along that will be difficult to get later (eg baseline data against which to measure progress). Try and get the person planning and implementing the project to complete the log frame. If this is not possible consider the implications of the objectives, indicators and verification on field staff and partners. Develop a problem tree. By turning the problems into objectives, this should help clarify the goal, purpose, outputs and activities of your project. Find a mentor with experience of writing log frames who can offer you advice and assistance. If you get stuck, don't panic - move on to the next stage and come back to the tricky bit later. Work in pencil so you can erase things and make amendments easily, and use a large sheet of paper with plenty of room for 'thinking' then reduce it down to A4 later. It is difficult to get it right first time, keep reflecting and revising until you are satisfied that the project is workable and the log frame is clearly logical! Use the log frame as the basis of the funding application and then throughout the project Lifecycle to track progress and adapt to changing situations. You can use the log frame to review assumptions and implications, and to keep donors and other stakeholders informed of significant changes.

About Bond

BOND is the network of over 270 UK- based non- governmental organisations (NGOs) working in international development and development education. BOND aims to improve the extent and quality of the UK and Europe's contribution to international development, the eradication of global poverty and the upholding of human rights. The Guidance Notes Series aims to provide 'how-to' information on a variety of topics for the development sector. This edition also provides signposts to resources for those keen to pursue the topic further.

Disclaimer

BOND's Guidance Notes aim toencourage good practice through practical advice, however, BOND cannot be held responsible for the outcome of any actions taken as a result of the information contained in the Guidance Notes series.

References

Akroyd, D. (1995). 'Steps toward the adoption of Logical Framework Approach in the African Development Bank: some illustrations for agricultural sector projects', Project Appraisal, 10 (1), 19-30.

Akroyd, D. (1995). 'The Logical Framework Approach and the post-evaluation of health sector projects by the, African Development Bank', Project Appraisal, 10 (4), Pages 210-222.

Argeetey, E. (1998). 'Consultative processes in community development in Northern Ghana', Community Development Journal, 33 (4), 301-33.

Cordingley, D. (1995), 'Incorporating the Logical Framework into the Management of technical co-operation projects', Project Appraisal, 10 (2), 103-112.

Cracknell, B. (1996), 'Evaluating Development Aid', Evaluation, 2 (1) 23-33.

DFID Chapter on Logical Frameworks in their Tools for Development series: http://62.189.42.51/DFIDstage/FOI/tools/chapter_05_frame.htm

Eggers, H. W. (1998), 'Project Cycle Management Revisited', The Courier May/June 1998, 69-72, Brussels: European Commission

Gasper, D. (1999), 'Problems in the Logical Framework Approach and the challenges for Project Cycle Management', The Courier, Jan/Feb 1999, 173, 75-77. Brussels: European Commission

Gosling L and Edwards M (1995). "Toolkits: a practical guide to assessment, monitoring, review and evaluation" Development Manual 5 pp. 178-192 Save The Children, £6.95 (New edition due in March 2003 with updated information on Logical Frameworks)

http://62.189.42.51./DFIDstage/faqs/files/guidelines_on_humanitarian_assistance_v5.0a.doc

http://iucn.org/themes/eval/english/lfa.htm

Jackson, B. Designing Projects and Project Evaluations using the Logical Framework Approach:

Nancholas, S. (1998). 'How to (or not to do)... A Logical Framework', Health Policy and Planning, 13 (2), 189-193.

The Civil Society Challenge Fund Guidelines 2002 include an Annex C on The Logical Framework: http://62.189.42.51/DFIDstage/Pubs/files/cscf_guide.htm

The DFID "Guidelines on Humanitarian Assistance" published in May 1997 includes guidance on preparing log frames, and is available on the DFID web-site.

Chapter 18

Supply Chain Management in Enterprise Building and Business Strategy

Manas Mohan Adhikary and Sankar Kumar Acharya

Supply Chain Management

Supply chain management (*SCM*) is the process of planning, implementing, and controlling the operations of the supply chain as efficiently as possible. Supply Chain Management spans all movement and storage of raw materials, work-in-process inventory, and finished goods from point-of-origin to point-of-consumption.

Supply Chain Management encompasses the planning and management of all activities involved in sourcing, procurement, conversion, and logistics management activities. Importantly, it also includes coordination and collaboration with channel partners, which can be suppliers, intermediaries, third-party service providers, and customers. In essence, Supply Chain Management integrates supply and demand management within and across enterprises/companies.

Some experts distinguish Supply Chain Management and logistics, while others consider the terms to be interchangeable.

Supply Chain Management is also a category of software products.

Supply chain event management (*abbreviated as SCEM*) is a consideration of all possible occurring events and factors that can cause a disruption in a supply chain. With *SCEM* possible scenarios can be created and solutions can be planned.

Supply Chain Management Problems

Distribution Network Configuration

Number and location of suppliers, production facilities, distribution centers, warehouses and customers.

Distribution Strategy

Centralized versus decentralized, direct shipment, Cross docking, pull or push strategies, third party logistics.

Information

Integration of systems and processes through the supply chain to share valuable information, including demand signals, forecasts, inventory and transportation etc.

Inventory Management

Quantity and location of inventory including raw materials, work-in-process and finished goods.

Cash-Flow

Arranging the payment terms and the methodologies for exchanging funds across entities within the supply chain.

Activities/Functions

Supply chain management is a cross-functional approach to managing the movement of raw materials into an organization, certain aspects of the internal processing of materials into finished goods, and then the movement of finished goods out of the organization toward the end-consumer. As organizations strive to focus on core competencies and becoming more flexible, they have reduced their ownership of raw materials sources and distribution channels. These functions are increasingly being outsourced to other entities that can perform the activities better or more cost effectively.

Several models have been proposed for understanding the activities required to manage material movements across organizational and functional boundaries. SCOR is a supply chain management model promoted by the Supply Chain Management Council. Another model is the SCM Model proposed by the Global Supply Chain Forum (GSCF).

Supply Chain Activities can be Grouped into Strategic, Tactical, and Operational Levels of Activities

Strategic

- ✰ Strategic network optimization, including the number, location, and size of warehouses, distribution centers and facilities.
- ✰ Strategic partnership with suppliers, distributors, and customers, creating communication channels for critical information and operational

improvements such as cross docking, direct shipping, and third-party logistics.

- Product design coordination, so that new and existing products can be optimally integrated into the supply chain, load management
- Information Technology infrastructure, to support supply chain operations.
- Where-to-make and what-to-make-or-buy decisions
- Aligning overall organizational strategy with supply strategy.

Tactical

- Sourcing contracts and other purchasing decisions.
- Production decisions, including contracting, locations, scheduling, and planning process definition.
- Inventory decisions, including quantity, location, and quality of inventory
- Transportation strategy, including frequency, routes, and contracting.
- Benchmarking of all operations against competitors and implementation of best practices throughout the enterprise.
- Milestone payments

Operational

- Daily production and distribution planning, including all nodes in the supply chain.
- Production scheduling for each manufacturing facility in the supply chain (minute by minute).
- Demand planning and forecasting, coordinating the demand forecast of all customers and sharing the forecast with all suppliers.
- Sourcing planning, including current inventory and forecast demand, in collaboration with all suppliers.
- Inbound operations, including transportation from suppliers and receiving inventory.
- Production operations, including the consumption of materials and flow of finished goods.
- Outbound operations, including all fulfillment activities and transportation to customers.
- Order promising, accounting for all constraints in the supply chain, including all suppliers, manufacturing facilities, distribution centers, and other customers.

Supply Chain Management

In Peter Drucker's (1998) management's new paradigms, this concept of business relationships extends beyond traditional enterprise boundaries and seeks to organize entire business processes throughout a value chain of multiple companies.

However, with the complicated interactions among the players, the network structure fits neither "market" nor "hierarchy"categories (Powell, 1990).

Traditionally, companies in a supply network concentrate on the inputs and outputs of the processes, with little concern for the internal management working of other individual players. Therefore, the choice of an internal management control structure is known to impact local firm performance (Mintzberg, 1979).

Supply Chain Business Process Integration

Successful SCM requires a change from managing individual functions to integrating activities into key supply chain processes. An example scenario: the purchasing department places orders as requirements become appropriate. Marketing, responding to customer demand, communicates with several distributors and retailers, and attempts to satisfy this demand. Shared information between supply chain partners can only be fully leveraged through process integration.

- Customer relationship management
- Customer service management
- Demand management
- Order fulfillment
- Manufacturing flow management
- Supplier relationship management
- Product development and commercialization
- Returns management

Other key critical supply business processes combining these processes stated by Lambert such as:

a) Customer service management
b) Procurement
c) Product development and commercialization
d) Manufacturing flow management/support
e) Physical distribution
f) Outsourcing/partnerships
g) Performance measurement

a) Customer Service Management Process

Customer Relationship Management concerns the relationship between the organization and its customers. Customer service provides the source of customer information. It also provides the customer with real-time information on promising dates and product availability through interfaces with the company's production and distribution operations. Successful organizations use following steps to build customer relationships:

- ☆ Determine mutually satisfying goals between organization and customers
- ☆ Establish and maintain customer rapport
- ☆ Produce positive feelings in the organization and the customers

b) Procurement Process

Strategic plans are developed with suppliers to support the manufacturing flow management process and development of new products. In firms where operations extend globally, sourcing should be managed on a global basis. The desired outcome is a win-win relationship, where both parties benefit, and reduction times in the design cycle and product development are achieved. Also, the purchasing function develops rapid communication systems, such as electronic data interchange (EDI) and Internet linkages to transfer possible requirements more rapidly. Activities related to obtaining products and materials from outside suppliers requires performing resource planning, supply sourcing, negotiation, order placement, inbound transportation, storage, handling and quality assurance, many of which include the responsibility to coordinate with suppliers in scheduling, supply continuity, hedging, and research into new sources or programmes

c) Product Development and Commercialization

Here, customers and suppliers must be united into the product development process, thus to reduce time to market. As product life cycles shorten, the appropriate products must be developed and successfully launched in ever shorter time-schedules to remain competitive. According to Lambert and Cooper (2000), managers of the product development and commercialization process must:

d) Manufacturing Flow Management Process

The manufacturing process is produced and supplies products to the distribution channels based on past forecasts. Manufacturing processes must be flexible to respond to market changes, and must accommodate mass customization.

e) Physical Distribution

This concerns movement of a finished product/service to customers. In physical distribution, the customer is the final destination of a marketing channel, and the availability of the product/service is a vital part of each channel participant's marketing effort.

f) Outsourcing/Partnerships

This is not just outsourcing the procurement of materials and components, but also outsourcing of services that traditionally have been provided in-house. The logic of this trend is that the company will increasingly focus on those activities in the value chain where it has a distinctive advantage and everything else it will outsource. This movement has been particularly evident in logistics where the provision of transport, warehousing and inventory control is increasingly subcontracted to specialists or logistics partners.

g) Performance Measurement

Experts found a strong relationship from the largest arcs of supplier and customer integration to market share and profitability. By taking advantage of supplier capabilities and emphasizing a long-term supply chain perspective in customer relationships can be both correlated with firm performance.

According to experts internal measures are generally collected and analyzed by the firm including

- ✰ Cost
- ✰ Customer Service
- ✰ Productivity measures
- ✰ Asset measurement, and
- ✰ Quality.

Supply Chain Management Components Integration

The Management Components of SCM

- ✰ The SCM components are the third element of the four-square circulation framework. The level of integration and management of a business process link is a function of the number and level, ranging from low to high, of components added to the link (Ellram and Cooper, 1990; Houlihan, 1985).
- ✰ Planning and control
- ✰ Work structure
- ✰ Organization structure
- ✰ Product flow facility structure
- ✰ Information flow facility structure
- ✰ Management methods
- ✰ Power and leadership structure
- ✰ Risk and reward structure
- ✰ Culture and attitude

Baziotopoulos (2004) suggests the following supply chain components:

For Customer Service Management

Includes the primary level component of customer relationship management, and secondary level components such as benchmarking and order fulfillment.

For Product Development and Commercialization

Includes the primary level component of Product Data Management (PDM), and secondary level components such as market share, customer satisfaction, profit margins, and returns to stakeholders.

For Physical Distribution, Manufacturing Support and Procurement

Includes the primary level component of enterprise resource planning (ERP), with secondary level components such as warehouse management, material management, manufacturing planning, personnel management, and postponement (order management).

For Performance Measurement

Includes the primary level component of logistics performance measurement, which is correlated with the information flow facility structure within the organization. Secondary level components may include four types of measurement such as: variation, direction, decision and policy measurements. More specifically, in accordance with these secondary level components, total cost analysis (TCA), customer profitability analysis (CPA), and asset management could be concerned as well.

For outsourcing

Includes the primary level component of management methods, and the strategic objectives for particular initiatives in key areas of information technology, operations, manufacturing capabilities, and logistics (secondary level components).

References

Haag, S., Cummings, M., McCubbrey, D., Pinsonneault, A., and Donovan, R. (2006). Management Information Systems for the Information Age (3rd Canadian Ed.), Canada: McGraw Hill Ryerson.

Handfield and Bechtel, 2001; Prater *et al.*, 2001; Kern and Willcocks, 2000. Bowersox and Closs, 1996; Christopher, 1992; Bowersox, 1989.

Kaushik K.D., and Cooper, M. (2000). Industrial Marketing Management. Volume29, Issue 1, January 2000, Pages 65-83

Chapter 19

Micro Planning and Network Analysis in Enterprise Management

Manas Mohan Adhikary and Sankar Kumar Acharya

The 'myths' and 'realities' of green revolution, as experienced by the soil and society of India, opened a new chapter of discourses and some which could be termed as 'grand discourses'. The so-called scale neutrality of technology intervention had come out to be grim reality for making "rich, richer and driving the poor, poorer". The decade of eighty's went on enthralling the philanthropic intellects all over the world on devising means and methodologies of technology generation vis-à-vis rural development process purported to the cause and confidence of the poorest of the poor. Robert Chambers (*Putting the last first*, 1983) came down heavily on the pernicious flamboyancy of rural development tourism, which to his mind had and had been developed and directed self-deceptive approaches in the conventional rural development programmes.

In search of a 'socialized methodology', imbued with humane techniques, social scientists endeavored time and again to redefine and refine rural peoples pristine commend on folk knowledge, wisdom, experience, intrinsic ability of estimation properties of process and matter in a given rural system. This had been only a beginning, across the globe on the quest of exploring and esteeming time tested methodology, have been nurturing, evolving within the people themselves.

Redefined Participation in Planning Process

The concept of participation's had a long way to evolve and mutate over the myriad variation of social context. People's participation in decision making process and their empowerment in the form of creating their own plans had been conceptualized in terms of availing *social choices, entitlement to resources, capability building and realizing empowerment.*

The conventional idea of participation may pay off little and sometimes go coercive instead of creating real benefit to the people. Prof. A. Sen (*Poverty and Famine*, 1984) while defining "real democracy", he stressed on people's participation resulting in building of capability among poorest of the poor to access social privileges. In his concept of techniques of social choices, he basically implied on generation of skills and capability with a view to attaining and upgrading quality of life.

The ultimate goal of providing people's participation in the programme is imparting the community with shills, strategies, opportunities and support to enhance its own functioning towards self-sufficiency and self-development. In concrete terms, the goal of people's participation could be helping people to build their own organization and institution equipped with an capable acceleration their own development.

With this backdrop, the importance and implication of micro-planning can be perceivable in terms of its methodology, techniques, objectives and goals and the following will keep on delineating some of the alternatives of micro planning towards people's participation in the planning process.

Micro-Planning

Micro-planning is basically a participatory planning process for a relatively smaller geographical distribution in order to accommodate and address highest possible micro-variation.

The earlier plans (1950 onwards), as developed and implemented, had a clear dent for blanket coverage (CDP, IADP, HYVP), no matter whether they had been destined for a comprehensive rural development or agricultural revolution. The history and evolution of planning process have got both the socio-political and techno-economic contexts. The limitations of conventional planning approach are:

Top-down approach (carried with the idea of trickledown effect)Blanket coverage –smaller, tiny and trivial variations were purposefully ignored; Subject to certain biases on: Resource, project, people, season; Excessive reliability on secondary sources of information; No room for objective evaluation – multiple realities were not esteemed.

1. Domination of bureaucracy and hence, every likelihood of red-tapism;
2. Planning has been a long and protractile process;
3. No such scope for making people empowered.

Advantages of Micro-Planning Process

Micro-planning is a historical development in the entire process of planning. The success of micro-planning depends on the following factors:

1. Organizing group,
2. Building goal clarity,
3. Experience profile of the participants
4. Climate setting
5. Group dynamics and group communication.

6. Planning skill
7. Resolving ability and leadership
8. Motivation.
9. Team building
10. Mutual faith and confidence.

Steps in Micro-Planning

1. Unfreezing and generation of information;
2. Assigning roles,
3. Analysis of problem and deriving its nature, distribution and impact;
4. Prioritization of problems;
5. Formulation of objectives based on needs of the people;
6. Designing intervention plans;
7. Facilitating the implementation;
8. Participatory evaluation;
9. Micro-level workshops and resetting objectives.

Some Tools and Techniques of Micro Planning

The following are some of the tools and techniques for evaluating designing and implementing micro-plans.

- ☆ LFA (Logical Framework Analysis)
- ☆ ASA (A Systematic Approach)
- ☆ Net work analysis
- ☆ Stake holder analysis

Activity Planning and Network Analysis

Every rural development project has to be implemented with the help of a time schedule and managed effectively to minimize cost escalation and avoid time overruns. The planning of these activities in the project life-cycle is know as implementation planning. In the process of management it is generally served by functions of organizing, staffing and directing. A proper implementation planning exercise helps developing a timetable for efficient management of the project.

Activity Planning

The first participatory management function is planning. Therefore, once a project has been accepted for implementation, it becomes imperative to plan for each activity of the project. The planning is carried out in a sequential manner. The steps are given below :

Step 1

Prepare a list of all activities that need to be executed under the project. Sometimes a project may require the help of a number of departments/sectors/agencies for its execution. It is most desirable to list the activities according to agencies or departments or sectors. This list will provide the Work Breakdown Structure.

Step 2

Specify a logical sequence in which the activities need to be executed. This will indicate (1) the order in which activities will follow; and (2) activities, which can be undertaken simultaneously.

Step 3

Estimate the specific duration of time required for each activity in the project. If you have executed similar projects in the past under identical conditions, you can specify the duration based on past experience. Otherwise, estimate the duration, using your judgment and statistical measures.

Step 4

Assemble the activities in the form of a flow diagram. This is known as Network Plan.

Step 5

Analyze the flow diagram. This is called Network Analysis.

The first step is to list the various activities as given below:

- ✰ Project plant approval
- ✰ Recruitment of staff
- ✰ Purchase of machinery
- ✰ Obtaining power supply line
- ✰ Training of staff
- ✰ Installing machinery
- ✰ Electrical wiring
- ✰ Point run

Coding of activities

Activity Codes	Activity Description
A	Project Plan Approval
B	Staff Recruitment
C	Machinery Purchase
D	Obtaining of Power Supply Line
E	Training of Staff
F	Installation of Machinery
G	Electrical Wiring
H	Pilot Run

The next step is to specify a logical sequence in which the activities need to be undertaken. For this we need to consider each activity separately and examine whether it has to wait for any other activity in the package. Then we will indicate the immediate predecessor or previous activity in a separate column.

Time Optimization

Development of Network Analysis (Especially CPM and PERT)

CPM was developed in 1956 at the E.I. du Pont Nemours and Co., USA to aid in the scheduling of routine plant overhaul, maintenance and construction work. This method differentiates between planning and scheduling.

Programme Evaluation and Review Technique or PERT was first developed in 1958 for use in defence projects specifically in the development of Polaris fleet ballistic missile programme. But now this technique is very popular in the hands of project planner land controller. PERT, now assists a business manager in planning and controlling a project. It allows a manager to calculate the expected total amount of time, the entire project will take to complete, at the stage of formulation and planning a project and at the same time highlights the critical or the bottleneck activities in the project.

Some Basic Terms

Network

A network is a series of related activities which result in some product (or services) which in turn contributes to the goals of a business or industrial or similar other organization.

Activity and Event

An activity represents some action and as such is a time consuming effort necessary to complete a particular part of the overall project. An event represents the start or completion of some activity and as such consumes no time. Hence, an event is a specific instant of time that denotes the beginning and end of an activity.

Network Diagram

Network diagram is a pictorial presentation of the various events and activities concerning a project. In it each event is represented by a circle and arrows are used to indicate activities.

Merge and Burst Events

It may be pointed out that it is not necessary for an event to be the ending event of only one activity but an event can be the ending event of two or more activities in which case the said event is technically described as *Merge event*. Similarly, if the event happens to be the beginning event of two or more activities it is technically called as the '*Burst event*'.

Preceeding, Succeeding and Concurrent Activities

Activities which must be accomplished before a given even can occur are termed as *Preceding activities*; activities which cannot be accomplished until an event has occurred are termed as *Succeeding activities* and activities which can be accomplished concurrently are known as con-current activities.

Dummy Activities

Some times we use Dummy activities in the preparation of network diagram. Such activities are to designate a precedence relationship and in the network diagram are shown as broken lines.

Path and Critical Path

A path is continuous chain of activities through a network, which connects the first event to the last event. Critical path consists of the sequence of those events and connected activities that require the maximum time in the completion of the project.

Critical Activities

All the activities associated with the critical path are called as critical or bottleneck activities.

Eartliest Star Time (EST)

EST for an activity is the earliest possible time an activity can begin on the assumption that all activities preceding to it started at the earliest possible times.

Eartliest Finish Time (EFT)

EFT is the sum of the earliest start time and the estimated time to perform the concerning activity.

Latest Finish Time (or LFT)

LFT for an activity is the latest possible time an activity can finish without delaying the project beyond its dead line on the assumption that all the subsequent activities are performed as planned.

FLOAT (Total Interfering, Independent and Free FLOATS)

Float may be understood as total float, interfering float, free float and independent float. Total float is the duration by which an activity can be delayed without delaying the project and can be worked out as either (LST.-EST.) or (LFT.-EFT.). Interfering float: is that part of the total float, which causes a reduction in the float of the successor activity or activities.

Free float is that portion of the total float within which an activity can be manipulated without affecting the float of subsequent activities.

Slack

The term 'Slack' normally associated with events.

Slack of an event = (latest occurrence time of the event – earliest occurrence of the event or (LT – ET).

Network Analysis

Activity Event

Conducting a survey to identify beneficiaries Survey conducted

Selection of sites for sinking bore well Sites identified

Appointment of staff appointed

The time taken by each set of activities along a path is just the sum of the durations of the individual activities. It can be written as below

Path	Time taken
A-F-I	7+5+10=22
B-E-I	6+3+10=19
C-D-E-I	8+0+3+10=21
C-G-H-I	8+4+5+10=27
C-G-J	8+4+16=28

The activities along all the paths will have to be completed for the project to be completed. The completion time of the project corresponds to the path which taken the longest time

Event 1: E = 0 fixed

Event 2: Only activity A terminates here, Therefore the EFT of A=0+7=7

Event 3: Activity C terminates, As above, E3 = 8

Event 4: B and D terminate here

EFT of B = 0+6=6

Eft of D = 8 Earliest event time for event 4 will be the larger of the two and E4 = 8

Event 5: Only G terminates with EFT = 8+4 = 12

E5 = 12

Event 6: Activities F, E and H terminate here

EFT of F = 7+5 = 12

EFT of E = 8+3 = 11

EFT of H = 12+5 = 17

The largest is 17 and E6 = 17

Event 7: Activities I and J terminate here

EFT of I = 17+10 = 27

EFT of J = 12+16 = 28

E7 = 28

Event 7: L=E=28 fixed

Event 6: Only activity I emanates from here Latest event time L6=18

Event 5: Activity H and J emanate from here

LST of H = 18-5=13

LST of J = 28-16=12

The lower of the two is 12 and L4 = 12

Event 4: Only activity E emanates from here

LSt of E = 18-3=15, L4 = 15

Event 3: Activities D and G emanate from here

LST of D = 15 – 0 = 15

LST of G = 12 - 4 = 8

L3 = 8

Event 2: Activity F emanates

LST of F = 18-5=13

L2 = 13

Event 1: Activities A, B and C emanate

LST of A = 13-7=6

LST of B = 15-6 = 9

LST of C – 8-8 = 0

L1 = 0

Chapter 20

Project Life Cycle and Market Led Extension: An Overview

Kaushik Pradhan

Part I: Project Life Cycle

A project is a unique endeavour to produce a set of deliverables within clearly specified constraints of time, cost and quality. Projects are different from business operations in terms of uniqueness, time scale, budget, resources, risk and change.

- Uniqueness: Every project is different from the last, whereas operational activities typically involve repetitive (if not identical) processes.
- Timescale: A project has clearly specified start and end dates within which deliverables are produced to meet the customers' requirements.
- Budget: A project has a maximum limit to the expenditure within which the deliverables must be produced, to meet the customers' requirement.
- Resource: A project is allocated a specified amount of labour, equipment and materials at the start.
- Risk: A project entails uncertainty and therefore carries business risk.
- Change: The purpose of the project is typically to improve an organisation throughout the implementation of the business change.

Why organize an activity or job as a project?

- It allows to better structure and organize the tasks that need to be performed
- Well developed approaches and tools are available for managing projects
- Easy-to-use software is available for scheduling and budgeting projects
- Experience has shown that the work/job can be done faster, cheaper, and better when managed as a project.

A project life cycle is a collection of project phases that defines:

- What work will be performed in each phase.
- What deliverables will be produced and when.
- Who is involved in each phase.
- How management will control and approve work produced in each phase.

The way in which projects are planned and carried out follows a sequence that has become known as the project cycle. The cycle starts with the identification of an idea and develops that idea into a working plan that can be implemented and evaluated. Ideas are identified in the context of agreed strategies. It provides a structure to ensure that stakeholders are consulted and relevant information is available so that informed decision can be made at key stages in the life of a project. The generic project cycle has six phases namely programming, identification, formulation, financing, implementation and evaluation. The details of what occurs during each phase differ between institutions reflecting differences in procedures. However, within all institutions the cycle shares three common themes:

- The cycle defines the key decisions, information requirements and responsibilities at each phase.
- The phases in the cycle are progressive – each phase needs to be completed for the next to be tackled with success.
- The cycle draws on evaluation to build experience from existing projects into the design of future programmes and projects.

The phases of the project life cycle can be described as follows:

Programming

During the Programming phase, the situation at national and sectoral level is analysed to identify problems, constraints and opportunities which development cooperation could address. This involves a review of socio-economic indicators, and of national and donor priorities. The purpose is to identify and agree the main objectives and sectoral priorities for development cooperation, and thus to provide a relevant and feasible programming framework within which projects can be identified and prepared. For each of these priorities strategies will be formulated that take account of the lessons of past experience.

Identification

During the Identification phase, ideas for projects and other development actions are identified and screened for further study. This involves consultation with the intended beneficiaries of each action, an analysis of the problems they face, and the identification of options to address these problems. A decision can then be made on the relevance of each project idea (both to the intended beneficiaries and to the programming framework), and on which ideas should be further studied during the Formulation phase.

Formulation

During the Formulation phase, relevant project ideas are developed into operational project plans. Beneficiaries and other stakeholders participate in the detailed specification of the project idea that is then assessed for its feasibility (whether it is likely to succeed) and sustainability (whether it is likely to generate long- term benefits for the beneficiaries). On the basis of this assessment, a decision is made on whether to draw up a formal project proposal and seek funding for the project.

Financing

During the Financing phase, project proposals are examined by the funding agency, and a decision is taken on whether to fund the project. The funding agency and partner country agree the modalities of implementation and formalise these in a legal document which sets out the arrangements by which the project will be funded and implemented.

Implementation

During the Implementation phase, the project is mobilised and executed. This may require the tendering and award of contracts for technical assistance or works and supplies. During implementation, and in consultation with beneficiaries and stakeholders, project management assesses actual progress against planned progress to determine whether the project is on track towards achieving its objectives. If necessary the project is re-oriented to bring it back on track, or to modify some of its objectives in the light of any significant changes that may have occurred since its formulation.

Evaluation

During the Evaluation phase, the funding agency and partner country assess the project to identify what has been achieved, and to identify lessons that have been learned. Evaluation findings are used to improve the design of future projects or programmes. Although in the generic cycle the evaluation phase comes after implementation, it is common practice also to conduct a mid-term evaluation during implementation, to identify lessons that can be applied during the remaining life of the project.

The separation of the cycle into six phases provides the minimum basis for effective project preparation, implementation and evaluation. The separation of the Identification and Formulation phases is particularly important. Project preparation takes place in a social and political context, where expectations are raised and often-conflicting demands and aspirations must be reconciled. By adhering to the identification phase, the relevance of project ideas can be systematically established before the preparation process is too far advanced for the idea to be dropped. During the formulation phase, project ideas can then be fully developed in the knowledge that they are based on real beneficiary needs and are sufficiently 'owned' by the main stakeholders. In practice, the project cycle might differ according to the type of programme being operated. Nevertheless, it is very useful to reconcile the current practice within your area o f work with the steps of the project cycle as outlined here. The following blank table provides you with a framework to do so.

Aid cooperation and partnership programmes with non-member states involve often-complex processes that require the active support of many parties. Experience among the main donors has shown that too many decisions concerning projects have been taken without sufficient consultation with beneficiaries and stakeholders, and without the necessary information. The purpose of the project cycle is to ensure that the stakeholders take the decisions, and that decisions are based on relevant and sufficient information. As a result the project life cycle requires a proper identification of project ideas and stakeholders.

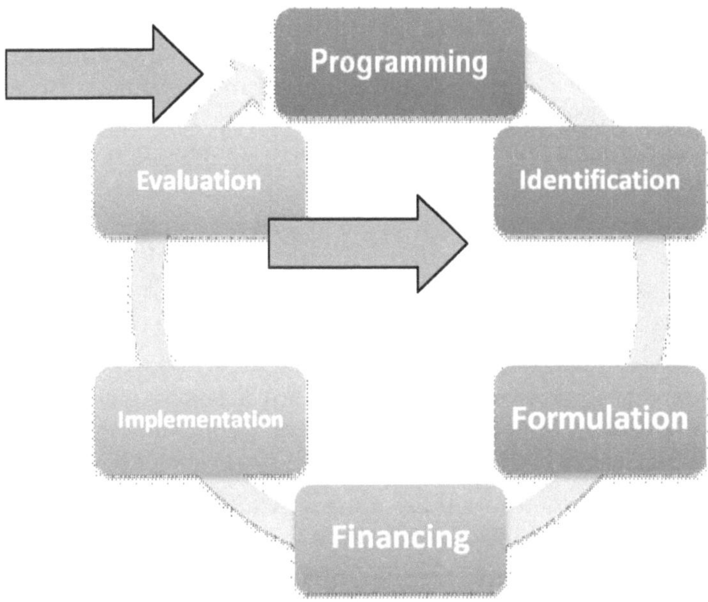

Aid cooperation and partnership programmes with non-member states involve often-complex processes that require the active support of many parties. Experience among the main donors has shown that too many decisions concerning projects have been taken without sufficient consultation with beneficiaries and stakeholders, and without the necessary information. The purpose of the project cycle is to ensure that the stakeholders take the decisions, and that decisions are based on relevant and sufficient information. As a result the project life cycle requires a proper identification of project ideas and stakeholders.

Selection of Project Idea

The project idea can be selected with the help of some numeric and non numeric methods. The non numeric methods of project selection are

1. Sacred Cow method (President's pet project),
2. Operating/competitive necessity method
3. Comparative benefits method.
4. Factor scoring method and
5. Weighted factor scoring method

The numeric methods are

1. Payback period method and discounted cash flow method (NPV).

Stakeholder Analysis

Stakeholder Analysis (SA) is a simple and important tool to identify the stakeholders related with the project building, implementation and its carrying over from a short term tenure to a protracted period. A stakeholder is any person/s or institutions/s who has/have a stake (interest) in any process or project. The stakeholders are of primary, secondary and Tertiary in nature and approach. The respective roles, contribution and impacts of stakeholders can be accessed through a structured format. But, one facilitator should go careful and meticulous in exercising this format to identify different stakeholders from a target group or community. Triggering of quality group dynamics as well as group communication would certainly breed on fostering the process.

Stakeholder analysis is a term that refers to the action of analyzing the attitudes of stakeholders towards something (most frequently a project). It is frequently used during the preparation phase of a project to assess the attitudes of the stakeholders regarding the potential changes. Stakeholder analysis can be done once or on a regular basis to track changes in stakeholder attitudes over time.

A stakeholder is any person or organization, who can be positively or negatively impacted by, or cause an impact on the actions of a company, government, or organization. Types of stakeholders are:

- ✰ Primary stakeholders: are those ultimately affected, either positively or negatively by an organization's actions.
- ✰ Secondary stakeholders: are the 'intermediaries', that is, persons or organizations who are indirectly affected by an organization's actions.
- ✰ Key stakeholders: (who can also belong to the first two groups) have significant influence upon or importance within an organization.

A stakeholder analysis can help a project identify:

- ✰ The interests of all stakeholders who may affect or be affected by the project;
- ✰ Potential conflicts or risks that could jeopardise the initiative;
- ✰ Opportunities and relationships that can be built on during implementation;
- ✰ Groups that should be encouraged to participate in different stages of the project;
- ✰ Appropriate strategies and approaches for stakeholder engagement; and
- ✰ Ways to reduce negative impacts on vulnerable and disadvantaged groups.

Stakeholder analysis is also an appropriate time to explore whether or not gender will be a factor in the elaboration and implementation of future efforts. It is well documented that discrimination by gender is likely to diminish the impact and effectiveness of projects and policies. Furthermore, the inclusion of women as stakeholders has the potential to achieve both better management of the resource base and improved community welfare. Gender analysis involves the assessment of:

- ✰ The distribution of tasks, activities, and rewards associated with the division of labour at a particular locality or across a region;
- ✰ The relative positions of women and men in terms of representation and influence;
- ✰ The benefits and disincentives associated with the allocation of tasks to women and men.

There are a number of ways of undertaking a stakeholder analysis. Workshops, focus groups and interviews are three common approaches. During the course of the project cycle you may use all three, matching the technique to the evolving needs of the project. Whatever approach is used, there are three essential steps in stakeholder analysis: 1) Identifying the key stakeholders and their interests (positive or negative) in the project; 2) Assessing the influence of, importance of, and level of impact upon each stakeholder; and 3) Identifying how best to engage stakeholders.

Stakeholder Analysis Matrix

Stakeholder	*Level*	*Activity/Role*	*Value (V)*	*Power (P)*	*Impact (VXP)*	*Rank*
	Primary					
	Secondary					
	Tertiary					

Stakeholder Analysis Grid

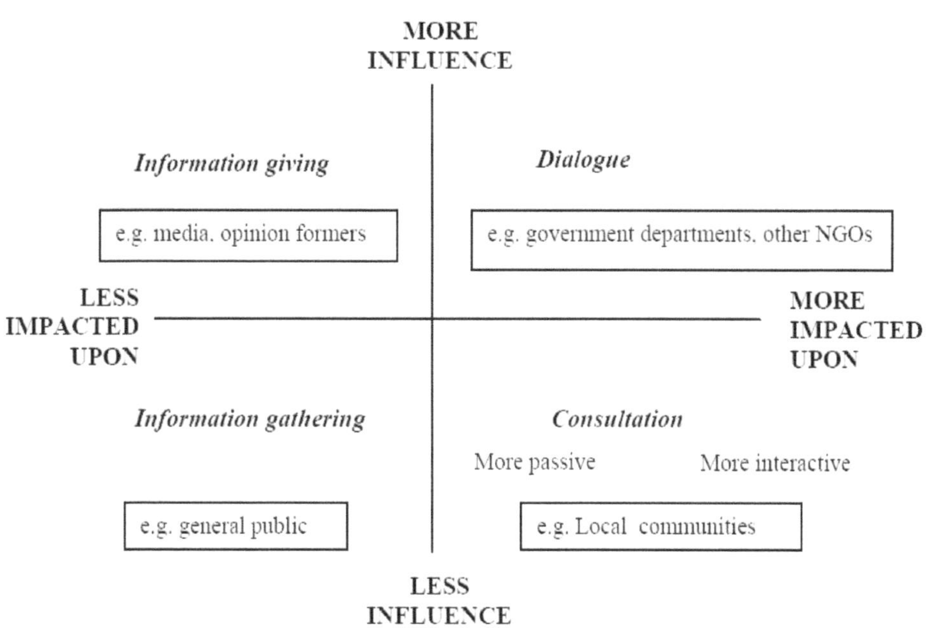

The Logical Framework Approach – A Project Design and Analysis Tool

The Logical Framework Approach is the main tool used for project design during the Identification and Formulation phases of the project life cycle. Using the LFA during Identification helps to ensure that project ideas are relevant, while during Formulation it helps to ensure feasibility and sustainability.

The approach is split into two phases:

- Phase 1 - the **Analysis Phase** during which the existing situation is analysed to develop a vision of the 'future desired situation' and to select the strategies that will be applied to achieve it
- Phase 2 – the **Planning Phase** during which the project idea will be developed in operational detail.

Analysis Phase

Projects are designed to address the problems faced by beneficiaries. A properly planned project addressing the real needs of the beneficiaries cannot be achieved without an analysis of the existing situation. There are three stages to the. Analysis Phase: Problem Analysis; Analysis of Objectives; Strategy Analysis.

Problem Analysis

Problem analysis identifies the negative aspects of an existing situation and establishes the 'cause and effect' relationships between the problems that exist. It involves three steps:

1. Identification of the stakeholders affected by the proposed project
2. Identification of the major problems faced by beneficiaries
3. Development of a problem tree to establish causes and effects

Analysis of Objective

While problem analysis presents the negative aspects of an existing situation, analysis of objectives presents the positive aspects of a desired future situation. This involves the reformulation of problems into objectives.

Strategy Analysis

The final stage of the analysis phase involves the selection of the strategy (ies) which will be used to achieve the desired objectives.

Planning Phase

The main output of the LFA is the log frame matrix. The log frame sets out the intervention logic of the project (if activities are undertaken, then results will be achieved, then project purpose, etc.) and describes the important assumptions and risks that underlie this logic. This provides the basis for checking the feasibility of the project. For management and supervision of projects, the logframe defines the tasks to be undertaken, the resources required, and the responsibilities of management. In the second and third columns (objectively verifiable indicators, and sources of verification), the logframe provides the framework against which progress will be monitored and evaluated.

Log Frame Matrix

The log frame matrix is the main output of the LFA. By bringing stakeholders together during the analysis phase, to discuss problems, objectives and strategies, the LFA encourages people to consider what are their own expectations, and how these might be achieved. By stating objectives clearly and setting them out in a 'hierarchy of objectives', it then provides a means of checking the internal logic of the project plan, ensuring that activities, results and objectives are linked. Planners are forced to identify the critical assumptions and risks that may affect project feasibility, and to specify the indicators and sources of information that will be used to monitor and evaluate the project. All of this key information is brought together in one document that provides a useful summary. The Logframe itself consists of a table, or matrix, which has four columns and (in its most basic form) four rows. The vertical logic identifies what the project intends to do clarifies the causal relationships and specifies the important assumptions and certainties beyond the project manager's control. The horizontal logic relates to the measurement of the effects of, and resources used by, the project through the specification of key indicators of measurement, and the means by which the measurement will be verified.

In the log frame matrix the horizontal logics are:

- ☆ **Goal** : The **Goal** of the project should explain why the programme is important to society, in terms of the longer-term benefits to beneficiaries and the wider benefits to other groups. It should also show how the programme fits into the regional/sectoral policies
- ☆ **Purpose** : The **Purpose** should address the core problem, and be defined in terms of the benefits to be received by the project beneficiaries or target group as a result of utilising the services provided by the project.
- ☆ **Output**: **Output** describes the services to be delivered to the intended beneficiaries or target group, and it should be possible for project management to be held accountable for their delivery. The results should address the main causes of the problems the target group faces. To ensure relevance of results, the problem analysis should therefore have identified a beneficiary demand for project services.
- ☆ **Activities** - how the project's goods and services will be delivered?

One of the keys to using the logframe successfully is understanding what the definitions mean in operational terms, and in particular the relationship between Outputs and Project Purpose.

The vertical logics of log frame matrix can be explained as:

Narrative Statements

These are the statements for the horizontal logics.

- **Objectively verifiable indicator (OVI):** Objectively Verifiable Indicators describe the project's objectives in operationally measurable terms, and provide the basis for performance measurement. The specification of OVIs acts as a check on the viability of objectives and forms the basis of the project monitoring system. Once the indicator has been identified, it should then be developed to include brief details of quantity, quality and time (QQT).
- **Means of verification (MOV):** When indicators are formulated, the source of information and means of collection should be specified. This will help to test whether or not the indicator can be realistically measured at the expense of a reasonable amount of time, money and effort. The MOV should specify the **format** in which the information should be made available (*e.g.* progress reports, project accounts, project records, official statistics etc.) who should provide the information.
- **Assumptions/Risks :** It will have become apparent during the Analysis Phase that the project alone cannot achieve all objectives identified in the objective tree. Once a strategy has been selected, objectives not included in the intervention logic and other external factors remain. These will affect the project's implementation and long-term sustainability but lie outside its control. These conditions must be met if the project is to succeed, and are included as assumptions in the fourth column of the Log frame.

Table 20.1 An Example of a Completed Log Frame

	Narrative Statement	*Objectively Verifiable Indicators*	*Means of Verification*	*Assumptions*
Goal	Infant and maternal mortality rates reduced	Mortality rates reduced for under-1s, under-5s and pregnant and nursing mothers from X to Y by 19xx	Dept of Health statistics, analysed ex-ante, mid-term and ex-post	
Purpose	Health status of pregnant and nursing mothers, infants and babies improved.	Incidence of post-partum and neo-natal infection within health centres reduced from X to Y by 19xx rates of infectious diseases (polio, measles, tetanus) among under-5s reduced from X to Y by 19xx Incidence of acute b irth complications reduced from X to Y by 19xx	Hospital and clinic records, analysed ex-ante, mid- term and ex-post Sample survey of target group conducted and analysed in years 1, 3 and 5	Incidence of infectious diseases in the household reduced.

Contd...

Table–*20.1 Contd...*

	Narrative Statement	*Objectively Verifiable Indicators*	*Means of Verification*	*Assumptions*
Outputs	Functioning primary healthcare service established at district level.	Number of villages provided with regular PHC services increased from X to Y by 19xx	Clinic attendance records, analysed quarterly	Mothers willing to attend clinics
		Proportion of under-5s vaccinated against polio, measles and tetanus increased from X to Y by 19xx No. birth complications diagnosed and successfully assisted increased from X to Y by 19xx	Clinic vaccination records, analysed quarterly	Household nutrition improved through increased seasonal availability of high protein foods, and increased proportion of breast-fed babies.
		Increased patient satisfaction with quality of services provided	Client satisfaction survey, conducted annually	Pregnant and nursing mothers able to access cash to pay for treatment.
Outputs	Quality and efficiency of secondary health-care improved.	Number of patients treated increased from X to Y by 19xx Average cost of treatment per patient reduced from X to Y by 19xx	Hospital records, analysed quarterly	
		Increased patient satisfaction with standards of care	Client satisfaction survey, conducted annually	
Activities	Rehabilitate/replace mobile clinic vehicles and equipment	Technical Assistance	Lump sum costs	Department of Health maintains level of funding at pre-project levels in real terms.
	Design and implement mobile clinic programme	Equipment	Reimbursables	Suitably qualified staffwilling to work in rural areas
	Recruit and train village birth attendants	Medical supplies		

Contd...

Table–*20.1 Contd...*

	Narrative Statement	Objectively Verifiable Indicators	Means of Verification	Assumptions
	Rehabilitate staff housing			
	Recruit new staff			
	Design and implement new procedures for: Personnel management Vehicle and equipment maintenance Drug storage and control Financial control and cost recovery			
	Design and implement staff training programme for: Patient care Hygiene maintenance Basic accounting Data collection and analysis			

Part II: Market Led Extension

Indian agriculture has made rapid progress in the last half century by augmenting the annual food grain production from 51 millions tones in the early fifties to 209 million tones in 1992-2000 and steered the country to a status of self sufficiency. It has been successful in keeping pace with the rising food demand of a growing population.

Food grain production quadrupled in the last 50yrs while population trippled from 350 million to billion during this period. Significantly the extension system had played its role untiringly in transfer of production technologies from lab to land besides the agricultural scientists, farmers and marketing network.

Though the production has increased dramatically, not so much bothered about remunerative prices, small and marginal farmers generally prone to sell produce on "as is where is basis" due to several constraints like repayment of personal hand loans and to meet domestic needs.

With globalization of market, farmers have to transform themselves from mere producers seller in domestic market to producers cum sellers in wider market sense to best realize the returns for his investments, risks and efforts. This to be achieved, farmers need to know answers to questions like what to produce, when to produce, how much to produce, when and where to sell, at what price and whom to sell his produce. Farmers received most of the production technologies from extension system. Extension system now needs to be oriented with knowledge and skills related to market.

An efficient marketing system is essential for the development of the agricultural sector. The marketing system contributes greatly to commercialization of subsistence farmers. Effective linkages of production system with marketing. Agro-processing and other value added activities would play an increasingly important role in the diversification of agriculture (MOA,GOI 2001)

The government provides much of the infrastructure required for efficient marketing. One of the most important is the information and extension services to farmers besides transport and communication facilities, public utility supply, like water, electricity, fiscal and trade administration and public storage, market and abattoir facilities (FAO, 2000).

Enhanced Roles of Agricultural Extension Personnel in Light of Market Led Extension

- **SWOT analysis of market**: Strengths (Demand, high market ability, good rice etc), Weaknesses (inadequate staff, poor skills and information) Opportunities (Export, appropriate time of selling etc) and Threats (imports and perishability of the produce etc.) need to be analyzed about the markets. Accordingly, the farmers need to be made aware of this analysis for planning of their production and marketing.
- Organization of Farmer's Interest Groups (FIGs) on commodity basis and building their capabilities with regard to management of their own farm enterprise.
- Enhancing interactive and communication skills of the farmers to exchange their views with customers and other market forces for getting feedback.
- Establishing marketing and agro-processing linkages between farmer's groups, markets and private processors.
- **Advice on product planning:** Selection of crops to be grown and varieties suiting the land holding and marketability of produce will be stating point of agri-enterprise.
- **Educating the farming community:** To treat agriculture as an entrepreneurial activity and accordingly plan various phases of crop production and marketing.
- **Direct marketing:** farmers need to be informed about the benefits of direct marketing.
- Capacity building of FIGs in terms improved production, post harvest operations, storage and transport and marketing.
- Acquiring complete market intelligence regularly on various aspects of marketing.
- Regular usage of internet facility through computers to get updated on market besides internet.

- **Organization of study tours of FIGs :** To the successful farmers/FIG's for various operations with similar socio-economic and farming systems as the farmers learn more from each other.
- Production of video films of success stories of community specific farmers.
- Creating of websites of successful FIGs in the field of agribusiness management with all the information to help other FIG achieve success.

Challenges to Market Led Extension

- Gigantic size of Public Extension System of the public extension in the country.
- Generation of data on market intelligence would be a huge task by itself.
- The present extension system suffers from several limitations of stationery, mobility, allowances, personal development of cadres etc.
- Infrastructure challenges.
- Storage and preservation of produce
- Minimization post harvest losses.
- Capacity building of extension personnel and farmers.

Suggestions to MLE

- Training programme for extension worker
- Basic concepts and issues in agriculture markets and marketing (and links w/agricultural extension)
- Collection, processing and dissemination of market information
- Helping farmers access markets: collective action, business linkages and other (based on needs assessment during mentoring phases)
- Development of local market and promoting local consumption
- Entrepreneurial training to farmers
- Linking farmers to markets
- Market reforms
- Formation of Market-Led Extension Policies and its implementation

References

Cameron, B.G., T. Seher, E.F. Crawley (2010). "Goals for space exploration based on stakeholder network value considerations." in: *Acta Astronautica* doi:10.1016/j.actaastro.2010.11.003.

Design, monitoring and evaluation of technical cooperation programmes and projects. A training manual. International Labour Office, Bureau of Programming and Management.

Development Co-operation of the European Union and the Process of Enlargement. Brussels, 2003.

Evaluation and Aid Effectiveness. No. 1 - Guidance for Evaluating Humanitarian Assistance in Complex Emergencies. OECD/DAC - Working Party on Aid Evaluation, 1999.

Evaluation and Aid Effectiveness. No. 3 - Donor Support for Institutional Capacity Development in Environment: Lessons Learned. OECD/DAC - Working Party on Aid Evaluation, 2000.

Evaluation and Aid Effectiveness. No. 5 - Evaluation Feedback for Effective Learning and Accountability. OECD/DAC - Working Party on Aid Evaluation, 2001.

Evaluation and Aid Effectiveness. No. 6 - Glossary of Key Terms in Evaluation and Results Based Management. OECD/DAC - Working Party on Aid Evaluation, 2002.

Evaluation Guidelines. Ministry of Foreign Affairs, DANIDA, Denmark, 1999

Fletcher, A., *et al*. (2003). "Mapping stakeholder perceptions for a third sector organization." in: *Journal of Intellectual Capital* 4(4): 505 – 527.

Guide de l´évaluation. Ministère des Affaires étrangères. France, 2003

Guide to the NGO Project Facility. CIDA, Canadian Partnership Branch, NGO Division, 2001.

Guidelines for Completing the Project Concept Document. World Bank. 1999.

Guidelines for Evaluators including the Project Evaluation Information Sheet. UNDP, World Development, Central Evaluation Office. August 1993.

Guidelines for Project Formulation and the Project Document Format. UNDP. 1988.

Hemmati, M., Dodds F., Enayti, J.,McHarry J. (2002). "Multistakeholder Procesess on Governance and Sustainability. London Earthscan

How to Perform Evaluations. 5 - Evaluation Workplans. CIDA, Performance Review Branch, 2000.

How to Perform Evaluations. 7 - Evaluation Reports. CIDA, Performance Review Branch, 2002.

Mendelow, A. (1991). 'Stakeholder Mapping', Proceedings of the 2nd International Conference on Information Systems, Cambridge, MA (Cited in Scholes,1998).

Mitchell, R. K., B. R. Agle, and D.J. Wood. (1997). "Toward a Theory of Stakeholder Identification and Salience: Defining the Principle of Who and What really Counts." in: *Academy of Management Review* 22(4): 853 - 888.

Savage, G. T., T. W. Nix, Whitehead and Blair. (1991). "Strategies for assessing and managing organizational stakeholders." In:*Academy of Management Executive* 5(2): 61 – 75.

Turner, J. R., V. Kristoffer, *et al.*, Eds. (2002). *The Project Manager as Change Agent.* London, McGraw-Hill Publishing Co.

Weaver, P. (2007). *A Simple View of Complexity in Project Management. Proceedings of the 4th World Project Management Week. Singapore.*

www.manage.gov.in

Chapter 21

Decision Support System (DSS) in Transforming Extension Education in India and World towards Enterprise Selection

Manas Mohan Adhikary and Sankar Kumar Acharya

The Basic Concepts

A **decision support system** (**DSS**) is a computer-based information system that supports business or organizational decision-making activities. DSSs serve the management, operations, and planning levels of an organization and help to make decisions, which may be rapidly changing and not easily specified in advance.

DSSs include knowledge-based systems. A properly designed DSS is an interactive software-based system intended to help decision makers compile useful information from a combination of raw data, documents, personal knowledge, or business models to identify and solve problems and make decisions.

Typical information that a decision support application might gather and present are:

- Inventories of information assets (including legacy and relational data sources, cubes, data warehouses, and data marts),
- Comparative sales figures between one period and the next,
- Projected revenue figures based on product sales assumptions.

DSS help executives make better decision by using historical and common data from internal information systems and external sources.By combining massive amounts of data with sophiscated analytical models and tools, and by making the system easy to use,they provide much better source of information to use in decision making process.DSS are a class of computerised information system that support decision making activities.DSS are interactive computer-based system and subsystem intended to help decision makers use communications technologies, data, documents, knowledge or models to successfully complete decision process task.

Decision Support System (DSS)

A decision support system or DSS is a computer based system intended for use by a particular manager or usually a group of managers at any organizational level in making a decision in the process of solving a semi structured decision (Figure 21.1). The DSS produces output in the form of periodic or special report or the results of mathematical simulations (Raymond, 1990). It is difficult to pinpoint that are completely structured or unstructured. The vast majorities are semi structured. This means that the DSS is aimed at the area where most semi structured decision is needed to be made.

The concept of DSS is extremely broad, and its definitions vary, depending on the author's point of view. To avoid exclusion of any of the existing types of DSSs, we will define them roughly as interactive computer-based systems that aid users in judgment and choice activities. Another name sometimes used as a synonym for DSS is knowledge-based systems, which refers to their attempt to formalize domain knowledge so that it is amenable to mechanized reasoning. For the last twenty years, different kinds of information systems are developed for different purposes, depending on the need of the business. Transaction Process Systems (TPS) function in operational

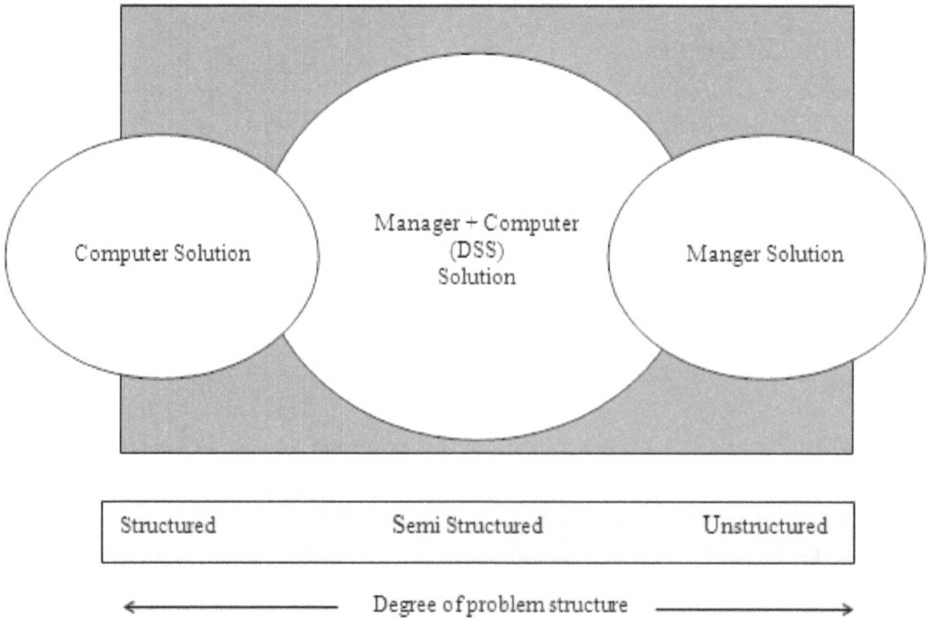

Figure 21.1: The DSS Focuses on Semi Structured Problems (Raymond, 1990)

level to process large amount of data for routine business transactions of the organization, Office Automation Systems (OAS) support data workers and Knowledge Work Systems (KWS) support professional workers. Higher-level systems include Management Information Systems (MIS) and Decision Support Systems (DSS). Expert System (ES) applies the expertise of decision makers to solve specific, unstructured problems. At the strategic level of management, there is Executive Support Systems (ESS). Group Decision Support Systems (GDSS) and the more generally described Computer Supported Collaborative Work (CSCW) systems aid group level decision making of a semi structured or unstructured decision. In the present article the authors discuss two kinds of information systems, namely, MIS, and DSS, and then their characteristics, interrelationship and their relations with decision-making process in an organization. Computerized decision support systems became practical with the development of minicomputers, timeshare operating systems and distributed computing.

The history of the implementation of such systems begins in the mid-1960s. In a technology field as diverse as DSS, chronicling history is neither neat nor linear. Different people perceive the field of Decision Support Systems from various vantage points and report different accounts of what happened and what was important (cf., Arnott and Pervan, 2005; Eom and Lee, 1990b; McCosh and Correa-Perez, 2006; Power, 2003; Power, 2004a; Silver, 1991). As technology evolved new computerized decision support applications were developed and studied. Researchers used multiple frameworks to help build and understand these systems. Today one can organize the history of DSS into the five broad DSS categories explained in Power (2001; 2002; 2004b), including: communications-driven, data-driven, document driven, knowledge-driven and model-driven decision support systems. A variety of DSS have been postulated for describing thecharacteristics of DSS. These frameworks are helpful in organizing and identifying the relationships of DSS.

The identification of DSS applications is important in planning organization strategy for the deployment of information technology. DSS is defined as the use of computer to: (i) Assist managers with their decision process in semi-structure tasks; (ii) To support, rather than replace managerial judgment, and (iii) To improve the effectiveness of decision making rather than its efficiency. While definitions of e-government by various sources may vary widely, there is a common theme. E-government involves using information technology, and especially the Internet, to improve the delivery of government services to citizens, businesses, and other government agencies to interact and receive services from the federal, state or local governments twenty four hours a day, seven days a week. E-government involves the use of information and communication technologies (ICTs) to support government operations and provide government services.

Characteristics of DSS

The characteristics of the DSS are as follows :

- ☆ DSS focus on towards providing help in analyzing situations rather than providing right information in from of various types of reports.

- ☆ DSS is individual specific. Each decisions maker can incorporate his own perceptions about the problem and analyze its effect.
- ☆ DSS incorporate various mathematics, statistical and operations research models.
- ☆ DSS is only supportive in nature decisions makers still retain their supremacy. It does not thrust its outcomes on the decision maker.
- ☆ DSS is effective in providing assistance to solve semi-structured problems at all levels. It is used at first line, middle level and top level management.
- ☆ DSS needs an effective database management system. It is extensively uses database.
- ☆ DSS helps decision makers to carry out **'what- if'** analysis.

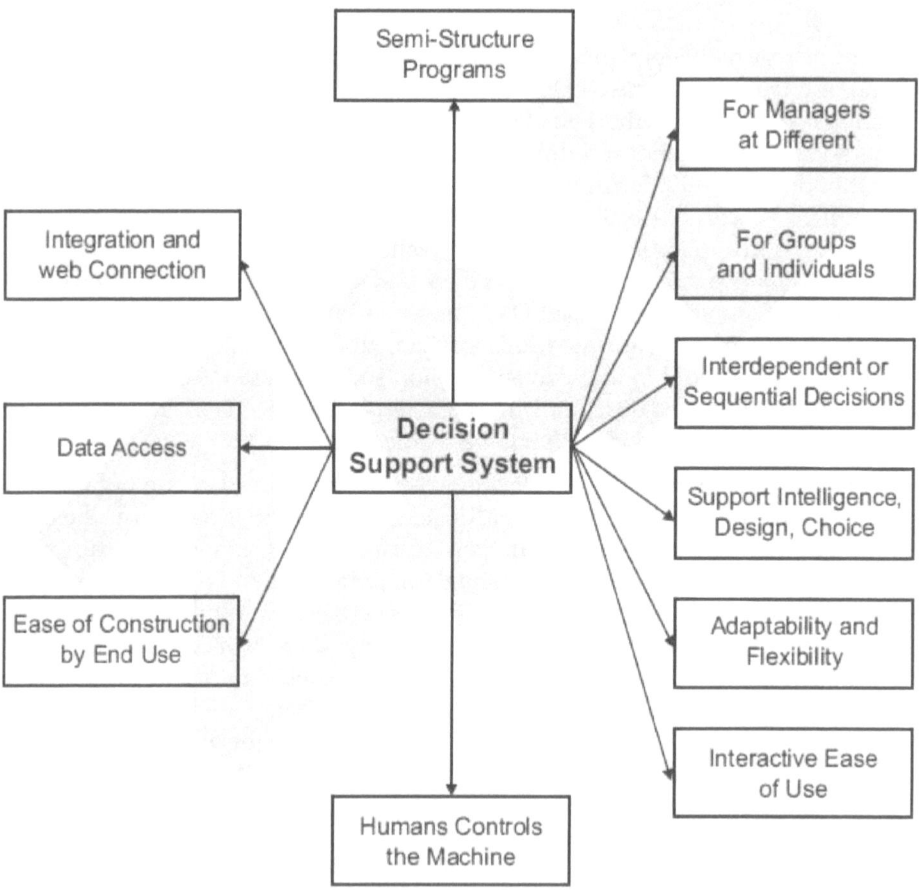

Operational Objectives of DSS

The objectives of the DSS are stated below:-

1. Provide assistance to decision makers in situations which are semi-structured.

2. Identify plans and potential actions to resolve problems.
3. Rank among the solution identified, those which can be implemented and provide list of viable alternatives.

Needs of DSS

DSS have become necessary for today's manager because of following reasons:-

1. **Fast computation:-** A decision maker can perform a large number of computations very quickly and that too at a low cost with the help of computer support systems.
2. **Enhanced productivity:-** Decision support system can enhance the productivity of support staff and also enable the group members to discuss the problems among themselves as a distance.
3. **Better decisions:-** Computer support system can help a decision-maker in arriving at a better decision. For examples, more alternatives can be evaluated, risk analysis be perform quickly, and views of experts from differ places can be collected quickly and at a lower cost.
4. **Data transmission:-** Sometimes the data, which may be stored at different locations, may be required to be transmitted quickly from distant locations. Computer support system can search, store and transmitted the required data quickly and economically.

DSS Benefits

That said, even if DSS do not necessarily result in quantifiable, tangible benefits, they do provide identifiable ones. The following list is derived from about 30 DSS studies. Only benefits mentioned in at least five case studies are included. A few typical illustrations or quotes are given for each category on the list:

1. Increase in number of alternatives examined
2. Better understanding of the business
3. Fast response to unexpected situations
4. Ability to carry out ad hoc analysis
5. New; insights and learning
6. Improved communication
7. Control
8. Cost savings
9. Better decisions
10. More effective teamwork
11. Time savings
12. Making better use of data resource

These categories add up to a concept of productivity. It is these often qualitative aspects of effectiveness that managers value. The operating assumption of Decision

Support is that improving communication, flexibility,' learning and responsiveness leads to better decision making.

DSS in Farm System

The model will help the farmers in increasing their productivity by raising the yield/hectare in food grains: thus, leading to their economic growth. This system has been developed to keep track of farmers all type of information related to crops.

Certain applications that are successfully developed using this database are:

- ✰ Farmers can manage their cash flow through the DSS system in a more predictable and efficient way. It is a more common problem with the farmers to manage the cash received at time of harvesting the crop.
- ✰ They can avail full benefit of their cash management by co-relating it with the loans and advances.
- ✰ The administrator can add information to the database without stopping the application.
- ✰ If implemented at Village, District and State level, the model will provide valuable information to other agencies and panchayats in particular.

Growing population and demands for improved watershed management, there is an obvious need to implement sustainable resource use that best serves the communities and the nation. To satisfy this need, the DSS is developed to aid decision – makers and various stakeholders in identifying and assessing options for resource uses. The DSS applies an integrative approach, combining biophysical data, perceptions and socioeconomic conditions of the farmers in the given area. The DSS attempts to stimulate the farmer's behaviour in selecting farming systems given relevant constraints and then aggregating up to the node.

A large number of database queries can be generated according to Crop, Water Availability and Requirement, Socio-economic constraints and so on. Design and Development of this database is purely based on Relation Database Management System Model, so the large volume of queries can be easily handled.

DSS for Agro-Technology Transfer

- ✰ The Decision Support System for Agro-technology Transfer (DSSAT) is a software package integrating the effects of soil, crop phenotype, weather and management options that allows users to ask "what if" questions and simulate results by conducting, in minutes on a desktop computer, experiments which would consume a significant part of an agronomist's career. It has been in use for more than 15 years by researchers in over 100 countries.
- ✰ DSSAT is a microcomputer software product that combines crop, soil and weather data bases into standard formats for access by crop models and application programs. The user can then simulate multi-year outcomes of crop management strategies for different crops at any location in the world.

- DSSAT also provides for validation of crop model outputs; thus allowing users to compare simulated outcomes with observed results. Crop model validation is accomplished by inputting the user's minimum data, running the model, and comparing outputs. By simulating probable outcomes of crop management strategies.
- DSSAT offers users information with which to rapidly appraise new crops, products, and practices for adoption. The release of DSSAT Version 4 incorporates changes to both the structure of the crop models and the interface to the models and associated analysis and utility programs.
- The DSSAT package incorporates models of 27 different crops with new tools that facilitate the creation and management of experimental, soil, and weather data files. DSSAT v4 includes improved application programs for seasonal and sequence analyses that assess the economic risks and environmental impacts associated with irrigation, fertilizer and nutrient management, climate change, soil carbon sequestration, climate variability and precision management.
- DSSAT is one of the principal products developed by the International Benchmark Sites Network for Agro-technology Transfer (IBSNAT) project supported by the U.S. Agency for International Development from 1983 to 1993. It has subsequently continued to be developed through collaboration among scientists from the University of Florida, the University of Georgia, University of Guelph, University of Hawaii, the International Centre for Soil Fertility and Agricultural Development, Iowa State University and other scientists associated with ICASA.

Minimum Data

The minimum data set (MDS) refers to a minimum set of data required to run the crop models and validate the outputs. Validation requires:

1. Site weather data for the duration of the growing season,
2. Site soil data, and
3. Management and observed data from an experiment.

A. MDS Weather Data

The required minimum weather data includes :

- Latitude and longitude of the weather station,
- Daily values of incoming solar radiation (MJ/m^2-day),
- Maximum and minimum air temperature (°C), and
- Rainfall (mm).

Accessory data sets, such as daily dry and wet bulb temperatures and wind speed, are optional. The period of weather records for validation must, at a minimum, cover the duration of the experiment and preferably should begin a few weeks before planting and continue a few weeks after harvest so that "what-if" type analyses may be performed as desirable.

B. MDS Soil Data

Desired soil data includes soil classification (*e.g.* USDA/NRCS), surface slope, soil colour, permeability, and drainage class. Soil profile data by soil horizons include:

- Upper and lower horizon depths (cm),
- Percentage sand, silt, and clay content,
- 1/3 bar bulk density,
- Organic carbon,
- Ph in water,
- Aluminium saturation, and
- Information on abundance of roots.

C. Management and Experiment/Observed Data

Management data includes information on planting date, dates when soil conditions were measured prior to planting, planting density, row spacing, planting depth, crop variety, irrigation, and fertilizer practices. This data are needed for both model validation and strategy evaluation. In addition to site soil and weather data, experimental data includes crop growth data, soil water and fertility measurements. This data are needed for model validation.

System Requirements

- Personal Computer running MS-Windows
- Pentium2 with 128MB RAM minimal
- Pentium4 with 512MB RAM recommended
- CD-Rom Drive
- Hard Disk Space requirement 300MB of free space

Components

The Cropping System Model (CSM) released with DSSAT Version 4 represents a major departure from previously released crop models in DSSAT, not in function, but in design. The computer source code for the model was restructured into a modular format in which components separate along lines of scientific discipline and are structured to allow easy replacement or addition of modules. CSM now incorporates all crops as modules using a single soil module and a single weather module. The new cropping system model now contains models of 17 crops derived from the old DSSAT CROPGRO and CERES models.

The major modules are:

- **Land Module**
- **Management Module**

- **Soil module** -A soil water balance sub-module and two soil nitrogen/ organic matter modules.
- **Weather module** - Reads or generates daily weather data
- **Soil-Plant-Atmosphere module** - Deals with competition for light and water among the soil, plants, and atmosphere
- **CROPGRO plant growth module**
  - Grain Legumes - Soybean, peanut, dry bean, chickpea, cowpea, velvet bean, and faba bean
  - Vegetables - Pepper, cabbage, tomato
  - Grassws-Bahia, brachiaria

Problems with DSS Development and Adoption

Although these systems seem to have many benefits to producers, they have not widely been taken up by them (Wilde, 1994; Lynch *et al.*, 2000). Cox (1996), Campbell (1999), and Lynch *et al.* (2000) have examined reasons behind the low adoption of DSS and other intelligent support systems within agriculture.

Some of the reasons they suggest are as follows:

- Limited computer ownership among producers,
- Lack of field testing,
- No end user input preceding and during development of the DSS,
- DSS complexity and possibly considerable data input,
- No reason seen for changing current management methods,
- Distrust for the output of a DSS because producers do not understand the underlying theories of the models,
- Mismatch of the DSS output with the decision-making style of the producer because the producer's conceptual models are excluded, and Unclear definition of the beneficiaries (*e.g.*, scientists, primary producers, and technology transfer agents).
- Availability of back up and software.

Some of these issues are discussed in the following sections :

Ease of Use

Features of the DSS will affect adoption (*i.e.*, that DSS are not useful to the user and(or) are not easy to use). Perceived ease of use and perceived usefulness issues has been discussed by Keil *et al.* (1995) and are best described through Figure 21.1. Perceived usefulness is a measure of how well the DSS will enhance a user's decision-making capability, that is, the amount of time it might take to perform a certain task. Alternatively, perceived ease of use is a measure of the reduction (or increase) in physical or mental effort to use the DSS. It is important to balance emphasis and effort between ease of use and usefulness. Keil *et al.* (1995) suggested that software that rates low in ease-of-use and low in usefulness will be rejected. Software that is high in ease-of use and low in usefulness they term "toys." Users may embrace this software initially but there is little chance of lasting acceptance. Software that is low in ease of use but high in usefulness will only be used by what they term "power users," very competent computer users. Most users will avoid this type of software because the time and effort required to learn how to use it outweighs the potential benefits. The aim is to develop software that rates high in ease of use and high in usefulness. They also suggest that designing for ease of use "must begin with some type of task analysis that goes beyond the typical considerations of ergonomics and user interface."

Problem Statement and Solution Methodology

The timely information to the farmers is closely linked to the Agricultural Development and well being of the rural communities. Quick information transfer between the researchers and the farmers has specific importance. Hence this research work is focused on the problem of developing a decision support system for farm entrepreneur. As the sharing of programme is beneficial, if used through Internet, hence the project is planned to be web based.

The additional advantages that will accrue are :

- Low cost of operation.
- Quick access to all relevant information.
- Direct and personalized.
- Integrated.

Conclusions

DSS practice, research and technology continue to evolve.. Decision support systems research and development will continue to exploit any new technology developments and will benefit from progress in very large data bases, artificial intelligence, human-computer interaction, simulation and optimization, software engineering, telecommunications and from more basic research on behavioural topics like organizational decision making, planning, behavioural decision theory and organizational behaviour. Trends suggest that data-driven DSS will use faster, real-time access to larger, better integrated databases. Model-driven DSS will be more complex, yet understandable, and systems built using simulations and their accompanying visual displays will be increasingly realistic. Communications-driven DSS will provide more real-time video communications support. Document-driven DSS will access larger repositories of unstructured data and the systems will present appropriate documents in more useable formats. Finally, knowledge-driven DSS will likely be more sophisticated and more comprehensive. The advice from knowledge-driven DSS will be better and the applications will cover broader domains.

Current researchers should remember that Decision Support Systems pioneers came from a wide variety of backgrounds and faced many challenges that they successfully overcame to demonstrate the value of using computers, information technologies and specific decision support software to enhance and in some situations improve decision making. The DSS pioneers created particular and distinct streams of technology development and research that serve as the foundation for much of today's interest in building and studying computerized decision support systems. The legacy of the pioneers must be preserved.

The association with e-governance is very important for government to interact with people and business transactions. E-government proffers a huge potential to find innovative way to satisfy need of people.This can be helpful to the farmers as well. Progress of new technologies allows electronic services to be applied in e-government. So, DSS must be integrated with e-government managerial levels. DSS is a very helpful tool for all e-government enterprises

References

Alavi, M., and Joachimsthaler, E. A., "Revisiting DSS Implementation Research: A Meta-Analysis of the literature and suggestions for researchers," MIS Quarterly, 16, 1, 1992, 95-116.

Alter, S.L. Decision Support Systems: Current Practice and Continuing Challenge. Reading, MA: Addison-Wesley, 1980.

Alter, S.L., *"A Study of Computer Aided Decision Making in Organizations,"* Ph.D. dissertation, M.I.T., 1975.

Alter, S.L., *"Why Is Man-Computer Interaction Important for Decision Support Systems?"*, Interfaces, 7, 2, Feb. 1977, 109-115.

Armstrong, M. P., Densham, P. J. and Rushton, G., *"Architecture for a microcomputer based spatial decision support system,"* Second International Symposium on Spatial Data Handling, 120, 131 International Geographics Union, 1986.

Arnott, D. and G. Pervan, *"A critical analysis of decision support systems research"*, Journal of Information Technology, 20, 2, 2005, 67-87.

Arnott, D. and G. Pervan, *"A critical analysis of decision support systems research"*, Journal of Information Technology, 20, 2, 2005, 67-87.

Baskerville, R., and Myers, M., *"Information Systems as a Reference Discipline"*, MIS Quarterly, 26, 1, 2002, 1 -14.

Baskerville, R., and Myers, M., *"Information Systems as a Reference Discipline"*, MIS Quarterly, 26, 1, 2002, 1 -14.

Berners-Lee, T., *"The World Wide Web: Past, Present and Future,"* August 1996, URL http://www.w3.org/People/Berners-Lee/1996/ppf.html, last accessed March 5, 2007.

Bhargava, H. and D. J. Power. Decision Support Systems and Web Technologies: A Status Report. Proceedings of the 2001 Americas Conference on Information Systems, Boston, MA, August 3 - 5, 2001.

Bhargava, H. K., R. Krishnan and R. Müller, *"Decision Support on Demand: Emerging Electronic Markets for Decision Technologies,"* Decision Support Systems, 19:3, pp. 193-214, 1997.

Bonczek, R. H., C.W. Holsapple, and A.B. Whinston. Foundations of Decision Support Systems, New York: Academic Press, 1981.

— *Part IV* —

Empirical Enterprise: The Approach, Method and Application

Chapter 22

Entreprise Building on Herbals and Nutraceuticals: A New World of Hope

Amit Baran Sharangi

Today's health-conscious consumers around the world prefer natural materials as food, flavour and colour. The same is true for prevention and treatment of diseases over the synthetics of equal efficacies (and toxicities), since genes and immune system of human beings are already exposed to numerous natural toxins during evolution. The value of herbs, spices and medicinal plants have been recognized for centuries throughout the globe.Such crops were especially common in India, China and many Eastern countries. Today,Western people are also opting for these natural healing and flavouring alternatives. An overview in this area with latest approaches has its direct implication on health and wellbeing of human beings by creating a synergy between the growers, food-scientists, breeders, doctors, and consumers equally. The tenet "Let food be thy medicine and medicine be thy food" was put forward by Hippocrates, the father of modern medicine. The "food as medicine" concept in the past decades witnessed a growing global interest on food and health and there was massive explosion of research on health products especially nutraceuticals and functional foods. Herbs and spices, the predominant flavouring, colouring and aromatic agents in foods and beverages could contribute a lot to health and wellness due to their multiple health beneficial effects and as sources of natural colour, flavour, anti- microbial properties and anti oxidants. Biologically active compounds in various herbs are the secondary metabolites and they include alkaloids, glycosides, flavones, saponines, gums and volatile oils.

Herbs: Importance and Scopes

Botanically herb is a plant with no woody stem above ground distinguished from a tree or a shrub. But, in general terms, any parts of the vegetable species (roots, leaves, bark, fruits, flowers, seeds, stem or any part of the plant) that can be used for medicine, cosmetic, culinary or such purposes are known as herbs.

Many of the chemicals of present day have been derived from the herbs, studied and then such molecules have been synthesized. The humankind has relied on herbs for thousands of years. The information collected from the experience has been passed on from generations to generations. Medicinal herbs are of much demand. Herb gardens are taking place of ornamental gardens.

Growing herbs is not very difficult; with little understanding of gardening, many herbs can be grown easily in backyard of the kitchen. Natural herbs are herbs found in the nature, such as found in the forest, along hillside etc. Healing herbs have healing properties, some times which looks like a miracle. Drying herbs is very easy and often done in sunlight. Weight loss herbs will make a person slim, only thing is that one must be prudent on how to use it. Bulk herbs and herbal supplements, herbal tea and much other herbal medicine are available through many herb stores. Ayurveda is ancient art and science of healing with herbal supplements. Ayurvedic medicine containing Indian herbs have long been used for health and long life. Herbs are products of nature, they are not synthetic chemicals manufactured in the laboratories or chemical plants. Human body is very complex in nature, the effect, reaction or response to herbs may vary from human to human.

Current Status of Herbal and their Future Perspectives in India

India is one of the 12 mega biodiversity centres having 45, 000 plant species; its diversity is unmatched due to the 16 different agroclaimatic zones, 10 vegetative zones, and 15 biotic provinces. The country has a rich floral diversity (Table 22.1).

Table 22.1: Floral Diversity in India

Numbers	*Species*
15, 000 - 18, 000	Flowering plants
23, 000	Fungi
25, 000	Algae
1,600	Lichens
1,800	Bryophytes
30 million	Microorganisms

Traditional medicine is the synthesis of therapeutic experience of generations of practicing physicians of indigenous systems of medicine. It comprises medicinal plants, minerals and organic matters etc. Herbal drug constitutes only those traditional medicines that primarily use medicinal plant preparations for therapy. The ancient record is evidencing their use by Indian, Chinese, Egyptian, Greek, Roman and Syrian dates back to about 5000 years (Table 22.2). About 500 plants with medicinal use are mentioned in ancient texts and around 800 plants have been used in indigenous systems of medicine. Indian subcontinent is a vast repository of medicinal plants that are used in traditional medical treatments (Chopra *et al.*, 1956), which also forms a rich source of knowledge (Nadkarni, 1982; Jone, 1984). The various indigenous systems such as Siddha,

Ayurveda, Unani and Allopathy use several plant species to treat different ailments (Rabe and Staden, 1997). In India, around 20,000 medicinal plant species have been recorded recently (Dev, 1997), but more than 500 traditional communities use about 800 plant species for curing different diseases (Kamboj, 2000). Currently 80 per cent of the world population depends on plant-derived medicine for the first line of primary health care for human alleviation because it has no side effects (Veale *et al.*, 1992). Plants are important sources of medicines and presently about 25 per cent of pharmaceutical prescriptions in the United States contain at least one plant-derived ingredient. In the last century, roughly 121 pharmaceutical products were formulated based on the traditional knowledge obtained from various sources (Anesini and Perez, 1993).

Table 22.2: Plant Derived Ethnotherapeutics and Traditional Modern Medicine

Sl.No	*Drug*	*Basic Investigation*
1.	Codeine, morphin	Opium the latex of *Papaver somniferum* used by ancient Sumarians. Egyptaians and Greeks for the treatment of headaches, arthritis and inducing sleep.
2	Atropine, hyoscyamine	*Atropa belladona, Hyascyamus niger* etc., were important drugs in Babylonium folklore.
3	Ephedrine	Crude drug (astringent yellow) derived from *Ephedra sinica* had been used by Chinese for respiratory ailments since 2700 BC.
4	Quinine	*Cinchona spp* were used by Peruvian Indians for the treatment of fevers
5	Emetine	Brazilian Indians and several others South American tribes used root and rhizomes of *Cephaelis spp* to induce vomiting and cure dysentery.
6	Colchicine	Use of Colchicum in the treatment of gout has been known in Europe since 78 AD.
7	Digoxin	Digitalis leaves were being used in heart therapy in Europe during the 18th century.

Modern Medicine from Higher Plants

Medicinal plants play a major role for the development of new drugs. There are many plant-based new drugs eg., deserpidine, reseinnamine, reserpine, vinblastine, vincristine, ectoposide, E-guggulsterone, teniposide, nabilone, plaunotol, Z-guggulsterone, lectinan, artemisinin and ginkgolides, paciltaxel, toptecan, gomishin, irinotecan etc. Plant based drugs provide outstanding contribution to modern therapeutics. For example: serpentine isolated from the roots of *Rauwolfia serpentina* in 1953, was a revolutionary event in the treatment of hypertension and lowering of blood pressure. Vinblastine isolated from the *Catharanthus rosesus* (Farnsworth and Blowster, 1967) is used for the treatment of Hodgkins, choriocarcinoma, non-hodgkins lymphomas, leukemia in children, testicular and neck cancer. Vincristine is recommended for acute lymphocytic leukemia in childhood advanced stages of hodgkins, lymophosarcoma, small cell lung, cervical and breast cancer. (Farnsworth and Bingel, 1977). Phophyllotoxin is a constituent of *Phodophyllum emodi* currently used against testicular, small cell lung cancer and lymphomas. Indian indigenous tree of *Nothapodytes nimmoniana* (*Mappia foetida*) are mostly used in Japan for the treatment of cervical cancer (Table 22.3). Plant derived drugs are used to cure mental illness, skin diseases, tuberculosis, diabetes, jaundice, hypertension and cancer. Plant derived drugs came into use in the modern medicine through the uses of plant material as indigenous cure in folklore or traditional systems of medicine. More than 64 plants have been found to possess significant antibacterial properties; and more than 24 plants have been found to possess antidiabetic properties (Arcamone *et al.*, 1980), antimicrobial studies of plants (Perumal Samy and Ignacimuthu, 1998; 1999 and Perumal Samy *et al.*, 2006), plant for antiodotes activity - *Daboia russellii* and *Naja kaouthia* venom neutralization by lupeol acetate isolated from the root extract of Indian sarsaparilla *Hemidesmus indicus* R.Br (Chatterjee, *et al.*, 2006). Which effectively neutralized *Daboia russellii* venom induced pathophysiological changes (Alam *et al.*, 1994). The present investigation explores the isolation and purification of another active compound from the methanolic root extract of *Hemidesmus indicus*, which was responsible for snake venom neutralization. Antagonism of both viper and cobra venom and antiserum action potentiation, antioxidant property of the active compound was studied in experimental animals. Recently, Chatterjee *et al.* (2004) from this laboratory reported that an active compound from the *Strychnus nux vomica* seed extract, inhibited viper venom induced lipid peroxidation in experimental animals. The mechanism of action of the plant derived micromolecules induced venom neutralization need further attention, for the development of plant-derived therapeutic antagonist against snakebite for the community in need. However, the toxicity of plants has known for a long period of time, and the history of these toxic plants side by side with medicinal ones are very old and popular worldwide, they considered the major natural source of folk medication and toxication even after arising of recent chemical synthesis of the active constituents contained by these plants (Adailkan and Gauthaman, 2001; Heinrich, 2000; Pfister *et al.*, 2002).

Table 22.3: Some of the Important Medicinal Plants Used for Major Modern Drugs for Cancer

Plant Name/Family	*Drugs*	*Treatment*
Cathranthus rosesus L. (Apocynaceae)	Vinblastine and vincristine	Hodgkins, Lymphosarcomas and children leukemia.
Podophyllum emodi Wall. (Beriberidaceae)	Podophyllotaxin,	Testicular cancer, small cell lung cancer and lymphomas.
Taxus brevifolius (Taxaceae)	Paciltaxel, taxotere	Ovarian cancer, lung cancer and malignant melanoma.
Mappia foetida Miers.	Comptothecin, Irenoteccan and topotecan	Lung, ovarian and cervical cancer.
Comptotheca acuminata	Quinoline and comptothecin alkaloids	Used in Japan for the treatment of cervical cancer
Juniperus communis L. (Cupressaceae)	Teniposide and etoposie	Lung cancer

Teniposide and etoposide isolated from *Podophyllum* species are used for testicular and lung cancer. Taxol isolated from *Taxus brevifolius* is used for the treatment of metastatic ovarian cancer and lung cancer. The above drugs came into use through the screening study of medicinal plants because they showed less side effects, were cost effective and possessed better compatibility.

Market Potential of Phytomedicine

The estimation of total phytomedicine sale reported in country wise European Union was about US$ 6 billion in 1991 and $ 4 billion in 1996, of which almost half were sold in Germany $ 3 billion, in France $ 1.6 billion, in Italy $ 0.6 billion and in Japan $ 1.5 billion. The present global market is said to be US 250 billion (Brower *et al.*, 1998). In India the sale of total herbal products is estimated at $ 1 billion and the export of herbal crude extract is about $ 80 million, of which 50 per cent is contributed by Ayurvedic classical preparations. Plant derived drugs are important in Germany and Russia. Particularly, herbal drugs are imported by several countries for their usage of traditional medicinal preparation from various parts of the country (Table 22.5). Some of the important Indian medicinal plants exported to various countries are reported (Table 22.6).

Table 22.5: Percentage of Herbal Drugs Imported by Various Countries for Drug Preparation

Country	*Percentage of Herbal Drugs Imported*
China	45 per cent
USA	15.6 per cent
Australia	10.5 per cent
India	3.7 per cent
South Korea	1.4 per cent
Taiwan	1.7 per cent
Indonesia	8.1 per cent

Table 22.6: Medicinal Plant Parts Exported from India, Importing Medicinal Plants and their Parts

Exporting of Herbals		*Importing of Herbals*	
Botanical Names	*Parts Used*	*Botanical Name*	*Parts Used*
Acorus calamus	Rhizome	*Aloe vera*	Dried leaf
Argemone mexicana	Fruit	*Adhatoda vasica*	Whole plant
Curcuma amada	Rhizome	*Cinnamomum iners*	Bark and leaf
Curcuma longa	Rhizome	*Curcuma aromatica*	Rhizome
Curcuma aromatica	Wild turmeric	*Garcinia indica*	Fruit
Cassia lanceolata	Leaves	*Gloriosa superba*	Tuber and seed
Glycyrrhiza glabra	Root	*Juniperus communis*	Fruit
Withania somnifera	Vegetable rennet	*Myrica nagi*	Bark

Contd...

Table 22.6–*Contd...*

Exporting of Herbals		*Importing of Herbals*	
Botanical Names	*Parts Used*	*Botanical Name*	*Parts Used*
Myrica nagi	Leaf	*Strycnos nux-vomica*	Bark and seed
Piper longum	Fruit	*Phyllanthus amarus*	Fruit
Rubia cordifolia	Madder root	*Ricinus communis*	Seed
Symplocos racemosa	Bark	*Rauvolfia serpentina*	Root
Swertia chirata	Whole plant	*Ocimum sanctum*	Leaf and essential oil
Terminalia chebula	Bark and seed	*Tylophora purpuria*	Root
Zingiber officinale	Rhizome	*Vinca rosea*	Leaf, seed and stem
–	–	*Wedelia calendula*	Leaf and root
–	–	*Withania somnifera*	Tuber

Important medicinal plants and their parts used for the preparation in indigenous systems of Indian medicines are reported in Table 22.7.

Table 22.7: The Major Traditional Sector of Pharmaceutical Companies

S.No	*Traditional Sector Pharmas*	*Modern Sector Pharmas*	*Standardization of Phytomedicine (Formulation)*
1.	Himalaya	Ranbaxy	Chromatography techniques
2.	Zandu	Lupin	Thin Layer chromatography (TLC)
3.	Dabur	Allembic	UV - Spectrophotometer
4.	Hamdard	–	High Performance of Liquid Chromatography (HPLC)
5.	Maharishi	–	Nuclear Magnetic Resonance spectroscopy (NMR)

Role of World Health Organization (WHO) in Phytomedicine Medicine

In 1991 WHO developed guidelines for the assessment of herbal medicine, and the 6th International Conference of Drug Regulatory Authorities held at Ottawa in the same year ratified the same. The salient features of WHO guidelines are: 1). **Quality assessment:** Crude plant materials or extract plant preparation and finished product. 2). **Stability:** Shelf life. 3). **Safety assessment:** Documentation of safety based on experience and toxicological studies. 4). **Assessment of efficacy:** Documented evidence of traditional use and activity determination (Animals and human).

Standardization of Phytomedicine

In the traditional system of medicine, the drugs are primarily dispensed as aqueous or ethanol extract. Fresh plant juice or crude extract are a rarity rather than a rule. The medicinal plants should be authentic and free from harmful materials like pesticides, heavy metals, and microbial and radioactive contamination. The medicinal plant should be single solvent extraction once or repeatedly or aqueous extract or as described in the ancient texts. The extract should be then checked for biological activity in experimental animal models. The bioactive extract should be standardized on the basis of active compound (Table 22.7). The bioactive extract should undergo limited safety studies.

New Sources of Tribal Medicine for Future Investigation

In most of the countries, where ethnomedical treatment is frequently used to treat cut wounds, skin infection, swelling, aging, mental illness, cancer, asthma, diabetes, jaundice, scabies, eczema, venereal diseases, snakebite and gastric ulcer, provide instructions to tribal or local people as how to prepare medicine from herbal (Puspangadan and Atal, 1984; Waller, 1993; Perumal Samy and Ignacimuthu, 1998-1999; Perumal Samy *et al.*, 2006). They keep no records and the information is mainly passed on verbally from generation to generation (Dhar *et al.*, 1973; Sofowara, 1982). World Health Organization (WHO) has shown great interest in documenting the use of medicinal plants used by tribals from different parts of the world (Kaido *et al.*, 1987). Many developing countries have intensified their efforts in documenting the ethnomedical data on medicinal plants. Research to find out scientific evidence for claims by tribal healers on Indian herbs has been intensified. Once these local ethnomedical preparations are scientifically evaluated and disseminated properly, people will be better informed regarding efficacious drug treatment and improved health status (Manandhar, 1987).

Nutraceuticals

- ✰ The term "Nutraceutical" was coined from "Nutrition" and "Pharmaceutical" in 1989 by Stephen DeFelice,MD, Founder and Chairman of the Foundation for Innovation in Medicine (FIM).
- ✰ Other words used in the context: Dietary supplementation, Functional, Multi-functional Foods, etc.

Nutraceutical can be defined as

"A food or part of food or nutrient, that provides health benefits, including the prevention and treatment of a disease."

A Global Picture of the Nutraceutical Industry

The global picture of the nutraceutical industry is very much interesting.The functional food and nutraceutical industry represents more than $75.5 billion US industry (Just-food, 2007) with a potential of growing to $167 billion by 2010 (Just-food, 2004). Operationally, the industry relies upon a network of supportive stakeholders with a vested interest, in one form or another in providing consumers with alternative health products with potential to prevent diseases resulting from nutrient deficiencies or with products that have beneficial physiologic effects beyond those simply attributed to their nutrient content. Although popularity of products marketed as functional foods and nutraceuticals is highly variable and often dependent upon historical allegiances, the industry as a whole has an international presence and operates in many countries.

Table 22.8: Nutraceuticals Available in Market

Brand Name	*Components*	*Function*
Betatene	Carotenoids	Immune function
Xangold	Lutein esters	Eye health
Lipoec	α-lipoic acid	Potent antioxidant
Generol	Phytosterol	CHD reduction
Premium probiotics	probiotics	Intestinal disorder
Soylife	Soyabean phytoestrogen	Bone health
Ztrim	Wheat	Zero calorie fat replecor
Linumlife	Lignan extract flax	Prostate health
Fenulife	Fenugreek galactomannon	Control blood sugar
Teamax	Green tea extract	Potent antioxidant
Marinol	ω 3 FA, DHA, EPA	Heart health protection
Clarinol	CLA	Weight loss ingredient
Cholestaid	Saponin	Reduce cholesterol

The United States of America (USA) currently possesses the largest and most rapidly expanding functional food and nutraceutical market in the world (World Nutraceuticals, 2006). In 2006 value of the industry was $21.3 billion (Datamonitor, 2007). Its strong domestic market supports major imports from Japan, North and South Korea, China, India, Brazil, the European Union (EU), Australia, New Zealand and other parts of the world (World Nutraceuticals 2006). For the USA it has been suggested that about 50 per cent of its multi-billion dollar food market can be related to use of nutraceuticals and functional food products (Belem, 1999). By comparison the Canadian nutraceutical and functional food market is relatively young and is growing. In 2003 Canadian trade in nutraceuticals and functional foods represented 3 per cent of the global market compared to the USA (35 per cent) and EU (32 per cent) (D'Innocenzo, 2006).

Nutraceutical and functional food markets in the EU have grown over the past eight years, from about $1.8 billion out of a $5.7 billion global market in 1999 (Kleter *et al.*, 2001) to $8 billion (Datamonitor, 2007) out of a global market of $75.5 billion (Just-food, 2007) in 2006. While growth of functional food markets within the EU was estimated to be only 2-5 fold in 2001 (Kleter *et al.*, 2001), it reached more than 10 fold in 2006 (World Nutraceuticals 2006). A number of factors have contributed to restricted growth within the industry throughout the EU. Strict regulations governing food labelling, product formulation, food processing, packaging, marketing, registration and licensing details are all strictly monitored in the EU and have been identified as restricting the size of the consumer market in these countries (Breithaupt, 2004; Kleter *et al.*, 2001; Moon and Balasubramanian, 2003; World Nutraceuticals, 2006). Major trading partners with the EU are the USA, Japan, south, south-east, far east and middle east Asia and Pacific regions (Japan's Nutarceuticals, 2003; World Nutraceuticals, 2006).

Eastern cultures have a long history of use of traditional medicines associated with health foods in forms of recognized nutritional foods, food supplements, medicinal herbs, and crude powdered drugs derived from plant, animal and marine sources (Datta Banik and Basu, 2002; Dhanukar, Kulkarni and Rege, 2000). India and China are the two most important countries known for their production of traditional functional food products and nutraceuticals. In India the most common forms of functional foods and nutraceuticals are available as traditional Indian Ayurvedic Medicines (IAM); these are marketed under different brand names (Patwardhan *et al.*, 2005). India is the home of a large number of medicinal herbs, spices and tree species that have a substantially large domestic market with no major foreign competition at present (Datta Banik and Basu, 2002; Patwardhan *et al.*, 2005). However, it is important to note that there are no strict pharmaceutical regulations on Ayurvedic and nutraceutical health products in India. Most of these products are available to consumers directly over the counter without need for a medical prescription (Patwardhan *et al.*, 2005). India has a large share of the international functional food and nutraceutical market, and exports products to the far east, south-east, west and middle east Asia as well as to parts of north Africa and the EU. However, India's major export destination is the USA and Japan (Patwardhan *et al.*, 2005). Labelling, and strict control over formulations and branding are still not required for most products. In addition cost of production typically is low, making this produce highly competitive in Asian and African markets (World Nutraceuticals, 2006). Estimated value of the industry is $10 billion per annum with exports of $1.1 billion per annum making a significant contribution to the export market (Singh, Singh and Khanuja, 2003).

A similar profile for the industry can be developed for China. Both functional foods and nutraceuticals are part of the traditional Chinese diet and are also a large component in Traditional Chinese Medicine (TCM). The industry is represented by around 1000 small to medium sized enterprises located throughout China (AAFC, 2002; Patwardhan *et al.*, 2005). There is a steady demand for functional foods and nutraceuticals in the country and a friendly business environment. Branding and labelling are not under stringent control and ironically, foreign-made preparations of Chinese medications have been reported to have better health and nutritional qualities than those produced in China. However, because of less stringent regulations, cheap labour, lower production costs and the enormous market involved, China has the potential to emerge as a leader in the international market (AAFC, 2002; Patwardhan *et al.*, 2005). Japan, Hong Kong, Korea and Singapore are major importers of TCMs and represent 66 per cent of Chinese plant drug exports (Report of the Chinese Ministry of Commerce, 2003). The annual Chinese herbal drug production is estimated at $48 billion, with estimated exports of $3.6 billion (Handa, 2004).

Japan is the second largest market in the world for nutraceutical products after the United States. Its nutraceutical market has exhibited a steady average growth rate of 9.6 per cent per annum for the past decade, and in 2006 its functional food industry was estimated to have a value of $27.1 billion (Functional Food Japan, Project Report,

2006). The Japanese people invented use of modern functional foods in the early 1970's and ever since the Japanese functional foods' industry has been a leader within the global marketplace. The percapita consumption of nutraceuticals by the Japanese is actually higher ($166.00 per annum) than that seen in the USA ($136.00 per annum) and in the EU ($92.00 per annum). Two types of functional foods have been approved by the Japanese government; *i.e.*, those with approved health claims or FOSHU (Foods for Specified Health Use) and foods that may provide health benefits (without any health claims).

Other large, emerging international markets in south and south-east Asia are seen in Taiwan, Sri Lanka, Thailand, the Philippines, Vietnam, Lagos, Kampuchea, Indonesia, Malaysia, North and South Korea (World Nutraceuticals, 2006). In addition Australia and New Zealand are emerging quickly as international competitors. Investment in research and development of high quality functional food and nutraceutical products along with promotional support from both the industry and government has helped in rapid establishment of the industry in these two countries and in their catching a share of the expanding global market (Heasman, 2004; FRST, 2003).

Potential markets also can be found in oil rich middle-eastern or gulf countries like Saudi Arabia, the United Arab Emirates, Qatar, Oman and Kuwait (World Nutraceuticals, 2006). It is estimated that demand for functional food and nutraceutical products will grow internationally by about 6 per cent per annum through 2010, and that China and India will be the fastest growing markets, while the USA will continue to be the largest, followed by Japan. African markets are still not well organized, although functional food and nutraceuticals are part of the African diet and culture. Scattered opportunities also have been identified in parts of northern and sub-Saharan Africa and in some southern African nations (Global Market Review of Functional Food, 2005).

Nutraceuticals and Pharmaceuticals and Healthcare in INDIA

[Source: http://www.indiabizclub.com]

- ADITHYA PHARMACEUTICALS PVT. LTD........... Bangalore
- ADVANCE SERVICE PROVIDERS.......... New Delhi
- ALEMBIC LIMITED.......... Vadodara
- ATRA PHARMACEUTICALS LIMITED.......... Aurangabad
- BIOSOURCING.COM (P) LTD.......... Bhubaneswar
- BRLB INTERNATIONAL.......... Hyderabad
- GLOBELA PHARMA PVT LTD.......... Surat
- KAWARLAL AND CO.......... chennai
- MAHIMA HERBALS.......... Newbombay
- MORSEF PHARMACEUTICALS PVT. LTD........... Mumbai
- NUTRICARE LIFE SCIENCES.......... Saharanpur
- ORCHID CHEMICALS and PHARMACEUTICALS LTD........... Chennai

Table 22.9 Important phytochemicals (nutraceuticals), their corresponding plant sources and medicinal properties

Phytochemicals: Chemical groups	Plant sources	Medicinal property
	Alkaloids	
Quinine	Cinchona	Anti-malarial
Tropane alkaloids	Solanaceous members : Deadly night shade, Datura	In treatment of heart ailments
Morphine	Opium poppy	Antidepressant, pain killer
Ergot alkaloids	Fungus: *Claviceps purpurea*	Abortifacients
Vincristine	Periwinkle	Antineoplastic
Vinblastine	Periwinkle	Antineoplastic
Coumarin,	Fenugreek	Hypoglycaemic
Scopoletin	Fenugreek	Hypoglycaemic
Fenugreekine	Fenugreek	Hypoglycaemic
Trigonelliine	Fenugreek	Hypoglycaemic
	Carotenoid terpenoids/ Isoprenoids	
α-carotene	Carrots	Antioxidants, anticarcinogenic
β-carotene	Fruits & vegetables	Antioxidants
β-cryptoxanthin	Oranges & tangerines	Antioxidants, anticancer
Lutein	Vegetables (kale, spinach, watercress, parsley)	Reduce risk of macular degeneration, protect against colon cancer
Zeaxanthin	Corn, avocado	Protects eye from macular degeneration and cataracts
Lycopene	Tomatoes, pink grapefruit, watermelon, guava, papaya	Reduces risk of prostrate cancer in males
	Non-carotenoid terpenoids	
Perillyl alcohol	Cherries & mints	Anticancer
Saponins	Legumes (Chicks, peas, fenugreek, all pulse crops)	Reduces cholesterol levels in blood
Terpenol	Carrots	Anticancer
Terpene limonoids	Peels and membranes of citrus fruits	Anticarcinogenic
	Flavonoid polyphenolics	
Anthocyanins	Stawberries, raspberries, cherries, cranberries, pomegranate, apples, red grapes	Antioxidants
Betacyanins	Beet root	Antioxidant
Catechins	Tea	Antioxidant
Flavonones	Citrus fruits	Antioxidant
Flavones	Fruits & vegetables	Anticancer
Isoflavones	Soybean	Anticancer
Hesperetin	Citrus fruits	Antioxidant
Naringin	Grapefruit	Reduces cholesterol
Rutin	Asparagus, buckwheat & citrus fruits	Protects against cardio vascular ailments
Quercitin	Red onions, buckwheat, red grapes, green tea, apple skins	Anti-sitaminic, antioxidant
Silymarin	Artichoke & milk thistle	Anti-atherosclerotic
Tangeretin	tangerines	Anticancer
Tannins	Cranberries, pomegranate, cocoa & tea	Reduces blood cholesterol,
	Phenolic acids	
Ellagic acids	Strawberries & raspberries	Prevents colon cancer
Chlorogenic acids	Blueberries, tomatoes, grapes & bell peppers	Antioxidant
p-coumaric acids	Red and green ball peppers, legumes	Antioxidant, anticancer
Phytic acids	Legumes and whole seed grains	Lowers blood glucose
Ferulic acids	Seeds of brown rice, whole wheat and oats, apple, artichoke, orange, peanut & pine apple	Antioxidant, anticancer
Vanillin	Vanilla bean	Antioxidant, anticancer
Cinnamic acid	Cinnamon, balsam tree resins	Antibacterial, antifungal
Hydroxycinnamic acid	Grapes, blueberries & blackberries	Antioxidant, anticancer
	Non-flavonoid polyphenolics	
Curcumin	Curcuma	Anti-microbial, anticancer, antioxidant
Resveratrol	Grapes	Anti-inflammatory, anticancer
Lignans	Plant cell walls	Reduces skin cancer
	Glucosinolates	
Isothiocyanates	Horseradish, radish & mustard	Anticancer
Phenethyl isothiocyanate	Watercress	Anticancer
Sulforaphene	Broccoli	Anticancer
Indoles	Broccoli	Anticancer
Thiosulfonates	Garlic & onions	Anticancer, antimicrobial, reduces blood pressure and blood cholesterol
Phytosterols (plant sterols)	Peanuts, cashews, almonds, peas, kidney beans & avocados Anticancer, blocks cholesterol absorption	

Table 22.9 contd...

	Anthraquinones	
Senna	Legumes and pulses	Purgative,
Barbaloin	Aloe	Laxative, anti-helminthic
Hypericin	St. John's wort	Analgesic
Capsaicin	Capsicum (hot peppers)	Anticancer, anti-inflammatory, anti-apoptotic
Piperine	Black peppers, jalapeno peppers	Helps in digestion
	Terpenes	
Menthol (Monoterpene)	Plants of mint family	Topical pain reliever & anti-pyretic
Borneol (Monoterpene)	Pine oil	Disinfectant
Santonin (Sesquiterpene)	Wormwood	Photosensitizer
Gossypol (Sesquiterpene)	Cotton	Contraceptive

- ✰ PIONEER ENTERPRISE.......... Mumbai
- ✰ RAMPAL.......... Madurai
- ✰ SALIUS PHARMA PVT. LTD........... Navi Mumbai
- ✰ SHAGUN PHARMACEUTICALS.......... Indore
- ✰ SHAMBHAVI ENTERPRISES........... Thane
- ✰ SSS BIOTIC PVT. LTD........... Bhubaneswar
- ✰ SSSBIOTIC.COM.......... Bhubaneswar
- ✰ TULSI AMRIT PVT. LTD........... Indore
- ✰ VENKAT PHARMA LIMITED.......... Hyderabad

[Source: Basu et al., 2007]

Conclusions

Determining the biological properties of plants used in traditional medicine is helpful to the rural communities and informal settlements. Several efforts are going on to isolate the active compound(s) by bioassay-guided fractionation from the species that showed high biological activity during screening. The reduced reliance on chemicals is further influencing the overall scenario. With a view to find out a balance between traditional and modern medicines alongwith modern science revisit, refinement and restructuring of methods and environments are being made. Apart from technological backup, economic and social factors are also taken into consideration. Further research is deserved to isolate the compounds responsible for the observed biological activity.

References

AAFC, 2002. The nutraceuticals market in China. Available online at: http://ats.agr.ca/asia/3826_e.htm.

Adailkan, P.G., Gauthaman, K., 2001. History of herbal medicines with an insight on the pharmacological properties of Tribulus terrestris. The Aging Male 4, 163–169.

Alam, M.I., Auddy, B., Gomes, A., 1994. Isolation and partial characterization of viper venom inhibiting factor from the root extract of the Indian medicinal plant sarsaparilla (*Hemidesmus indicus* R.Br.). Toxicon 32, 1551–1557.

Basu, S. K.; Thomas, J. E. and Acharya, S. N. (2007) Prospects for Growth in Global Nutraceutical and Functional Food Markets:A Canadian Perspective. *Aust. J. Basic and Appl. Sci.*, **1**(4): 637-649.

Belem, M.A.F., 1999. Application of biotechnology in the product development of nutraceuticals in Canada. *Trends Food Sci and Technol*, 10(3): 101-106.

Bhakuni DS, Dhar ML, Dhar MM, Dhawan BN, Mehrotra BN (1969) Screening of Indian plants for biological activity: Part II. *Indian J. Exp. Biol*. 7: 250-262.

Breithaupt, H., 2004. GM plants for your health. EMBO Rep, 5(11): 1031-1033.

Chatterjee, I. Chakravarty, A.K., Gomesa A. (2006). *Daboia russellii* and *Naja kaouthia* venom neutralization by lupeol acetate isolated from the root extract of Indian sarsaparilla *Hemidesmus indicus* R.Br. *Journal of Ethnopharmacology* 106(1), 38-43.

Chatterjee, I., Chakravarty, A.K., Gomes, A., 2004. Antisnake venom activity of ethanolic seed extract of *Stychnos nux vomica* Linn. *Indian Journal of Experimental Biology* 42, 468–475.

Chopra, R.N., Nayar, S.L. and Chopra, I.C. (1956). In Glossary of Indian medicinal plants, Vol. I. *Council of Scientific and Industrial Research, New Delhi*, pp. 197.

Cox P, Balick M. The ethnobotanical approach to drug discovery. Scientific American, 1994, June, 82-87.

Cox PA, Ethnopharmacology and the search for new drugs. (Eds. D.J. Chadwick and J. Marsh). In Bioactive Compounds from Plants. Ciba Foundation Symposium 154, John Wiley and Sons, Chichester, 1990, pp.40-55.

D'Innocenzo, L., 2006. The future is ripe. Bio Business, XII: 20-24.

Dahanukar, S.A., R.A. Kulkarni and N.N. Rege, 2000. Pharmacology of medicinal plants and natural products. Indian J Pharmacol, 32: S81-S118.

Datamonitor, 2007. Functional food and drink consumption trends. Product code dmcm2982. Available online at: http://www.market-research-report.com/datamonitor/DMCM2982.htm.

Datta Banik S. and S.K. Basu (eds), 2002. Environmental perspectives and human responses. National Service Scheme 2000-01: Government of India. S. Graphics, Kolkata, India.

Dev S (1997). Ethnotherapeutic and modern drug development: The potential of Ayurveda. *Cur. Sci.* 73 (11): 909-928.

Dhar ML, Dhar MM, Dhawan BN, Mehrotra BN, Ray C (1968) Screening of Indian plants for biological activity: Part I. *Indian J. Exp. Biol*. 7: 232-247.

Farnsworth NR, Bingel AS. *Problems and prospects of discovery new drugs from higher plants by pharmacological screening*. In: H.Wagner and P.Wolff (eds.), New Natural products and plant drugs with pharmacological, biological and therapeutical activity, Springer Verlag, Berlin 1997: 1-22.

Farnsworth NR, Blowster RN, Darmratoski D, Meer WA, Cammarato LV (1967). Studies on *Catharanthus* alkaloids IV Evaluation by means of TLC and ceric ammonium sulphate spray reagent, *Lloydia* 27: 302-314.

FRST, (Foundation for Research Science and Technology), 2003. Draft investment signals and review updates. Foundation for Research Science and Technology. Wellington, New Zealand. Functional Food Japan, 2006. Project Report. Japan.

Global Market Review of Functional Foods, 2005. Available online at http://www.justfood.com/store/product.aspx?ID=44028.

Handa, S.S., 2004. Indian efforts for quality control and standardization of herbal drugs/products. Proc.1st Joint Workshop on Quality Control and Standardization of Traditional Medicine. January, 8-10.

Heasman, M., 2004. Food, health, functional foods and Australian rural industries. Rural Industries Research and Development Corporation (RIRDC), Government of Australia, Australia.

Heinrich, M., 2000. Plant resources of south-east Asia medicinal and poisonous plants 1. Phytochemistry 53, 619–620.

Just-food, 2004. Global market review of functional foods – forecasts to 2010. Aroq Limited. Available online at: http://www.researchandmarkets.com/reportinfo.asp?report_id=246286.

Just-food, 2007. Global market review of functional foods – forecasts to 2012. Aroq Limited. Available online at: http://www.just-food.com/store/product.aspx?id=44028 and lk=pop.

Kaido, T.L., Veale, D.J.H., Havlik, I., and Rama, D.B.K. Preliminary screening of plants used in South Africa as traditional herbal remedies during pregnancy and labour. *J. Ethnopharmacol.*, 1997; 55: 185-191.

Kamboj VP (2000): Herbal medicine. *Cur. Sc.* 78(1): 35-39.

Kandil, O., Radwan, N.M., Hassan, A.B., Amer, A.M.M. and El-Banna, H.R. (1994) Extracts and fractions of *Thymus captitatus* exhibit antimicrobial activities. *J. Ethnopharmacol.* 44, 19-24.

Kleter, G.A., W.M. van der E.J. Krieken, Kok, D Bosch, W Jordi and L.J.W.J. Gillissen, 2001. Regulation and exploitation of the genetically modified crops. Nature Biotechnol, 19: 1105-1110.

Latha PG, Evans DA, Panikkar KR, Jayavardhanan KK (2000) Inhibition of chemical carcinogenesis in mice by *Ixora coccinea* flower. *Pharmaceutical Biol.* 38(2): 152-156.

Manandhar, N.P. Traditional medicinal plants used by tribals of Lamjung District, Nepal. *Int. J. Crude Drug Res.*, 1987; 25 (4): 236-240.

Moon, W. and S.K. Balasubramanian, 2003. Is there a market for genetically modified foods in Europe? Contingent valuation of Gm and non GM breakfast cereals in United Kingdom. Agri Bio Forum, 6: 128-133.

Perumal Samy R, Ignacimuthu S (1998) Screening of 34 Indian medicinal plants for antibacterial properties. *J. Ethnopharmacol.* 62: 173-182.

Pfister, J.A., Ralphs, M.H., Gardner, D.R., Stegeleier, B.L., Manners, G.D., Panter, K.E., 2002. Management of three toxic Delphinium species based on alkaloid concentrations. Biochemical Systematics and Ecology 30, 129–138.

Puspangadan, P., and Atal, C.K. Ethnomedico-botanical investigation in Kerala I. Some primitive tribals of Western Ghats and their herbal medicine. *J. Ethnopharmacol.*, 1984; 11: 59-77.

Rabe T, Staden JV (1997) Antibacterial activity of South African plants used for medicinal purposes. *J. Ethnopharmacol.* 56: 81-87.

Sofowora, A. *Medicinal Plants and Traditional Medicine in Africa*. Wiley and Sons, Chichester, 1982; pp. 75-76.

World Health Assembly, Resolution, WHA 1977; 30: 49.

Patwardhan, B, D Warude, P Pushpangadan and N Bhatt, 2005. Ayurveda and Traditional Chinese Medicine: a comparative overview. eCAM., 2(4): 465-473 (doi:10.1093/ecam/neh140).

Report of the Ministry of Commerce, 2003. People's Republic of China.

Singh, J., A.K. Singh and S.P.S. Khanuja, 2003. Medicinal plants: India's opportunities Pharma Bioworld, 1: 59-66.

World Nutraceuticals, 2006. Industry Study with Forecasts to 2010 and 2015. The Freedonia Group, Cleveland, OH USA.

http://www.indiabizclub.com

Chapter 23

Dairy Enterprise a Great Stay in India's Globalization Journey

Asitava Sur

Indian Stay

Being the world's largest producer and consumer of dairy products, India represents one of most lucrative dairy markets. The sales of dairy products in India will nearly double its size from INR 2.6 Trillion (US$ 60 Billion) to around INR 5.1 Trillion (US$ 115 Billion) by 2016.

The dairy sector in the India has shown remarkable development in the past decade and India has now become one of the largest producers of milk and value-added milk products in the world.

India is the largest exporter of dairy products and exported 25,639.49 MT of dairy products to the world for the worth of Rs. 289.37 crores during the year 2011-12. Major Export Destinations being United Arab Emirates, Nepal, Singapore, Oman and Egypt Arab Republic.

Milk Production and International Trade

- India is the largest milk producer in the world.
- India produces 14 per cent of world milk production, but its share in international dairy trade is even less than 0.1 per cent
- In contrast, New Zealand and Australia produce 2.4 per cent and 1.7 per cent of world milk production, but their share in international dairy trade is 31 per cent and 15 per cent respectively.

Productivity

- In comparison with the world standards, the average milk yield (average milk per animal per year/lactation) in India is considerably low (less than 900 kg).
- In contrast, Europe has 7,000 kg, USA 8,000 kg and Israel 12,000 kg. of average milk yield per cow per year.

SWOT Analysis of Indian Dairy Industry

Strengths

- **Demand profile:** Absolutely optimistic.
- **Margins:** Quite reasonable, even on packed liquid milk.
- **Flexibility of product mix:** Tremendous
- **Technical manpower:** Professionally-trained, technical human resource pool, largest in the world.
- **Availability of raw material:** Abundant.

Weaknesses

- **Perishability:** Wider use of Pasteurization, Cooling, Microfilteration, Sterilization, Uperization Thermization, LP System, Vacreation, UHT Treatment, Irradiation, Pulsed HVEF, UHP Technique, IR induction, Bactofugation, Microwave Heating, Bacteriocins, Carbonization, Dielectric Heating, Probiotics, Ohmic Heating, Thermosonication, Bactotherm,
- **Lack of control over yield:** Use of developments like embryo transplant, artificial insemination and animal husbandry practices,
- **Logistics of procurement:** Bad roads and inadequate transportation facility make milk procurement problematic. Overall economic improvement is required.
- **Problematic distribution:** Require a cold chain linking the producer to the refrigerator at the consumer's home.
- **Competition:** With so many newcomers entering this industry, competition is becoming tougher day by day. But the market is large enough for many to carve out their niche

Opportunities

- **Planning Commission, India,** has targeted milk group output and focused on increasing productivity and spoilage management in the farm-consumer chain.
- **Mckinsey,** in its study of processed foods in India, has earmarked processed liquid milk as the highest growth product.
- **FAO** has short listed 26 ethnic Indian milk products for their high potential for international marketing in its "Tradition Reinvented" study on global foods.

- Efforts to exploit export potential are already on.
- **88 per cent** of the milk produced is yet to move up the Value chain.
- Large number of Indian consumers being vegetarian, vast potential for milk products in food habits/cuisine.

Threats

- Environment norms and energy costs warrants a change in packaging material and system.
- Urgent need for implementing hygiene/food safety systems.
- **Milk vendors, the un-organized sector**: Today milk vendors are occupying the pride of place in the industry. Organized dissemination of information about the harm that they are doing to producers and consumers should see a steady decline in their importance

Conclusion

- Strenghth and Opportunities far outweigh Weaknesses and Threats.
- Strengths and Opportunities are fundamental, Weaknesses and Threats are Transitory.
- Essential Ingredients :
  - Entrepreneurship.
  - Innovativeness.
  - Value Orientation.

Present Efforts to Improve Quality

- It is true that the low raw milk quality is everybody's concern and conscious efforts are being made by a few large dairy processors, NABARD and other dairy related institutions in selected area by providing bulk coolers, milking machines to farmers.
- But the problem of poor raw milk quality is so grave and so little and so inadequate is being done to control, it if not completely overcome it.
- These "knee-jerk" fashion measures that too in selective areas shall not bring in any conspicuous improvements in the overall situation unless the total integrated dairy activity is carried out in an organized way and in every district and Taluka of India

Longing for Long

Since long with lot of expectation from our research and extension activities, the obvious prescription has been in the use of newer knowledge and technologies in the following key areas.

- DEVELOP PEOPLE
- CHANGE MINDSET

- ☆ CHANGE RULES OF THE GAME/GOVERNANCE
- ☆ ENCOURAGE CREATIVITY
- ☆ GET RID OF WASTE
- ☆ GO FOR LATEST SYSTEMS
- ☆ GET LATEST TECHNOLOGY

Nothing Special in Management

For a turn around, more than the technicalities, the management lacunas need stressing, which the dairy industry has failed to do as of yet. 'Mind Tree' famous Mr. Jaitlee has indicated the reasons of excellence in WELCOME.

i) **W**orkplace
ii) **E**mpowerment and inclusion
iii) **L**earning and development
iv) **C**ommunication
v) **O**rganization
vi) **M**anagement team
vii) **E**quity

Chapter 24

Water Resources as Enterprise Generation and Enterprise Mentoring: The Journey and Success Stories

Anandamoy Puste

Introduction

What are the Natural Resources?

1. Virtually, global is endowed with vast natural resources in terms of : (i). Land, (ii). Water and (iii) Biodiversity.
2. India is blessed with natural resources with a great diversity from scares to endowed or full potentiality of all and
3. Among the biodiversity, natural calamities is also important, to be considered for any adaptation/conservation practices.

Diversification of Land Use in India

Actually, we have 6 agro-climatic regions, has been divided based on : (a) availability of precipitation, (b) land and soil condition, (c) type of vegetation grown and (d) prevailing of other climatic condition in those situations, are : i. Arid, ii. Semi-arid, iii. Sub-humid, iv. Humid, v. Wet and vii. Very wet situation. Beside rainfall pattern, temperature and a large/wide verify of soils spreading all over Indian sub-continent are derived from a wide range of geological rocks and parent,materials interesting with soil forming factors like (1) climate (2) Vegetation and (3) topography for varying period.

The total 329 m ha of India is covered with –

i. Red soils – 105.5 m ha
ii. Black soils – 73.5 m ha
iii. Alluvial soils – 58.4 m ha
iv. Laterite soils – 11.7 m ha
v. Hill and terai soils – 30.0 m ha

Water and its Characteristics

The word 'water' has derived from the Latin word *'aqua'* and anything, which is associated with or growing or living in or near water, characteristically prevails aquatic in nature or otherwise any living or biological entity contents water, may be little or more. Water is the key and inseparable primary input that occurs on the earth in all three states, *viz.* (i). liquid (moisture), (ii). solid and (iii). gaseous (vapour) in various degrees of motion. Virtually, our country is endowed with vast natural resources in terms of land, water and biodiversity.

Sources of Water

Truly precipitation is the prime source of water in the world. However, the nature gift us lot of water through many ways like rivers, dams, lakes, waterfalls, streams, rains and ground water. We, the people visualizing physically water in the different places, where it is in the form of stored for future use. Out of total rainfall, we are using only 10 per cent water in a calendar year monsoon, which is a natural gift to the country with great diversity, and rest water flows to the river and to various others water resources. For the conception in depth, we can divide it into 3 main heads, as (1). Mountain water, (2). Land water and (3). Ocean water

Global Water Cycle

As per the Anonymous (1998) - "The circulation of water between the surface of the earth and its atmosphere is driven by energy from the sun. Water vapour is lost from seas, rivers and lakes by evaporation and from plants by transpiration. The rising vapour cools and condenses into droplets of water, which forms clouds. Carried by the wind, clouds continue to absorb water until saturated, when the water droplets falls as rain, hail or snow (precipitation). In this way water is redistributed around the planet". The quantity of water in the earth is fixed; only the form is changed time to time in its natural consequences through its different ways including natural calamities (Figure 24.1).

World Scenario

Water is the key and integral input of the world, which is the most abundant compound of all living organisms. More than 70 per cent of the earth's surface is covered with water. The total resources of the earth in solid, liquid and gaseous forms are presented in Table 24.1. The total water resources of the earth are about 1,384 million km^3. About 97.16 per cent of total water in the world occurs in the oceans as salt water. Only 2.60 per cent is fresh water. The water in the oceans has a salt content

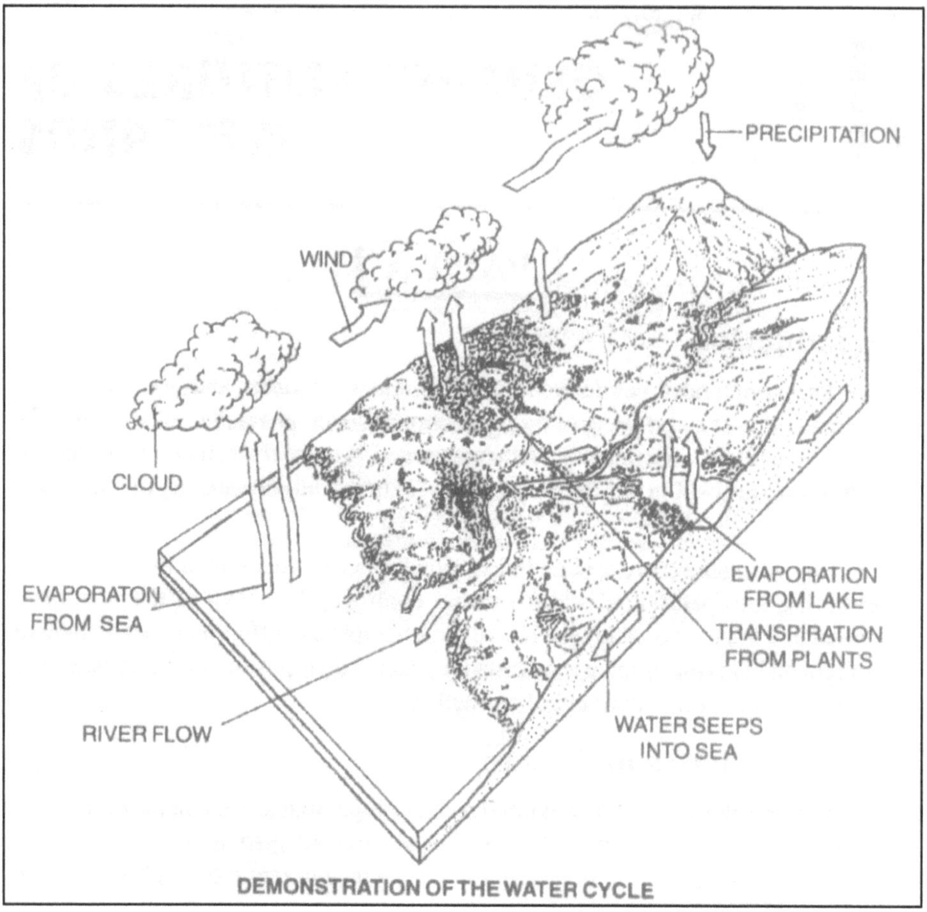

Figure 24.1: Global Water Cycle Water Resources of India.

Table 24.1: Water Resources of the Earth in Solid, Liquid and Gaseous Form Quantity of Water

Resources	*Form*	*Quality*	*Volume ('000 km³)*	*Percent of Total*
Oceans	Liquid	sl	1,348,000	97.16
Polar ice caps, ice-bergs and glaciers	Solid	f	27,820	2.01
Ground water and soil moisture	Liquid	f	8,062	0.58
Lakes and rivers	Liquid	f	125	0.009
Lakes and rivers	Liquid	sl	100	0.007
Atmosphere	Gas	f	13	0.0009
Total			1,384,120	100

Sl: Salt Water; f: Freshwater.

Source: Baugartner and Reichel (1975).

over 3 per cent and is not fit for direct use by humans, many plants, livestock's or in the industry. High costs, high-energy requirements and associated environmental degradation limit the possibility of mass desalinization except small quantities for domestic use. Even in the near future, there is no possibility to extract economically large quantities of fresh water from the seas (Reddy and Reddy, 1997).

India is blessed with a good amount of rainfall receiving in a given year. But these is a great variation in regional distribution of rainfall from <100 mm in extreme arid part of western Rajasthan to highest rainfall regional region (>10,000 mm per annum) in Cherrapunji (Mawsynram) and adjoining *Khasi* hills of Assam and Meghalaya in the country (Table 24.2). However, on the basis of availability of annual rainfall in an area, the regions may be classified accordingly –

Table 24.2: Based on the Availability of Rainfall the Zone/Region/Area may be Divided Broadly

Sl.No.	*Regions*	*Per cent of Cultivated Area*	*Rainfall Received mm/am*
1.	Arid	25	<250
2.	Semi arid	30	250-500
3.	Sub-humid	20	500-1000
4.	Humid	11	1000-1500
5.	Wet	9	1500-2000
6.	Very wet	5	>2000

However, on an average the country's annual rainfall is about 1194.0 mm or round about 1200 mm, which amounting or quantity (volume) of 400 million hectare meter (m ha m). Actually, we can estimate the total volume or quantity of rainfall that carries in the different resources, source in the country etc. based on this average rainfall value: - There are two types of estimation –

1st Method of Estimation

i) The country's annual rainfall is about 1194 mm. The country's geographical area is 329 million ha

Therefore, total rainfall in India = 1194 mm x 328 m ha

= 391632 million ha mm

= 391.632 m ha m or 392 m ha m

This may be rounded off 400 m ha m including snowfall.

ii) 2nd Method of Estimation

Generally our rainy months occurred 4 months in a year, staring from June to September. Our Bay of Bengal branch of monsoon current carries moisture about amounting to 770 m ha m and Arabian Sea branch monsoon current carries about 340 m ha m. So, total monsoon moisture carries by both the sources = 770 + 340 = 1110 m ha m.

On an average 27.5 per cent moisture carried by monsoon current, which is precipitated as rainfall in the country.

$$\text{So, total rainfall} = 1100 \times \frac{27.5}{100} = 305.25 \text{ m ha m}$$

For remaining 8 months, the precipitation is in the order of 100 m ha m. Therefore, the total rainfall according to the calculation = 305.25 + 100 = 405.25 m ha m or round about 400 m ha m. On an average, 130 days in a year is wet days in our country. Out of which, 75 days have rainfall <2.5 mm. It moistens a few mm of the top soil and than it evaporates rapidly. Rest 55 to 60 days have rainfall $>25 - 30$ mm, which is affected soil-environment in different ways. The amount is sufficient for saturated the upper and sub-soil column, if it is in optimum quantity. If this rainfall is more than optimum need, then the surplus amount may be added or will act as —

1. Ground water contribution,
2. Flow downwards and admixture with the adjoining river, tributaries, channels, low-lying areas and
3. Ultimately adds with the sea water.

However, if the situation is in severe form, particularly within a short time, then this excess water may create flood situation even sometimes in unprecedented form.

Water availability in the country

So, the total availability of water in the country is about 400 m ha m. This amount of water is utilized in different ways or forms round the year. On the basis of calculation, the total evaporation (loss) to the astrosphere is about 70 m ha m out of 400 m ha m. The other utilizable sites are :

400 m ha m
- → 70 m ha m — evaporation
- → 115 m ha m — run off
- → 215 m ha m — infiltrate into the soil.

Total Surface Flow in India

Irrigation commission of India (1972) has placed the total surface water flow in the country is about 180 m ha m through the major rivers including tributaries system in the country. The sources are as follows :

1. 115 m ha m contribution from rainfall,
2. 45 m ha m as regenerated flow from ground water as assessed from river flow during non-rainy months and
3. 20 m ha m brought in by streams and rivers from catchments lying outside the country (Bangladesh, Pakistan and China).

Utilization of Surface Flow

The total surface flow of water (180 m ha m) in the country is utilized through different ways, as follows:

1. 15 m ha m is utilized through diversion works and direct pumping (RLI),
2. 15 m ha m is stored in various reservoirs and tanks and
3. 150 m ha m goes to the sea and some adjoining countries.

This amount and forms vary from time to time and year to year. This picture is not rigid one.

Soil Moisture and Groundwater Recharge

On the basis of permeability of soils in different regions and rainfall pattern, it has been estimated that for the country as a whole about 12.5 per cent of the total precipitation infiltrates to the ground water table, then,

$$\frac{400 \text{ m ha m} \times 12.5}{100} = 50.0 \text{ m ha m}$$

We know that of the total amount of water in the country through rainfall is 400 m ha m, of which about 215 m ha m infiltrates in to soil. Out of 215 m ha m about 50 m ha m percolates to the water table through soil stratum and rest 165 m ha m is retained in the soil column as soil moisture. However, the total ground water of the country has been estimated as 67 m ha m, which is counting from the different ways/ source.

67 m ha m
- → 50 m ha m – from precipitation or rainfall
- → 5 m ha m — as flood flow
- → 12 m ha m — from irrigation system

Utilization of Groundwater

The total ground water of 67 m ha m is utilized though various ways —

1. 13 m ha m extracted from various uses,
2. 45 m ha m regenerated into rivers and
3. 9 m ha m is used or loss for E_T (evapotranspiration) and raising of ground water table.

In near future with the full exploitation and development of water resources, it is expected to be about 85 m ha m.

Loss or Utilization of Water as Transpiration

For maintaining turgidity, metabolic etc., a reasonable quantity of water is essential through the plants system. Major part is lost through the ways of transpiration. However, the amount of transpiration from different source is about 110 m ha m, as follows:

1. Irrigated crops – 13 m ha m
2. Non-irrigated crops – 42 m ha m
3. Forests and other vegetations — 55 m ha m

On full development, it will increase to 125 m ha m.

Potential Utilization of Water Resources

During 1976, the total utilization of water in the country was only 10 per cent of the water resources; it is likely to rise about 26 per cent. However, with improved technology and adaptation in future years, it is expected to increase over 30 per cent.

Reasons for Low Utilization of Water Resources

The causes low utilization of water resources in India, are – i) Rainfall in a limited time and erratic in nature, ii) Lacuna of construction of dams and watersheds and sedimented and over siltation in most of the dams, watershed areas, iii) Scarcity of arable land adjoining river basins *i.e.* lack of catchments area, and above all iv) Short length in between river and sea.

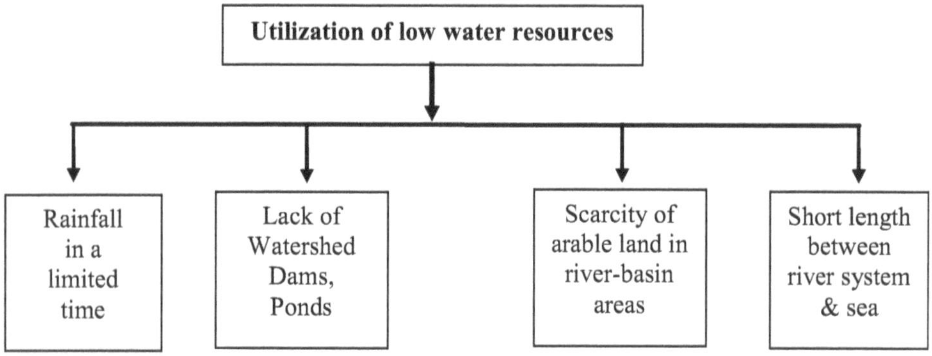

i) Rainfall in a Limited Time

The rainfall in most part of the country is confined to a few months in a year and sometimes, occurrence is scanty may be long or short time and sometimes heavy rains may be within short time. If it more within short time, then surplus rain water will flow downwards and maximum river flow will occur during that period. As the rainy season's flows can not be fully utilized during this short period, water has to be stored artificially in the reservoirs for subsequent use. *In situ* utilization is comparatively low from the sources during that time.

ii) Construction of Dams and Watersheds

Construction of suitable dams or reservoirs or watersheds are one of the mitigation options for storage this surplus rain water in a place. So, in suitable sites dams or reservoirs can be built for large storage of rain water. But these are very limited in the country, even it is non-functioning of the existing system due to so many causes. Some major rivers in the country do not have enough storage capacity for completely harnessing the river flow during rainy season.

iii) Scarcity of Arable Land Adjoining River Basins

Scarcity to arable land in some river basins leads to unutilized large quantity of river water in many adjoining river places. *e.g.* for the enormous size of the river *Brahamaputra* its valley in Assam is quite narrow having a mean width of only 90 km, *i.e.* not enough adjoining land to irrigate and

iv) Short Length in between River and Sea

Similarly, because of their short length to the sea, the west flowing rivers of south India (*Tapi* river), offer only limited scope for further development for irrigation.

Milestones in the Development of Irrigation in India

In the efforts to develop water resources for irrigation, the managements aspects of water resources for irrigation did not receive due attention. As a result of ineffective irrigation management, this is a gap of 9.8 million ha between the irrigation potential created and utilized by 1996-97 (Ready, 1998).

However, a substantial irrigation infrastructure had already been created during pre-independence period, and undivided India had one of the largest areas under irrigation in the world. But at the time of independence in 1947, India had to face the tragedy of partition and 1/3rd of its irrigated area went to Pakistan. However, after partition, the country was coupled to grow more food to support large population with relatively much less agricultural land and irrigation resources. At this time, large scale development of irrigation and power infrastructure was recognized as critical input for rapid agricultural growth. An immediate need was felt to develop several small, medium and large multi-purpose irrigation projects to tap available water resource in the country. A target of doubling the area under irrigation was set, which was achieved by 1974. As a result, a large irrigated potential, above 90 m ha achieved by the year of 1995 - 96 as against 22.6 m ha in 1950 - 51. During the first, 4 decades of planned development in independent India, the country spent about 450 billion on irrigation and the average growth of irrigation potential has been one of the highest in the world (Yadav and Singh, 2000).

Irrigated Ecosystem in India

Depending upon the natural water resources, each region has certain area under irrigated agriculture. But broadly considering, two distinct irrigated ecosystems exists *e.g.*

1. Indo-Gangetic plain region comprising of the states of Punjab, Haryana, plains of Utter Pradesh, Bihar, West Bengal and Plains of Jammu and Kashmir and
2. Coastal areas comprising of Andhra Pradesh and T. Nadu

The 7 states (Punjab, Haryana, UP, Bihar and West Bengal) in the *Gangetic* plain region and Andhra Pradesh and T. Nadu in South India are the potential areas, which has been developed for utilizing the natural resources like water potential based on the available resources of the Ganges.

Role of Water in Plant's Life

Soil-plant-water relationships relate to the properties of soil and plants that affect the (i) movement, (ii) relation and (iii) use of water. Soil provides the room for water to be used by the plants through the roots present in the same medium. Water, as such and also as a carrier of large amounts of nutrients, is required for the successful growth of crops. Plants grow on soils that provide them water and nutrients. They absorb the water from soils, mainly through their root system and use only 1.0 to 1.5 per cent of the volume of water absorbed for building their (i) vegetative structures and (ii) performing various physiological and biochemical activities. The rest of the water absorbed by the plants, is lost through transpiration. So, a certain quantity of water is needed by a plant to maintain its turgidity, even after loosing this amount. In fact, an excess or deficit of soil water hinders/reduces the plants growth and ultimately which will affect the crop yield. A close relationship, therefore, exists between soil water and plant to make it fruitful productive in nature.

Simply, we can say water plays a vital role in plant life. It is essential to plants in the following ways:

1. Water is a structural constituent of plants cells when water is plenty, cells are turgid and retains normal structural form. It contains 85-90 per cent of the body weight of young plants and 20-50 per cent of older or mature plants.
2. Water is a source of two essential elements, oxygen and hydrogen, which required for synthesis of carbohydrate during photosynthesis.
3. Water acts as a solvent of plant nutrients and helps in uptake of nutrients from soil.
4. Water also acts as a carrier for distribution of plants foods for maintaining their life, that manufactured by the plants.
5. Transpiration is a vital process in plant and it occurs at a potential rate as long as water is available in soil adequately. If it deficits, transpiration process is curtailed down seriously, which affect seriously plant growth and ultimate yield.
6. Adequate supply of water maintains the turgor pressure of guard cells, helping stomata to open fully. Water deficit on the other hand closes stomata partially or completely reducing water loss through transpiration.
7. Cells and tissues are formed and simultaneously, growth of plant occurs when an adequate amount of soil water is available. Water deficit slows down the growth processes.
8. Leaves get heated up with solar radiation, plants dissipate or transmit such heat by increasing transpiration to maintain its equilibrium. Water acts as a buffer against high or low temperature injury and

9. Water encourages good growth, development and yield of plants and quality of plant produce, when it is available in plenty. Conversely, plants die when water supply curtailed down.

Thus optimum quality of water needed by the plants, either it is met through the source of precipitation naturally or by means of artificial in the form of irrigation. So, a close relationship, therefore, exists between soil, water and plant and that should be clearly understood to decide upon: (i) the time (when to irrigate), (ii) depth of irrigation (how much to irrigate) and (iii) to make most efficient use of irrigation water (how best to irrigate).

Water in Crop Production

Water is a prime natural resource and important input for assured agricultural production. In absence of it, crop production is drastically hampered, sometimes crop may total failure, even other management variables are in optimum in the field. It has been observed through different experimentation that the yield of a crop may be doubled or even more if water supply is in (i) proper time (ii) with proper quantity. This proper time with required quantity may be performed either through natural resources like precipitation or through artificial means of irrigation. It may be concluded that water will continue to be the principal key factor in mitigating climatic constraints, enhancing agricultural production systems and ultimately augmenting farmer's income, in future as well.

Irrigation Potentiality

Recognizing the vital role of water as a critical input in crop production, huge investments have been made in the last decades since independence for harnessing water resources for various uses including agriculture. Irrigation at present accounts for nearly 84 per cent of water requirements. In future, non-irrigation requirements will be growing faster and still by 2025, the irrigation will account for 73 per cent of the total water requirements of the country.

Some Terms Related with Rainfed Dryland Agriculture

Concept of Dry Farming Areas

As water is the primary limiting factor in crop production in the problematic areas in India, hence a great deal of attention is to be paid on the study of the soil moisture status particularly in dry farming areas. Among the natural resources, rainfall, climatic factors and physical condition of the soil is obviously related with the crops that are grown, soil physical condition includes particle size/class of the soil, porousness, water holding capacity (WHC) etc. are the most important factors that limits the crop production and choice of crops is another most important factor particularly in dry farming areas (genetically adaptable varieties, those who are acquainted with this adverse climatic condition).

In simple, in dry farming areas there is more or less 3 distinct problems, *viz.*

1. Low (amount) and erratic rainfall (unusual),
2. High evaporative demands due to climatic condition (high temperature) and

2. low water holding and nutrient capacity of the soil due to soil physical condition.

Arid Region

Area where the potential water losses by evaporation and transportation (E_T) are grater than the amount of water supplied by precipitation.

Semi-Arid

Region intermediate between the true deserts and sub-humid areas that lying under semi-arid condition.

Contour

It is an imaginary line connecting points of equal/same elevation on the surface of the soil.

Figure 24.2: Imaginary Line Connecting Points of Equal/Same Elevation on the Surface of the Soil

Contour Farming

Sowing of crops along the contour of land. It is one of the best way to conserve soil moisture *in situ*. Flow of moving water checked by thick stand of crop stems, reducing its velocity and lessening erosion. Saves moisture for crop production, reduces soil loss and increases yield.

Contouring

Practice of driving the plough furrow, ridge or trench along the contour of the land to control erosion in areas of high rainfall. This practice is also effective to conserve all available moisture in areas of low rainfall; includes phases of water catching.

Contour Planting

The planting or sowing of crops in rows along the contour lines, especially it is done in comparative less rainfall as well as soil are mostly loose and friable in nature in mildly sloppy or slightly/mildy hilly areas.

Contour Ploughing

Ploughing a field on a contour or line of equal level of lands.

Contour Terracing

Series of properly spaced ridges or drainage channels formed along contours by construction of suitable mounds of earths – called level terracing when constructed

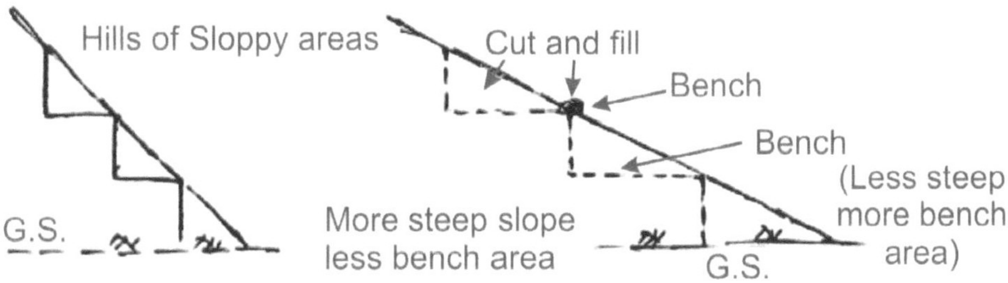

Figure: 24.3 Contour Terracing

along contour; but in case of graded terrace if given slight longitudinal gradient to promote safe discharge of excess water from the area.

Contour System of Planting

System of planting fruit trees or other plants usually followed on the hills with high slopes, where tree rows are planted along a uniform slope and, usually at right angles to the slope with the idea of reducing loss of top soil by erosion. Planting done just as in the square system.

Grass Waterway

A waterway means, which has been covered with grass to prevent gully erosion.

Grassland

Grassland means a land under grass, but not under cultivation.

Gully

1. A gash (deep cut) cut into a slope of soil or loose, unconsolidated sediments by concentration of rainfall runoff into a channel. One of the most destructive forms of soil erosion.
2. A furrow or channel cut by running water during and just after rains. When it is shallow in nature, called a rill (sheet/rill is the initial stage of soil erosion).

Gully Contour

The checking of gully erosion by mechanical or vegetal means.

Gully Erosion

The third and final stage of water erosion, closely followed rill erosion, the tiny grooves developing into wider and deeper channels or gullies assuming huge sizes; gullies trend to deeper and widen with every heavy rainfall, cutting up larger fields into smaller fragments, and in course of time, making them unfit for cultivation.

Gully Head

The upper extremity of gully, which is usually the place where the water falls.

Rainfall Pattern and its Distribution in India

The average annual rainfall in the country is 1200 mm, which amounts to 400 m ha m of precipitation falling over the country's geographical area. Over 80 per cent of the annual rainfall of the country is received advancement of monsoon across the country is a regular phenomenon. Annual variability with respect to onset and its distribution is noticed throughout the country. The variability increases with decrease in rainfall.

However, we have an uneven distribution of rainfall throughout the country in respect to amount and type. We have an extreme variation in respect to amount of average rainfall in the India, as for example, in Cherrapunji (Mawsynram) and adjoining *Khasi* hills of Meghalaya and Assam, the average rainfall is >10,000 mm, whereas in extreme arid part of western Rajasthan it is only <250 mm (even it is only round about 100 mm).

Normally the onset of monsoon occurs on the first of June over Kerala coast and advances steadily further into different parts of the country. By the end of June, it covers the whole country expect the extreme parts of western Rajasthan where it enters by mid of July.

The withdrawal of monsoon (termination of main rainy season) begins from west Rajasthan by 1st week of September and extends rapidly over western India up to West Uttar Pradesh, West Madhya Pradesh and Gujarat plain, NE India, eastern Madhya Pradesh, Vidarbha and other adjoining areas of Maharastra and interior Peninsula. The NE monsoon rains merge with the withdrawing SW monsoon along the coastal Andhra Pradesh, Tamil Nadu and Kerala. There is also a considerable variability in the timing of onset and withdrawal of NE monsoon.

What is Dry Farming

System of farming in regions of low, uncertain or erratic and variable rainfall *i.e.* not in a systematic manner. Or otherwise it may be said that agriculture practicing is totally dependent on the available on the variability of rainfall. It is estimated at about 1/3rd of total arable land our country is under rainfed and in agricultural practicing is about 2/3rd of the total arable land. Simply, otherwise we can say, an improved system of cultivation in which maximum amount of moisture is conserved in soil from low and untimely rainfall for the production of optimum quantities of crops on a sustained basis.

Among the management part of dryland ecology, some measures are very important and these are:

1) Mechanical Measures

a) Selection of proper land for different dryland crops,

b) Contour bunding,

c) Contour cultivation and

d) Bunding, terracing etc.

2) Crop Management Practices

a) Use of improved varieties adaptable for dry land ecology (low-water requirement crops)
b) Adoption of lower seed rate
c) Wider spacing of crops/less plant population
d) Use of organic matters and fertilizers to improve soil structure, supply of plant foods
e) Practicing of contour stripping
f) Fallowing
g) Crop rotation etc.

3) Materials Used for Reducing E_T from Soil as well as from Plant Surface

a) Mulching and
b) Use of artitranspirants.

What is Watershed and its Management

Watershed is the integral part of any area (dry, arid, semi-arid, sub-humid, humid, wet and even very-wet areas), in which excess or surplus rainwater is possible to store at its maximum level and this are effective as irrigation and other uses as recycling way to meet up the crop demands *in situ* or succeeding crops. Its management is important so that it could be served as sustainable manner. The watershed is a manageable hydrological unit that makes a harmonious use of the (i) prevailing climate, (ii) soils, (iii) water, (iv) locally available materials and (v) human resource towards stepping up crop yields or,

Watershed management or protection implies the wise use of soil and water resources within a given geographical area so as to enable sustainable production and to minimize floods.

Watershed management in terms of its physical components is very synonymous with soil and water conservation – with an emphasis on optimum production, rather than a maximizing crop production.

Watershed management practices are those changes in (i) land use, (ii) vegetative cover and (iii) other structural and non-structural actions that are taken in a watershed to achieve specific watershed management objectives. The objective of the plan will have to improve the situation through some step of actions *e.g.* a cutting of vegetation will increase discharges in streams. Therefore, a programme of management will often mean a compromise between two objectives —

The overall objectives of all watershed management are —

- ☆ to increase infiltration into the soil
- ☆ to control damaging excess runoff
- ☆ to manage and utilize runoff for useful purpose (Tideman, 1999).

(WARASA – 'Watershed Areas Rainfed Agricultural System Approach', it is work plan or procedure for the development of watershed areas).

Runoff Harvesting, Storage and Recycling for Soil and Water Conservation

Harvesting of runoff at a micro–level for storage and recycling is necessary for better utilization of rainfall in an area. Such operations are effective for control of erosion and providing life-saving irrigations to crops during dry spells in the monsoon season and also for growing a second crop in succeeding *rabi* season in many regions of the country. This may widely be applicable from humid areas of Dehradun to arid and semi-arid regions of Rajasthan and other dry areas of the country.

Small ponds and tanks are essentially micro–reservoirs during the intense rainy periods. But they are subject to seepage and evaporation losses at every moment. These can be controlled to some extent applying various control measures. In the ponds constructed of alluvial soils (Agra and Vasad regions) heavy seepage losses have been observed in so many studies. On the other hand, seepage losses from dug–out ponds in black–clay soils at Bellary are negligible. Seepage contour techniques for farm ponds are still in the experimental stages in so many areas, as it is subject to infiltration/percolation losses mostly loose and friable soils. At present, the comparison with the cost of water recourses development by large reservoirs or by minor irrigation works, the cost of lining (*pacca* construction by concrete) in small farm ponds is done with bricks and cement. However, cement concrete appears to be justified (although initial cost is high enough), particularly in areas where no other source of irrigation water is available (Dhruba Narayan *et al.*, 1997), as agriculture is totally dependent on the variability of rainfall.

Non-Agricultural Lands

These lands have many limitations, like as steep-slopes, excessive erosion, stoniness, rockiness, shallow soils, wetness, flooding or waterlogged etc. Such lands are, therefore, unsuitable for normal agriculture. The work done at research centers has shown that these lands can be put to productive use as positives and forests lands.

Role of Vegetation Cover: The Study

Establishment of vegetative cover is one of the most effective ways of soil and water conservation. It has been demonstrated that natural fallow *i.e.* with grasses or shrubs or both give the least and negligible runoff and soil loss when compared with the practice of cultivated fallow, which gives maximum rate of runoff and soil loss (Table 24.3).

Trees intercept rainfall and reduce its velocity and the force with which raindrops strike the soil surface. The interception value for *Acacia* (at Chandigarh) is about 15.5 to 20.0 per cent and for *Shorea robusta* at Dehradun at about 14.0 per cent (Tejwani *et al.*, 1975).

Micro-Level Water Harvesting and Wise Use

In principle, water harvesting consists of using water, derived from an area that has been treated to increase runoff of precipitation in order to supplement soil moisture in an adjacent area, situated generally at a lower elevation. In may dry regions, in which contemporary precipitation (low rainfall) alone is not sufficient to ensure a crop. These harvested water that accumulated in the soil will be sufficient, if a proper ratio between donor area and recipient area is established.

Table 24.3: Runoff and Soil Loss on 9.0 per cent Slope Under Grass and Fallow at Dehradun Regions

Sl.No.	Treatment	Monsoon Rains (June to Oct) Rainfall 1250 mm		Winter Rains (Nov to March) Rainfall 173 mm	
		Runoff as (per cent) of Rainfall	Soil Loss (t/ha)	Runoff as (per cent) of Rainfall	Soil Loss (t/ha)
1.	Grass cover (*Cynodon dactylon*)	27.1	2.1	11.4	0.2
2.	Bare fallow	71.1	42.4	58.7	3.6
3.	Bared and ploughed	59.6	156.0	45.6	5.8
4.	Natural grasses	21.2	1.0	4.1	0.0

It is one of the greater example of management of rainwater at micro-level, may be controlled within the field itself, which may be possible only through some manipulations. In this system, one row of crops alternately may be skipped-off and would be covered up the skipped row by plastic mulching or any other means, from where harvested water would be available to the rows where plants has been grown. Plants population would be maintained as per the availability of rainfall in the area.

Methodological Approaches

(i) Micro Watersheds

Cropping system may be developed in which runoff water from parts of a field are concentrated in to the strips, in which plants are planted. The crops are sown in narrow strips between wide intervals that are ridged as artificial miniature watershed. These later are compacted to increase runoff of water to the crop rows. The relative

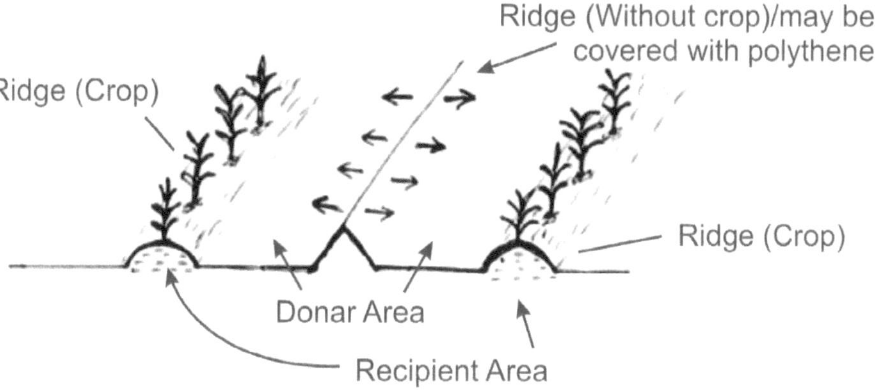

Figure: 24.4 Micro-Level Water Harvesting Structure in Field Condition.

widths of the water-shedding strips and of the crop producing strips depend on the amount of annual precipitation that can be expected. The two approaches for increasing the runoff, used are –

a) Ground covers with plastic films, rubber or metal sheet materials aiming to minimum evaporation loss from the soil surface and may be performed by
b) Dispersing the soil colloids and sealing soil pores by spraying sodium carbonate (Na_2CO_3) @ 45 kg/ha. About 70 per cent runoff water of rainfall was obtained by using sodium carbonate (Myers, 1967). The main disadvantages of the various films are the high cost and susceptibility to damage by winds.

(ii) Mechanical Measures: It includes –

a) Contour farming,
b) Contour bunding,
c) Contour graded bunding,
d) Contour conservation ditches and
e) Conservation bench terraces

Mechanical Measures

These measures form the second line of defense in agricultural scarce areas, which are subject to relatively severe erosion hazards and runoff loss in the problematic areas. The mechanical measure includes i) contour and graded bunds, ii) contour ditches and iii) bench terracing on relatively steep slopes in hilly or undulated areas.

Contour Farming

Figure shows that contour farming is an effective soil conservation measure for gently agricultural sloppy lands. If has been seen that contour cultivation alone can reduce runoff and soil loss from agricultural watershed areas even in high rainfall regions.

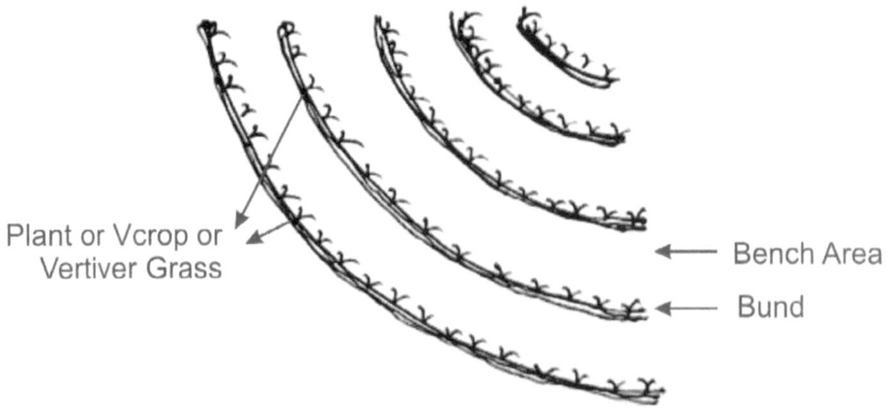

Figure: 24.5 Contour Farming as Mechanical Measure for Practicing of Crops in Sloppy Areas.

Contour Bunding

Among the mechanical measures, field bunding is a common practice for moisture conservation and erosion contour in agricultural watersheds. At Dehradun, an agricultural watershed treated with contour/field bunding reduced the runoff volume and peak discharge to the extent of 62 per cent and 40 per cent, respectively.

Contour bunding can be adopted on almost all the soils of the country receiving an annual rainfall up to 600 mm and having adequate infiltration rates. Contour bunding in agricultural watersheds of many regions is found to reduce surface runoff and soil erosion considerably. In the Dehradun regions, runoff observations on a 55 ha agricultural watersheds treated with contour/field bunds have shown that runoff volume and peak were reduced to 62 per cent and 40 per cent, respectively of the values from fields before treatment. (Ram Babu *et al.*, 1980).

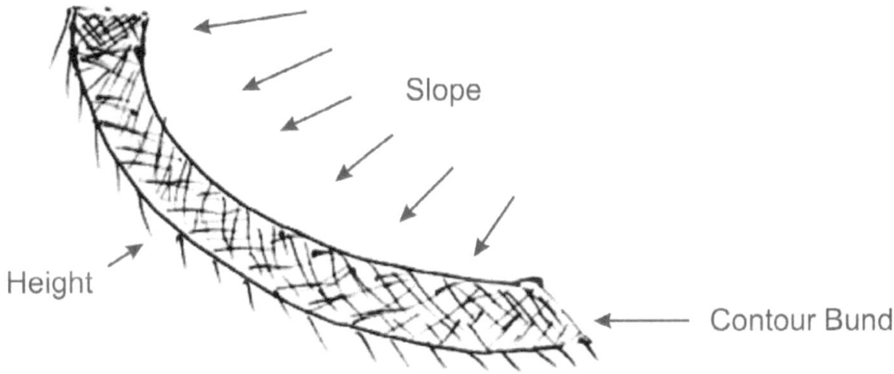

Figure: 24.6 Contour Bunding is Effective for Soil and Moisture Conservation in Sloppy and Hilly Areas.

Graded Bunding

A broad-based channel type terrace at Bellary (Karnataka). In black soils, broad beds on 0.3 per cent grade have been found useful in conservation more moisture and giving 24 per cent and 8 per cent increased yields of sorghum and safflower, respectively.

Graded bunding or drainage type terrace is generally practiced in areas receiving more than 60 cm of annual rainfall irrespectively of the soil texture. It is useful for a conservation practice especially in the deep black soils where the annual rainfall is less than 60 cm (600 mm). The graded bunds are constructed on grades varying from 0.2 to 0.4 per cent and the excess runoff water is drained to the grassed waterways which lead to farm ponds.

Conservation Ditches

In deep black soils, which crack on drying, conservation ditches as an alternative to contour and graded bunds, have been found to harness scare rainwater effectively. The stored runoff water in the ditch is recycled for crop production during the lean period.

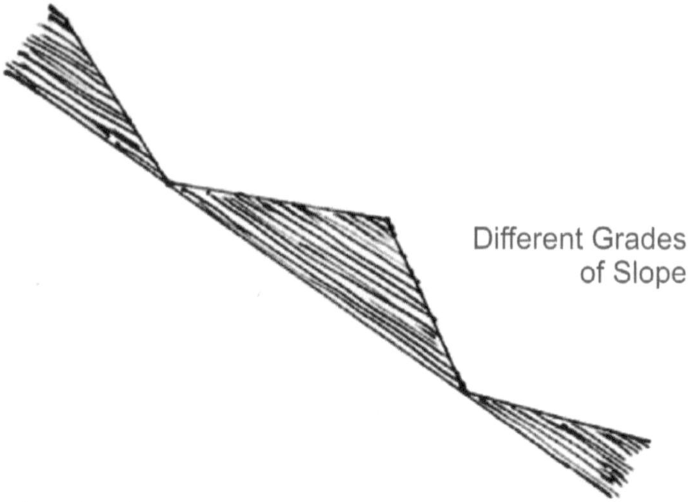

Figure: 24.7 Graded Bunding is Effective for Soil and Moisture Conservation in Slppy and Hilly Areas.

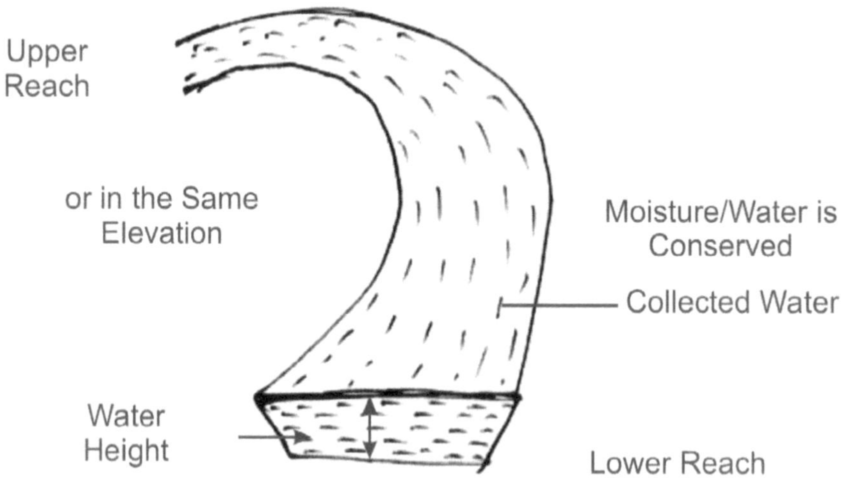

Figure: 24.8 Conservation Ditches is Effective for Soil and Moisture Conservation in Sloppy, Hilly Areas.

In deep black soil regions of Karnataka, conservation ditching is being tried as an alternative to contour and graded bunds. The contour ditches, trapezoidal in shape are laid on contours at the same vertical interval as the contour or graded bunds *i.e.* 0.9 m. These conservation ditches serve as mini-water storage ponds for each field. The stored runoff water is recycled for crop production during lean spells. Such contour ditches are relatively free from the limitations of cracking etc., observed in the bunds constructed in deep black soils.

Conservation Bench-Terraces

On steep sloppy lands, intensive farming can be practiced only with the developing of bench terracing. A terrace system for semi-arid regions have been designed by Zingg and Hanser (1959), that not only reduces the soil and water erosion but also spreads runoff water over a levelled bench, where it is started in the soil for use by the crop itself.

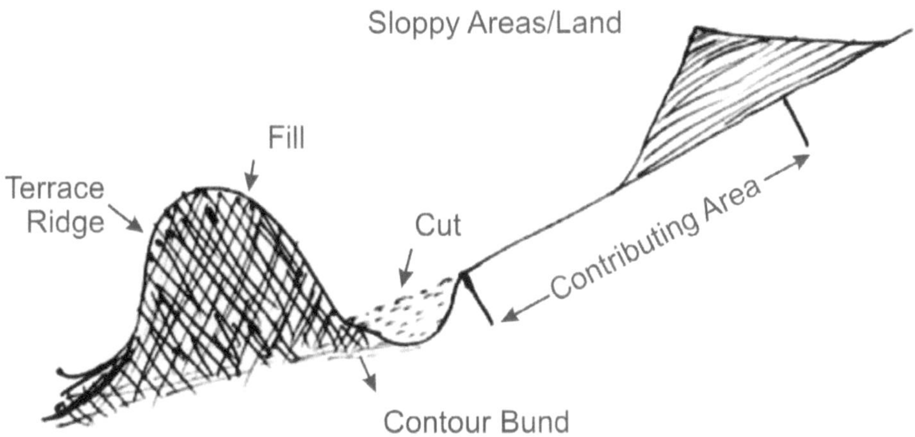

Figure: 24.9 Conservation Bench-Terracing is Effective for Soil and Moisture Conservation in Sloppy, Hilly Areas.

Experience in the field with these terraces shows that good results can be achieved with this method, provided that –

a) Precession levelling is carried out
b) There are adequate terrace outlets to handle excess runoff water and
c) Fertilizers are applied so that the stored water is used efficiently

However in rainfed area, bench terracing can be recommended only on slopes ranging from 6 to 30 per cent. Studies on specifications for bench terraces where conducted areas of Nilgiri in Southern India. It was observed that on areas with an average slopes of 20 per cent, bench terrace with length of about 100 meter, longitudinal grades in the range of 0.2 per cent to 0.8 per cent and inward grades greater than 1 per cent conserved the soil and soil moisture more effectively. These specification can therefore, be greatly adopted in the Nilgiri hills and its adjacent areas.

Recent Advances in Dryland Agriculture in Respect to Soil and Water Conservation

Concept

Indian agriculture is predominantly a rainfed agriculture, under which both dry farming and dryland agriculture are included. Dry farming was the earlier concept, where amount of rainfall (< 500 mm per annum) remained the deciding factor for more than 50 years.

In modern concept, dryland areas are those where the balance of moisture is always on the deficit side (shortage of moisture). In other words, annual E_T exceeds precipitation that received in the areas, is considered as dryland agriculture. There is no consideration on the amount of rainfall receives in a given year, *e.g.* suppose in an area where an amount of 1100 mm or more rainfall annually receiving in an area, but there loss of water from the cropped field (E_T) is more, should fall in the category of dryland agriculture under this concept. To be more specific, the average annual rainfall of Varanasi is around 1100 mm and the annual potential E_T loss is 1500 mm. Thus, the average moisture deficit so created comes to 400 mm annually. This deficit of moisture is bound to affect the crop production under dryland situation, and ultimately resulting into total or partial crop failure. Accordingly the production is either low or extremely uncertain and unstable, which are the real problems of dryland in the country.

In that situation either practices of water conservation and/or management of field, crops etc. would be in such a way, so that some expectation of crop yield should be ascertained without any drastic crop failure. The concept of conservation agriculture are much emphasized concerning all management practices together with for greater outcome but without any degradation.

What is 'Water Balance': The Pre-Concept for Conservation ?

The success of crop production in dryland areas depends on the amount and distribution of rainfall, as this influence the stored soil moisture and moisture used by the crops. The amount of water used by the crops and stored in the soil is governed by the **'Water balance'** -

Water Balance Equation

Water balance equation : $E_T = P - (R + S)$. When the balance of the equation shifts towards right, precipitation (P) is higher than ET, so that there may be waterlogging or it may even lead to run-off (R) and flooding. On the other hand, if balance shifts to the left ET becomes higher than the precipitation, resulting in drought of various severity. In fact, the balance of equation is controlled by weather, season, crops and cropping pattern in a place.

where,

E_T: Evapotranspiration in an area (from soil and crop)

P: Precipitation or rainfall in that area

R: Runoff water exceeding S

S: Used by the crop + stored soil moisture

Example

1) Category 1

ET = 1000 mm in a given time

P = 1500 mm in a given time

R = 200 mm in a given time

S = 100 mm in a given time

Then, 1000 = 1500 – (200 + 100)

= 1500 – 300 = 1200 mm ('+ve' site, as ET loss 1000 mm is less than

P = 1500 mm)

2) Category 2

Suppose, ET = 1500 mm

P = 1100 mm

R = 000 mm

S = 200 mm

Then, 1500 = 1100 – (0 + 200)= 1100-200 = 900 mm (-) ('-ve' site, as ET loss 1500 mm is less than P = 1100 mm)

Main Problems in Rainfed Dryland Areas

In dryland and/or rainfed agriculture, scarcity of water is the main problem. Apart from this –

1. Low and erratic rainfall
2. High evaporative demands from the surface and
3. Limited water holding capacity of the soil, are the principal constraints in crop production in dry farming areas.

Mitigation Options in Dryland Farming System for Soil and Moisture Conservation: Improved Technology

The characteristics of crop production in dryland areas, is that maintaining a certain moisture regime in the soil as well as within the crop itself. The improved techniques and practices, which have so far been generated and recommended for achieving the objective of increased and stable crop production in dryland areas. These are as follows:

1. Crop planning and seed manipulations (low seed rate/wider spacing, seed soaking etc.)
2. Planning for aberrant weather (crop, field management related with weather)
3. Crop substitution or contingent crop planning
4. Efficient cropping system (intercropping/crop sequence)
5. Addition of organic matter for improving soil property (nutrient status, WHC etc.)
6. Need-based addition of fertilizer
7. Rainwater management for better utilization

8. Watershed approach for resources improvement and utilization
9. Micro-watershed (in the field system)
10. Improvised cultivation practices (tillage, sowing and intercultural operations)
11. Alternate land use system
12. Use of efficient implements for reducing moisture loss
13. Moisture conservation by mulching
14. Drought stress and controlled by mechanical means
15. Developed and use of drought tolerant genotypes

Cropping Systems Based on Availability of Rainfall under Rainfed Ecosystem

In rainfed sub-regions, the length of water availability period determines the crop and cropping systems at a location. This in turn depends upon –

i) Start of the rainy season,
ii) Distribution of rainfall during the rainy season,
iii) End of the rainy season,
iv) Water holding capacity of the soil and
v) Evaporative demand.

However, as per the development programme on dryland farming system, some specific areas/pockets (AESR, Agro-ecological sub-regions) have achieved for irrigation facilities, creating and development of watershed basins, ponds etc. (may be lined bottom or *pacca*) efficient conservation of rainwater (runoff).

Agronomic-cum-Field Management for Soil and Water Conservation

Management Practices for Increasing Soil Water Conservation vis-à-vis for Increasing Yields Potentials

Objectives

There are so many mechanical measures including crop management in the field, can be adopted suitably for increasing soil water potential *in situ* as well as it ensures maximization of crop production particularly in dry areas of the country. Successful crop production in region of frequent drought requires specific methods and practices for providing or maintaining sufficient available water for crop growth. Among these methods and practices are selection of crops that evade (escaping drought periods) or endure (tolerance at its maximum level of this adverse situation) periods of insufficient moisture utilization of cultural and soil management techniques that increase soil water storage or decrease the rate of use of the limited soil water supply and development of methods that reduce evapo-transpiration (E_T) : The operations are –

1. Bunding
2. Contour ploughing
3. Addition of organic matter
4. Low seed rate with wider spacing
5. Crop rotation/cropping sequence
6. Strip Cropping
7. Fallowing
8. Water harvesting and recycling
9. Scooping
10. Closing of soil cracks
11. Laddering
12. Minimum tillage and wheel track planting
13. Selection of planting dates
14. Methods of sowing
15. Gap filling
16. Thinning
17. Mulching
18. Use of artitranspirants etc.

1) Bunding

Bunding is mainly done for (i) moisture conservation, (ii) for maintaining fertility of the soil and (iii) for effectiveness in soil conservation. In dry farming areas, since the rainfall is irregular with low concentration and comes suddenly, the loss of water occurs through runoff. This may check effectively by giving bunds and by adapting this, water will be conserved *in silu* at a maximum rate. Bunds are generally given across the slope of the land. Moreover, as the excess rainwater drains out as runoff from the crop field in huge quality, these is every possibility to erode soil with the runoff water. Thus soil erosion is also checked by given bunds. Top soils ensures a good state of soil fertility thereby is checked or controlled by applying this mechanical measures like bunds in the sloppy areas, particularly where the soil is too erode in nature. Bunds may be performed by any materials available in the areas, by which we can check runoff water as well as soil particle from the areas. This may be done by earths, stones, gravels, concrete, trees, shrubs or vertiver grasses etc.

2) Ploughing

Ploughing of shallow, medium in light soil should be omitted as far as practicable. Generally in these areas water holding capacity of soil is very low and it is necessary to check evaporation of moisture as much as possible. If ploughing is done frequently, the sub-soil will be stirred which will favours evaporation. At the same time loosening of sub-soil may help greatly for soil erosion. Mechanically this should be controlled by giving hoeing twice instead of ploughing operation frequently.

3) Addition of Organic Matter

Addition of organic matter helps to improve (i) soil texture, (ii) soil structure, (iii) drainage condition as well as (iv) fertility status of the soil. It is noted that higher temperature lowers the organic matter contents in the soil, so it is obvious that in dry areas organic matter is comparatively less. To compensate with this losses as well as to increase the water holding capacity of the soil it is an integral part to incorporate more and more organic matter with the soil due to improve its physical (texture, structure, BD, WHC) and chemical environment (nutrient, pH etc.) as well.

4) Lower Seed Rate with Wider Spacing

It is wise and most desirable that less plant population in dry rainfed tracts, which ensure less quality of water would be lost from the plant body itself through transpiration. This has, however, been aggravated if the soil contains less amount of water during seeding and post-seeding period. For this, lesser quantity of seed material should be used for obtaining less plant population in the field, which can be obtained by maintaining wider spacing and low seed rate.

5) Crop Rotation

Suitable crop rotations are to be followed under different dry regions of the country. This should be based on the soil and climate condition of that area. It is to be advocated always to incorporate any of the leguminous plants/crops in the rotation. It will not only for addition of organic matter (bio-fertilizer) but also improves the soil condition due to more canopy development. Moreover, its root system will adhere more soil volume which will ensure to hold more soil in the field and it will obstruct to erode soil from the field also. Due to more canopy development as well as very close to the ground it will act as vegetation mulch over the field. As for example :

Jower – Jower - Groundnut/gram

More or less all the crops in the rotation are of low water demand. Hence, these are to be incorporated in the rotation.

6) Strip Cropping

Simplest, this is a technique or practice of growing soil erosion resisting and soil permitting crop in strips one after another for achieving a good result in regard to control soil erosion as well as conserving more moisture *in situ*. Strip cropping is practiced by growing different types of crops, - each kind in a particular strip. This is very effective to check the soil erosion by water and this is a very good practice in dry farming areas. More over, as we need all types of crops in our dietary system, so it is essential to incorporate all sorts of cropping systems, like cereals, pulse, oilseeds etc.

7) Fallowing

Fallowing means keeping the land without any crop for a particular period. There are different types of fallowing practiced in the field for the different type of activities :

- **(a) Uncultivated :** After harvesting of one crop the land is kept without any ploughing operation. This is further divided into two categories, - *viz.*

(i) **Weedy fallow :** Without controlling weeds in the field *i.e.* undisturbed

(ii) **Unweedy fallow** – Where weeds are controlled by the application of herbicides only, without the field disturbed by any mechanical means.

(b) Cultivated : Land is ploughed to control the weeds and conservation of soil moisture by using repeated laddering operation etc. to compact the soil for conserve more moisture in the field. This operation will restrict loss of evaporation from the field.

Pre-Requisite for Ploughing

Actually some criteria would be available and performed for practicing of fallowing in an area/land.

i) Sufficient rainfall would be available,

ii) Well distributed rainfall *i.e.* in timely available,

iii) Weed control may be performed for lesser loss of water as transpiration and

iv) Medium type of soil instead loose and friable erode in nature.

Efficiency of Fallowing

Efficiency of fallowing (E.f.) means the objective would be fulfilled by the practicing of fallowing *i.e.* assessment of moisture content in the soil after fallowing.

E. F. =

$$\frac{\text{Moisture content in soil after fallowing}}{\text{Moisture content before fallowing} + \text{Rainfall received during fallowing}} \times 100$$

If the E.f. value is less that 25, there is no use of fallowing for moisture conservation.

Factors Affecting E.F.

1) Soil type – Medium type of soil is best.

2) Soil depth – More the soil depth, more value of E.F.

3) Well distributed and reasonable intensity of rainfall means more E.F.

4) Season – There are two fallowing practices

   i) Summer fallowing

   ii) Winter fallowing

In winter fallowing the E.A. is more as because evapo-transpiration is less during period as compared to summer season.

Practices during Fallowing

(i) Ploughing

(ii) Intercultural operation like seeding advantages of fallowing

Advantages of Fallowing

(i) Moisture conservation

(ii) Weed control

(iii) **NO_3 formation :** Nitrification is influenced due to fallowing. In a semi arid winter rainfall area, fallow increased available nitrogen from 5 to 6 times over recent cropping or continuous cropping system.

8) Water Harvesting and Recycling

Already discussed

9) Scooping

It is a very good practices in dry farming areas. With a machine/scooper some patches/pits are formed on the field before rain comes. When rain comes these particular patches holds more moisture than the unscooped area and the plants or seeds are then sown in the mid portion of the scooped areas to grow a good crop. And moisture would be available from the sub-soil scooped areas, as there is more moisture holds during the rainy time.

10) Closing of Soil Cracks

In dry areas there is cracking in the field due to lack of moisture. This situation is generally created during the drier months of the year and more accurate particularly in heavy or compact type of soil (black cotton soil also in Maharastra, Gujarat etc.). These cracks should be closed by suitable methods before rainfall occurrence, otherwise water will be lost by deep percolation through these cracks.

11). Laddering

It is an effective field operations to conserve more moisture in the field. After cultivation of land, repeated laddering is done with a view to compact the soil surface, otherwise moisture will be lost by evaporation and it will be more if the soil is loose or friable in nature after plough operations. This operation is done by *'Sohaga'* or 'ladders', mainly made with wood.

12) Minimum Tillage and Wheel-Track Planting

Minimum tillage can be defined as a method aimed at reducing tillage to the minimum necessary but obviously will perform the favourable condition for (i) ensuring a good seed bed, (ii) rapid germination, (iii) satisfactory stand and (iv) favourable growing conditions in the field. One example of the several 'minimum tillage' methods of the several minimum tillage developed is 'wheel-track planting'.

Wheel Track Planting

In this method, the field is ploughed as usual, bet instead of using the conventional methods of seedbed preparation, the seeds are planted in the compacted soil by the wheels of the tractor. Experience has shown that tractor wheel make a firm satisfactory seed-bed even on cloddy soil, ensuring rapid and even germination. The soil between the rows remain rough, loose and is therefore, better able to absorb more moisture and to reduce rain-off. Wheel-track planting was found to save about 40 per cent of tillage costs. Minimum cultivation has to be associated with chemical weed control.

13) Scheduling of Planting Dates

The planting dates should be adjusted to the pattern of precipitation over year a particular area (to observe LTA value of meteorological data of the area). The humid season of the semi-arid regions can be divided into three principal periods (Cochemen and Franquin, 1967).

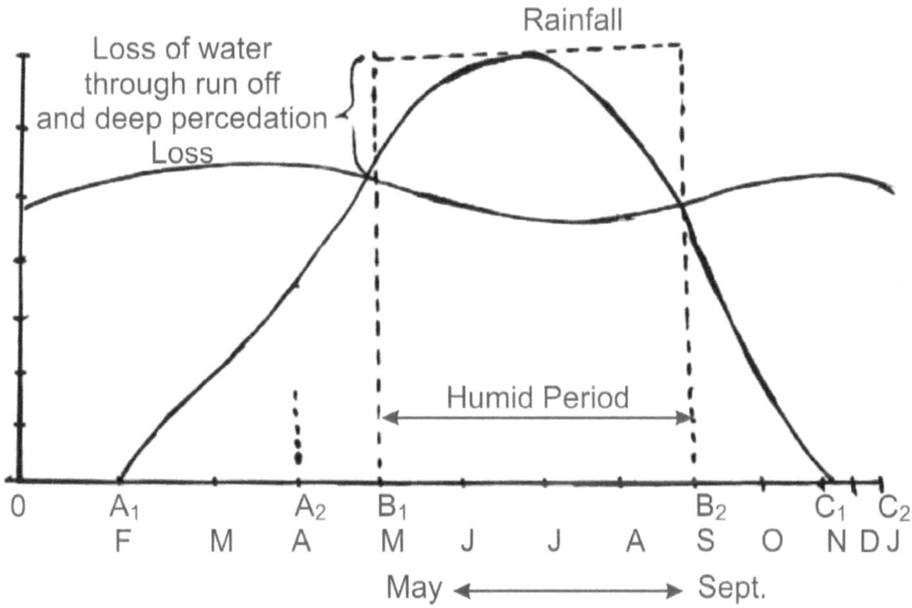

Figure 24.10 Schedule of Planting Dates

Sub-division of the Cropping Season in a Semi-arid Climate.

a) Pre-humid period
b) The humid period
c) The post-humid period

A_1 – beginning of the rainy season
A_2 – beginning of transition period (ppt = half of PET)
B_1 – End of transition period è beginning of humid period
B_2 – end of humid period
C_1 – End of rainy season
C_2 – exhaustion of moisture, reserves in the soil

14) Methods of Sowing

i) **Kera methods of sowing**: After land preparation, dragging the plow and make furrows the land. Put seeds in these furrows and laddering are done immediately after sowing operation. If soil moisture is high enough then sowing may be practiced by following Kera methods, which is done relatively in upper layer of the soil surface.

ii) **Pora methods of sowing:** If the soil moisture in upper layer is comparatively less, than Pora methods of sowing is practiced. In this method laddering is not done. Immediately after plough operation, seeds are put in furrows, comparatively in deeper layer and covered the seeds by the soil and compact it.

15) Gap Filling

If there is patchy germination in the field, the gaps are filled up by the germinated seeds. Immediately after putting the germinated seeds in the furrows some watering is to be provided for further emergence and growth of the seedlings.

16) Thinning

Thinning operation is very essential and it is done according to the moisture content of the soil, particularly in drought prone areas. More or less it is adjusted the plant population as per the availability of rainfall over the area

17) Use of Antitranspirants

Control on Evapotranspiration (E_T)

Reducing water losses due to evapotranspiration from the soil as well as from the plant system can be controlled through some mechanical means. Evaporation from the soil can be controlled by (A) mulching and by (B) chemical treatment of the soil.

Water losses due to transpiration can be reduced by plant treatments aimed at directly decreasing transpiration and indirectly by management practices such as weed control, maintaining optimum planting density and planting patterns etc.

In true sense, dryland and rainfed farming is the integrated and rational system of growing purely rainfed crops under **low** and **erratic** rainfall condition (40-120 cm rainfall annually) and erratic means irregular.

In principle, transpiration can be reduced in the field in four ways.

a) Increasing leaf resistance to water vapour loss by applying materials that (this may be possible by the use of antitranspirants)
b) Reducing top growth by growth retardants
c) Increasing the air resistance to water vapour transfer, by using wind breaks.

Stomata Closure

Two types of gas exchange between plant and atmosphere transpiration and CO_2 assimilation for photosynthesis are intimately linked. Stomata resistance slows down simultaneously both process; the loss of water to the atmosphere and entry of CO_2 into the plants.

Objectives of Antitranspirants

The main objective of antitranspirants is that, by reducing transpiration loss, they might cause an increase in leaf temperature to critical levels. It has lower ever, been estimated that under field condition, transpiration can not reduce the temperature of leaves by more than 5 per cent and that under most conditions, the best antitranspirants reduce transpiration by 30 – 40 per cent at the most.

Inducing Stomata Closure

Soupy materials used for various purposes such as certain herbicides, fungicides, metabolic inhabitants and growth hormones, have been found to cause the closure of guard cells of the stomata and thereby reduce transpiration.

The main practical problem is to obtain materials that induce stomata closure without toxic effects and that are long-lasting. For example, one of the initially most promising materials phenyl, mercuric acetate (PMA), a fungicide that also causes stomata closure, was found to cause direct inhibitions of mesophyl photosynthesis. Now, it has been made it unsuitable for use as an antitranspirant.

Use of Antitranspirant

An antitranspirant is any material applied to transpiring plant surfaces with the aim to reducing water losses from the plant. Approximately, 99 per cent of the water taken up by the plant roots is transpired to the atmosphere through stomata pores in the leaves.

Types

a) Reflecting type
b) Film-forming type and
c) Systemic type.

a) Reflecting Type

When these antitranspirants are used on leaves and stems, it helps to reflect the sunlight and lowers the heat load. This will check the transpiration loss of water, eg. Kaolin, $CaCO_3$ etc.

b) Film-forming Type

These chemical form a thin film on the leaf surface and ultimately reduce the transpirant, *e.g.* Mobileaf. Davenport *et al.* (1972) used this commercially available was emulsion manufactured by the mobil oil company, on peach trees and found resistance to water vapour diffusion from the leaves and also the water potential of those leaves increased.

c) Systemic Type

Stomata closure by chemical sprays has been investigated and one the most reliable inhibiters is phenyl mercuric acetate (PMA). This chemical closes the stomata opening and reduces transpirant. PMA has several desirable properties which make it useful as the chemical:

1. Closes stomata when sprayed on leaves at low concentrations,
2. Affects only the stomata in the sprayed area and is not likely to be translocated and
3. Will inhibit stomata opening for at least 4 days without apparent toxicity in many species.

18) Mulching

Definitions

A mulch is any material at the soil surface that was grown and/or maintained in place; any material grown, but modified before placement and any material processed or manufactured and transported before placement. Some examples of the materials included are : crop residues, leaves, clippings, bark, manures, paper, plastic films, petroleum products and gravels etc. These mulching materials are used for the purpose.

In broad sense it is the aim of any dryland agriculture is to conserve soil and water more in the field by applying mulch materials over the soil surface with the aim of creating obstructions or barriers for the loss of evaporation from the soil surface. For this article, dryland agriculture is defined as agriculture in areas where potential evaporation (E) and transpiration (T) are greater than the amount of water supplied/available by precipitation over the area.

Manipulations

A) Reducing of water loss through evaporation from the soil surface can be controlled by mulching. This can be done by 3 ways :
   i) Soil mulch :
      a) Removal of weed (manual)
      b) Chemical weed control

Drawback of Soil Mulch

(i) Frequent stirring may destroy soil structure and (ii) Soil is subject to erosion.

(ii) **Straw mulch :** This can be done by the using of stubbles, straw but the only drawback is the attacking of termites or habitants of insect pest, diseases etc.

(iii) **Polythene mulch :** Now-a-days mulching with various polythene are using, especially in dry and scares areas. Polythene mulches are also popularity used for this purpose, especially when the crop is grown as line sowing or any other systematic manner. Polythene mulching can not be used if the crop is sown as broadcasting.

B) **By chemical treatment**, the evaporation from the soil can be reduced. Hexagonal compounds mixed with the surface ¼th of an inch soil, reduced the evaporation by about 43 per cent.

Manipulation of climate to any great extent is not possible at present. Mulches have been tried to reduce evaporation from the soil with very limited success. Studies have shown that evaporation reduction is probably attained by surface mulching with plant residues only in the top few inches of soil and during the time when the soil is wet. Beyond this point, intrinsic soil properties appear to dominate evaporation, generally a heat beyond this level. A perfect mulch must have an artificial capillary break or a poor capillary conductivity. Straw mulches absorb measurable water and do not meet the criteria for perfect mulch. Artificial mulches are not more effective, partly because as soils dry a dust mulch usually forms.

Effects of Mulches on some Soil Properties and Conditions

Numerous soil properties and conditions are affected by mulches favourably or unfavourably, either directly or indirectly. Among those affected are (i) soil water increases through runoff control, (ii) increases infiltration, (iii) decreased evaporation and (iv) weed control etc. Apart from these, increased soil temperature through radiation shielding heat conduction and trapping and evaporative cooling.

It also improves the nutrient status of the field through addition of organic matter, differential nitrification and mineral solubility. It improves soil structure, soil biological regime through organic matter additions, microbial and soil fauna populations and more over plant root proliferation and distributions within the soil. Mulching decreases soil erodability and soil salinity through leaching and evaporation control in the area. Probably of greatest importance of mulching, however, for agriculture in dry regions are improves soil water, temperature, soil structure and salinity.

[**SWOT**: Strengths, Weakness, Opportunities, Threats. It is the pre-requisite to know the basic information (*e.g.* primary and secondary data) in an area, particularly effective for considering any developmental plan/programme and subsequent implement, may be agriculturally, socially or environmentally etc.]

Conclusion

As water is the key and integral elements/part of natural resources, it would obviously be utilized in precise and wise way for benefits of mankind, even in sub-humid to humid areas and sustainability in the world, either may be ground, surface and/or wetland sources. In the country, in addition with this, vast wetland ecosystem may effectively be utilized through the cultivation of so many aquatic crops, fish variables which not only valued to human beings but also imperative for the upliftment of resource poor rural economy as well, who are inextricably linked with the system. Besides, wetlands are continuously enriched by adding large quantities of biomass, enriched in sequence, which has possible to utilize this enriched biomass effectively for production of arable crops as well as improvement soil nutrient status in consequence in the zones.

Bedsides, in our country, greater part of agricultural land is under dryland ecosystem, may be more effective for management the ecosystem if managing it could be able to use more efficiently and precisely for the optimization of production system for making their livelihood in better way. There are so many management options pertaining to the environment to mitigate the adverse situation, would be overcome in a sustainable manner without losing soil and water loss. Water harvesting structures are the pathways, wherein emphasis would be paid to conserve more runoff, even in wet areas.

References

Anonymous (1998). *In*: Illustrated Oxford Dictionary, published by Dorling Kindersley Limited and Oxford University Press : 942.

Dhruba Narayan V. V., Sastry, G. and Patnaik, U. S. (1990). *In*: Watershed Management, published by Publications and Information Division, ICAR, New Delhi.

Ram Babu, Srivastava, M. M., Sastry, G. and Puri, D. N. (1980). Studies on hydrological behaviour of small watersheds in different land uses. Annual Report 1980, CSWCRTI, Dehradun.

Tajwani, K. G., Gupta, S. K. and Mathur, H. N. (1975). *In*: Soil and Water Conservation Research. ICAR, New Delhi.

Tideman, E. M. (1996). *In*: Watershed Management, published by Omega Scientific Publishers, New Delhi.

Chapter 25

Apiary: A Rural Enterprise for Livelihood Generation and Ecological Health

Shantanu Jha

The Flower as a Food Source

- ☆ More than a hundred million years ago the plant kingdom discovered that so far as its relationships with insects were concerned, fair exchange was no robbery. Insects could be duped into rendering a service, *i.e.* pollination, as long as they were offered a suitable reward.
- ☆ So began one of the longest partnership in biological history and over the intervening millennia the beneficiaries have become ever more closely attached. In return for its services, the pollinating insect receives nectar and pollen, and plants have evolved every means to heighten the allurement that their flowers present.

Nectar and Nectaries

- ☆ Flowers produce two of the most vital ingredients in an insect's diet: carbohydrate and protein. Obligatory flower feeders, such as bees receive virtually all their nourishment from this single source.
- ☆ It has been estimated that a honey-bee can only be sustained in flight if its muscles receive a supply of 10mg. of sugar every hour. Bees generally prefer nectars with concentrations in the range 30-50 per cent.
- ☆ To produce 450g pot of pure honey, apxly. 17000 honey bee foraging trips during which 10 million individual flowers will have been visited. An average foraging trip takes in about 500 flowers and lasts about 25 minutes.

In all therefore, 450g of pure honey represents an investment of 7000 bee-hours of labours.

Bee Keeping

- Bee keeping or apiculture (from latin word apis, bee) is the maintenance of honey bee colonies, commonly in hives, by humans.
- A bee keeper (or apiarist) keeps bees in order to collect honey and other products of the hive, to pollinate crops or to produce bees for sale to other bee keepers.
- A location where bees are keept is called apiary of bee yard.

Wild Honey Harvesting

- Collecting honey from wild bee colonies is one of the most ancient human activities and still practiced by aboriginal societies.
- Honey collecting entailed the destruction of the entire colony when the honey was harvested. This was destructive and unhygienic.
- In settled societies the destruction of bee colony meant the loss of valuable resource.

Modern Hives

- They are all rectangular or square, movable wooden frames.
- Consist of a floor, brood box, honey super, crown board and roof.
- Also use queen excluders between the brood box and honey supers to keep the queen from laying eggs in cells next to those containing honey intended for consumption.

Beekeeping: As a Source of Livelihood

- There are many different entry points for projects to strengthen livelihoods with beekeeping, such as including trees for bees within planting schemes to improve pollination and increase crop harvests, assisting honey hunters through beekeeping or making and marketing honey wines or beeswax cosmetics.
- Beekeeping projects have been started in many developing countries and are frequently supported by international organizations, governments or NGOs. Beekeeping fits in well with other interventions and is often incorporated as one of a number implemented. Some minimum resources, however, should normally be available to people.

Natural Resources: Indigenous Species

Beekeeping projects can improve the potential for beekeeping by planting melliferous vegetation. Indigenous honeybees have evolved and survived successfully under local conditions and will be better suited to them than introduced bees. The European honeybees introduced into many countries and African bees introduced into Central and South America currently form the basis of successful beekeeping industries.

Human Resources: Beekeeping Skills, Training and Extension

Beekeeping is a widespread activity with a wealth of existing local knowledge and skills. The addition of a little technical information, however, can lead to greatly improved harvests of honey and beeswax. There are many ways to assist honey hunters or beekeepers to build on their resources to create more income by harvesting and processing honey more skillfully, and to obtain better prices by saving and selling beeswax and by making secondary products.

Beekeepers and trainers often lack appropriate training materials - most of the literature discusses keeping European bees in temperate zone conditions. Training is often theoretical rather than practical, placing emphasis on changing the type of hive used without providing practical guidance and follow up. New beekeepers need training in how to work with bees, how to maintain honey quality, how to separate honey from beeswax, how to render beeswax, how to manufacture secondary products and how to make beekeeping clothes and equipment.

Physical Resources: Equipment and Transport

Limited access to transport is the main reason why beekeepers in remote areas receive the lowest prices for their products. Projects can do much to alleviate this problem. Rural people can find it difficult to obtain equipment, containers and packaging. The answer is not merely to donate the items but to train local people to make their own equipment and find access to good containers and packaging, and credit with which to acquire them.

The equipment needed for beekeeping can be simple: the humble plastic bucket is one of the most essential items. Recommending good-quality, lidded, stackable plastic buckets may not bring great professional kudos to the beekeeping expert, but such buckets are useful for beekeepers living in remote places who need to keep their honey clean until they are able to sell it. Honey of excellent quality can be harvested as long as clean buckets are available, along with cotton or baskets for sieving honey and containers for melting wax and packaging the honey and other products.

The appropriate equipment for harvesting and processing honey and beeswax depends on the quantities to be processed and the type of product required. In some areas, beekeeping using traditional local hives is practiced on a large scale and justifies the provision of relatively sophisticated, large-scale processing equipment capable of dealing with honey in bulk. Where a cooperative has established a honey-packing unit, a few specialized items often have to be imported, such as effective taps for use on honey containers, special gauze for filtering honey and refract meters to measure water content.

Financial Resources: Credit

In poor societies, lack of credit is a major constraint to everyone concerned with selling and buying honey. Beekeepers with honey to sell expect to receive cash from honey-collection centres or private-sector traders; otherwise they prefer to sell their honey in small quantities in markets to obtain an instant but low cash return. People buying honey need access to credit during the honey season. Lack of credit leads to insignificant volumes of honey being available for sale, no interest from traders and a stagnant industry.

Social Resources: Sector Support and Marketing

In poor countries, there are usually government officers responsible for training and extension in beekeeping. Often, however, they have little relevant training and lack access to transport and other resources. National policies are needed to promote apiculture and protect pollinators. A national NGO is a considerable advantage and able to represent the interests of beekeepers, establish communication between producers and traders and facilitate marketing

In many developing countries, much can be done to increase retail honey sales, for example, by improving and diversifying packaging, especially for small volume markets. Marketing initiatives can involve promoting honey in the media, interacting with consumers and traders to increase honey consumption and sales, and creating links with packaging suppliers. Honey consumption increases according to living standards; people are keen to buy honey when it is well presented and they have more confidence in the product.

The first aim of a marketing initiative should be import replacement, which means ensuring that local honey is packaged and presented as attractively as the imported brands. Only when the local need for honey is satisfied should export be planned, as inexpensive honey is readily available on the world market. In some countries, producers have benefited from having their honey or beeswax certified as organic or produced according to fair-trade criteria. This type of certification can help small-scale producers to find niche markets that pay premium prices. Honey export to the European Community requires expert knowledge of trade rules and import requirements (Brad-bear, 2001).

Project Planning

- ☆ Establish the most effective entry point for a beekeeping intervention, for example, by providing assistance with technical aspects of beekeeping or product marketing.
- ☆ Ascertain that the planned intervention is appropriate for the people concerned.
- ☆ Recognize that beekeepers are often the poorest people in local society and may live in remote places; focus the intervention to reach them.
- ☆ Determine that the producer groups will have market access, if more honey and beeswax are harvested.

Project Evaluation

Many beekeeping projects have involved the distribution of hives and equipment and the provision of technical training. Donors and local leaders might be satisfied with the outcome of such projects when shown convincing numbers of new hives installed in new apiaries. Closer examination, however, often reveals that hives of a newly introduced technology are not always used efficiently. Good training and follow up are essential. Beekeeping is a seasonal activity, and it cannot be learned in the classroom. The true test of success in any beekeeping development project should not be "How many hives were distributed?" but "Were people's livelihoods strengthened?"

Small interventions such as beekeeping projects are not always popular with donors. In poor societies, however, large beekeeping projects with high capital input are frequently doomed to failure. In far too many beekeeping projects, a well-meaning donor has allocated a significant budget to a project, a large portion of which is inevitably intended for equipment. This has led to equipment being introduced that is not always appropriate to the resources available, such as imported woodworking machinery for making hives that may become unusable as soon as a spare part is needed. Training is sometimes irrelevant to local resources and knowledge.

Frame-hive equipment should not be used unless the infrastructure exists for manufacturing it locally. Frame-hive beekeeping is practised in all industrialized countries and many projects have tried to introduce this type of beekeeping in poor countries. It is essential that a community have the physical, human, natural and financial assets to support this type of beekeeping, if such a project is to succeed.

Case Study

Learning

Input	*Outcome*	*Impact*
Merger financial resources	Started operation with simplified technology	Limited to the technology supplier
Capacity Building input from different support agencies	Worked on it. Skill enhancement	Gaining confidence, shaping of enterprise. Started networking
Migratory Beekeeping	Technology diversification	Innovations experimented and fine tuned

Learning

Input	*Outcome*	*Impact*
Technology innovations during resource constraints	Refinement of Technology	Product diversification and cost reduction
Multiple bee hive products	Reaching segmented market	Income enhancement
Product diversification	Reaching segmented market	Income enhancement

Learning

Input	*Outcome*	*Impact*
Technology innovation	Product diversification	Income distribution throughout the year
Technology innovation	Reduction of cost of production	Productivity through hugine
Market intelligent	Finding different buyers	More bargaining power

Learning

Input	*Outcome*	*Impact*
Value addition	Product diversification	Income enhancement through resource optimization
Business plan on the past experience	Leveraging different resource unit	Scaling up

Chapter 26
Value Addition in Vegetables Based Enterprises

Tapan Kumar Maity

Importance

Agriculture in general is the prime sector of Indian economy contributing 22 per cent of GDP while horticulture contributes about 28.5 per cent of GDP in agriculture and 52 per cent of export share in agriculture from cultivated area of 8.5 per cent. In the recent past, Horticulture continuing becomes the prime sector of Indian economy. After the macro economy policy changes, India is in the restructuring and privatization phases of this economy transition. Domestic environment is becoming increasingly competitive. Entrepreneurs are feeling the benefits of liberalization. However, as they consolidate their position in gaining domestic benefits, they should now look for greater scale of such benefits since any measures of economic benefit would be incomplete without entering the International area.

Due to the development in production technology in Horticulture, vegetable production in India has now reached to the level of 101.43 million tonnes from 6.75 million hectares during 2004-05. Horticulture sector has witnessed a growth rate of 2.5 per cent during the last two decades. Although the value of horticultural products has increased significantly, their share in the total exports of the country has in a declined trend. Major horticultural products of exports are spices, cashew, fruits and vegetables. India has the advantage of diversified agro-climatic conditions which provides unlimited scope for producing large quantities of every known economic species of fruits, vegetable, spices, plantation crops and ornamental plants. India is the second leading producers of fruits and vegetables in the world. India has a greater abundance of man power in all categories than any other nation in the world. Our technology options are good. If the available potentials are properly utilized, the productivity can increase two fold within the next five years. The country has several agro-based industries. If the productivity of horticultural crops is increased in these industries, it is not only possible to fully meet the growing demands of the country but also to export substantial quantities to the world market. Development of human resources is an area of prime importance for the food processing industry today, owing to the explosive potential of the agro-based sector, and the consequent spurt in its investments. In India, food has become a high-profit industry by reason of the scope it offers for value addition, particularly with the food processing industry getting recognised here as a high-priority area in the liberalised era. In order to achieve the expected returns on investment through good quality and right value addition, application of high technology has become imperative, leading to an unprecedented demand for scientists, technologists and other professionals who can manage the emerging challenges of the food processing industry effectively.

About 30-35 per cent gets of India's 60 million tones of fruits and vegetable gets wasted due to lack of adequate storage and other facilities. In value terms, the wastage is estimated at Rs 58,000 corers, which is estimated at more than the total production of fresh fruits and vegetables in Great Britain.. Now the question is

- ✰ Will this quantity be good enough to cater for requirement of 1050 million populations in a given year?
- ✰ Does our produce meet with quality parameters to match the International market through the proposed agri-export zone?
- ✰ How much of the processable varieties of horticultural products are available to satisfy the requirements of the food processing units and food parks?

Obviously, we are far too short of horticultural produce to satisfy each of these requirements. It is a known fact that more than 80 per cent of our farmers are small and marginal with fragmented land holdings. In addition, the woes of the small farmers are far too many. They raise multiple products in small quantities with heavy wastages. Farmer's organizations are either non-existent or dis-organised, as against the unorganized traders, and there is exploitation by commission agents/traders. Thus, there is an urgency to strengthen the network of post-harvest management and cold-chain infrastructure (grading/packing centres, pre-cooling units, refrigerated vans, cold storages etc.) both at the areas of production and at the retail outlets.

It is alarming to note that due to perishable nature of these high value horticultural produce in general and vegetables in particular as well as lack of proper infrastructures and management practices in post production period, the quantum of loss has been estimated to be very high. In order to remove these constraints and minimize the post production losses, appropriate strategies should be taken care of such as

- ✰ Adoption of market driven production system
- ✰ Planned truck farming
- ✰ Appropriate post harvest handling through promotion of pack house concept including storage, processing and value addition

- ☆ Creation of adequate transportation infrastructure and organized marketing system

The horticultural produce in general and vegetables in particular faces a glut situation when it arrives at the market and the farmers resort to distress sale due to perishability. Then comes the scarcity and vacuum in the market. For this purpose, in addition to the cold chain infrastructure, there is a need to produce new technologies like controlled and modified atmosphere stores and ultra low oxygen technologies etc. to meet the globalization and trade liberalization requirements. Therefore, every horticultural produce calls for value addition and product formulation for the purpose of

1. Table and raw consumption
2. Liquids, beverages (alcoholic and non-alcoholic)
3. Health drinks
4. Pulp and other semi-liquids
5. Culinary products
6. Pharmaceuticals, oleoresins, cosmetics, confectioneries etc.
7. Recycling of horticultural waste

Food Processing – Pathway for Value Addition

Food processing involves any type of value addition to agricultural or horticultural produce and also includes processes such as grading, sorting, and packaging which enhance shelf life of food products. The food processing industry provides vital linkages and synergies between industry and agriculture/horticulture. The Food Processing Industry sector in India is one of the largest in terms of production, consumption, export and growth prospects. The government has accorded it a high priority, with a number of fiscal relief's and incentives,

- ☆ To encourage commercialization and value addition to agricultural/horticultural produce
- ☆ For minimizing pre/post harvest wastage
- ☆ Generating employment
- ☆ Leads to diversification and commercialization of the agri-/horticultural produce
- ☆ Enhancing the income of the farmers
- ☆ Creating surpluses for export
- ☆ Overall for comprehensive development of the rural economy.

The broad-based development of the food processing industry will improve both the social and physical infrastructure of rural India. The food processing industry is one of the largest industries in India - it is ranked fifth in terms of production, consumption, export and expected growth. The industry is worth Rs. 8,60,000 crores including Rs. 1,80,000 crores worth of value added products. Food processing is a large sector that covers activities such as agriculture, horticulture, plantation, animal husbandry and fisheries. It also includes other industries that use agriculture inputs for manufacturing of edible products. Processed food industry in India contributes 6.3 per cent of the GDP, and accounts for 13 per cent of export and 6 per cent of the capital investment. India produces about 600 million tonnes of farm produce annually.

India is one of the largest producers of raw materials for the food processing industry in the world but the industry itself is not fully developed. Less than 2 per cent of fruits and vegetables are processed in the country, as compared to 30 per cent in Thailand, 70 per cent in Brazil, 78 per cent in the Philippines and 80 per cent in Malaysia. The value addition in the food sector is only 7 per cent. The level of processing in food processing sector in India is presented in Table 26.1. There were 7,521 regulated markets and 27,294 rural periodic markets in the country for handling part of this produce. There are an estimated 40,000 food processing units in India. The export of processed food from India is also growing at more than 10-12 per cent per annum. According to the Reserve Bank of India, the actual inflow of foreign direct investment (FDI) in the food and food-processing sector has been over US$ 711.4 million (Rs. 3187 crores) till March 2004. Nearly 30 per cent of FDI in this sector comes from EU countries such as Netherlands, Germany, Italy and France.

Table 26.1: Level of Processing in Food Processing Sector in India

Item	*Level of Processing in Organized Sector*	*Level of processing in Unorganized Sector*	*Total Processing*
Fruits and Vegetables	1.2 per cent	0.6 per cent	1.8 per cent
Dairy products	15 per cent	22 per cent	37 per cent
Meat	21 per cent		21 per cent
Poultry	6 per cent		6 per cent
Marine fisheries	1.7 per cent	9 per cent	10.7 per cent
Shrimps	0.4 per cent	1 per cent	1.4 per cent

Source: Cygnus Report, Indian Food Processing Sector, 2005.

India has arable land of 184 million hectares and produces annually 150 million tonnes of fruits and vegetables (second largest). India's agricultural production base is huge. However, processing level is very low *i.e.* around 2 per cent in fruits and vegetables. The share of India's export of processed food in global trade is only 1.5 per cent. Hence, there is immense potential for investment in this sector. To facilitate the prompt growth of food processing industry, the Government has implemented the scheme for infrastructure development comprising a food park scheme, establishing packaging centres, integrated cold chain facility, value added centres, Irradiation Facilities and Modernized Abattoir.

Advantages of Food Processing in India

1. India is one of the largest food producers in the world

2. India has diverse agro-climatic conditions and has a large and diverse raw material base suitable for food processing companies
3. Investment requirement of around US$ 15 billion exists in the food processing sector
4. India is looking for investment in infrastructure, packaging and marketing
5. India has huge scientific and research talent pool
6. A largely untapped domestic market of 1000 million consumers
7. 300 million upper and middle class consume processed food
8. 200 million more consumers expected to shift to processed food by 2010
9. Well developed infrastructure and distribution network
10. Rapid urbanization, increased literacy, changing life style, increased number of women in workforce, rising per capita income- leading to rapid growth and new opportunities in food and beverages sector
11. 50 per cent of household expenditure by Indians is on food items
12. Strategic geographic location (proximity of India to markets in Europe and Far East, South East and West Asia)

Indian Vegetable Processing Industries– Road Map for Future

India is the second largest producer of vegetables in the world (ranks next to China) and accounts for 15 per cent of the world's production of vegetables. The current production level is over 71 million MT and the total area under vegetable cultivation is around 6.2 million hectares which is about 3 per cent of the total area under cultivation in the country. TIFAC study has focused on 12 select vegetables which account for about 65 per cent of the total production in India. It is estimated that around 20-25 per cent of the total vegetables is lost due to poor post harvesting practices. Less than 2 per cent of the total vegetables produced in the country is commercially processed as compared to 70 per cent in Brazil and 65 per cent in USA. Around 150,000 MT of vegetables is sold as processed products.

India exported 68,500 MT of processed vegetables worth Rs.1.6 billion in 1993. Export of processed vegetables has registered a compounded annual growth rate of 16 per cent in volume and 25 per cent in value in recent times. Onions account for about 93 per cent (in volume) of the total export of fresh vegetables from India. The other major items of export are potato, tomato, brinjal, beans, carrots, chillies, capsicum etc. The major export markets are Gulf countries, UK, Sri Lanka, Malaysia and Singapore. Though India ranks second in the vegetable production in the world, the average yield for various vegetables are low compared to those experienced in other countries of the world. Land ceiling has been a major deterrent for large scale cultivation of fruits and vegetables especially in the organised sector. The small captive orchards are insufficient to meet the requirements of the fruit processing industry.

In case of vegetables, potato, tomato, onion, cabbage and cauliflower account for around 60 per cent of the total vegetable production in the country. Vegetables are typically grown in India in field conditions; the concept is opposed to the cultivation of vegetables in green houses as practices in the developed countries for high yields. The vegetables sector also suffers from lack of availability of good quality planting material and low rise of hybrid seeds. Poor farm management and manual harvesting practices also apply to these vegetables cultivation. The expert opinion survey has inferred tile following trends in the future for vegetables sector.

Vegetables would continue to be harvested manually in the future. While small land holdings and non-availability of good quality planting material have been the major issues of concern, it is expected that quality of planting material would improve in the long run due to selection, hybridization, breeding and tissue culture. For poor farm management practices, there exists strong need for extension education and training for the growers. Cooperative and contract farming may solve the problems for small land holdings towards improved yield and quality in the long run. Application of fungicides/pesticides and chemical preservatives would be phased out and would be replaced by more environment friendly technologies in the long run. While the level of processing would however around 5- 1 0 per cent in the next 1 0 years, 15-20 per cent of fruits and vegetables may he processed in the long term. The share of sectoral consumption for processed fruits and vegetables in the long term would be as follows:

- Domestic = 30 per cent
- Institutions = 40 per cent (including defence)
- Exports = 30 per cent

While the small scale processing units would dominate in the short term, an advent of large/medium scale units is likely in the long term. The summary of future technology and business scenario for the vegetables sector is presented in the Table 26.2.

The installed capacity of fruits and vegetables processing industry has increased from 1.1 mn tonnes in January 1993 to 2 mn tonnes in 2000 and further to 2.1 mn tonnes in 2006. The prominent processed items in this segment are fruit pulps and juices, fruit based ready-to-serve beverages, canned fruits and vegetables, jams, squashes, pickles, chutneys and dehydrated vegetables. Some recent products introduced in this segment include vegetable curries in retortable pouches, canned mushroom and mushroom products, dried fruits and vegetables and fruit juice concentrates. The fruits and vegetable processing industry is highly decentralized, and a large number of units are in the cottage/household and small scale sector, having small capacities of up to 250 tonnes/annum. Since 2000, the industry has seen significant growth in ready-to-serve beverages, fruit juices and pulps, dehydrated and frozen fruits and vegetable products, pickles, processed mushrooms and curried vegetables, and units engaged in these segments are export oriented (Table 26.3). The domestic industry is yet to change its preference in favour of processed foods. Consumption of value added fruits and vegetables is low compared to the primary processed foods, and fresh fruits and vegetables. The inclination towards processed foods is mostly visible in urban centers.

Table 26.2: Summary of Future Technology and Business Scenario for the Vegetables Processing in India

Aspect	*Short Term*	*Medium Term*	*Long Term*
Mechanization in Harvesting	Essentially manual as land holdings are likely to remain small Semi-mechanized means for bulk crops like potato and onion Use of hand tools for fruits	Will continue to be manual Increased use of farm tools, implements and tractors	Limited use of mechanized harvesters for potato and onion to be cultivated on large farms
Post Harvest Treatments	Selective use of pre-cooling for high value Use of irradiation not likely	Increased use of pre-cooling techniques like ✰ Mobile Pre-cooling ✰ Ventilated Pre-cooling ✰ Evaporative Cooling Use of broad spectrum chemicals to continue	Extensive use of mobile pre-cooling Limited use of irradiation for potato, onion and spices Use of broad spectrum chemicals to be curtailed
Storage	No significant change from currently followed practices in storage at field level Less than 5 per cent of produce would be channelized through cold chainUse of CA/MA storage only for high value perishable produce	5-10 per cent of fresh produce to be channelized through cold chain Very limited use of CA/MA Storage	20-25 per cent of fresh produce to be channelized through cold chain Extensive use of CA/MA Storage
Transportation	Mainly by trucks and by cart load	Increased use of Reefer trucks	Reefer trucks and CA/MA containers for transport
Packaging	Traditional packaging methods – (Jute bags, Rattan baskets, modern boxes) to continue Use of CFB Cartons, shrink wrapping for exports	Traditional methods to continue	Traditional methods to continue but use of plastic packaging to grow at the cost of wood based packaging Use of CA/MA packaging to increase significantly Emergence of packing stations
Marketing and Distribution	No significant change from current practices Middlemen would continue to play dominant role	Increased role of cooperatives and private institutions	Institutionalisation of marketing and distribution network

Contd...

Table 26.2–*Contd...*

Aspect	*Short Term*	*Medium Term*	*Long Term*
Processing	Level of processing (as per cent of total production of fresh product) would about 5-8 per cent Consumption to be dominated by institutional/defence and export segments	Level of processing about 8-12 per cent	Level of processing would be about 15-20 per cent. Focus to shift from exports to domestic consumption Increase in share of consumption by household sector
Industry Structure	Small scale sector to dominate		Small scale units to emerge as subsidiaries/ancillaries to large scale units
Processing Technology			
Dehydration	Sun drying and forced air drying		Freeze drying
Concentration	Predominantly batch pan concentration in the small scale sector	Use of low temperature vacuum evaporation	High temperature short time evaporation will also be used
	Low temperature vacuum evaporation for large processing units		
IQF	*IQF*	*IQF*	
Freezing	Plate Freezing		Plate Freezing to decrease
Thermal Processing	Canning to dominate Introduction of aseptic filling	Canning as well as use of aseptic filling	Use of aseptic filling
Packaging of Processed Products	No significant change from current practices	Increase in use of food grade plastics Aseptic packaging to gain popularity	Use of food grade plastic and aseptic laminates
Exports	Will account for about 1-2 per cent of world trade	To account for about 2-3 per cent of world trade	To account for about 3-5 per cent of world trade
R and D	Thrust on extension services	Thrust on developing new high yielding varieties	Increasing role of private sector in R and D
	Thrust on development of varieties with extended shelf life	Fruits and Vegetables with specific characteristics suited for processing to be developed	Increasing role of private sector in R and D.

Table 26.3: Exports of Processed Fruits and Vegetables (Quantity in MT, Value in Rs Mn)

Processed Products	2001-02		2004-05		CAGR	
	Qty.	Value	Qty.	Value	Qty.	Value
Dried and Preserved Vegetables	209157.80	5371.5	351034.3	7657.5	18.8	12.5
Mango pulp	76735.18	2413.4	90988.6	3008.6	5.8	7.6
Pickles and Chutney	38758.97	1203.4	67193.29	1205.8	20.1	0.1
Other processed Fruits and Vegetables	61332.39	2017.4	80760.5	2755.3	9.6	10.9
Total	**385984.3**	**11005.7**	**589976.7**	**14627.2**	**15.2**	**9.9**

Source: Ministry of Food Processing Industries, Annual Report 2005-06.

A significant thrust can be given to this sector by strengthening linkages between farmers and processors. The weak linkage between farmers and markets, as well as, farmers and processing companies has brought about inefficiencies in the supply chain and encouraged the involvement of middlemen. The Government of India's National Agriculture Policy (NAP) envisages the participation of the private sector through contract farming and land leasing arrangements which not only assures supply of raw material for processing units, but also a market for agriculture produce, accelerate technology transfer and capital inflow into the agriculture/horticulture sector. Contract farming in Tomatoes practiced in Punjab by Pepsi Foods Ltd is one of the successful examples of contract farming in India, which changed the farming landscape and promoted the cultivation of processable variety of farm produce. Such innovative

Table 26.4: Major Food Processing Companies in India

Major MNCs	Major Indian Companies	Indian MNCs likely to Enter
1. Nestle	1. ITC	1. Reliance
2. Pepsi	2. Dabur	2. Bharti group
3. Coke	3. Britania	3. Tatas
4. Kelloggs	4. Parle	4. Wipro
5. Conagra	5. Amul	5. Thapars etc.
6. Unilever	6. Haldiram	
7. Perfetti	7. Godrej	
8. Glaxo	8. Veny's	
9. Smithkline		
10. Heinz		
11. Wyeth		
12. Ajinomoto		
13. Nissin met		
14. Walmart		

practices will power the fruits and vegetables processing industry. Apart from such initiatives, fiscal incentives and tax concessions will also give impetus to the sector. The five-year 100 per cent tax exemption announced by the Government in FY05 was one such incentive for upcoming fruits and vegetable processing units.

The entry of multinational companies (Table 26.4) has increased competition in the food processing industry. At the same time, these companies are facing tough competition from strong Indian brands. These levels of competition have increased innovations, facilitating a sustained growth of the sector and also improve global competitiveness. The emerging new growth phase of the sector is just in its initial stages with the potential for India to emerge as a leading food supplier to the world.

Processing of Vegetables

Since vegetables are highly perishable foods they need to be preserved and processed in various forms. After harvest the biological produce can be either preserved or processed. **Value addition** is a terminology used to define the processing of biological produce. Through processing the value of the commodities can be increased by converting it to different products by using conventional or modern processing techniques, thereby the storage life of the produce is enhanced. The vegetable are processed into more stable products that can be stored for extended period of time by canning, freezing or through chemical preservatives etc. Actually processing includes all the steps begins after harvest and end before consumption of food. Normally handling, transportation, reception, temporary storage, washing, sporting, peeling, size reduction *i.e.*, slicing, blanching followed by either canning, freezing or drying.

1. **Canning:** Heat processing helps in destruction of microorganism and hermetic sealing of cans prevent from further recontamination.
2. **Dehydration:** Water or moisture removal by dehydration prevents the growth or reproduction of decay causing micro-organisms and minimizes deteriorative reactions. It also brings substantial reduction in weight and volume, minimizing, packaging, storage and transportation cost and products can be stored at room temperatures.
3. **Freezing:** Freezing of vegetables results in substantial reduction of microbial activity as well as biochemical changes. Frozen products should be stored at -18^{0} C temperature till use. It can be stored for more than one year.
4. **Chemical preservation:** Pickles, ketchups, sauce are made using chemical; preservatives like potassium metabisluphite, sodium benzoate, acetic acid, salts etc.
5. **Freeze Dehydrated vegetables** : The conventionally dried vegetables lose their structural integrity and have poor rehydration characteristics, freeze-dried or dehydrated ones retain their nascent texture and colour, and reconstitute fairly rapidly even in cold water owing to their porous structure. Freeze-dehydration can be thought of as an up scaling on lyophilization, a technique which is widely used in pharmaceutical and biotechnological industry to dehydrate and thereby conserve and preserve certain pharma and biological materials and formulations like enzyme preparations,

antibiotics, vaccines, various drugs, hormone preparations, microbial cultures, blood plasma and other products. Industrially freeze dehydration runs in two cycles - freezing and drying. A whole generation of vegetables can be freeze dried in particulate form or in puree or pulp/paste form. Vegetables and their products with initial moisture contents of 85-95 per cent take about 24-30 hours to freeze dehydrated to a final moisture content of about 2 per cent. As these products have very low moisture contents, there is no threat of residual microbial growth. The products remain shelf-stable under ambient conditions for more than a year without spoilage when adequately packed. Usually polymeric films with aluminium foil having adequate barrier protection against oxygen; water and lightness are most suitable for packaging freeze dehydrated vegetables.

6. **Improved Dehydrated Vegetables:** Currently most of the dehydrated vegetables are produced by hot air dehydration which is the simplest and most economical; among the forced air drying techniques. However, conventionally, hot air dried products often lose their original shape and size at the alter thermoconvective stresses, showed delayed and unsatisfactory rehydration and loose consumer appeal. Owing to rather poor quality of traditionally dried vegetables, their demand has remained static and mostly confined to catering to the requirements of our defence force. Improved dehydration techniques offer a highly effective and practical means of preserving vegetables to reduce post harvest losses and offset the shortage in supply.
    a. Treating blanched cauliflower with 3 per cent salt and 6 per cent sucrose for 12-16 hours at 4^0C markedly reduces shrinkage and improves rehydration with out affecting palatability besides increasing the shelf life of the dried products.
    b. A short soak treatment involving soaking first in 20 per cent sucrose for 1 hour followed by 4 per cent salt for 1 hour also brings about similar improvement. Cabbage, brinjal, bitter gourd and small gourd also showed improvements in their rehydration ratios and reconstitution times, when pretreated similarly with varying concentrations of sucrose and salt.
7. **IM Vegetables:** Intermediate Moisture (IM) foods technology offers a simple, inexpensive and excellent alternative to traditionally used dehydration and canning process for preserving vegetables. Moist infusion (desorption) of fresh vegetables in soak solution containing ingredients such as glycerol (humectant), sucrose, salt, potassium sorbate (antimycotic), potassium/ sodium metabisluphite, dextrose, pectine coupled with or without partial dehydration is intrinsic to IM technology in obtaining acceptable shelf-stable products through osmotic process. Recycling of residual infusion solution entailing filtration, decolourization, vacuum concentration and restoration of humectants, antimycotic and preservative concentration is also central to attaining economy in IM food manufacturing. This technology is indeed quite appropriate and suited to salvaging the current colossal post harvest losses in vegetables in cost effective manner. The IM carrot slices and

cauliflower florets with moisture content of 47-49 and 0.80-0.83 per cent having freshly cooked appearance have also been developed.

8. **HT Stabilized Vegetables:** Hurdle Technology (HT) stabilization is another paradigm shift in preserving vegetables, offering a value alternative to canning and dehydration. The HT processing is a combination of processing technique in which several technological approaches and unit operations such as freezing, refrigeration, dehydration, irradiation, pulse electric fields, high pressure processing, use of chemical preservatives etc are used in unison in a synergistic fashion to bring about desired microbiological safety and sensory acceptability. This technology provides a simple process for preserving surplus and seasonal vegetables with value addition. It can be easily adopted by any new entrepreneur for commercial exploitation at cottage, small, medium or large scale. The technique is also less energy intensive, requiring less capital investment. The products have fairly long shelf life and can be stored in flexi-polymeric packs without refrigeration. Convenience, safety, microbiological stability and fresh-like appearance are qualities of HT products.
9. **Minimally Processed Vegetables:** Minimal processing falls in the realm of an art by which trimmed and cut vegetables preserved without causing significant changes in the fresh like properties of the produce. Such products are quite suitable for salads, curries and custard like preparations. Chopping, cutting, slicing and peeling, all injures the plant materials and lead to higher respiration activity and ethylene production and enhanced senescence besides providing extra sites for pathogen entry and substrates for their growth. Minimal processing cut vegetables entails certain additive pretreatments, infusion treatments and ionizing radiations that control the spoilage microorganisms and check the onset of deteriorative physio-chemical reactions when stored at refrigeration temperatures. In minimal processing, the trimmed, peeled, cut pieces and shreds of vegetables are non-thermally treated by dipping them in a soak solution, generally, consisting of potassium metabisluphite (0.01 – 0.09 per cent), sodium benzoate (0.03 – 0.08 per cent), citric acid (0.5 – 1.2 per cent), ascorbic acid (200-300 ppm) and salt (3.0 – 6.0 per cent) at 25-30 ^{0}C for a few hours or, at 5 + 1^{0}C for several hours. Sweet potato, yam, bottle gourd, ridge gourd, cabbage and carrot could be stored for up to 40- 60 days depending on the nature of the produce under refrigeration without spoilage retaining their fresh-like characteristics. However, cowpea, bitter gourd, cluster bean seeds showed shelf life of only 20-25 days at the same temperature. Above all, minimal processing encompassing various non-thermal soak pretreatments including ionizing radiations offer a promising technology that can be used to improve the safety of ready-to-eat cut fresh vegetables.

Principle of Vegetable Processing

The principle of vegetable processing include the following

1. Harvesting maturity and methods
2. Packaging
3. Post harvest treatment
4. Storage
5. Canning
6. Marketing and transport

Post-harvest Value Addition Includes

- Primary, secondary, and tertiary processing
- Operations performed on farm produce
- To provide longer shelf life
- To maintain/improve quality
- To enhance form, space and time utility of the produce for food, feed, fibre, fuel and industrial purposes
- Farm produce includes food crops, horticultural produce, sugarcane, plantation crops, spices and condiments, medicinal and aromatic produce, livestock and fish produce, fibre, fuel crops and others
- The post harvest operations include on-farm handling, cleaning, grading, moisture conditioning, milling, extraction, cooling, freezing, roasting, puffing, flaking, retort processing, packaging, transport and storage

Value Addition

The market for processed foods in 2014-2015 in comparison to 2003-2004 is expected to go up **11 times for fruits and vegetables**, 4 times for dairy, 15 times for poultry, 4.3 times for buffalo meat, 1.4 times for milled rice, 1.5 times for milled wheat, 16 times for ready-to-eat (RTE) foods, 5.4 times for marine products, 9.3 times for sugar and sugar based products, 4.8 times for alcoholic beverages, 2.5 times for aerated cold beverages, twice for pulses and 4.5 times for spices. The growth per annum in processed foods would be about 10 per cent, primary processed foods 7 per cent, and **value added foods 15 per cent. The share of value added products would increase from the existing 38 per cent to 58 per cent by 2014-2015**. The level of processing in organized sector is expected to be in fruits and vegetables (15 per cent), dairy (30 per cent), buffalo meat (45 per cent), poultry (25 per cent), and marine products (20 per cent).

Fresh Steps in Value Addition

1. Food Policy is proposed to support projects for reduction in post-harvest losses, enhance the level of processing, encourage value addition, employment generation and ensure remunerative price to farmers. In the Government of India Budget 2006-2007, Food Processing has been placed as a priority sector for bank credit.

2. National Bank for Agriculture and Rural Development (NABARD) has set aside Rs. 1,000 crore, especially for agro-processing infrastructure and market development.
3. The Ministry of Food Processing Industries has already approved establishment of 47 Food Parks in different parts of the country having facilities for cold storing, IQF, packaging and so on at one place.

Quality Standards

The quality standards that govern the food industry include;

1. Prevention of Food Adulteration Act, 1954 (PFA), Vegetables Product Order, 1967 (VPO), Food Products Order, 1955 (FPO), Agri-produce (Grading and Marketing) Act, 1937, Bureau of Indian Standard (BIS), Export (Quality Control and Inspection Act) (1963)
2. There are few standards for unprocessed/raw food materials. Good Manufacturing Practices (GMP), Good Hygiene Practices (GHP), Hazard Analysis Critical Control Point (HACCP), Codex, International Standards Organisation (ISO) 9000 are some other certification procedures applicable to processing of foods and products.
3. Codex covers 234 foods, 32 functional groups (additives), 36 flavouring agents, 6 labeling standards, 49 codes of practices, and 41 guidelines.
4. BIS has about 700 Indian Standards for use in the area of agricultural produce and value added products.
5. The key issues addressed by these standards include preventing adulteration, regulate hygienic conditions, inform consumers about the product, manufacturer etc., provide product specifications as well as export specifications
6. An integrated food law *viz.*, Food Safety and Standards Act 2005 has come into existence that consolidates most of the food related laws in India to meet the international standards.

Organic Products

1. Out of about 200 products under "Organic" certification, the major items produced in India by the year 2005 included chilli, okra, brinjal, garlic, onion, tomato, potato, etc. Globally, the area under organic cultivation has been growing rapidly. India, with 5147 certified "organic farms" in 2005, ranked 33 in the world having 0.114 million ha under organic cultivation out of total 31 million ha in the world.
2. Lack of quality testing and certification laboratories, certified seeds and planting materials, cool chain facilities including reefer vans, and poor road conditions have been the major problems faced by the sector that need priority attention.

Employment Generation

1. The food processing industry ranks fifth in its contribution to value addition but tops the list in terms of employment opportunities with approximately 15 lakhs employed consisting of 19 per cent of the total investment in the industrial sector but contributes 18 per cent to the GDP.
2. Employment potential in post-harvest and value addition sector is considered to be very high. Every Rs. 1 crore invested in fruits and vegetable processing in the organized sector generates 140 persons per year of employment.
3. The same level of investment in Small Scale Investment (SSI) units creates 1050 person day of employment per year. The SSI unit in food industry employs 4, 80,000 persons, which accounts for 13 per cent of all SSI units employed

Value Added Products in Vegetables

Vegetables typically constitute an essential part of the daily diet in India and they are in great demand round the year from most sections of the population. The commercial value of vegetables in terms of direct consumption besides, processing as well as trade has risen substantially in recent years. Their economic importance has also increased and high labour intensity in the production of most vegetables makes them important from the employment angle as well. The growing food processing industry is poised to dominate the palate of Indian households in the near future. India's middle class is estimated at 250 - 300 million people, the majority of whom earn between $1,200 and $3,600 US per household per year, an amount that is short of what many deem to be the income required to purchase value-added food products ($5,000 US). Yet, about 30-40 million Indians earn $20,000 US or more a year, a huge market by any standard, but scattered all over India in major cities. The consumption and living patterns of this emerging middle class is comparable to those in western market economies during the 1960s. Due to globalization, with the Government of India providing incentives and an easier working environment for foreign investors, both the Indian corporate sector and Multi National Corporations (MNCs) have jump-started the food processing industry, with participation from farm to factory. Primary food processing is a major industry with hundreds of thousands of ricemills/ hullers, flour mills, pulse mills and oil-seed mills. There are several thousands of bakeries, traditional food units and fruit/vegetable/spice processing units in unorganized sector. Size of the semi-processed and ready to eat packaged food industry is valued at over Rs. 4000 crores (US $0.8 billion) and is growing at over 20 per cent annually. As a result of the continuing reform process the food processing industry has drawn a considerable amount of Foreign Direct Investment (FDI) in the last few years. As a result of several policy initiatives undertaken since liberalization in August 1991, the industry has witnessed fast growth in most of the segments. The various value added products in vegetables are presented in Table 26.5.

Table 26.5: Processed Vegetable Products

Sl.No.	Vegetable	Value Added Products
1.	Tomato	Canned whole or in form of pulp, puree, paste or juice, sauce or ketchup
2.	Onion	Dehydrated (flakes, granules, powder), paste
3.	Peas	Canned, frozen, dehydrated
4.	Okra	Canned, frozen, dehydrated
5.	Watermelon	Ready to serve beverage
6.	Cauliflower	Frozen, dehydrated
7.	Carrot	Frozen, dehydrated, juice
8.	Beans	Canned, frozen, dehydrated
9.	Pickles	Mixed vegetables like cucumber, cauliflower, carrot, peas etc.

Enterprises for Processed Vegetable Products

1. **Dehydrated Vegetables (Technology Source: CFTRI):** Vegetables are seasonal and perishable. Dehydration is one of the methods to preserve them and make available throughout the year in hygienic conditions at reasonable cost. The dehydrated vegetables are easy to transport and cater to the needs of large catering establishments. They can be used in various preparations at any season of the year. Traditional sun drying is time consuming, less hygienic and climate dependent. The process for controlled dehydration of vegetables consists of grading/sorting, washing, peeling/trimming, size reduction, blanching, Chemical treatment, dehydration and packing in unit can be established.
2. **Prepackaging of Vegetables (Technology Source: CFTRI):** This simple technique involves cleaning, trimming, cutting of the fresh produce and packing the same in unit packages in polyethylene bags. Bean, carrot, brinjal, green chilli, root crops, leafy vegetables, can be prepackaged to obtain 1 to 2 times extension in shelf life in polyethylene bags under normal conditions without any refrigeration. The prepacked produce presents better consumer appeal, longer shelf life and has considerable handling advantages in transport and marketing. The major facility required at the farm level or in orchards is a packing shed.
3. **Wax Emulsion for Vegetables (Technology Source: CFTRI):** A large number of units in tiny sector can be set up for improving the shelf life of fresh vegetables in villages where they are grown for marketing in the urban areas. The wax emulsion is diluted with cold water and used for dipping fruits and vegetables. It enhances the shelf life, protects vegetables from fungal attack, and reduces desiccation and weight loss during storage. The emulsion is harmless and imparts a gloss to fruits and vegetables. The process is quite simple and economical.
4. **Evaporative Cool Stores (Technology Source: CFTRI):** Evaporative cooling system is a simple and cheap method for short term storage of vegetables.

It does not depend on conventional energy sources. A 170 litre capacity EC chamber has been designed and tested. The special features are dripping system for water, compactness, easy loading and unloading through door and recycling of water. During summer months the system maintained a 20-25°C temperature and 90-95 per cent RH as compared to ambient conditions of 23 –38°C and 30-75 5 RH. The shelf life of commodities is extended up to 20 per cent in EC storage. The freshness is well maintained due to the high humidity environment. EC stored tomatoes develop better colour and uniform ripening. This system can also be used for various other purposes like horticulture, plant propagation, mushroom cultivation, fermentation, sericulture, etc.

5. **Pickles and Chutneys (Technology Source: CFTRI):** Pickles and chutney have a great important in the Indian menu and have now become essential items in any feast and lunch. Pickling of vegetables is an old art. A large variety of these items are method of preparation varies. The basic method is salt curing of fruits and vegetables, acidifying, addition of vinegar/oil and the spices.

6. **Instant Pickles (Technology Source: CFTRI):** The traditional pickle making involves elaborate and time consuming process. A new concept has been worked out for making an instant pickle mix of lime and mango, which can easily be reconstituted, into a tasty pickle by the simple addition of oil and water. The dry mix has the features of convenience, better keeping quality, relatively low cost of production, reduced packaging and transportation cost. The ever expanding market for pickles, both in domestic and export sectors can be exploited by adopting suitable market strategies. The process is simple and consists of the selection of fully matured commodities for picking, washing, and cutting matured commodities for pickling, washing and cutting, brine curing at optimal conditions, drying, addition of dry spice mixes and packing in unit packages.

7. **Potato Flour (Technology Source: CFTRI):** Many ready –to -eat products are prepared from potato. Units based on potato products can easily be established in rural areas and the market can be exploited in urban and semi urban centres. Potato flour, granules and mash are used in the preparation of instant foods, soups etc. as binding materials and also for preparing kheer, tikki, chops, pakoda, cutlets, stuffed parotha, kofta and other products. Production of potato flour can be taken up easily using indigenous equipment's. The process involves peeling of potato, cutting, pre-treatment with salt and permitted preservatives, soaking, granulating and drying. The dried product is ground and packed. The flour obtained by this process can be easily reconstituted with boiling water to get the mashed potato and used for making a variety of products.

8. **Potato Chips (Technology Source: CFTRI):** Potatoes are grown extensively in the northern region of India. During the glut season, growers do not get remunerative prices. Therefore it is advisable to utilize the surplus produce

by preparing potato chips under hygienic conditions.The product based on developed sun drying technology is superior to the conventional one both in colour and quality. The process is simple and can be easily adopted at rural areas

9. **Potato Wafer (Technology Source: CFTRI):** Potatoes are grown extensively in the northern eastern region of India. During the glut season, growers do not get remunerating prices. Therefore it is advisable to utilize the surplus produce by preparing potato chips under hygienic conditions. The product based on developed sun drying technology is superior to the conventional one both in colour and quality. The process is simple and can be easily adopted at rural level.
10. **Tomato Products (Technology Source: CFTRI, RRL-Jmu):** Tomato is extensive grown in India and used for the preparation of puree, paste, ketchup, sauce and ready- to- eat products. There is a good domestic and export market. Since the fast food sector is expanding rapidly the demand, particularly for tomato ketchup and sauces, is also increasing

Vegetable Varieties Suitable for Processing

The various vegetable varieties suitable for value addition in the processing industries are presented in Table 26.6.

Table 26.6: Vegetable Varieties Suitable for Processing

Sl.No.	*Crop*	*Varieties/Hybrids*	*Important Processed Products*
1.	Tomato	Pusa Gaurav, Pusa Uphar, Roma, Punjab Chhuara, Arka Saurav, Pusa Hybrid 2, Pusa Hybrid 4	Sauce, ketchup, chutney etc.
2.	Carrot	Pusa Keshar, Pusa Meghalli	Shreds, halwa
3.	Onion Pusa White Flat	Pusa Red, Pusa White Round,	Flakes, powder
4.	Bitter gourd	Pusa Vishesh, Pusa Hybrid 1	Pickles
5.	Pumpkin	Pusa Vikas	Sauce
6.	Ash gourd	Pusa Shakti, Co 1	Petha sweet
7.	Vegetable Amaranth	Pusa Lal Chauli	Soup
8.	Garden pea	Pusa Pragati, Arkel	Canned, freezing
9.	Chilli	Pusa Sadabahar, Punjab lal, Pant C1	Sauce
10.	Cauliflower	Pusa Snowball KT1, Pusa Snowball KT25	Pickle, dehydrated
11.	Cabbage	Hari Rani	Shreds
12.	French bean	Pusa Parvati, Contender	Canned
13.	Potato	Kufri Chipsona 1, Kufri Chisona 2	Chips and French fry

Value Addition in some of the Important Vegetables

The flow chart for processing of some important vegetable products are presented as follows

A. Pickles and Sauerkraut

Reception
↓
Control
↓
Temporary storage
↓
Grading
↓
Washing
↓
Small holes
↓
Receptacle filling
↓
Salt solution
↓
20-30 per cent aerobically Addition (6.1 NaCl) in cucumber skin
↓
Fermentation
↓
Storage

B. Shredded Sauerkraut

Reception
↓
Control
↓
Temporary storage
↓
Removal of External leaves
↓
Add of 2 per cent of salt solution
↓
Fermentation 1.5 per cent lactic acid pH 4.1

Processing of Peeled Tomatoes

Reception
↓
Washing
↓
Sorting
↓
Scalding
↓
Cooling
↓
Peeling
↓
Filling
↓
Closing
↓
Pasteurization
↓
Cooling
↓
Labelling
↓
Storage

C. Processing of Tomato Pulp

Reception
↓
Washing
↓
Sorting
↓
Draining
↓
Cutting
↓
Extraction

Concentration
↓
Filling
↓
Capping
↓
Labelling
↓
Cooling
↓
Pasteurization

D. Processing of Dried Carrots

Reception
↓
Cleaning
↓
Washing
↓
Scrapping
↓
Control
↓
Cutting
↓
Cooling
↓
Sulphiting
↓
Drying
↓
Sorting
↓
Blenching Control Packing and Storage

E. Dried Leeks

Reception
↓
Temporary
↓
Sorting
↓
Washing
↓
Cutting
↓
Blanching
↓
Drying
↓
Sorting
↓
Control
↓
Packaging
↓
Storage

F. Processing of Dried Onions

Reception
↓
Cleaning
↓
Size grading
↓
Removal of tops roots/outer leaves
↓
Pre washing
↓
Cutting
↓
Washing

↓
Draining
↓
Drying
↓
Sorting
↓
Packing
↓
Control and Storage

G. Processing of Dried Tomatoes

Reception
↓
Washing
↓
Trimming
↓
Sorting
↓
Draining
↓
Slicing
↓
Drying
↓
Packing
↓
Labelling
↓
Storage

H. Tomato Juices

Tomatoes
↓
Washing
↓

Control

↓

Cutting

↓

Preheating (60^{0}C)

↓

Juice Extract Reaction

↓

Refining

↓

Conc. (10-12 ^{0}Brix)

↓

Filling

↓

Closing

↓

Pasturization

↓

Cooling

↓

Labelling

↓

Storing

I. Tomato Ketchup or Sauce

Tomatoes

↓

Sorting

↓

Wholesome red Tomatoes

↓

Soaking and Washing

↓

Coring and Slicing

↓

Heating and Crushing

↓

Pulping
↓
Tomato Pulp Cooking
↓
Spice, Sugar, Salt, Onion, Garlic, Vinegar added
↓
Finisher
↓
Preservatives
↓
Filling
↓
Low pasturization
↓
Labeling and packing

J. Processing of Dried Green Beans

Reception
↓
Washing
↓
Cutting
↓
Blanching
↓
Cooling
↓
Sulphiting
↓
Drying
↓
Packing
↓
Storage

K. Vegetable Juices

Vegetable juices may be grouped into 6 classes:

1. Juices prepared from normally acid products such as Tomato, Rhubarb etc. may be processed at relatively low temperature
2. Vegetable juices or their blends are acidified with such highly acid prodcts as citrus, pineapple, tomato, sauerkraut and Rhubarb juices
3. Vegetable juices are acidified with organic or mineral acids so that they can be processed at relatively low temperature
4. Excess juices obtained from fermented vegetables are often used in their natural state or they may be heat processed at low temperature. Sauerkraut juice is the important commercial product in this group.
5. Juices are freshly extracted from non-acid vegetables at health food store immediately before consumption. They are neither heated nor acidified.
6. Vegetable juices or their blends when not acidified must be processed at relatively high temperature to kill spores of spore forming microbes.

General principle adopted for juice processing have shown as under

Clear → Juice Separation of coarser
Clear → Pressing the juice
↓
Heat Treatment → High Temperature to inactive enzymes
↓
Sterilization → High Temperature
Sterilization → Low Temp
↓
Acidification → Add of organic acids
Acidification → Fermentation
↓

Tomato processing industry in India

India is not a dominant player in the world market, with a production of 7,430 thousand MT as against worlds production of 84,873 thousand MT. India's share in world production is only 5.7 per cent. However, Indian tomato processing industry prides itself on being the largest in south Asia. Infact, the output of the Indian industry is more than twice that of Japan, and considerably greater than Thailand and Taiwan. Another encouraging trend has been that India's production level of processed tomato has risen by 50 per cent.

Table 26.7: Production of Tomato in India

Sl.No.	*Particulars*	*1999*	*2000*	*2005*
1.	Production quantity (1000 tonnes)	4,603.45	7,430.00	7,600.00
2.	Yield per hectare (kg/ha)	158.59	161.52	140.74
3.	Area harvested (1000 ha)	290.28	460.00	540.00

Source: FAOSTAT.

Recently, India has started the export of processed tomato. India's exports mainly consist of tomato ketchup/sauce, paste/*chutney*. In the export of processed products tomato paste contributes a large share and touched the level of 68 thousand tonnes as against the target of one lakh ton, this is a substantial jump from the base of 30 thousand tonnes. The increased export of about 127 per cent during last year encourages the growers. In terms of value, exports do show an increase of about 122 per cent as reported. The level of exports has been rising in value and volume. India exports processed tomato paste and *chutney*, to the USA, Saudi Arabia and Japan. Other forms of preserved tomatoes are exported in small quantities to Kuwait, Sri Lanka and the UAE. Processed tomato is a major trade item in the world. India has been exporting processed tomato in the form of tomato paste and ketchup. India does possess moderate export competitiveness on the world market for tomato paste. The domestic market is expanding, and there is also world demand for this product. Tomato paste is another item in which India possesses export competitiveness.

Quality of Raw Material Suitable for Processing

The quality of a tomato product is judged by its colour, which is dependent on the redness of the tomatoes used. In fact, the red pigment (lycopene) can be used an index of the amount of tomato actually present in a product. Tomatoes should be ripe, red, and firm to soft, free of all mould growth (by cutting out infected parts) and free of stems, leaves, dirt and other soils (by washing). It is less important if the tomatoes have surface blemishes or splits/cracks (provided these are not infected) and in most processes they will be cut or pulped. Tomato crop raised in winter season is superior in quality because they contain more total solids.

Value Added Products of Tomato

The following products can be prepared from tomato

1. Tomato paste
2. Tomato juice
3. Tomato ketchup
4. Tomato pickle and
5. Tomato sauce

Working Costs of Preparation of Tomato Ketchup

It is observed from the table that, total working cost amounted to Rs. 60,782.96/tonne of tomato ketchup. Expenditure towards total raw material cost has maximum share of 87.84 per cent, followed by tomato (20.36) and glass bottle (16.45). Table 26.8 indicates working cost of preparation of tomato ketchup.

Fixed Costs of Preparation of Tomato Ketchup

Total fixed cost accounted for Rs. 10, 75,000. Among the fixed costs, area under building has maximum share of 41.86 per cent, followed by machineries and equipments has 37.21 per cent as shown in the Table 26.9.

Table 26.8: Working Costs of Preparation of Tomato Ketchup (Rs./tones)

Sl.No.	Particulars (Rs./tonnes)	Percent to Total Working Cost
A. Expenses on Human labour		
1.	Plant manager 120.00	0.20
2.	Plant supervisor 96.00	0.16
3.	Technical assistants 180.00	0.30
4.	Skilled artisans 144.00	0.24
5.	Office assistants 36.00	0.06
6.	Unskilled workers 300.00	0.49
7.	Plant manager 120.00	0.20
	Total human labour cost 876.00	**1.44**
B. Expenditure on raw materials and Packing		
1.	Tomato 12375.00	20.36
2.	Sugar 4000.00	6.58
3.	Salt 175.00	0.29
4.	Onion 1500.00	2.47
5.	Ginger 450.00	0.74
6.	Garlic 405.00	0.67
7.	Red chilli powder 1125.00	1.85
8.	Cinnamon 3250.00	5.35
9.	Cardamom 5750.00	9.46
10.	Cumin 3000.00	4.94
11.	Black pepper 4500.00	7.40
12.	Clove (Nos.) 1500.00	2.47
13.	Vinegar(Lit) 600.00	0.99
14.	Sodium benzoate 84.50	0.14
15.	Glass bottle (500g. Capacity) 10000.00	16.45
16.	Plastic Cork 1000.00	1.65
17.	Plastic Cap 1000.00	1.65
18.	Corrugated box @12 bottles/box 1530.00	2.52
19.	Expenditure towards electricity@ 140 units 910.00	1.50
20.	Water charges 240.00 0.39	0.39
21.	Total raw material cost 53394.50	87.84
22.	Interest on working capital @ 12 per cent 6512.46	10.71
	Total variable cost 60782.96	**100.00**

Table 26.9: Fixed Costs of Preparation of Tomato Ketchup (in Rs.)

Sl.No.	Fixed Cost (Rs.)	Per cent	
1.	Land (500 esq.)	90000.00	8.37
2.	Area under building (150 m2)	450000.00	41.86
3.	Machineries and equipments	400000.00	37.21
4.	Installation and erection charges @ 15 per cent cost of equipments and machineries	60000.00	5.58
5.	Office expenses (Furniture and chair)	75000.00	6.98
	Total fixed cost	**10,75,000.00**	**100.00**
	Grand Total cost	**11,35,782.96**	

Cost of Production of Tomato Ketchup

The details of cost of production of tomato ketchup are given in the table.10. The cost of production of tomato products are worked out for a tonne of output. It is observed from the table that cost of production was Rs. 62,882.96 per tonne of Tomato ketchup.

Table 26.10: Cost of Production of Tomato Ketchup (Rs./tonne)

Sl.No.	Particulars	Per cent
A.	**Cost of production (Rs./tonnes)**	
1.	Recurring expenses 60782.96	96.66
2.	Depreciation on building (SLM) 490.00	0.78
3.	Depreciation on machineries and equipments (SLM) 437.00	0.69
4.	Interest on fixed capital @ 12 per cent 1173.00	1.87
5.	Total 62882.96	100.00
B.	**Return :** Gross return of Tomato ketchup (1000 Kgs.) 85000.00	
C.	**Cost Economics** Net return 22117.04	

Recurring expenses has maximum share of 96.66 per cent of the cost of production. Gross return of tomato ketchup was Rs. 85000/tonne and Net return was Rs.22117.04/tonne.

Problems Post Harvest Value Addition of Vegetables in India

The sector has been facing the problems of

1. Erratic and inadequate supply of quality raw materials
2. Inadequate infrastructure
3. Inadequate investment in organized sector, fragmented research and development (R and D)
4. Lack of adequately trained human resource
5. Lack of quality testing and certification laboratories

6. Long chain of intermediaries contributing to high costs and inefficiencies
7. High costs of carrying inventories, working capital, and taxation
8. Overshadowing of consumption patterns by cultural practices
9. Very heavy and multiple taxation for processing industry
10. No arrangements for processing waste utilization
11. The local entrepreneurs have to compete with multinationals in terms of quality, hygiene, appearance and standards
12. The bulk handling facilities are not available in all ports
13. Refrigerated warehousing and transportation facilities are also inadequate and costly.

Priorities of Vegetable Processing

The priority areas in vegetable processing are:

1. Processing of special vegetable and canning s well as storage facility
2. Emphasis on cottage industry involving village women for the manufacture of vegetable products
3. Popularization of low cost engineering storage structures.
4. Starch production from potato and tuber crops
5. Strengthening of research base with adequate financial support
6. Emphasis on production of value added products from locally available vegetables

There are many other areas of processing aspects which should be given priority. Once the processing, research and industry programme picks up many other outlets shall come up automatically. Processing industry has very good employment potential.

Strategies for Value Addition to Vegetables

Recent trend in human diet is towards the larger intake of vegetables. A large number of vegetables are not available throughout the year because of their seasonal nature of production, warranting the need of their preservation through storage and processing in order to ensure their availability in off season. Indian farmers get only about 28.5 per cent share of the payment made by consumer for his produce. This necessitates value-addition to his produce so as to increase the income of the farmers and improve the quality of produce. The value addition strategies can be planned at different levels.

Farmers' Level

A vegetable producer can increase the market value of his produce by adopting various measures such as:

- ✰ Pre-cooling
- ✰ Washing
- ✰ Grading

- Post harvest treatment (hot water treatment with or without fungicides)
- Waxing
- Packaging
- Proper transportation

Farmers' Organizations Level

The support of farmers' organizations is necessary because it is not possible for an individual farmer to arrange all the required infrastructures. Attention should be paid to the development of village based processing such as preservation of vegetables in various forms such as puree, paste, ketchup, chutney etc.

Provision of Infrastructure Facilities

A large number of infrastructural facilities are required for quality and safety against health hazards for vegetable processing. Depending of the vegetables, infrastructural facilities include:

- Pre-cooling facility
- Packaging and grading chambers
- Refrigerated vans
- Cold storage/controlled atmosphere storage
- Pesticide analyzing laboratories
- Good roads
- Efficient communication system

Providing Credit and Insurance Cover

Since more and costly inputs are required to produce vegetable processing, it is necessary to ensure credit availability at reasonable interest without much hassles is very critical in sustaining competitiveness. Likewise, an insurance cover will be necessary to safeguard farmers against natural calamities such as outbreak of pests and diseases, hailstorm, frost and heavy rains etc.

Information Dissemination

This is an information era signifying the role of information and communication technologies in transforming various value added products in vegetables. Providing farmers with relevant, timely information about:

- Different vegetable varieties
- Planting densities
- Integrated water and nutrient management
- Integrated pest management
- Maturity indices
- Market intelligence etc can help a great deal in increasing their productivity and profitability.

Adoption of Hi-tech Horticulture

In order to increase the quality of value addition in vegetables for competition in global market, our vegetables which will make imperative the adoption of various technologies of hi-tech horticulture like?

- ✰ Genetically modified crop varieties
- ✰ Micro propagation
- ✰ Fertigation
- ✰ Integrated pest management
- ✰ Cold-chain
- ✰ Modified atmosphere storage
- ✰ Application of bio-fertilizers and bio-pesticides etc.

According to G C P Ranga Rao, Organising Secretary, FoodPlus-2005, the strategies for vegetables processing in India are as follows:

1. In order to reduce the huge post-harvest losses in the fruit and vegetable sector, results of research carried out at institutional levels, should be transferred to the end user on mission mode
2. Policies should be formulated in order to encourage value addition to horticulture produce
3. Areas that need to be critically looked into include availability of quality seeds, introduction of varieties that are suitable for processing
4. Organic farming
5. Contract farming
6. Cooperative farming
7. Establishment of packing houses and cold chain infrastructure
8. Market access and availability of market intelligence to producers
9. Elimination of middlemen by amending APMC act
10. Elimination of taxes and surcharges on fresh and processed fruits and vegetables and on inputs that go into their production, etc.
11. If India has to become a preferred destination for global outsourcing of fruit and vegetable products, negotiations with key countries such as China, USA and Japan should be expedited to lift quarantine restrictions

SWOT Analysis of Vegetable Processing Industry

Strengths

1. Abundant availability of vegetable based raw material
2. Priority sector status for agro-processing given by the central Government
3. Vast network of manufacturing facilities all over the country
4. Vast domestic market

Weaknesses

1. Low availability of adequate infrastructural facilities
2. Lack of adequate quality control and testing methods as per international standards
3. Inefficient supply chain due to a large number of intermediaries
4. High requirement of working capital.
5. Inadequately developed linkages between R and D labs and industry
6. Seasonality of vegetable based raw material

Opportunities

1. Large crop and material base offering a vast potential for agro processing activities
2. Setting of SEZ/AEZ and food parks for providing added incentive to develop Greenfield projects
3. Rising income levels and changing consumption patterns
4. Favourable demographic profile and changing lifestyles
5. Integration of development in contemporary technologies such as electronics, material science, bio-technology etc. offer vast scope for rapid improvement and progress
6. Opening of global markets

Threats

1. Affordability and cultural preferences of fresh food
2. High inventory carrying cost
3. High taxation
4. High packaging cost

Policy Initiatives for Processing Industries

In order to promote investment in the food processing sector, several policy initiatives have been taken during recent years. These include:

1. Food processing industry has been declared a priority area. So it qualifies for a number of fiscal relief and incentives to encourage commercialization and value addition to agricultural products
2. Full repatriation of profits and capital is allowed
3. Almost the entire sector is delicensed freeing it from bureaucratic hassles
4. Automatic approvals for foreign investment up to 100 per cent, except in few cases, and also technology transfer
5. Zero duty import of capital goods and raw material for 100 per cent export-oriented units

6. Agro based 100 per cent export oriented units are allowed sales up to 50 per cent in domestic tariff area
7. All processed fruits and vegetable products are exempted from central excise duty
8. Government grants given for setting up common facilities in Agro Food Park
9. Full duty exemption on all imports for units in export processing zones
10. Vision 2015 adopted by the Ministry of Food Processing Industries envisages:
    - Trebling the size of the processed food sector
    - Increasing level of processing of perishables from 6 per cent to 20 per cent
    - Value addition to increase from 20 per cent to 35 per cent
    - Share in global food trade to increase from 1.5 per cent to 3 per cent

Conclusion

Because of the liberal government and other developmental measures being taken the future of the vegetable processing Industry looks very bright. The production base is being enlarged, modern methods of cultivation are being adopted thus improving the productivity and cutting the per unit cost. To some extent cold chain is being provided, which will help in retaining quality, freshness and reduce post-harvest losses. With the new hybrid varieties being added the production season is also being extended. These developments shall result in the greater availability of quality raw materials to the industry thus resulting in better capacity utilization and producing a wider range of products and of international quality. The quality is now the watchword for success. The multinationals now entering the food industry have an international marketing network and have their brand loyalties all over the world. This will enable the Indian products reaching all over the world in the form and packing required. With the rise in the per capita income particularly of the middle class a drastic change in the food habits has been noticed. This will lead to an increased domestic consumption of processed foodstuffs. ?A strong and dynamic food processing sector could play a vital role in diversifying agricultural activities, improving value addition opportunities and creating surplus for exports of agro food products. The desired growth of horticulture sector in general and economic prosperity of rural farmer in particular is achievable only with the effective integration and synergy between horticulture and agro based industries. Value addition ensures better market and remunerative price of the horticultural produce by making it more marketable, creating better entrepreneurship for the unemployed youth.

References

D.S.Ksurdiya (2003). Promoting Export of Value added products in Indian Horticulture, April-June.

FOOD PROCESSING INDUSTRIES – ROAD MAP FOR FUTURE

G C P Ranga Rao (2005). Needed: 4 more CFTRI's to turn around country's food fortune, Saturday, December 03, 2005, *The author is Organising Secretary, FoodPlus-2005 and Hon. Secretary, AFST(I)*

J.L Mangal and S.Siddiqui (2000).Post harvest technology of vegetables: Present Status and Future Strategies.Challenges in *Vegetable Production in Twenty First Century. In Emerging Scenario in Vegetable Research and Development. Ed.G.Kalloo and K.Singh,* pp.336-34

Policies, Useful Web Links, Ministry of Food Processing Industries Post Harvest Technology and Value Addition, Industry and Services

S. Arunajatesan and S. Balaji (2004) :Agro industries: the lure of value addition,Online edition of India's National Newspaper The Hindu, Monday, Feb 02, 2004

S. Biswas, T. Chandrasekhar, Y.S. Rajan, D.N. Singh, A. Amudeswari (1999). Long-term Technology Vision: An Indian Experience Overview of Indian Food Processing Industries, Indian Express Front Page

T MAT, Value Added Products, Dr. A. Malaisamy, Ph.D,Assistant Professor and R. Parimalarangan, Senior Research Fellow, Department of Agricultural Economics, Tamil Nadu Agricultural University, Coimbatore, Tamil Nadu,India - 641 003.

The Latest Technology - Modified Atmospheric Packaging of Fruits and Vegetables (MAP) and Other Value Added Products developed by Central Food Technological Research Institute (CFTRI), Mysore. North East India CSIR Technology, The Fourth EU-India Business Summit, November 28, 2003, New Delhi

Vegetable Processing: Principle and its Practical Application: Paper presented by S.P. Singh on Winter School on Recent Advances in Vegetable Production Technology at IIVR, Varanasi

Virtual University for Agricultural Trade – A project by the department of Agriculture, Govt. of Kerala : Processing and Value addition

Chapter 27

Livestock Enterprises Towards Generation of Income and Livelihood for Economic Sustainability

Subhransu Pan

Entrepreneurship is not only the ability to successfully predict future needs of the people but also produce goods and services accordingly to satisfy these needs. Further, it needs constant evaluation of the changing social and economic order. It also requires highest level of imagination, creativity, and innovation. By and large, there is no dearth of the people possessing these properties. In developing countries, however, a large number of people with potential to develop entrepreneurship either sit idle or wait for a job to come their way. Becoming an entrepreneur to earn a living and thereby generate employment for others and also create assets for the whole economy is not an option for them. An Entrepreneur is a person who organizes and manages a business undertaking, assuming the risk for the sake of profit. Any person (any age) who starts and operates a business is an entrepreneur.

Livestock production provides a robust support to the economy of Indian farmer by ensuring farm level diversification and flexibility in agri-business. They serve either as complement to crop enterprises or independent enterprises and supply products of food value and materials of commercial importance. The livestock production business in India is mainly in the hands of small and marginal farmers and agriculture labour who maintain livestock as means to meet part of their family food requirements and to earn supplementary income. Scheduled castes, scheduled tribes, backward castes, women and other weaker sections of the society have a greater stake in livestock sector in this regard. Animals serve as a savings account and ready source of cash. Poultry are often the "coins", sheep and goats the "small bills", and large animals -if available -are sold to meet large expenditures, such as building a house. Where there is no efficient banking system and the rate of inflation is high, this is very rational.

The National Commission on Agriculture, therefore, laid emphasis on the role of animal husbandry development in the context of food production and income generation for the rural poor. In fact, livestock keeping forms an important component of IRDP and other rural development programmes. Thus, it is important that the components, complexities, planning and management of livestock enterprises of different scales are clear to the relevant planners, administrators, bankers, technical personnel and all others concerned with development of this sector. It is only on the basis of such knowledge, one can strive to improve the programme content and develop strategies for its successful implementation in the context of rural development.

Advantages of Entrepreneurship

- Entrepreneur is his own boss.
- Entrepreneur enjoys the profits from his efforts.
- Entrepreneur senses pride of his business.
- Offers flexibility in his work schedule.

Disadvantages of Entrepreneurship

- Will need to put in long hours.
- Need money to start.
- Have to keep up with government rules and regulations.
- May have to mark hard decisions (hiring, firing, etc.).
- May lose money.

Three major aspects of entrepreneurship are:

- The identification/recognition of market opportunity and the generation of a business idea (product or service) to address the opportunity
- The marshalling and commitment of resources in the face of risk to pursue the opportunity
- The creation of an operating business organization to implement the opportunity-motivated business idea.

Livestock Production Enterprises in Livelihood of Rural Poor

The role of livestock production in the lives of the small farmers and the village poor is many fold as shown below.

1. It buffers the risks due to crop failure, un- and underemployment.

2. Enhances family nutrition status, allows some domestic consumption of essential nutrients.
3. Provides regular cash income.
4. Renders services – draft, manure, fuel, sports and tourism.
5. Helps enhance socio-economic status.

But the small farmer faces several constraints in developing his livestock enterprise. Some such constraints are listed below.

1. They possess meager resources.
2. They have limited access to resources, services, technology and market, due to their low literacy and socio-economic status.
3. Theirs is a subsistence farming/livelihood.
4. They are poorly organized rather unorganized.
5. They are distributed over a wide area, with no/poor transport and communication facilities.
6. They are generally tradition/taboo/apathy bound to allow big changes in technology towards.

Selection of Suitable Livestock Enterprise

Every farmer at every locality of the country cannot effectively use all types of livestock enterprises. Based on our experiences and researchers, the scope of different livestock enterprises for different regions of the country can be as follows.

Cattle	:	Milk – Milk-shed areas, near towns and cities.
		Female calf rearing – Milk-shed areas, near towns and cities.
Buffalo	:	Milk – Milk-shed areas, near towns and cities.
		Female calf rearing – Milk-shed areas, near towns and cities.
		Male calf rearing – Around slaughterhouses.
Work Animals	:	Bullocks – for hiring, anywhere.
		Camels – for hiring, N-W India.
		Horse- sports, festival, social ceremony.
		Donkeys – for hiring, anywhere
Sheep	:	Wool – Himalayan, N-W region
		Meat – Any area.
Goats	:	Milk – Any area.
		Meat – Any area.
Pigs	:	Pork – Near cities and specific areas and people.

Poultry	:	Eggs – Poultry-shed areas, near cities and towns.
		Meat – any area
Rabbits	:	Meat – Any area
		Fur –Hilly areas
Yak and Mithun	:	Milk and Meat – N.E. region
Allied Activities	:	Feeds and Fodder – Production/collection anywhere. Processing – Crop by products, in any area. Hatcheries – in selected areas. Cold Stores – for egg and meat storage in selected areas. Feed mills – Compound feed preparation anywhere. Medicines and vaccines- Any area.

Normally livestock are used for alleviation of poverty in rural areas via certain specifically designed schemes or programmes *e.g.*, Integrated Rural Development Programme with livestock components. The considerations to be kept in mind in organisation of livestock sector for beneficiaries in an area are as follows.

1. Select appropriate livestock enterprise for the target micro area. Some considerations desirable for this are -
   (a) Enterprises whose products have local/nearby demand/market.
   (b) Enterprises generally needing low/intermediate (not intensive) technology;
   (c) Enterprises generally producing less/not perishable items ;
   (d) Enterprises that fit into local "likes" and "dislikes".
2. Adopt "growing in" rather than "going in" strategy for the sake of the small entrepreneurs.
3. The livestock owners need be organized.
4. Input services need to be organized:
   (a) Provision of capital/loan/subsidy;
   (b) Supply of "improved" birds/animals;
   (c) Supply of feed/fodder seeds/tree saplings (Subabul);
   (d) Creation/improving of infrastructure facilities for -
      a. Health cover
      b. Feed processing technology
      c. Slaughter houses
      d. Egg and meat handling/storage facilities
      e. Technology transfer
      f. Marketing facilities, etc.

Service and Input Requirements

It can be seen from the above that organisation of supply of inputs and services constitute the main plank of livestock enterprises in the rural areas. The broad guidelines for development of essential inputs for livestock sector particularly for the small farmer recommended by the Food and Agriculture Organisation are as follows:

1. Improvement of fallow land for grazing through fencing controlled grazing, cultivation, fertilization.
2. Augmenting nutritive value of agricultural by-products through simple technologies: pre-soaking, ensiling with Urea, Urea-Molasses-Block, compounded feeds etc.
3. Diffusion of less costly, simple, improved techniques of stock management.
4. Correction of mineral deficiencies by use of mineral mixture supplementation salt-bricks.
5. Control of infertility, infectious and elementary diseases.
6. Resorting to group approach in services and marketing. Some possible ways are – clusters of beneficiaries, complexes (poultry, dairy), cooperatives etc.

Planning and Management of Livestock Enterprises

The time has now come, not only for management of livestock enterprise in a logical and planned manner, but also to plan and manage livestock development in a given area. In view of the wide variation in the geo-agro-climatic patterns, socio-economic backgrounds of people, type/breed and species of livestock etc. between regions, it is necessary to organize livestock enterprises specific to each micro area (block, *tehsil*, district). Only then the aims and objectives of any livestock development plan for an area will succeed, most likely.

Constraints and Conditions

Before going into the mode of planning for livestock enterprises for a given micro-area some constraints encountered and conditions to be fulfilled in planning for livestock sector in India in general may be considered. Some such constraints and conditions are as follows.

1. Need for intermediate technology:
   (a) Low or no demand on capital or other resources (subsidies and low interest loans can be helpful).
   (b) Gradual stock improvement without removal or slow removal of useless stock.
   (c) General manipulation in management practices for improving animal performance.
   (d) Improving the quality of crop byproducts as feed.
2. Need for government role in providing inputs/services. Some voluntary organizations are coming up.
3. Need for organisation of farmers and establishment of enterprises in clusters.
4. Need for simultaneous development of marketing facilities for livestock and livestock products including product storage facilities.

5. Need for operating input supply services and extension (transfer of technology) services as two faces of the same coin.

Step 1 – Stock Taking

The first step in entrepreneurship planning is to take stock of existing situation of the livestock sector in an area as a prelude to its development.

For this the sources of information can be – (a) Livestock census, (b) Agriculture statistics, (c) Demographic data, (d) Data on revenue, marketing etc, (e) Reports of previous surveys, (f) Records of Districts/Zilla Parishads/Blocks/Panchayats etc. (f) Sample surveys. From such information the operation in the livestock sector in the region may be deciphered on the following lines.

1. Livestock numbers (per thousand humans; per holding) –
   (a) Cattle (milch, dry, bulls, bullocks, young stock, male and female);
   (b) Buffalo (milch, dry, bulls, bullocks, young stock, male and female);
   (c) Sheep (Male and female, young and adult stock);
   (d) Goats (male and female, young and adult stock);
   (e) Pigs (male and female, young and adult stock);
   (f) Work animals (bullocks, camels, donkeys, ponies, horses);
   (g) Poultry (chicken, ducks, geese, quails and others).
2. Quality/type/breed of different livestock –
   (a) Productivity – average yield of milk, meat, wool, work, eggs etc.
   (b) Type: cattle – dry, draft, mixed; Buffalo – same as cattle; sheep-wool, meat; Goats – dairy, meat; poultry – layer, broiler.
   (c) Breed: Non-descript, indigenous breeds, exotic breeds, crossbreeds.
3. System of farming –
   (a) Specialized farming (dairy, poultry etc.);
   (b) Diversified farming (within specialized farming, *e.g.*, poultry farming turkeys, quails etc.)
   (c) Mixed farming (crop farming + dairy etc.)
   (d) Contract farming;
   (e) Others.
4. Feed resources available –
   (a) Pasture, grazing lands available.
   (b) Green fodder.
   (c) Dry fodder (straws, stovers, begasse, pods)
   (d) Concentrate feeds (brans, oil cakes, coarse, grains, pulse, molasses etc.)
5. Intensity of farming –
   (a) Intensive: high inputs – high returns, quick, *e.g.*, Poultry in cages;
   (b) Extensive: low inputs – low returns, slow, *e.g.*, poultry on ranges.

   (c) Semi-intensive: Intermediate level inputs – intermediate level outputs, reasonably quick, *e.g.*, Poultry on deep litter.

6. Categories of holdings possessing livestock –

   (a) Landless, (b) Marginal, (c) Medium, (d) Big, (e) Large.

7. Extent of use of resources (Land, Livestock, Labour, Capital, others) in livestock sector fully, substantially or partly –

   (a) All, (b) Some, (c) None;

8. Existing infrastructure facilities and magnitude of provision of technical inputs –
   (a) Veterinary hospitals, dispensaries, stockmen centres etc.
   (b) Semen banks, A. I. Centres and sub-centres
   (c) Milk plants, Chilling centres, collection centres.
   (d) Feed plants
   (e) Extension services
   (f) Slaughterhouses, wool centres, by product utilization centres, egg and meat storage and processing facilities.
   (g) Marketing facilities
   (h) Manpower availability
   (i) Service organizations (Coops, Boards, Corporations)
9. Production of milk, meat, eggs, wool, other products per annum, per 1000 humans, per holding.
10. Contribution of livestock to domestic income (household, village, block, district) –
    (a) Only domestic consumption
    (b) Les than 5 per cent of total income
    (c) 5.1 to 10 per cent of total income
    (d) 10-20 per cent of total income
    (e) more than 20 per cent of total income.
11. Source of family income generation-
    (a) Sole source
    (b) Principal source
    (c) Subsidiary source
    (d) Down-stream business

Step 2 – Determining Requirements – Targets

Once the facts and figures about the existing situation is known, the next step in planning is determining production targets and input requirements. This may involve the following:

1. *Determine the annual requirements of products* – milk, meat, wool, eggs, etc. required in the area. The requirements can be developed as per –
   (a) Needs/head as per standards, *e.g.*, Nutritional Society of India recommendations
   (b) Actual market needs of last year extrapolated to this year.
   (c) Decisions regarding plant capacity, export to other regions etc.
2. Determine the quality/type/breed of livestock and their numbers required for sustaining the desired level of production over and above the present stock.
3. *Determine the feed requirements.* Feed is the main (70 per cent of the total costs) input –
   (a) For the purpose of calculation, it is better to reduce the total number of all categories of grass eating livestock into Adult/Ruminant/Livestock Units (AU or RU or LU). One common constant for such conversion is to consider as follows.
      (a) Cattle above 3 years age = 1.00 LU
      (b) Buffalo above 3 years of age = 1.10 LU
      (c) Older calves (heifers) = 0.65 LU
      (d) Younger calves = 0.35 LU
      (e) Sheep and goats = 0.35 LU
      (f) Pigs = Provide only for concentrates
      (g) Poultry = Provide only for concentrates
   (b) Then calculate feed requirements as per the rough rate – 30 kg greens, 5 kg dry fodder and 1 kg concentrate per LU per day.
   (c) The concentrates, typically, may contain 50-60 per cent coarse grains/ millets; 20-25 per cent coarse pulses (gram), 10-15 per cent oil cakes and 10-15 per cent brans + 2 kg salt and 1 kg mineral mix per quintal feed.
   (d) For poultry and pigs only concentrate feed be provided @ 15 kg per day for 100 birds or for 10 pigs, presuming that there will be some scavenging too by them.
   (e) Green fodder availability –
      (a) *Grazing :* Under Indian conditions about 8 to 10 kg greens per day (3 to 4 kg dry matter) can be grazed per LU from natural grasslands on an average day. (It can be more in some season and at some places and less in other seasons or other places). On this basis grazing land needs can be found out. Note that grass yield from grazing land can be improved by better pasture management. Also land for grazing can be known based on the carrying capacity of land as 3-5 LU per ha/year.

(b) *Yield of cultivated green fodders:* Data available for the region under consideration may be obtained from local institutions or even farmers.

(f) Yield of dry fodder may be considered as – (a) 1.6 times to grain yield/ha for large grain crops (rice, wheat), (b) 3 times the grain yield/ha for small-grain crops (Bajra, Ragi), (c) 0.02 times of cane yield in case of sugarcane tops, and (d) 2.5 times of grain yield in case of gram/pulse crops, important in sheep and goat feeding.

4. Determine infrastructure requirements:
5. Determine strategy of developing enterprises -
   (a) A single enterprise (poultry, dairy) for the whole area/community.
   (b) A single enterprise for part of the area/community
   (c) Several enterprises for several parts of the area/community
   (d) General area plan

   This can be done based on specific needs of each part of community/area and scope for developing individual enterprises in each part of area/community.
6. Develop, accordingly, general plans and plans (projects/schemes) for individual parts of area/community.
7. Determine priorities of (a) items, (b) species, (c) areas, (d) group/community for development.
8. Determine finances needed and sources of finances if possible. Common sources of funds are – government departments (central and provincial) and organizations, banks etc. Plans have to be prepared in the form of full-fledged Schemes or Projects in the format that the financial organizations prescribe and within rules and regulations.

Step 3 – Deciding on Services and Inputs

Once a plan is carefully developed for particular entrepreneur in an area, the various groups of activities, the components of livestock development have to be organized. From purely practical point of view, the following may be considered as the minimal activities that need be implemented in the target area by way of livestock development.

1. *General improvement:* Improving the quality of livestock so that they may produce more milk, eggs, meat etc. The steps include -
   (a) Supply of improved animals/birds of superior breeds; Artificial insemination with semen of good bulls.
   (b) Removal of scrub/low grade animals through castoration and slaughter.
   (c) Herd recording for knowing production worth of animals, selection of superior animals and their propagation.

2. *Feed resource augmentation:* Low production of our stock is mostly due to poor feeding. Feed cost constitutes 60-70 per cent of production costs. Hence need for developing economic feeds in adequate quantity and of quality.
   (a) Supply of improved fodder seeds/saplings;
   (b) Increasing area under fodder;
   (c) Pasture/grassland development;
   (d) Conversion of "lush" season greens as hay and silage;
   (e) Improving feeding value of straws and stovers by use of new techniques: (i) urea + molasses treatment of straw/bagasse/stovers; (ii) ensiling straws with urea ; (iii) soaking, etc.
   (f) Feed mills – Feed pellets; combined/compound feeds; poultry feeds;
3. *Stock management:* To keep them in comfortable condition –
   (a) Housing/shelters – renovation/improvement of the existing ones: construction of new ones.
   (b) Hygiene and sanitation – clean surroundings.
   (c) Protection against hot (wallowing, showering, body wetting) and cold (bedding, curtains, jackets) weather conditions.
   (d) Proper schedules of working (draft animals), milking, feeding, shearing slaughter etc.
4. *Health cover:* Prevention of occurrence of diseases as well as treatment of sick animals.
   (a) *Infrastructure development and expansion:* Hospitals, Dispensaries, First Aid Centres/Stockmen Centres, A.I. Centres, Semen Banks, Disease Investigation Centres etc.
   (b) *Man power development:* Veterinarians, Stockmen, Componders, Extension workers.
   (c) *Organisation of preventive measures:* (i) Vaccinations; (ii) Deworming campaigns; (iii) Testing for disease carriers; (iv) Fertility surveys and camps.
5. *Product handling and marketing:* Unless the products are handled and sold without loss in quality the income will be less.
   (a) Milk collection centres, Chilling centres, milk plants.
   (b) Slaughter houses, by product utilization centres.
   (c) Egg and meat storage and transport facilities.
   (d) Wool grading and marketing centres.

Development of Livestock Entrepreneurship

A prospective (livestock) entrepreneur must be evaluated for the following qualities to be successful in the venture.

(a) Whether he is a self-starter?

(b) How does he feel about other people?
(c) Can he lead other?
(d) Can he take responsibility?
(e) How good organizer he is?
(f) How good worker he is?
(g) Can he make decisions?
(h) Can other people trust him?
(i) Can he stick with?
(j) How good he is in keeping records?

According to the Small Business Administration, successful entrepreneurs should have five characteristics:

- Drive, which is defined as the most important attribute. Entrepreneurs can expect long hours, high stress and endless problems, as they launch a new business.
- Thinking ability, or the characteristic that encompasses creativity, critical thinking, analytical abilities and originality.
- Aptitude for human relations. This characteristic recognizes the importance of the ability to motivate employees, sell customers, negotiate with suppliers and convince lenders. Personality plays a big part in success in this area.
- Communication skills, or the ability to make themselves understood.
- Technical ability speaks to the need of the entrepreneur to know their product and their market. They must consider the long- and short-term implications of their decisions, their strengths and weaknesses, and their competition. In short, they need strategic management skills.

Livestock enterprises are the oldest forms of enterprise of rural India. People are traditionally set to these forms of enterprises. Rural livestock enterprises are heavily dependent on natural resources. Unfortunately, not much scientific attention has been paid to make these enterprises either more remunerative or less dependent on depleting natural resources. Though very encouraging success has been achieved in certain areas like poultry yet much has to be done in other areas.

Chapter 28

Women Entrepreneurship through Floral Art and Floristry

Nilimesh Roy Chowdhury

Flower is the best medium to express all the emotions of your heart: joy, happiness, sense of celebration, sorrow, romance. The aesthetic value of flowers and ornamental plants, their use in social events, overall satisfaction in working with them and high income generating power are attracting modern entrepreneurs to invest money in the floriculture industry. The demand for flowers and ornamental plants for different needs like religious, official ceremonies, parties, house decoration, wedding, funerals, etc. is on the rise. This demand for fresh flowers and plants will increase world wide over the coming years.

Table 28.1: Export of Floriculture Products from India (Rupees in Million)

Item	*1991-92*	*1992-93*	*1993-94*	*1994-95*	*1995-96*
Cut Flower Fresh	3.99	10.90	9.96	29.98	100.33
Dried Flowers	64.35	74.31	109.21	195.78	364.56
Live Plants	40.15	30.56	30.52	60.43	81.48
Dried Plants	23.99	25.45	23.72	10.35	35.83
Bulbs, Tubers etc.	12.06	7.83	14.95	11.83	19.21
Total	**144.54**	**149.05**	**188.36**	**308.37**	**601.41**

Source: FICCI.

Globally more than 145 countries are involved in the cultivation of ornamental crops and area under these crops is increasing steadily. And due to globalization and its effect on income, there is growing per capita consumption in most of the countries. India, bestowed with diverse agro-climatic and ecological conditions, abounds in ornamental plants and it is the place of origin of many plants of floricultural importance. In our country a large number of people earn their livelihood wither by producing or marketing of flowers.

According to a report of the APEDA, the total area under flower crops was estimated around 34,000 hectares, which included 24,000 hectares under traditional flowers such as marigold, jasmine, aster, rose, chrysanthemum, tuberose and 10,000 hectares under modern flowers like cornation, rose, gerbera, gladiolous, anthurium.

Returns from floricultural products were estimated at Rs.205 Crores, which included Rs.105 Crores from traditional and Rs. 100 Crores from modern flowers. The data collected by National Horticultural Board.

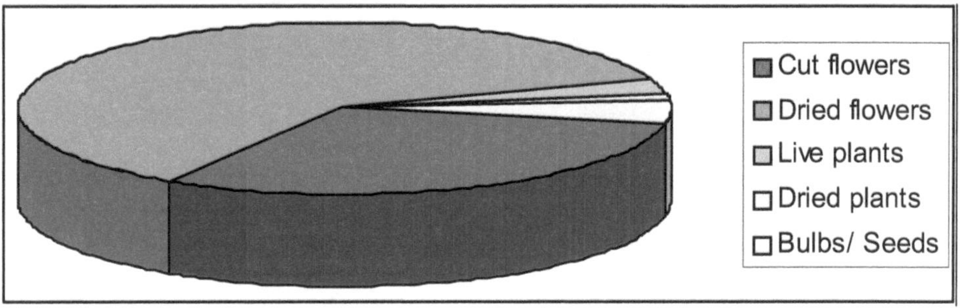

Figure 28.1: Per cent Contribution of Different Items to Total Floricultural Export in the Year 2003-2004.

The total business of flouriculture products in India in 2005 was Rs.8174 Lakhs while it increased to Rs.10117 Lakhs by April,in 2006. There were more than 300 export-oriented units in India. More than 50 per cent of the floriculture units are based in South zone mainly in Karnataka, Andhra Pradesh, Tamil Nadu. Also West Bengal, Maharashtra, Rajasthan have large areas under floriculture. The domestic flower production goes on increasing annually. Technical collaborations with foreign companies have been approved for India, in order to increase total share in the floriculture world trade.

Past trend analysis (Table 28.1 and Figure 28.1) clearly depicts that the amount of foreign exchange earned by export of floricultural item is steadily increasing in India, too.

To estimate of domestic market size of floriculture, no quantitative data is available. Apart from the growing in the traditional market for flowers, there is also an increasing trade for flowers used for decorations and gifts, which are specially called as " Florist's Item". However, at the retail level the domestic floriculture market is worth about Rs.500 crore

As a cascading effect of Globalization, value addition has become inevitable in every sphere of our life, thus effecting floriculture, too. Now packaging and arrangement can bring a huge difference in price of a flower of medium quality. On the other side of the coin, even the best quality rose may not be able to fetch a minimum price if not arranged or displayed properly. Thus different methods of value addition have become extremely important for the survival of farmers as well as florists.

The easiest and fastest way to add value is by preparing floral sticks/spikes, bouquets and other floral arrangements. Besides, preparation of Garlands, *Gajra, Veni* etc. from loose florets is also an important avenue. Further preparation of dry flower, essential oils and bio dyes are also coming up well and engulfing a major share of floricultural export (Figure 28.1)

Floral Art and Flower Arrangements

Floral art is not just about using flowers, foliage and containers. At another level, it uplifts the mood, gives confidence by inculcating a sense of achievement. It stimulates the inner power and can be a healing source as well. Every flower arrangement and every colour has a different meaning. A bouquet of flowers makes the recipient smile. And it is pleasurable to make flower arrangement, if you will, or in designing a piece of floral art-all this, surely, is within the reach of all of us. We have heard how the floral art works as a color therapy. It refreshes the mood, there is liveliness and freshness. Sometimes, art forms depict a philosophy of life; at other times, they reflect the psychology. Floral art also heals and relaxes through aroma and it is a good source of inspiration, motivation and creativity. It is an art that has a discipline about it, leading to self-awareness and a way of expressing and pleasing. All this are inherent in the concepts of designing and composing in the concepts of designing and composing in the concepts of designing and composing in the world of modern and contemporary floral art. The floral art piece unfolds life in its different aspects. Take the bud; it holds a promise for the future. The full bloom may not last long but what a thing of beauty it is for the beholders, while it lasts. The foliage expresses a long lasting companionship.

Florists either design arrangements per customer request or sell and design arrangements according to standard designs. Their job also entails floral decoration of buildings, cost and price consulting, and training others in floriculture. Most traditional retail florists are affiliated with national wire services that handle long-distance orders by telephone. The receiving florist gets a commission from the sending florist depending on the size of the order. Often, customers are charged an additional service fee by the sending florist.

When it comes to give someone flowers, there are no ground rules. With flowers, it really is the thought that counts! However, to select the ideal flowers for a person or occasion, consider the types of flowers that best reflects the personality, gender or theme of the intended recipient or venue. A florist can help you customize your arrangement of flowers for an array with a personal touch. Flowers help to define mood, establish a sense of style and create texture. Florists either design arrangements per customer request or sell and design arrangements according to standard designs. Their job also entails floral decoration of buildings, cost and price consulting, and training others in floriculture. Most traditional retail florists are affiliated with national wire services that handle long-distance orders by telephone. The receiving florist gets a commission from the sending florist depending on the size of the order. Often, customers are charged an additional service fee by the sending florist.

In order to increase service, computers were used to send customers' orders to other towns. Some consumers used toll-free numbers to reach the florist who actually sent the product. Others used various florists, who in turn used various wire services like Florist Transworld Delivery, Teleflora, and AFS to send the flowers. Wire services offering memberships give florists advertising and promotional materials and assistance, as well as training in management, marketing, and point-of-purchase aids. Standard arrangements were designed to make it easier for the customer to order. Other factors that helped this industry included telephones and credit cards. These allowed direct contact between retailers and growers ' unlike the old days of telegrams 'and provided a better way to pay, thereby protecting both parties even though they did not know each other. Flowers are a great gift for any occasion. Nobody wants to pay more for a flower product that has been over-valued. If you find a good flower shop online, go for it. Flower delivery online really can be useful in saving time and hopefully, money. Flowers are a gift of beauty. Although they are beautiful, flowers have many different meanings. Sending flowers with special meanings can be appropriate for almost any occasion - birthdays, funerals, weddings, anniversaries, new baby, get well, graduation, new home etc. Whatever the event, you want flowers that have a deeper meaning.

Ikebana

Japanese way of flower arrangement is known all over the world and is appreciated. Japanese art of arranging flowers is called ikebana. In contrast to the decorative form of flower arranging in western countries, Japanese flower arrangement creates a harmony of linear construction, rhythm, and colour; in themselves, decorative requirements. The Japanese emphasize the linear aspects of the arrangement. They have developed the art to include the vase, stems, leaves, and branches, as well as the flowers. The entire structure of a Japanese flower arrangement is based on three main points that symbolize heaven, earth, and man.

Quality Assurance to be Offered by a Florist

- Every flower arrangement that is delivered is made fresh. All Gerberas and Tulips are wired for extra support and life span.
- Even if there is extreme weather and no one is home to collect the bouquet, their courier will place it in water for the customer.

What is Dry Flowers?

The term "Dry Flowers" includes a vast range of dried plant material in any form starting from traditional flowers to many other plant items like spikes of wheat, rice and sorghum, flowers of sugarcane and moog, various kinds of weeds foliage, gourds, raffia, nuts, moss, fibers and ribs which is presentable as dried and coloured (optional), falls under the domain of Dry Flowers.

Through this dried plant parts various **house decorative materials** and **household fancy utility materials** are prepared which are sold in the urban market. As products are house decorative materials, the manufacturing activities demands skill in designing, finishing, and product diversification to suit different client demands. The capacity building part of the enterprise demands extensive training in product design and product diversification.

Different End Products

- ☆ Potpourri
- ☆ Decorative floral crafts for interior decoration and commercial exploitation.
- ☆ Greetings card.
- ☆ Flower arrangements.
- ☆ Dried Flower Bouquet
- ☆ Dried Flower Wreath
- ☆ Dried Flower Craft
- ☆ Flower Arrangement Crafts
- ☆ Twisted Twig
- ☆ Terracotta Artificial Fruits
- ☆ Artificial Flowers Sola
- ☆ Artificial Flower Arrangement
- ☆ Dried Floral Garland
- ☆ Dried Flower Arrangement
- ☆ Dried Leaves Artificial Foiliages
- ☆ Palm Dried Flower Swage
- ☆ Dried Floral Arrangement

Techniques for Drying Flowers

1. Air drying
2. Pressing
3. Embedded drying/Embedding

Air Drying

In the attics or dark corners of unused rooms where there is an adequate circulation of air, is ideal for hanging the plant materials in an inverted position for drying. Time required for drying varies from a week to several months depending on-

- ☆ The type of material
- ☆ Harvesting stage and time.
- ☆ Relative humidity of the drying.
- ☆ Air circulation of the room.

The commonest choice for this technique is members of the *Compositae* family (eg. Straw flowers, helipterums and xeranthemum), several seed heads, members of the *Umbelliferae* family (angelica, fennel) large spike seed heads (hollyhock), flowers composed of many tiny blossoms (gypsophylla, achillea), *Helichrysum*, etc. Air drying is the simplest and cheapest among all the methods of dehydration. However, considerable shrinkage of most of the flowers and dependence on weather are the major constrain for this method.

Pressing

Plant materials are placed between pieces of newspaper or blotting paper and some weight is kept in the form of heavy books. Commercially the whole bundle or stack of plant materials and newspaper or blotting paper in the plant press and its screws are tightened. After 24 hours, the bundle is removed to an electric hot air oven for 24 hours at 40 -45 degree Celcius, when most of the flowers and leaves get dried. The main problem of press drying is that the fininshed product is rather two dimensional than three dimensional and very brittle. Any flat structure especially leaves of thuja, taxodium,marigold, gravellia, rose,various ferns, casuarinas and grasses can be press dried successfully.

Embedding

To maintain the three dimensional structure of the finished product, the plant materials are dried in a desiccant by embedding. During desiccation the moisture of the plant material is absorbed by the surrounding desiccants, *viz.* silica gel, borax, alum, sand or yellow cornmeal. After embedding the containers can be kept in room or can be exposed to sun on a regular basis or can be kept in oven (electric oven or micro-wave oven) for faster or rapid drying. For embedding, bottom of the containers are filled up to 5cm. with the embedding materials, upon which stalks of the flowers are inserted upright. The entire material is covered with desiccants and during the process sufficient care should be taken to protect the floral structure.

Sl.No.	Flower	Time (min.) taken for Embedded Drying in Microwave Oven
1.	Snapdragon, chrysanthe-mum, bougainvillea (Partha, Subhra), gerbera, yellow gladiolus, pride of India and narcissus	3.00
2.	Larkspur, strawflower, phlox and statice	2.30
3.	Paper flower, gulmohar and ixora	2.00
4.	Combreatum and *Petrea*	1.00
5.	Water lily	4.00

Bleaching, Sulfuring and Coloring

Some flower species lose their colour and become unattractive on the course of drying that is why some people prefer to paint the dried flowers. Before coloring some prefers to remove the inherent colour by bleaching or sulfuring. Bleaching is an oxidative process where hypochlorite, chlorite and peroxide are most commonly used. Whereas sulfuring is a reductive process done by burning sublimed sulfur powder or sulfur-di-oxide in a sealed chamber prior to drying. These bleached materials can be coloured by using cationic aniline dyes of primary colours (red, yellow, blue), blended for enormous colour range. Car aerosol spray paints is quick drying process giving a high gloss finish. Brush work is slower, but gives better perfection.

- ☆ Gomphrena and Ceosia-Vat colours 0.25
- ☆ Bleaching:
  - Sodium Chlorite10 per cent solution at 70 °C takes 7 hours for pink gomphrena to remove colour
  - Hydrogen peroxide 30 per cent at 60° C - 7 hours for gomphrena
  - Common bleaching powder (calcium hypo chlorite) at 3-5 per cent concentration at 70° C is economic for bleaching reeds, grasses, ear heads etc
- ☆ The embedded flowers should be taken out by tilting the container.
- ☆ The dried flowers are either picked by finger or tweezers, cleaned by inverting them and tapping the stem with finger slowly and gently.
- ☆ Any remaining desiccant is finally removed with the help of fine hair painting brush.
- ☆ The dried flowers should be stored in desiccators or glass jars or plastic jars in which blue silica crystals are kept at the bottom. This is done to prevent them from spoilage and their future utilization.

Market Potential

During the last 16 years, Indian economy, integrated with the global market, has paved the way for increased export of floricultural products and the contribution of dried flowers is enormous. The organized research and production of Dry flowers in the country by making use of rich biodiversity and cheap labour would further trigger the growth of floriculture industry. In view of the unorganized nature of the business, it is difficult to estimate the size of flower trade, both in terms of volume and value. During last five years or so this business has really boomed in India, which is reflected in the number of new florist outlets in all cities and increase in the purchase of flowers as gifts. This has put the current trade at several times than earlier estimate. A recent study of Delhi market alone put the value of flowers traded on wholesale as Rs. 500 million.

During that period the growth of total floriculture is 4 times where as the Dry flower export has grown 5.5 times. During 1999-2000 the reported export of floriculture is 1167 million Rupees. During the same period the export component of Dry flower has also grown sharply and the demand of Indian dried flowers and plants has sharply increased world-over.

- **Goodwill Nursery**

  Contact Name: Mr. Shekhar Dey

  Address: 14, Mansatala Lane, Kidderpore City, Kolkata - 700 023, West Bengal, India

  Phone: 91-33-24595907

- **A Global Dried Flowers**

  Contact Name: Mr. Rakesh Jain

  Address: 31, Shakespeare Sarani # 301A, Kolkata - 700 017, West Bengal, India

  Phone: 91-33-22873901

- **Allied Exports**

  Address: 3A, Ripon Street, Kolkata - 700 016, West Bengal, India

  Phone: 91-33-22296336

- **Floral Solutions P. Ltd**

  Address: 139/1, Basement, Humayun Pur, Safdarjung Enclave, New Delhi - 110 029, New Delhi, India

  Phone: 91-11-26191981

- **Floral Solutions P. Ltd**

  Address: 139/1, Basement, Humayun Pur Safdarjung Enclave, New Delhi - 110 029, India

  Phone: 91-11-26191981

Grassroot Entrepreneurship

Spreading the "Dry Flower' enterprise in the grassroots will enhance the income of women entrepreneurs with the potentials of diversified grassroots level innovation to stay afloat in the capital intensive competitive market. Women can earn a good amount from their home without compromising the normal household level activities.The cutting edge in bargaining power through networking at the production line will strengthen the sustainability issue

Networking

Number of entrepreneurs joining the Grower's Cooperative is the key issue for product diversity and quantity production. A minimum number of 200 BPL women are necessary for a viable economic entity. Active partnership with panchayats and its downstream organizational structure like self-help groups, farmers club

The entire enterprise model:

- **Collection of raw materials**: Initially from the wholesalers of Barabazar, Newmarket, Gobordanga. Afterwords from natural resourses of forest areas. Of course in the later part even these materials will be prepared by the entrepreneurs that will be sold as such as well as serve the purpose of raw materials in floral arrangements. However, the total cost of raw materials of an arrangement should not exceed **20 per cent of the product market price**

- ☆ **Labour wage per item: It should be 45 per cent of the product market price**
- ☆ **Cost of assembling+ rent of common storage place/collection centre +Transportation cost (in between producer and cooperative+Cooperative to market) = 10 per cent of the product market price**
- ☆ Share of cooperative for advertisement, branding, networking and marker linkage= **20 per cent of the product market price**
- ☆ **Loan repayment/miscellaneous : 5 per cent of product market price**

A very close market survey revealed that a standard branded product having the market price of **Rs. 60/-** contains raw material of **Rs. 10 to 12/-.** Considering the upper limit, if we fix it to Rs **12/-** the share of raw material stands to 20 per cent. In practice, middleman enjoys the lions share of the product margin and the producers get practically nothing (10 per cent or less). Here we have given maximum share to the producers of fixed **45 per cent** resulting to **Rs. 27/-.** Cooperative's share is a nominal 20 per cent along with other costs **(10 per cent and 5 per cent).**

Now from entrepreneur's point: Earning **Rs 27/-** per bouquet. If she can produce 20 bouquets per month **she can earn Rs. 540/-per month (spending 3 hrs a day)**

From cooperatives point of view : If one entrepreneur is producing 20 arrangements per month then 100 women will produce 8000 items per month through which the cooperative will have the share of **Rs. 24000/- per month.**

Summary

- ☆ Being a low-cost and high-demand industry, new generation of professional florists are coming forward with great *razzle-dazzle.*
- ☆ High income-generating power of flowers can be utilized elegantly in this manner also.
- ☆ This industry as a whole can contribute immensely to area development and, in turn, to the overall development of the nation.

Chapter 29

Biodiversity Conservation through Participatory Monitoring: A Case Study from People's Protected Area Dhamtari, Chhattisgarh

Raktima Mukhopadhyay

Biodiversity plays a significant role in maintaining the life support system of the people and provide them with livelihood opportunities. Forests are the repository of biodiversity. Often such biodiversity resources face the threats of degradation and extinction due to over exploitation and lack of monitoring mechanism. People living in the forest fringe areas can play important role in conservation of biodiversity provided they are properly oriented, their capacity is built and they are empowered to check the drivers of degradation. The present paper focuses on how the floral biodiversity resources of Dugli Jabarra and Sankra Chandanbehera PPAs under Dhamtari Forest Division in Chattisgarh are conserved through active participation of the Joint Forest Management Committee members. Forest fringe dwellers are organized in the form of Joint Forest Management Committees to protect and conserve the biodiversity, sensitized to adopt non destructive harvesting practices of biodiversity, their capacity is built for resource survey to measure the status of biodiversity at a regular interval and take corrective action to reverse the process of degradation. Such empowerment of people has brought positive changes in the status of biodiversity and the substantial increase in the income of the people through biodiversity based livelihood.

Keywords: *PPA, JFM, Institutions, Livelihood, NTFP, Regeneration, Non Destructive Harvesting.*

Introduction

Biodiversity provides the life support system of human being, but there has always been a challenge to negotiate the dichotomy between conservation of biodiversity and development. The answer could be a resilient and adaptive management which can reconcile the dichotomy of threat perceptions arising out of conservation. The role and importance of biodiversity have been focused globally through the International Convention on Biodiversity (CBD). Through the agreement focus has been made to put effort on conservation of the biological species on the one hand and their sustainable use for creating livelihood opportunities on the other especially for the poor people living in the biodiversity rich forest fringe areas. International Union for Conservation of Nature (IUCN) defined six associated management categories of protected areas among which Category VI is defined as *Managed Resource Protected Area: protected area managed mainly for the sustainable use of natural resources* – area containing predominantly unmodified natural systems, managed to ensure long-term protection and maintenance of biological diversity, while also providing a sustainable flow of natural products and services to meet community needs. Chattisgarh Forest Department has established People's Protected Area (PPA) in Forest Divisions like Dhamtari, Jagdalpur, East Bhanupratapur, Bilaspur, Marwahi, Raigarh, East Sarguja, Korea and Durg. The underlying philosophy of the People's Protected Areas (PPA) is evolving Sustainable Livelihood Approach with Biodiversity Conservation (SLAB). The PPA programme focuses to ensure achievement of the broad goals of conservation and sustainable utilization of the forest resources by building community institutions and developing their capacity to promote in situ and ex situ conservation and propagation, developing standards and norms for non destructive harvesting for wild collection of biodiversity and periodic monitoring by the community institutions to measure the changes of biodiversity to take strategic corrective actions for conservation and develop options for sustainable livelihood. The present paper focuses on how through participatory monitoring the forest resources could be conserved in the PPA of Dhamtari Forest Division in Chattisgarh

Study Area

Dhamtari Forest Division

Dhamtari Forest Division is located between 31075'E and 32015'E longitude and 200 0'N and 200 40'N Latitude. The division stretches about 58 kms from North to South and 39 km from East to West. The total geographical area of Dhamtari district is 4,081sq kms, out of which the forest area comprises 21,25,540 km2 (52.1 per cent of total geographical area).

Forest Types of the Study Area

The forests of this division are of two type's *viz.* Sal and Mixed forests. The Sal (*Shorea Robusta*) forests are found concentrated in four ranges namely, Dugli, Nagri, Birgudi and South Singpur while those of the Dhamtari, Keregaon and North Singhpur ranges are of the mixed forest type.

Selection of Dugli-Jabarra and Sankra-Chandanbahera PPAs under Dhamtari Forest Division

To start with, the biodiversity rich areas have been delineated as Dugli-Jabarra and Sankra Chandanbahera. The PPAs are constituted in these two areas in consonance with IUCN chapter VI that is, i.e, *protected area managed mainly for the sustainable use of natural resources* to ensure long-term protection and maintenance of biological diversity, while also providing a sustainable flow of natural products and services to meet community needs. With the consent and the help of the local people, 1,000 ha area with density of 0.5 and above, with a potential of rich biodiversity, less biotic pressure and already existing community institution such as Joint Forest Management Committees (JFMC) for protection of forest, have been demarcated under each PPA unit for *insitu* conservation every year, over a period of five years. These 5,000 ha areas have been protected from all kinds of biotic pressure like grazing, fire and illicit felling with active participation of JFMCs. While seven villages namely Jabarra, Dinkarpur, Dugli, Kharka, Chargaon, Deogaon and Munaikera fall under Dugli Jabarra PPA there are eight villages *viz.*, Basin, Belarbahera, Barpahar, Tumribahar, Arjuni, Daur, Theni and Ghoragaon under Sankra Chandanbahera PPA. The Dugli-Jabarra PPA covers 27,457.24 ha forest area. Out of the total area under Dugli-Jabarra PPA, conservation and sustainable use of the species is carried on in 16, 194.30 ha and 11,262.94 ha is kept for Nistar purpose from where villagers can get fuel, fodder, small timber and bamboo for their bonafide domestic use. Under Sankra-Chandanbehera PPA 24,255.19 ha area is covered, out of which 3,024.12 ha area is kept for in situ conservation and 11,231.07 ha is for Nistar purposes. Each year, 1,000 ha area under each PPA is demarcated for protection and conservation purposes. Five such 1,000 ha areas are demarcated under each PPA which are named as P1, P2, P3, P4 and P5.

Methodology

The study is conducted based on the primary data collected from JFMC members on the social, institutional and economic aspects as well as the longitudinal data collected to capture the changes of forest vegetation through the grid survey by the survey team comprising of the JFMC members of Dugli – Jabarra and Sankra-Chandanbahara PPA was procured from the Office of the Divisional Forest Officer (DFO), Dhamtari. This was analysed to understand the impact of people's participation on the conservation of biodiversity. Interviews were conducted with the JFMC members in order to elicit their specific knowledge regarding the project objectives and activities, status of forest, benefits accrued by them etc. Retrospective recall method was used to get the past year data and status quo method was used to assess the current year data.

Sampling

A total of 1000 hectare of *in-situ* conservation area has been surveyed by the survey team members of Dhamtari forest division to identify the flora over a period of five years. The various steps involved in resource survey were as follows. 1. First of all, North – South and East – West grid lines were drawn at an interval of 3 cm X 3 cm on a map of *in-situ* area of 1:15000 scale. A total of 50 grid points were established and the points were given symbol in order to avoid confusion or duplicity in resource surveying. After reaching at the grid point, four diagonals were laid out in all four directions (North, South, East and West) at a distance of 22.40 meter. A straight line, measuring 31.40 meters, joined all four corners such achieved (Figure 29.1). Thus, the

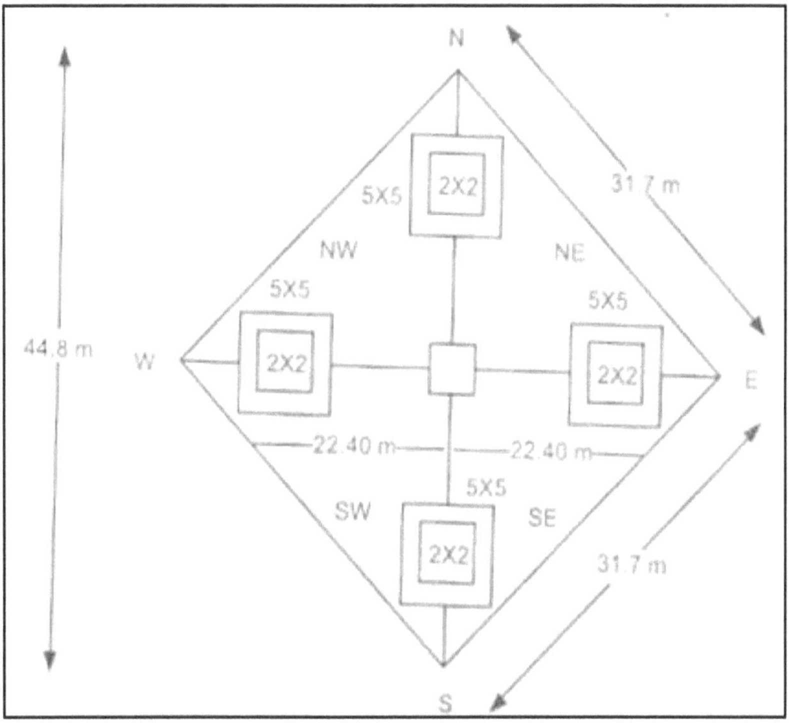

Figure 29.1: Figure of the Plot.

total plot area became 0.1 ha. Wooden pegs were fixed on the ground to mark the area sample plot. Now in the center cross point and center of the line on each direction, a plot of 2 meter x 2 meter was established, 9 (Nine) plots on each grid point, 5 plots of 2x2 meter and, 4 plots of 5x5 meter measurement. The plots of 2x2 meter measurement were termed as Regeneration plot and the plots of 5x5 meter measurement were termed as Medicinal plots. In the regeneration plot, all the established and new recruits of important species were counted. Similarly in the medicinal plots, number of all the medicinal plants has been counted (excepting *lantana, chhind* and the main spp. which have already been counted in the regeneration plot).

Regeneration Survey

The seedlings were counted in 5 plots of 2m X 2m each in 50 grids as mentioned above. All the regeneration calculations were done using the formula to identify the Regeneration Status = Established X 1 + Un established X 0.25 + One year old seedlings X 0.

Data Analysis

For the data so obtained, density, relative density and regeneration status were calculated using formulae given by Curtis and McIntosh (1951). Density is the number of individuals per unit area, which gives the numerical strength of the species. In general, density is the total number of individual of a species relative to the total area examined. Relative density is a proportion of total number of individuals of a species with the total number of all species with an area.

Density

Total no of individuals of species 'A' in all quadrates (D) stems/ha = * 10000 Total number of quadrates sampled X size of quadrates Relative Density Total no of individuals of species 'A' (RD) per cent = * 100 Total individuals of all species

Calculation of Species Conservation Value (adopted from Ganeshaiah *et al.,* 2001)

SCV =Where,

SCV = Species Conservation Values = Species Diversity or number of species in a given plot pi = population size of a species in a given plot

Results

In the PPA of Dhamtari Forest Division, the method adopted for forest regeneration is Assisted Natural Regeneration which is characterized by enrichment planting, improved silvicultural practices and forest floor management along with soil and water conservation measures.

Regeneration Status

Regeneration status of P2 *Dugli* and P1 *Sankra* is considered as they have the data for consecutive five years (Tables 1 and 2). Looking at the percentage increase in total regeneration we may infer that during the protection period maximum increase in regeneration has taken place in *Saja* followed by *Sal*. There has been an increase of 629.9 per cent and 294.4 per cent in the two species respectively. The presence of new recruits of all the species shows the signs of protection in the area. The comparative assessment of the regeneration status between 2002-03 and 2006-07 says that regeneration status of *Tendu, Sal* and *Saja* has been quite good over the protection period. The established regeneration in *Tendu, Sal* and *Saja* has been 257, 262 and 169 respectively. The new recruits coming up in the different species are the clear evidence of protection in the area.

Status of Medicinal Plants in P2 Dugli

Shrubs

In the shrubs category the Relative Density varied from 1 to 40. In the shrubs category the commercial species like *Charota* and Baibirang have had a greater increase in comparison to the non-commercial species like *Salparni,Gursukri* and *Bankapas*. The overall density of the shrubs increased from 1918 shrubs/ha to 6164 shrubs/ha.

Herbs

The relative density under the herbs category ranged from 0.08 to 27.38. Under the commercial species *Kaali musli, Safed musli,dasmul* and *Kaalmegh* were the main species present in the area. In the commercial category *Kaali musli* had the maximum density of 13532 individuals per hectare followed by *Bada Dash-mool* (13298

Table 29.1: Percentage Increase in Regeneration Status in P2 *Dugli*

Species	*Scientific Name*	*Established Regeneration*			*Total Regeneration*		
		2002-03	*2006-07*	*Per cent Increase*	*2002-03*	*2006-07*	*Per cent Increase*
Sal	*Shorea robusta*	30	123	310.00	31.50	124.25	294.10
Bija	*Pterocarpus marsupium*	27	69	155.60	27.25	69.25	154.01
Saja	*Terminalia tomentosa*	57	432	657.90	60.25	439.75	629.90
Lendiya	*Lagerstroemia parviflora* (roxb)	16	28	75.00	16.75	31.00	85.1
Dhawda	*Anogeissus Latifolia* wall	135	355	163.00	135.00	359.25	166.10
Tendu	*Diospyros melanoxylon* (roxb)	138	153	10.90	140.00	155.25	10.90
Amla	*Phyllanthus emblica*	0	19	0	0	19.50	0
Karra		0	67	0	0	70.25	0
Harra	*Terminalia chebula*	0	3	0	0	3.25	0
Others		188	788	391.10	199.75	806.25	309.8
	Total	**591**	**2037**	**244.70**	**607.05**	**2078.00**	**242.10**

individuals/ha). *Tikhur,* which though commercially important has not grown so well over the period. Under the non-commercial category species with highest density was Genji (9532 individuals/ha).

Climber

Under this category, *Karukandh* 1094.97 shows the highest increase in numbers over the span of 3 years from 378 individuals in 2002 to 2164 individuals in the year 2005. On the contrary, the maximum decrease is in the case of *Anantamul* with 125 in 2002 to 65 in 2006 showing a decrease of 4 per cent over the period. Some species that were absent in 2002 but are found in 2006 were *Baichandi, Satavar,Bhespar, Dotakandh, Gaypar, Jarkakand, Satavar* and *Suaokand*. Looking at the commercial importance of *Baichandi,* its growth over the span of 3 years is not so good hence it is important to promote its growth by weeding out the less useful species in the climber's category. The overall density of climbers has increased from 2538 individuals per hectare to 9372 individuals per hectare

Status of Medicinal Plants in P1 Sankra

Shrubs

In P1 of *Sankra,* 8 species of shrubs with medicinal value were observed. The relative density of the shrub species ranged from 50.32 to 1.48. The highest RD was recorded in *Hasiyadpar* followed by *Bibirang* [44.76] and *Ghirgholi* [38.55]. *Dhawai* has economic importance but its presence in the P1 area is quite low. It is important to promote its development in the area. *Gudsukri* had the highest percentage as earlier there was only 1 individual (2002-03) of the species present in the plot as compared to 55 in the next year [03-04] and 108 in the consecutive year [04-05] and 164 in 2006-07. The density of the NTFPs having medicinal value under shrub category in the P1 area has increased greatly from 400 stems/ha to 2708 stems/ha during the protection period. Species like Bankapas and *Hasiyadpar, Ghirgholi, Khuddu* were completely absent in the base year [2002-03] but then made good presence as 340 individuals in *Hasiyadpar.*

Herbs

The main species present in this category were *Safed musli, Bhuiamla, Kayukand,Kali musli, Tikhur, Chirayata, Lajjwanti* and *Jangli adrak*. The relative density of the herb species varied from 27.90 to 0.10. There were certain species which were absent in the base year (2002-03) but have come up in large number in the next year worth mentioning are, *Lajvanti* [0-237], *Teenpaniya* [154] and *Bada Teenpaniya* [303], There has been a promising increase in some species like *Bhui amla* [from 4 individuals to 1624individuals in the plot], *Kalmegh* [176 to 1966 individuals] Chirchira [62 to 643 individuals] and *Ban Adrak* [1-49 individuals]. The highest density was recorded in *Kaalimusli* with 6076 stems/ha, followed by Kalmegh with 3932 stems/ha and *Bhui Amla* with 3248 stems/ha and with 1200 stems/ha. In the rare species like *Kamraj, Tejraj,* and *Sarpdasa* there has not been any improvement.

Table 29.2: Percentage Increase in Regeneration in P1 *Sankra*

Species	Scientific Name	Established Regeneration			Total Regeneration		
		2002-03	2006-07	Per cent Increase	2002-03	2006-07	Per cent Increase
Senha		1	33	3200.00	1	34.75	3375.00
Dhawda	*Anogeissus Latifolia* Wall	6	87	1350.00	6	97.00	1516.67
Saja	*Terminalia tomentosa*	19	169	789.47	19	183.00	863.158
Sal	*Shorea robusta*	149	262	75.84	155.5	305.50	96.46
Bija	*Pterocarpus marsupium*	9	112	1144.44	9.25	118.75	1183.78
Tendu	*Diospyros melanoxylon* (Roxb)	81	257	217.28	82.25	306.00	272.03
	Total	**92**	**886**	**863.04**	**100.25**	**998.25**	**895.76**

Climbers

The relative density of the species in this category ranged from 21.58 to 0.15. In this category there were certain species which were absent in the base year but were found present in the next consecutive years. These species are namely *Dotokand, Gayapar* and *Satavar*. In the final year these species had 400, 228 and 118 individuals respectively. The density in the climbers ranged from 1144 stems/ha to 8 stems/ha. The highest density was recorded in *Maalkangini* (1144 stems/ha), *Karukand* (952), *Gaypar* (918 stems/ha) and *Dotakandh* (800 stems/ha). The lowest density was observed in *Baichandi* (8 stems/ha) followed by *Dokrakandh*. The result of protection is clearly evident in P2 area of Sankra where the overall density of the climbers with medicinal value has increased from 684 Stems/ha to 5300 stems/ha during the protection period.

Species Conservation Value

The species conservation value in both the Ranges has shown exponential increase within the year of inception of the PPA programme and continued to grow in the second year as well. However, in its third year the SCV started to erode. However, once such data was shared it was stabilised and improved (Figure 29.2). This trend has been observed in case of both the ranges. As shown in the Figure 29.2, it could be concluded participatory conservation programme are very effective and efficient way of conserving biodiversity. It also strengthen the fact that if allowed, due to the resilience of nature the biodiversity regenerated exponentially in the initial stages and then slows and saturates reaching climax of its inherent ecological state, following the typical S shape curve. This phenomenon was observed in both the cases that is, in Dugli as well as in Sankara. Comparatively it was observed that cumulatively Dugli Range SCV is higher than the Sankara Range (Figure 29.3). There could be various reasons for the same including that the SCV of Dugli was higher since inception of the programme (Figure 29.1).

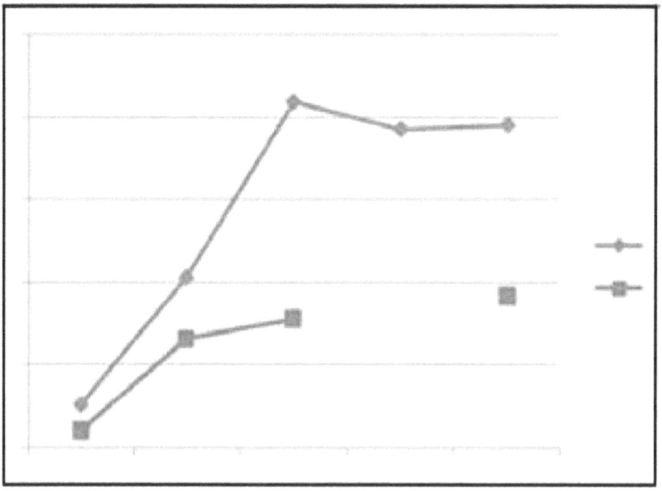

Figure 29.2: Species Conservation Value Trend in the Sampled Plots from 2003-2007of Dugli and Sankara Range.

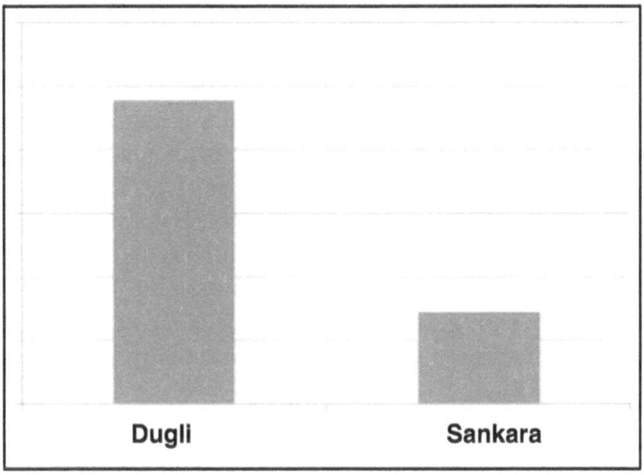

Figure 29.3: Cumulative Species Conservation Value of Dugli and Sankara Range.

Discussion

Role of Community Institution in Conserving Biodiversity *Developing and Strengthening People's Institutions*

The individual of human society interacts with the elements of forest ecosystem as per the social norm, structure and system. The social norms and cultural behaviour will determine the practices for conservation and harvesting of the forest resources. Once the community institutions and government departments institutional procedures are matched, popularly known as *Bilateral Matching Institutions* (Roy 1992), it is expected that the members of the community will develop and follow the norms of sustainable harvesting of biodiversity in agreement with the forest department that would facilitate the process of biodiversity conservation. The effective governance system developed by involving the local community, organized in the form of Joint Forest Management Committees to ensure protection of forest has yielded positive results in terms of species regeneration. JFMC members have been orientated to promote *in situ* conservation of the forest. The villagers were sensitized to protect the forest from fire, grazing, illicit felling and they have been trained in techniques like creating fire line, controlling fire when it occurs etc. As a result, in the PPA there has been no incidence of forest fire in the past years. They have also controlled grazing in the PPA areas.

Identification of Drivers of Degradation and Taking Corrective Actions to Reverse the Process

Norms and Practices for Non-destructive Harvesting of NTFPS for Off Take and Ecosystem Sustainability

Villagers are being sensitized to identify the major drivers of degradation. It has been found that destructive harvesting practices to earn quick return as one of the major drivers of degradation. Non-destructive harvesting of the species is essential for biodiversity conservation and maintenance of ecosystem function. To promote non-destructive harvesting, the villagers are sensitized to adopt the practice of 'sustainable use' of different species. Strong market forces can lead to a rapid loss of biodiversity through *commercial use,* whereas harvesting rates tend to be lower for *subsistence uses,* unless demand becomes very high for a particular product (or products). For example it has been found that over exploitation of Kalmegh by the people of Munaikera, has led to non-availability of the same species in the following year. The high demand for the plants in the market has caused over exploitation and near extinction of the species from the local forest. On the other hand, people of Jabarra has shared that though they used to collect Safed *Musli* for two years, they have stopped collecting it now, as there is no market demand and on the other hand processing cost is very high.

Thus, the lack of demand in the market has caused reduction of the extraction rate of the product. Thus consumptive and commercial uses of biodiversity without considering its ecological implications are most likely to lead to unsustainable harvests (Freese *et al.,* 1996; Freese 1997). Under the PPA programme, efforts are being made to promote non-destructive harvesting by maintaining off take and eco-system sustainability. The villagers have been sensitized for the timely collection of NTFPs. In the case of rhizome species like *Tikhur, Keokand* and *Baichandi,* they are given the option to choose one out of the three tubers, so that the regeneration is taken care of. Though uprooting of a whole tuber of *Tikhur* is easier than leaving 25 per cent in the ground, the repeated practice however difficult of collecting 75 per cent of the tuber, has increased the regeneration and it's sustained availability. Therefore, the locals are now practicing non destructive harvest of *Tikhur.*New rituals have been developed like *"Aonla Navmi"* and the villagers have been sensitized to harvest mature *Aonla* fruits only after *Aonla Navmi.* Mature fruits ensures better price. They have also been educated to leave some fruits in the plants for regeneration. Through this non-destructive harvesting, there has been a greater regeneration of NTFPs especially *Kalmegh* and *Aonla* in the forest. As a result, people are getting more NTFPs and better prices for them. People have also stopped the felling of *Achar* and *Aonla* trees. The incidence of fire in *Mahua* trees has been reduced.

Controlled Grazing and Livestock Management

While the PPA emphasised on forest protection which involve controlled grazing it has also provided a way out for regulatory grazing through rotational basis and introduction of better quality of fodder.

Capacity Building of the Villagers on Participatory Monitoring

There exist-number of scientific research methods of monitoring the biodiversity resources, particularly that comes from the forest and human resources but the challenging task is to develop such criteria and indicators for monitoring which can be understood and adopted by the local people themselves so that people and the Forest Department can also make use of the findings to support their decision making process (Roy 2004). Participatory Vegetation Monitoring is such an innovation in the utilitarian or technological order in the sense that it endures the villagers to determine what would be the judicious amount for harvesting the forest products so that sustainability is ensured (Roy *et al.*, 2000). Villager's capacity has been built so that they can use quadrate method for vegetation monitoring. Quadrate method is the most common method, which can be adopted if it is a homogeneous forest or a monoculture plantation. In Dugli and Sankra PPA the selected persons from the villages are being trained on techniques on grid survey, data collection and analysis. They are conducting grid survey in the PPAs each year. They are acting as resource persons and in turn are training other villagers for the survey. A team of 10 persons have been formed who are being trained to conduct grid survey for the five PPAs in Dugli Jabarra area.

Tradeoffs to Seek the Appropriate Balance Between, and Integration of Conservation and Use of Biological Diversity

Through the participatory monitoring the species having contemporary market demands are being identified and their present status is analysed. Based on the availability of the species and their market demand processing and value addition of certain species are planned. Local community, especially the women is organized in the form of Self Help Groups (SHGs). Forest department has organized training for them and has created the infrastructure facilities. Several processing units have been established in Sankra and Dugli. People are involved in preparation of herbal medicines, *Baichandi* chips, *Tikhur* powder, *Mahul* leaf plates, *Ketki* fibres etc. The net profit earned through *Mohul* leaf plate by the processing centre at Dugli, in the year 2002 – 2003 is Rs 1,811.75 in 2003-2004 it was Rs 12,583, in 2004-2005 it was Rs 11,173, in 2005-2006 it was Rs 29, 205.

It has been observed that there has been an increase in the rate of small plates from Rs 50.00 per 100 in the year 2004-2005 to Rs 80.00 per 100 in the year 2005-06. In the same way, the rate of medium sized plates has been increased to Rs 60.00 per 100 from Rs 40.00 per 100 and the bowls from Rs 25 per 100 from Rs 20 per 100. The JFM committees are provided with protection money at the rate of Rs 300 per ha to ensure protection of the PPA area from fire, grazing, illicit felling etc. This is unique in case of PPA whereas other JFM committees where PPA is not being practiced are not getting the protection money. The committee members decide about employing the watchers and fix the wage rate. They are also being empowered to utilize the funds received from the timber sharing as well as for development purposes as per the consensus of the committee.

Conclusion

The PPAs of Jabarra Dugli and Sankra Chandanbahera have been established in consonance with IUCN category VI where the focus has been on the management of the protected area to ensure long-term protection and maintenance of biological diversity, while also providing a sustainable flow of natural products and services to meet the community needs. The people in PPA have realized that to get the sustainable supply of any target species, they have to maintain other species as well and accordingly have developed social norms for conservation. Through *in situ* conservation protection against felling, grazing, and fire have been given to the forest area to conserve the rare and endangered species and increase their density in the forest area. Through the efforts made for *in situ* conservation visible changes have been found in the forest area in terms of regeneration status, density of the forest, species richness etc. JFMC members have been trained to conduct quadrate survey of the same sample plots on a yearly basis and prepare a resource inventory whereby the change in the status of the forest has been documented as well as areas of intervention has been identified especially for sustainable harvesting and management of NTFPs for sustainable livelihood development. It has become evident through the study that if local community members are involved through the process of participatory monitoring to measure the changes in the forest condition and status and are empowered to take corrective actions to reverse the process of degradation by identifying the drivers of degradation, the chances of conservation of forest increases. The regulatory mechanism designed, adopted and practiced to ensure non destructive harvesting of the species and equitable benefit sharing arrangements help conserving the forest on the one hand and exploring options for sustainable livelihood of the people on the other.

Recommendations

Though there has been encouraging results but to sustain the results efforts should be made to involve more members of the community for forest resource survey through sensitization on voluntary basis. A systematic market study should be conducted to explore market potentials of different non timber forest produces (NTFPs). It is also important to involve community members in preparation of livelihood plan in consistence with the working plan and microplan and its implementation through identification of Common Interest Groups which could be developed as Common Enterprise Groups (CEGs). Capacity of the people should be made to measure other associated ecosystem services that the forest provides through their conservation efforts along with the improvement and sustainable use of biodiversity.

Acknowledgements

The authors express their indebtedness to Chattisgarh Forest Department for giving the opportunity to conduct the study in PPA Dhamtari. They are also grateful to Range officials and Beat Guards of Dugli and Sankra for their support and cooperation to conduct the field work. It would not have been possible to prepare the case study without the sincere efforts and cooperation from the JFMC members belonging to Dugli Jabarra and Sankra PPAs and the authors express their sincere gratitude to them.

References

Convention on Biological Diversity. From www.cbd.int/convention/articles

Curtis JT, McIntosh RP 1951. An upland forest continuum in the prairie-forest border region of Wisconsin. *Ecology*, 32: 476-496.

Forest Genetic Resources: Status, Threats and Conservation Strategies. New Delhi: Oxford and IBH Publishing Co. Pvt. Ltd IUCN 1994. *Guidelines for Protected Areas Categories.*

Freese CH (Ed.) 1996. *The Commercial, Consumptive Use of Wild Species: Managing it for the Benefit of Biodiversity.* Washington DC, USA: WWF Freese CH 1997. *Harvesting Wild Species: Implications for Biodiversity Conservation.* Baltimore, USA:Johns Hopkins University Press

Ganeshaiah KN, Shaanker RU, Bawa KS (Eds.) 2001.

IUCN, Gland: CNPPA/WCMC. From < From www.iucn.org/wcpa>

Roy SB 1992. Bilateral matching institution: An illustration in forest conservation. *J of Indian Anthropological Society of India,* 27: 253-262

Roy SB 2004. *Participatory Vegetation Monitoring:Examples from West Bengal; Root to Canopy.* New Delhi: Winrock International India

Roy SB. Mahapatra TK, Yadav G 2000. Forest Protection and Management by Community in Midnapore district, West Bengal. In: *Joint Forest Management and Community Forestry in India – An Ecological and Institutional Assessment.* Delhi: Oxford and IBH Publishing Co. Pvt. Ltd.

Chapter 30

Corn as A Value-Added Crop

Koushik Brahmachari

What is Value Addition

Process of Changing or transforming a product or a commodity from its Original state to a more Valuable state.

Feasibility of Maize as C_4 Plant under Changed Climatic Conditions

- ☆ Carbon assimilation is high in C_4 plants even at very low CO_2 concentration
- ☆ Photo respiration is very low in C_4 plants where as it is high in C_3 plants
- ☆ Light saturation is high in C_4 plants compared to C_3 plants
- ☆ CO_2 compensation point is lower in C_4 plants compared to C_3 plants
- ☆ C_4 plants assimilate nitrate more efficiently than C_3 plants

Table 30.1: Area and Production of Maize in World and India's Share

Area (in 000'ha)			*Production (in 000't)*		
2006	*2007*	*2008*	*2006*	*2007*	*2008*
144376 (5.3)	158034 (4.9)	161017 (5.2)	695228 (2.12)	791795 (2.39)	822713 (2.34)

[Data in the parenthesis represent per cent share of India]

Source: *Agricultural statistics at a glance,* 2007, 2008, 2009 and 2010, Directorate of Economics and Statistics, Ministry of Agriculture, Government of India. (Website: http://www.dacnet.nic.in/eands)

Table 30.2: Trends in Maize Production (million tonnes) during Kharif and Rabi Season in India

Season	*2010-11*	*2009-2010*	*2008-2009*	*2007-2008*
Kharif	16.3	12	14.1	15.1
Rabi	4.9	4.7	5.6	3.9
Total	**21.2**	**16.7**	**19.7**	**19**

Source: Ministry of Agriculture.

Table 30.3: Maize Area (' 000 ha) and Production (' 000 tonnes) in West Bengal

Year	*Area*	*Production*
1980-81	52.6	55.6
1990-91	64.6	81.6
2000-01	35.3	88.3
2006-07	85.4	253.5
2007-08	77.2	244.5
2008-09	90.8	343.4
2009-10	97.7	385.2

Source: Directorate of Agriculture, Evaluation wing. Government of West Bengal.

Need of Value Addition

- To upgrade nutritional and health promoting potential
- To reduce nutrition related chronic disease mainly in the developed world
- Value addition leads to "Less tension agriculture"

Six Key Strategies for Adding Value

- Changing Physical State of Products (Milling wheat or Corn into Flour).
- Producing Enhanced Value Products (Growing Organic Crops).
- Differentiating Products (Selling Pre-seasoned Corn on the cob).
- Bundling Products (Green House growers sell Pre-planted Hanging Baskets).
- Producing more products that Improve Efficiency up the Supply Chain (Producing new Wheat Varieties that improve Milling and Baking efficiency).
- Owning Assets Up the Supply Chain (Producing Ethanol from Corn, Dairies market their own Organic Ice cream).

Present Use and Suggested Diversification of Field Crop for Better Domestic Utilization and Export Promotion

Field Crop	Present Value-added Products	Suggested Product Diversification
Paddy (Rice)	Milled raw and parboiled rice, flour and flakes and puffed rice.	Quick cooking rice, Sela Basmati rice and rice based extruded snack/breakfast cereals.
Wheat	Flour, Maida, Suji, Dalia and Noodels.	Puffed & flaked breakfast foods.
Maize	Flour and flakes, starch and dextrins and puffed corns.	Degermed maize flour and corn oil.
Sorghum	Flour, flakes and puffed sorghum.	Food and feed uses.
Millets	Flour and pearled millets.	Food and feed uses.
Oilseeds	Oil and cake.	Protein rich soybean flour, protein isolates and concentrates, dal anologs, soy-lecithin, fatty acids and derivatives, direct food uses of oilseeds.
Pulses	Dal, Besan, feed (husk and brokens) and roasted snack foods.	Quick cooking dal, mixed and speciality dal.
Cotton	Cotton and cotton yarn, cottonseed oil and cake.	Colour cotton, byproduct utilization and refinement of cottonseed oil and cake for food and feed.
Jute and Mesta	Gunny bags, carpets, ropes and bags.	For interior decortications, blended yarn, soft luggage, shoe uppers and disposable sleeper, non-woven fabric and geotextile.

Source : Ali, 1999.

Health Benefits of Corn/Maize

- ✰ The high amount of fiber present in corn helps lower cholesterol levels and also reduces the risk of colon cancer.
- ✰ Corn, if consumed in moderate quantities, has been seen to be beneficial for those suffering from diabetes.
- ✰ Being rich in folate, corn helps the generation of new cells, especially important before and during pregnancy.
- ✰ Those suffering from anaemia have shown positive effects after consuming corn.
- ✰ The pantothenic acid present in corn help the physiological functions of the body.
- ✰ Owing to the presence of thiamin, corn has been said to help in the metabolism of carbohydrates.
- ✰ Corn has been found to be helpful in treating kidney problems, including renal dysfunction.
- ✰ Regular consumption of corn, in moderate quantities, has been associated with better cardiovascular health.

- The beta-cryptoxanthin in corn makes it good for the health of the lungs and may even help prevent lung cancer.
- The insoluble fiber in corn makes it good for those suffering from common digestive ailments, like constipation and hemorrhoids.
- Corn is a very good source of Thiamine, Vitamin B1, which is very essential for brain cell function.

Nutritional Value of Corn Amount of Nutrients in 100 gm of corn

- Dietary fiber - 2.7 gm
- Fat – 3.60 gm
- Niacin - 1.8 mg
- Phosphorus - 348 mg
- Potassium - 270 mg
- Protein – 11.10 gm
- Sugars - 3.2 gm
- Iron – 2.0 mg
- Magnesium - 37 mg
- Carbohydrates -66.20 gm
- Calcium - 10 mg
- Mineral– 1.50 gm
- Energy – 342.00 Kcal
- Carotene -90.00 mg

Some Facts about Corn

- Being deficient in Niacin Vitamin (Nicotinic acid) and relatively poor in Protein Corn has lower Nutritive Values than Wheat.
- Pellegra is prevalent among people who rely upon corn for a large portion of their daily food as 50-60 per cent of the Niacin in corn occurs in bound form which is biologically unavailable.
- Treating with 0.5-1.5 per cent Lime (called Tortilla in Mexico) release the bound Niacin and make it available to the consumers.
- In some parts of Africa, traditional food prepared by spontaneous Fermentation of corn flour dough followed by boiling it in water (called Kenkey) as Niacin and Lysine are increased during Dough fermentation.
- Corn is also blended with Wheat flour upto 20 per cent for manufacturing Cakes, Biscuits and Snack food.

Types of Value Addition in Corn

It is of two types :

1. Dry Milling : For making various types of Food products.
2. Wet Milling : For producing Industrial products.

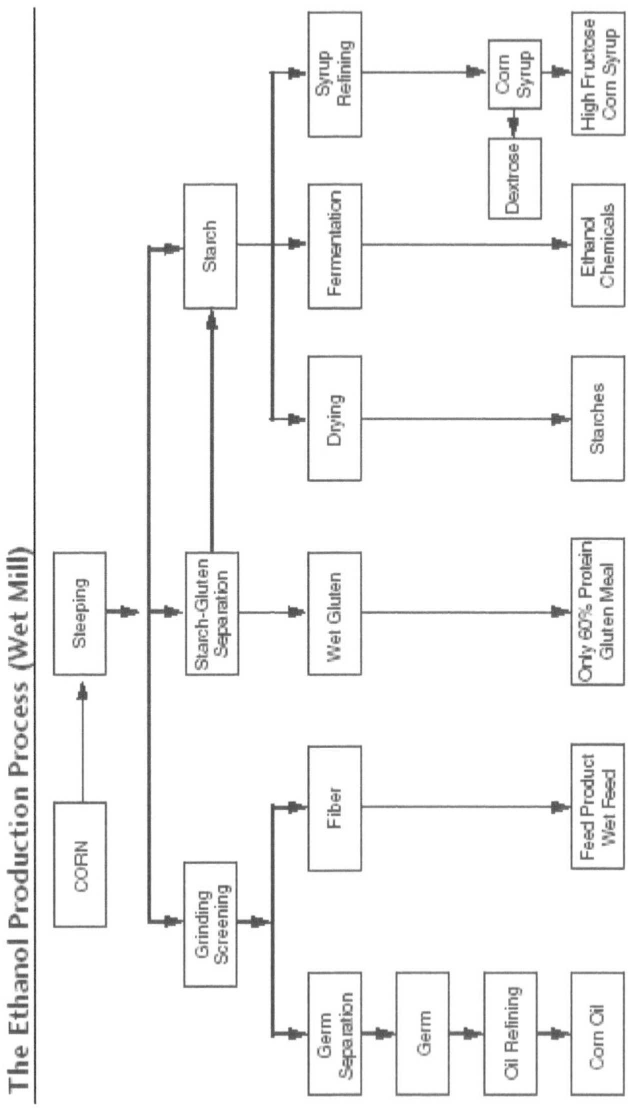

Figure 30.1: Versatile Products from Corn.

A. Food Products

1. Corn Meal: It is made by grinding whole corn. Coarsest meal is used for making Corn Flakes. Medium grade is used to make corn bread, deep fry batter and for dusting pizza dough. Finest grade is Corn flour used for Pancakes, and baby food etc.
2. Corn Starch: It is used as a thickening agent in soups and liquid-based products like, sauce, gravies and custards by mixing it with a cold liquid to form a paste or slurry.

Some Popular Dishes of Corn

1. **Popcorn:** Also known as popping corn, is a type of corn (*Zea mays var. everta*) that expands from the kernel and puffs up when heated. Corn is able to pop because its kernels have a hard moisture-sealed hull and a dense starchy interior. This allows pressure to build inside the kernel until explosive "pop" results. Some strains of corn are now cultivated specifically as popping corns.

Nutritional Value of Pop Corn

1. No cholesterol
2. Very low in saturated fat
3. Very low in sodium
4. Very low in sugar
5. High in dietary fibre
6. Good source of protein

Popcorn Nutrition (30 gm)

Content	*Amount (Typical)*
Calories	93
Calories from Fat	10
Carbohydrates	18.7g
Dietary Fibre	15g
Protein	3.1g
Iron	2.7mg
Cholesterol	0mg
Fat	1.1g

Source: about.com.

2. **Sweet Corn**: Unlike field corn varieties, which are harvested when the kernels are dry and mature (dent stage), sweet corn is picked when the crop is immature (milk stage) and eaten as a vegetable, rather than a grain. Cooked sweet corn contains ferulic acid which provides anti-cancer properties.

3. **Baby Corn**: Baby corn is a popular Asian vegetable that can be consumed cooked or raw due to its sweet and succulent taste. Baby corn is the immature ear of fully grown standard cultivars; ears are harvested two or three days after silk emergence, but prior to fertilization.
4. **Corn Flakes**: Corn flakes are a common breakfast cereal that delivers a crunchy texture and satisfying taste that's not overly sweet. One cup of Corn flakes contains 100 calories, and none of these calories are derived from fat.
5. **Corn Flour**: Unlike most other flours, cornflour blends to a smooth cream or slurry when combined with cold liquid without lumps forming. It can be also used a thickening agent in pakoda. Cornflour is also used, usually with other flours, in the making of cakes and baked goods, to create a lighter texture or to maintain a gluten free status as well as in certain desserts/pudding.

B. Industrial Uses

1. **Corn Syrup:** Made from corn starch, used in food to soften texture, add volume and enhance flavour. HFCS is created when corn syrup undergoes enzymatic processing.
2. **Corn Oil:** It is extracted from the germ of corn, main use in cooking, where its high smoke point makes refined corn oil valuable frying oil. Other industrial uses of corn oil include soap, paint, rust-proofing for metal surfaces, inks, nitroglycerine and in insecticide.
3. **Ethanol:** Ethanol fuel or Bio-fuel is made by distilled corn. Widely used in Brazil and in U.S.A. Regular gasoline powered cars can run on gas blended with up to 10 per cent ethanol. Corn is a renewable resource, so bio-fuels can be a replacement of Fossil fuel.
4. **Corn Silk:** Tea brewed from corn silk is used as a remedy for urinary tract infection as it has diuretic properties. The tea is also helpful from diabetes to cancer.
5. **Corn Steep Liquor:** It is a by-product of corn wet milling. It contains organic acids, yeast, gluten, and plenty of organic nitrogen, Perfect medium to grow large quantities of Penicillin.
6. **Corn Cobs:** Corn cobs might seem like the throwaway part of corn, but have their uses. Ground cobs are used for livestock feed. Traditional farm uses include animal bedding, landfill, and to make corn cob jelly. Also used as absorbents for oil and hazardous waste, insecticides, fertilizer, and grit for tumbling and blasting.

C. Fodder Use

Maize makes a greater quantity of epigeous mass than other cereal plants, so can be used for fodder. Digestibility and palatability are higher when ensiled and fermented, rather than dried.

Corn Silage

For silo corn production the whole plant is harvested when still not fully ripe, chopped and stored in silos. Within these silos, which are widely hermetic, lactic acid fermentation occurs. Chaff material is then converted into durable silage which is primarily used for animal feeding in winter.

Quality Protein Maize

- ✩ Quality Protein Maize (QPM) contains nearly twice as much usable protein as other maize (or corn) grown in the tropics and yields 10 per cent more grain than traditional varieties of maize. It was developed by Dr. Surinder Vasal and Dr. Evangelina Villegas at the International Maize and Wheat Improvement Center (CIMMYT) in the late 1990s. For their achievement, they won the World Food Prize in 2000.
- ✩ QPM produces 70-100 per cent more of lysine and tryptophan than the most modern varieties of tropical maize. These two amino acids allow the body to manufacture complete proteins, thereby eliminating wet-malnutrition. In addition tryptophan can be converted in the body to Niacin, which theoretically reduces the incidence of Pellagra.
- ✩ QPM hybrids have been developed and tested for varying climatic and growing conditions; QPM varieties are grown on roughly 9 million acres (36,000 km^2) worldwide. Meanwhile, QPM research and development have spread from Mexico to throughout Latin America and to Africa, Europe, and Asia.

Conclusion

It is clear from our discussion that corn has now already occupied a good position in the world market as a value-added agricultural crop. The developed countries including some south-east Asian countries have already put their foot prints on this pavement. Now, time has come how we would be able to enlist our country in this que.

Chapter 31

Value Addition and Processing in Medicinal and Aromatic Plants: The New Horizon of Agricultural Enterprises

Anjan Bhattacharyya

Introduction

Medicinal plants have proved their efficacy and safety in therapeutic treatment for alleviation or cure of human diseases since time immemorial. The evidences for the therapeutic action of herbal drugs are documented in Indian, Chinese, European and African system of medicine. Medicinal plants are utilized in two distinct areas of health management *i.e.* (1) Modern system of medicine and (2) traditional system of medicine. At present about 130 clinically tested prescribed drugs are used in modern system of medicine and they are solely derived from about 100 species of higher plants. The traditional systems of medicine still continue to carter the primary health care of 80 per cent world populations.

In India a number of Traditional systems of medicine are practiced such as Ayurveda, Siddha, Unani, Homeopathy, Yoga and Naturopathy for the total health care. The Global market for herbal products including medicines, health supplements, herbal beauty and toiletry products, is estimated at $62 billion and is growing at a rate of 7 per cent annually. The WHO's forecast is that the global market for herbal products would be of the order of $5 trillion by the year 2050. However India's share in the global export market of medicinal plants related trade is just 0.5 per cent. This is against India's rich biodiversity of 45,000 plant species spread across 16 Agro-climatic zones. The system of standardization should be established for every plant medicine in the market because the scope of variation in different batches of medicine is enormous. Knowledge of chemical components of a plant is essential for quality control analysis of a plant, extract or any formulation containing them. A compound or group of compounds present can serve as a "biomarker" and the presence and concentration of the same can be followed to decide on the geniuses of the drug/ formulations.

Any component other than the biomarkers present indicates adulteration. Not many plants are studied for these markers. Knowledge of this compound and their specific analytical methods will facilitate the herbal industry in checking adulteration and thus raise its standard. Therefore, it is essential that such uses of natural products be documented and studied for systematic regulation and wide spread application.

With this objective in view a regional quality control laboratory in the department of Agriculture chemicals, BCKV have already standardized the valuable medicinal plants grown in different agro climatic zones of West Bengal to ensure quality of crude materials by HPTLC finger print techniques and clearly identify and quantify the active principles through rapid chemical analysis.

Materials and Method

HPTLC finger printing techniques consists of the following steps:

- ☆ Extraction of plant material
- ☆ Sample application
- ☆ Chromatogram development
- ☆ Chromatogram evaluation

Extraction of Plant Material

The dried plant materials were extracted with suitable organic solvents. The extract is then filtered, concentrated and made up to 20ml volume.

Sample Application

Samples along with standards were on a HPTLC glass plate coated with 60GF 254 silica gel (20x20; 300µm layer thickness) prepared by HPTLC auto coater (Camag). Samples and analytical standards of known concentrations were applied on the glass plates as 8mm wide band, positioned 15mm from the bottom and 20mm from the side of the plate, using an automated TLC applicator linomat 5(Camag, Multenz; Switzerland) with nitrogen flow providing a delivery speed 150nl/s from the syringe.

Chromatogram Development

After completion of sample application, the plate was developed in a Camag twin trough chamber which was pre-saturated with the mobile phase. Different compositions of the mobile phase for HPTLC analysis were tested to obtain high resolution and reproducible peaks. The TLC runs were performed under laboratory conditions of 25±5°C and 50 per cent relative humidity.

Chromatogram Evaluation

After development plates were taken off and dried by drier. The plates were then scanned with Camag, TLCscanner-3 equipped with wincats software (Camag) under the following conditions: slit width 6x0.45, absorption and reflection mode, lamps used deuterium and tungsten.

Result and Discussion

HPTLC Fingerprint Profile of Kalmegh (*Andrographis paniculata*)

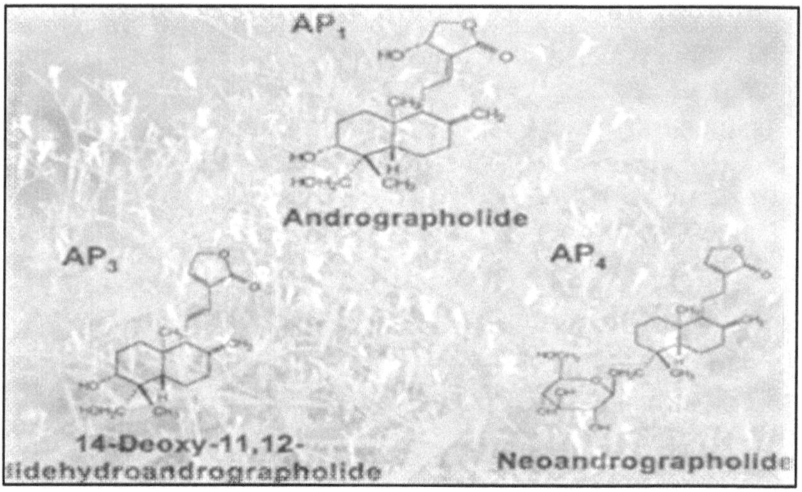

- ✰ The methanol extract of leaves along with andrographolide solution was developed with Chloroform : methanol 7: 3 and scanned at λmax 230nm.
- ✰ In HPTLC finger print of *A. paniculata* (leaf) sample, six peaks were observed (Figure 31.1). Compound with R_f 0.36 was in highest concentration and was found to be identical with a diterpenoid Andrographolide, the chemical active of A. paniculata.
- ✰ Matching of UV spectra of standard Andrographolide and that of the extract confirms that they are the same substance.

Table 31.1: Andrographolide Content in Leaves of *Andrographis paniculata*

Source	*Per cent of Chemical Active Present*
Mohanpur	1.98
Canning	2.23
Tarokeswar	2.02
Asansol	2.49
Mecheda	2.18
Raghunathpur	2.11
Nabwadip	1.33

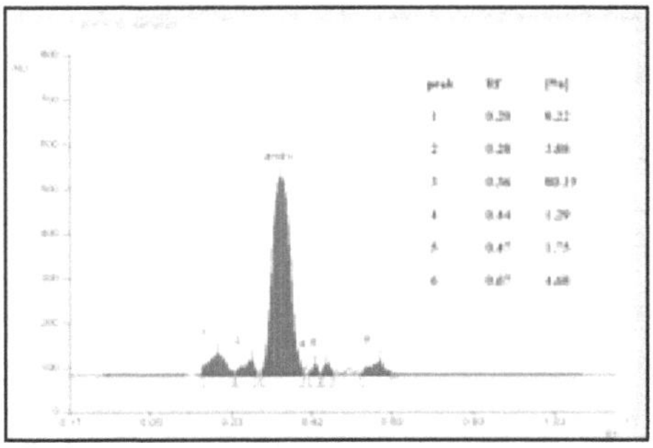

Figure 31.1: Fingerprint of Andrographolide Extract at λmax 230 nm

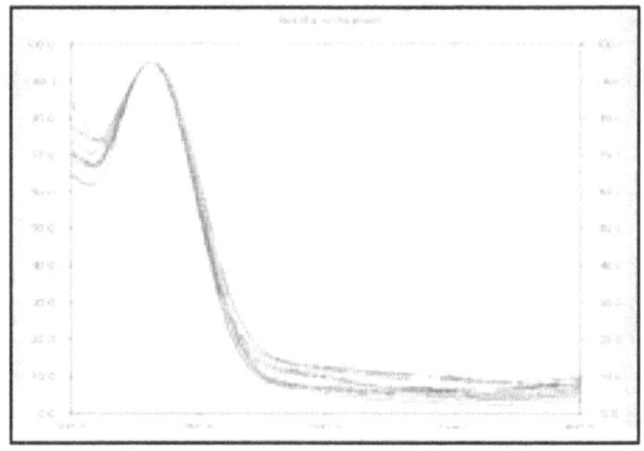

Figure 31.2: Supermposable UV Spectra of Andrographolide and its Extract.

The maximum concentration of Andrographolide was found in Asansol region

- ✰ Withania somnifera (fam: Solanaceae) is one of the most important herb used in the treatment of various ailment in Indian system of medicine.

- ✰ Roots are used in insomnia, tissue-building and nervous breakdown disorder.
- ✰ Leaves are used in fever, painful swellings, Opthalmities.
- ✰ Withaferin A, an unsaturated lactone isolated from W.somnifera is the most important withanolide to which the active properties of the plant are attributed.

HPTLC Analysis of *Withania somnifera* (Ashwagandha)

In HPTLC finger print of Ashwagandha root samples, six peaks were observed. Compounds with R_f 0.68 was found to be identical with withaferin A, confirmed by superimposing UV spectra of withaferin A with that of spot at R_f 0.68.

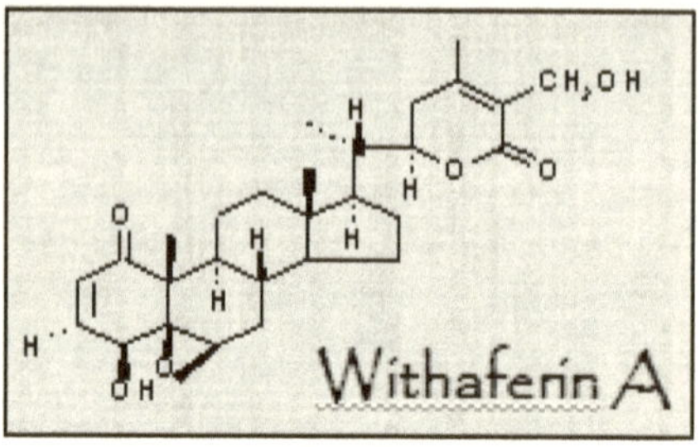

The extraction of roots was followed by approved method. Crude withanolide and the standard withaferin A was applied on a HPTLC plate, developed in chloroform: methanol (9:1) and scanned at λmax 220nm (Figure 31.4).

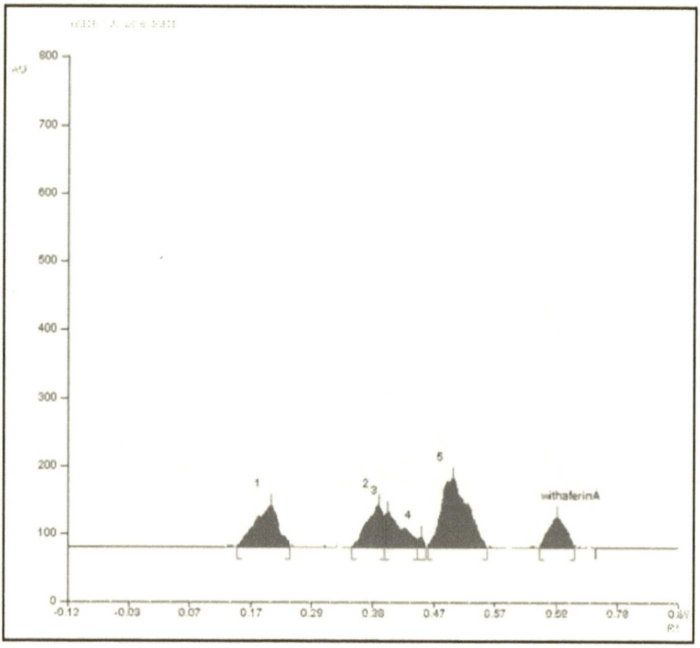

Figure 31.3

Table 31.2: The Per cent Alkaloid and Withaferin A in Roots of Ashwagandha

Plants Parts	*Alkaloid (per cent)*	*Withanolide (per cent)*	*Withafein A (per cent)*
Root	0.15	1.4	0.1
Stem	0.11	1.2	0.01
Leaves	0.51	2	0.53

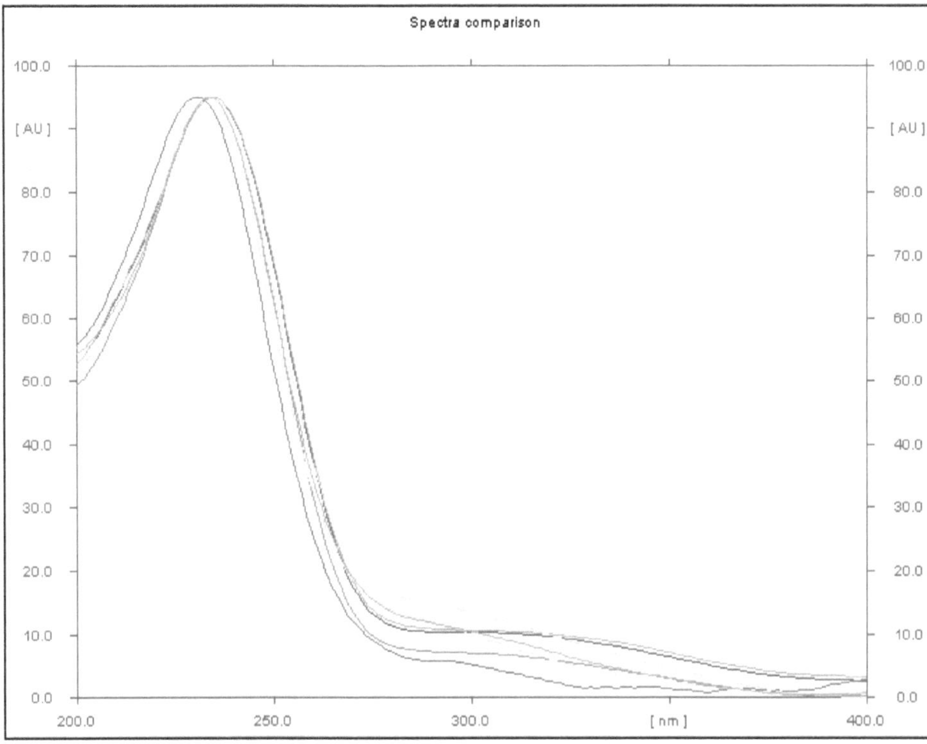
Spectra comparison
100.0
[AU]
80.0
70.0
60.0
50.0
40.0
30.0
20.0
10.0
0.0
200.0
250.0
300.0
[nm]
400.0

Figure 31.4

Cassia anguistifolia (Senna)

- ☆ A variety of polyherbal ayurvedic preparations,commercially available in India to relieve constipation were analyzed to quantify these two anthraquinones.

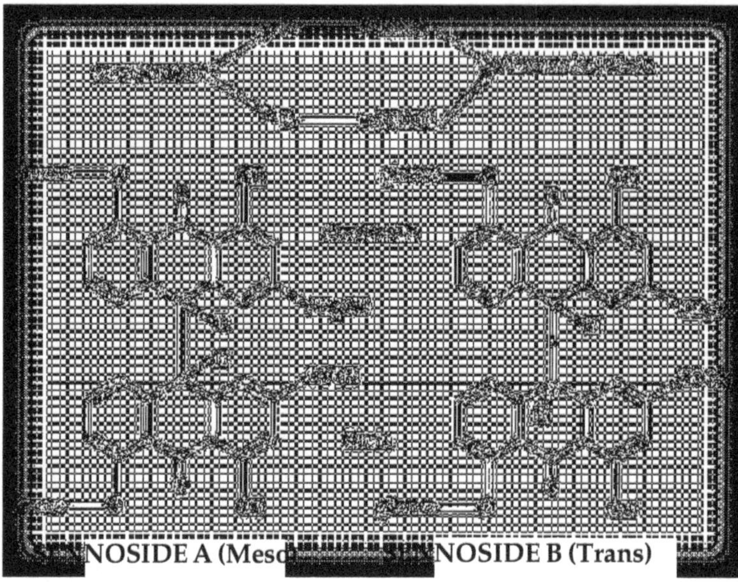

- ☆ Sennosides A and B, the two anthraquinones isolated from C. *angustifolia* (*Senna*) have the strong laxative properties.

The methanol (70 per cent) extract of the sample standard sennosides A and B were applied on a HPTLC plate, developed in a mobile phase containing 2-propanol: ethyl acetate: water: Formic acid (17:19:12:02.) and scanned at 350nm (Figure 31.5).

Validity of peak was confirmed by comparing UV spectra of standard and that of the standard within the same R_f window.

Table 31.3: Content of Sennoside A and Sennoside B in *Cassia angustifolia*

Plant Parts Used	*Per cent of Sennoside A*	*Per cent Sennoside B*
Leaves	1.8	2.3
Pods	2.2	2.5

***Aloe vera* (Linn.)** contain two major juice materials -

A yellow **exudate** which is used in traditional medicine as a cathartic agent and a **mucilaginous gel** that is utilized as cosmetic and as dietary supplement in beverages.

A HPTLC method was developed and validated to quantify the aloin content in leaves collected at different time intervals and in different marketed formulations (Table 31.5 and 31.6).

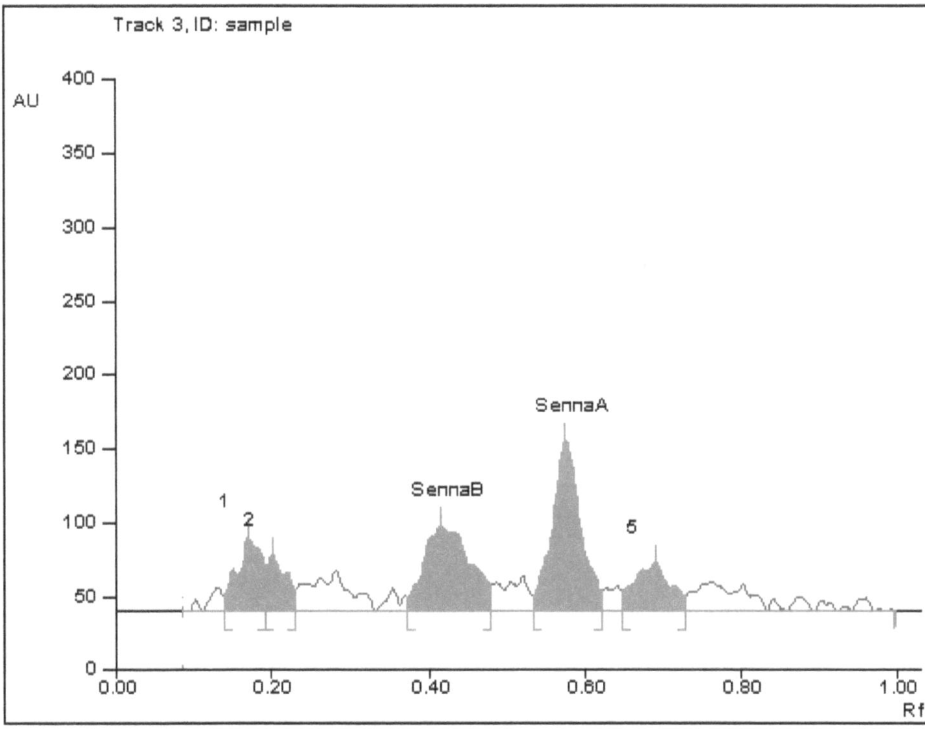

Figure 31.5: HPTLC Fingerprint of Senna Formulation.

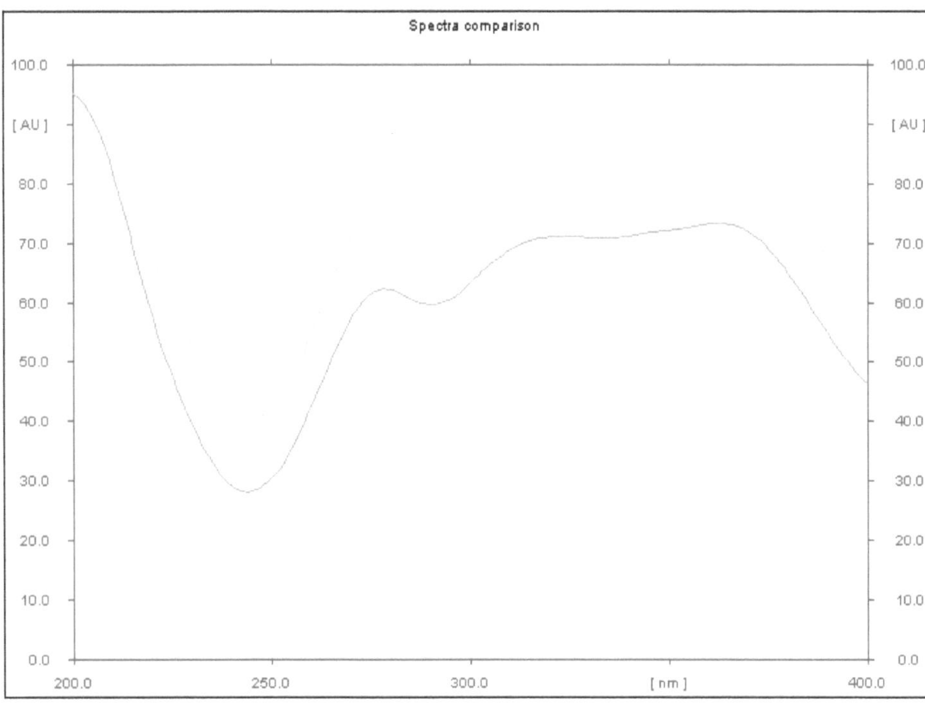

Table 31.4: Concentration of Senoside A and Senoside B in different Ayurvedic Formulation

Formulations	*Sennoside A [mg/g] of the Formulation*	*Sennoside B [mg/g] of the Formulation*
1	2.5	25.9
2	2.3	12.7
3	2.1	2.6
4	1.8	1.87
5	1.62	1.85
6	1.58	1.72
7	1.4	1.52
8	1.2	1.5
9	0.92	1.4
10	0.91	1.4

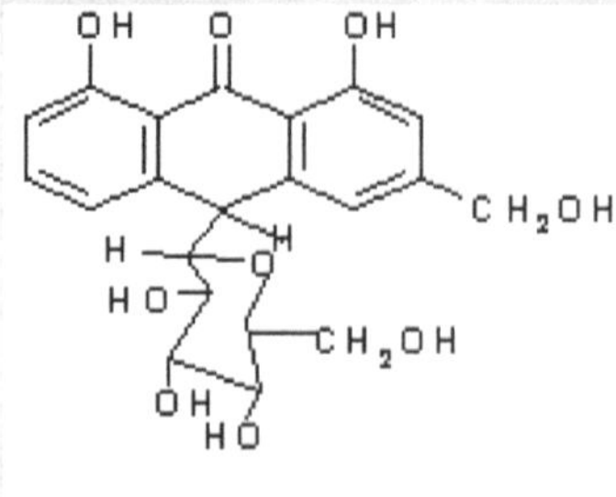

Aloin A

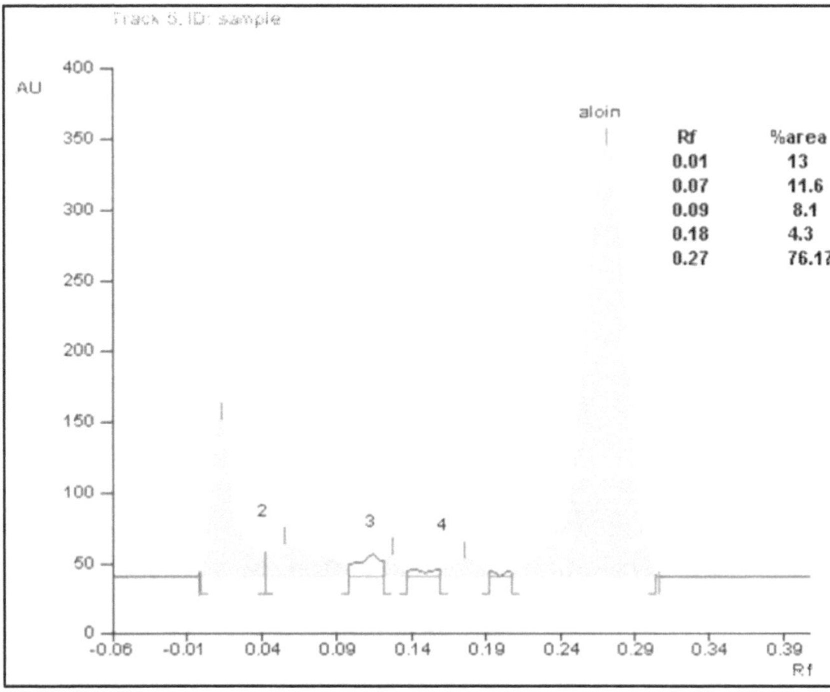

The methanol extract of sample and the standard Aloin was developed in a mobile phase containing Ethyl acetate: MeOH: H_2O = 200: 33: 27 at λmax: 370nm.

HPTLC finger print of Gritakumari showed five peaks.

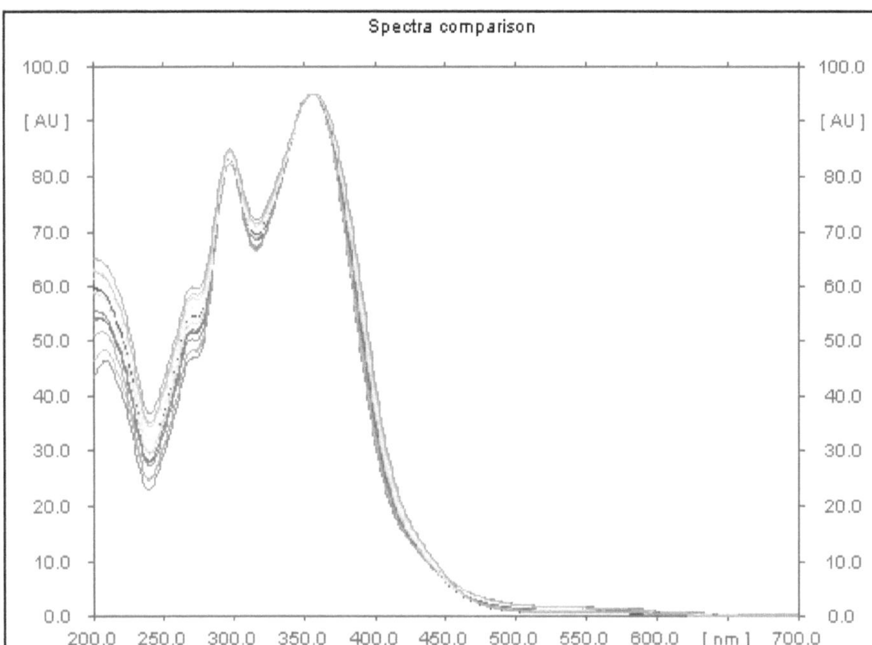

The major peak at R_f 0.27 was confirmed as aloin by superimposing UV spectra of aloin with that of the spot at R_f 0.27

Table 31.5: Aloin Content (per cent) in Leaf Exudates of *Aloe vera* at Different Harvesting Stages

Harvesting Time (Months after planting)	*(Per cent) Aloin Content in Leaf Exudate*			*Mean ± S.D.*
	R_1	R_2	R_3	
12	4.59	4.72	4.70	4.67 ± 0.07
13	4.81	4.78	4.78	4.79 ± 0.02
14	5.05	4.97	4.92	4.98 ± 0.07
15	5.31	5.28	5.22	5.27 ± 0.05
16	5.76	5.73	5.79	5.76 ± 0.03
17	5.80	5.82	5.81	5.81 ± 0.01
18	5.86	5.91	5.84	5.87 ± 0.04

Table 31.6: Content of Aloin in different Marketed Formulations

Aloe gel supplied	3mg/100gm of gel
Aloe gel+ Na benzoate supplied	3.5mg/100gm of gel
Gel collected from supplied leaf	4mg/100gm of gel
Whole leaf juice	45mg/100gm of whole leaf juice
Tonic	19.75
Capsules	9.75

Vasak (*Adhatoda vasica*)

Adhatoda vasica is a well known plant drug used in the treatment of various diseases and disorders particularly for the respiratory track ailments. Vasicine the main active quinazoline alkaloids have been shown to posses bronchodilatory and potent respiratory activities.

HPTLC Analysis of *Adhatoda vasica*

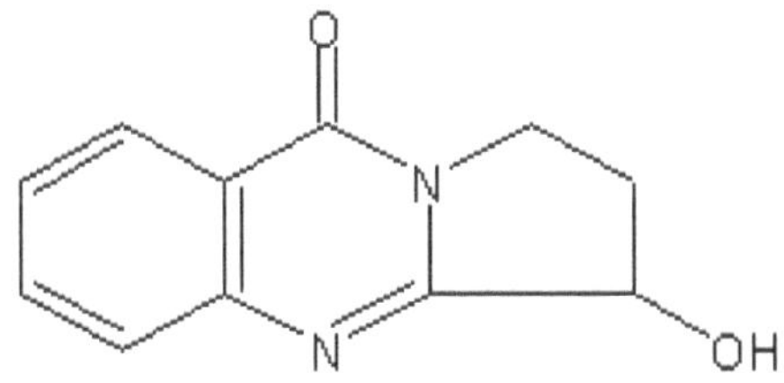

Molecular Formula = $C_{11}H_{10}N_2O_2$
Formula Weight = 202.209

Vasicine

- ☆ Samples and standard were developed in MeOH: Toluene: Dioxane: ammonia 2:2:5:1 and scanned at λmax: 270 and 281nm.

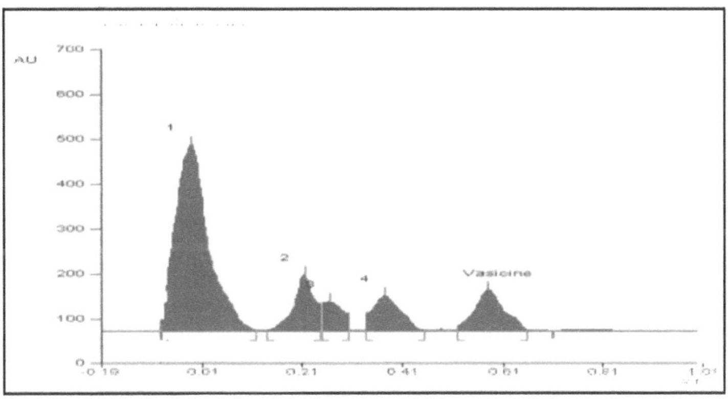

- ☆ HPTLC fingerprint of Vasak showed the separation of five peaks.
- ☆ Peak at R_f 0.61 was confirmed as vasicine by comprising UV spectra of sample with that of the standard at the same R_f window.
- ☆ Concentration of vasicine was maximum in leaves (Table 31.7).

Table 31.7: Content of Vasicine

Plant Parts	*Vasicine (per cent)*
Leaves	0.65
Stem	0.31
Root	0.062

Turmeric (*Curcuma longa*)

- ☆ Curcuma longa (turmeric), a yellow spice from the family Zingiberaceae have long been used in Ayurveda to treat a diverse range of conditions.

☆ These actions are mainly due to the phenolic fraction containing Curcumine, Demethoxycurcumine and bisdemethoxycurcumine, collectively known as curcuminoids.

HPTLC Analysis of *Curcumine*

A HPTLC finger print technique is developed to standardize the turmeric based formulations (Tables 31.8 and 31.9).

Table 31.8: The Curcuminoids Composition in *Curcuma longa* (Collected from Medicinal Garden, BCKV)

Curcumine (per cent)	*DMC (per cent)*	*BDMC (per cent)*
2.2	0.8	0.2

Table 31.9: The Curcuminoids composition of some commercial samples of turmeric

Sample	*Curcumine (per cent)*	*DMC (per cent)*	*BDMC (per cent)*
Source 1	70	25	8.5
Source 2	75	20	5.78
Source 3	79	16	3.9

Methanolic solutions of samples and standards were developed in chloroform: methanol (48:2) and scanned at 420nm.

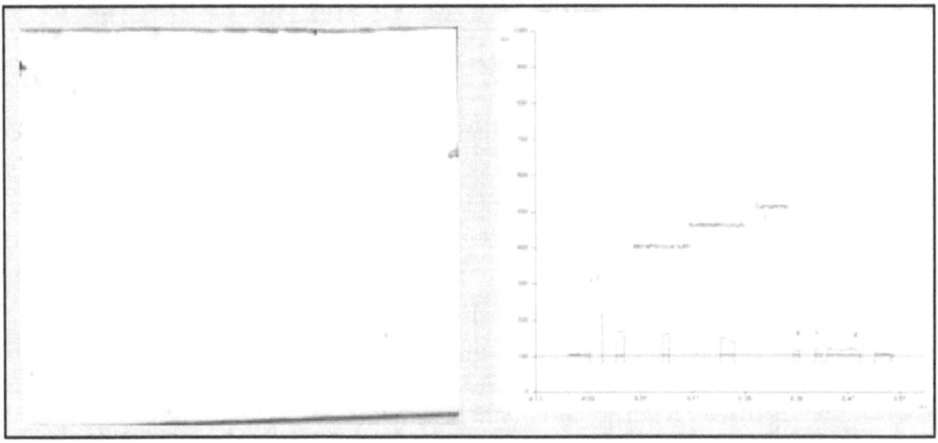

Possibilities and Potential

- In order to develop the medicinal plants sector in different agro-climatic zones across the country, there is a need for a coherent approach linking all stakeholders of the medicinal plants *viz.* growers, collectors, traders, manufactures, exporters and others.
- A very active R and D programme is also essential to fully convert the potential of our medicinal plants into economic wealth. The R and D has to cover all aspects relating to the spices from collection to utilization.

The following strategies are suggested to improve National as well as the International trade of Medicinal and their various products.

1. Crop-wise post harvest technology must be developed for high quality produce.
2. Medicinal plants must be domesticated and cultivated for consistent quality of the produce.
3. Toxicity (safety) and efficacy trials need to be conducted on plant drugs.
4. Materials of proven safety, effectively and quality need only to be exported.
5. Monographs for each of the medicinal plant that are currently used in formulations for domestic and export market, including information of physiological effects, efficacy, quality standard and references etc., are needed to be developed.
6. Cultivation of medicinal plants should be integrated with that of food crops by devising suitable cropping schedules.
7. Good Manufacturing Practices have been made mandatory for all drug manufacturing units to improve for standard of drugs for promotion of exports.

Conclusion

- ☆ The system of standardization should be established for every plant medicine in the market.
- ☆ Because the scope of variation in different batches of medicine is enormous. Adding to this variability is the fact that in herbal medicine several plants may be used together in the same preparation.
- ☆ This means there should be a quality test for the entire preparation to ensure quality of the preparation.
- ☆ Standard procedure should be developed for getting the final product.
- ☆ HPTLC finger print techniques of plant materials would help in lying down pharmacopoeia standards to check the identity of the medicinal plant and hence to control the quality of herbal medicine.

Chapter 32

Seed Enterprise: The Stay and Prospects for Farmers of Small and Marginal Resources

Asit Kumar Basu

Definition of Seed and Quality Seed

Botanically **seed** is the matured ovule, containing the small plant (an embryo) with enough food to start its development. In term of '**Seed Science**' seed can be defined as 'any plant parts (true seeds, leaf, stem, root, rhizome, corm, cormlet, tuber, bulb, cloves, fruit, suckers, crown, runner, etc.) used for plant propagation.

Seed is the starting point of agriculture and dictates ultimate productivity of other inputs. Good quality seed alone increases the yield by 15-20 per cent. **Quality seeds** are the seed which are produced by the experts in planned way taking all care to maintain all parameters of seeds. Grains cannot be considered as seed, but seed can be considered as grain if not treated with pesticide. The differences between seed and grain are being given in the Table 32.1.

Classes of Quality Seed

There are four classes of seed based on the generation system of quality seed multiplication in India. Each of the classes are being briefly discussed below.

Nucleus Seed

Nucleus seed is the seed produced by the breeder to develop the particular variety and is directly used for multiplication as breeder seed. At the time of release of a variety, small quantity of seed normally known as nucleus seed is available with the plant breeder. Commercial quantity of seed is produced after a series of multiplication steps. Starting from maintenance programme in which nucleus seed is multiplied in a generation system of multiplication as breeder, foundation and certified seed. The genetic purity and physical purity of nucleus seed crop should be maintained at 100 per cent.

Table 32.1

Parameter	*Seed*	*Grain*
Viability	It should be a viable one	Need not be a viable one
Food	It can be used as food if not treated with pesticide	Grain is always used for food
Production	Seed is produced by expert with well planned way	Grain can be produced by any cultivator
Purity	It should have maximum genetic and physical purity	Not so
Standard	It should satisfy the minimum seed certification standard	No such requirement
Certification	Should be certified through 'Seed Certification Agency'	No certification is required
Packing	Seed are packed in standard packet or container	No specifications for packing of grains
Seed treatment	It should be completely treated with pesticides to protect the seed against storage pests	Grain should not be treated with any chemicals as it is used for consumption
Storage	In ordinary condition, seed cannot be stored for longer duration	It can be stored comparatively longer duration
Biological activities	Respiration rate and other physiological and biological processes should be kept at low level during storage	No such specifications are needed

Breeder's Seed

Breeder seed is the seed or vegetatively propagated material directly controlled by the originating or sponsoring plant breeder of the breeding programme or institution. The breeder seed production is personally supervised by a qualified breeder. The Breeder seed is the source for the initial and recurring increase of foundation seed. The Breeder seed plot is subjected for inspection by ***monitoring team*** consists of a nominee from the Nodal Officer for breeder seed production, Deputy Director of Seed Certification, Area Manager, National Seed Corporation and the Producing breeder. The team will inspect the crop during flowering and maturity stages. The breeder seed after passing the field and seed standards should be bagged with breeder seed tag containg all details about crop, variety, lot number, date of test, physical and genetic purity, germination percentage. Grow out test to determine the genetic purity of the seed lot should be conducted for all the breeder seed produced by the breeder.

When seed multiplication rate is very low or breeder seed is extremely short supply production of **Breeder Seed Stage II** is permitted to continue the seed chain in an effective manner. **Breeder Seed Stage II** is produced from the **Breeder Seed Stage I**.

Foundation seed

Foundation seed is obtained from breeder seed by direct increase. Foundation seed is genetically pure and is the source of registered and or certified seed. Production of FS is the responsibility of NSC. Foundation seed is produced on government farms, at experiment stations, by agricultural universities or competent seed growers under strict supervision of experts from NSC. This class of seed should be produced in the area of adaptation of the concern variety.

Production of foundation seed stage-I and stage-II may thus be permitted, if supervised and approved by the Certification Agency and if the production process is so handled as to maintain specific genetic purity and identity. **Foundation seed stage-I** and **Foundation seed stage-II** are permitted when the seed multiplication rate of the particular crop is very low, for example, groundnut, potato etc.

Certified Seed

Certified Seed is produced from Foundation Seed. This is so known because it is certified by the State Seed Certification Agency. The certified seeds are usually produced by progressive farmers according to standard seed production practice. To be certified the seed must meet the prescribed requirements regarding purity and quality. These standards vary from one crop to another. Certified seed is available fro general distribution to farmers for commercial crop production. Its production is generally supervised by State Seeds Corporation but NSC also undertakes the supervision of certified seed production if required.

Truthfully Labelled Seed

It is the non registered class of seed. The Truthfully Labelled Seed shall be the progeny of foundation, certified, or labelled seed. Seed and field standard are equivalent to certified seed. Production procedure is same as certified seed. No Seed Certification Agency is necessary. Market monitoring is the only quality testing tool of the Truthfully Labelled Seed. Producer will declare and guarantee the quality attributes of Truthfully Labelled Seed.

Importance of Quality Seed

- Low cost as compare to other cost of cultivation
- High purity
- High germination
- High vigour
- Uniform size
- Absence of seed borne disease and insect
- Optimum moisture content

Classes of Seeds at Farm Level

The seeds which we have discussed above all are not reaching at field level. The seed used by the farmers, in general is for the production of crops and not for seeds. The seeds received by the farmers for crop production may be classified as Certified Seed, Truthfully Labelled Seed and Farm Saved Seed.

The farmers of India stand first in terms amount of seed production of major field crops, which may be realized from the Seed Replacement Rate (Table 32.2). A part of the crop is saved by the farmers for the seed of the next season or next crop is known as **Farm Saved Seeds**.

Table 32.2: Seed Replacement Rates of Important Crops in India

Crop	*Seed Replacement Rate*					
	2001	*2002*	*2003*	*2004*	*2005*	*2006*
Cereals						
Rice	19.22	19.31	19.16	23.28	24.35	25.10
Wheat	13.04	13.00	13.00	16.48	17.64	18.03
Ragi	17.50	18.00	17.36	19.10	–	–
Maize (variety)	20.98	21.35	24.41	31.35	35.39	36.23
Jowar	18.36	18.78	26.71	20.87	–	–
Bajra	45.92	48.47	51.02	44.90	55.36	56.12
Pulses						
Bengal gram	4.17	4.23	7.09	9.87	9.41	9.54
Black gram	16.55	17.06	20.48	17.24	15.70	16.21
Green gram	13.47	13.48	19.48	12.34	12.50	12.82
Red gram	8.71	8.84	13.60	9.80	10.48	10.88
Oil seeds						
Groundnut	5.20	5.50	11.00	7.11	6.89	7.03
Soybean	12.44	12.45	15.58	27.00	28.88	32.34
Sunflower	13.73	15.69	19.61	60.15	67.67	69.17
Commercial Crops						
Cotton	21.21	21.86	19.84	20.73	21.78	22.27
Jute	28.77	30.14	27.40	26.03	26.03	26.03

Seed Production Models

There are two models of seed production, namely formal system and informal system of seed production. The seed producers who follow the generation system of seed production come under the formal seed production system. This includes public sectors (Agricultural Universities, Seed Farms of State Agricultural Departments etc.), private sectors (national and international seed companies), NGOs, SHGs etc.

Informal seed production system includes Farm Saved Seeds, Exchanged Seeds from other farmers, Seeds from Local Markets whose seed source and quality parameters are unknown.

Village Based Seed Enterprise for Production of Quality Seeds

A village, wherein trained group of fanners are involved in production 'of seeds of various crops and cater to the needs of themselves, fellow fanners of the village and fanners of neighbouring villages in appropriate time and at affordable cost is called "a seed village". Following are the main objectives of the seed village concept:

- ☆ Organizing seed production in cluster (or) compact area
- ☆ Replacing existing local varieties with new high yielding varieties.
- ☆ Increasing the seed production
- ☆ To meet the local demand, timely supply and reasonable cost
- ☆ Self sufficiency and self reliance of the village
- ☆ Increasing the seed replacement rate

Justification of Requirement of Village Based Seed Enterprises

- ☆ Low seed replacement rate
- ☆ Gap between the requirement of quality seeds and their supply is large
- ☆ Gap may be reduced by involving farmers in production of quality seeds
- ☆ Area under University is very small to produce require quantity of quality seeds

Features of Village Based Seed Enterprises

- ☆ Seed is available at the door steps of farms at an appropriate time
- ☆ Seed availability at affordable cost even lesser than market price
- ☆ Increased confidence among the farmers about the quality because of known source of production
- ☆ Producer and consumer are mutually benefited
- ☆ Facilitates fast spread of new cultivars of different kinds

Establishment of Village Based Seed Enterprises

Seed village concept is to promote the quality seed production of foundation and certified seed classes. The area which is suitable for raising a particular crop will be selected, and raised with single variety of a kind. The area for seed village may be selected based on the following criteria:

- ☆ The area should have irrigation facilities
- ☆ Conducive climatic conditions for the particular crop whose seed to be produced
- ☆ Availability of labour
- ☆ Local farmers should have enough knowledge on that particular crop

- ☆ No occurrence or outbreak of pest and designated diseases
- ☆ Closeness to a urban area for easy movement of seed and other inputs

Seed Supply

Suitable area for seed production will be identified by the Scientists. The foundation/certified seeds or University labelled seeds will be supplied by the University through Krishi Vigyan Kendras (KVKs) and Research Stations at 50 per cent subsidy cost to the identified farmers in the area. The farmers will use these quality seeds and take up their own seed production in a small area (1 acre) for their own use. The crops are Rice, Pulses and Oilseeds.

Capacity Building

In order to harness the synergy between technologies and the community participation, special emphasis is being given to build farmer's capacity to produce quality seeds. Training on seed production and seed technology to the identified farmers for the seed crops grown in the seed villages will be given for technology empowerment of farmers. Duration of the training is 3 days

- ☆ ***First one day training :*** At the time of sowing
- ☆ **Training on :** Isolation distance, sowing practices, seed treatment, and other agronomic practices.
- ☆ ***Second one day training :*** During flowering
- ☆ **Training on:** Identifying offtypes and removal, maintenance of seed plots, plant protection measures, maturity status and harvesting methods.
- ☆ ***Third one day training :*** After harvest
- ☆ **Training on :** Seed cleaning, grading, seed treating, bagging and storage aspects, seed sampling and sending to seed testing laboratory for analysis.

A seed grower forum will be organized for further empowerment of technology and marketing.

Chapter 33
Framework of Entrepreneurial Venture

Dipak De

A farmer does not become an entrepreneur only by adopting a new agricultural technology but he becomes an entrepreneur only when he comes to be an operator of a farm business

De (1986)

Introduction

All round development of agriculture is possible only with effective exploitation of entrepreneurial behaviour skills as well as material resources. Our country is having material resources but abundant of human resources. So, we can identify individuals in all segments of the population, who have the requisite entrepreneurial behaviour skills. The entrepreneur is an economic man, who tries to maximise his profits by identification and adoption of innovations. However, the entrepreneurs are not simply innovators but they are the persons with a will to act to assure risk and bring about a change through organisation of human efforts. It plays a key role in economic development of the country. Importance of development of entrepreneurship as an ingredient of economic development has been recognised long time back (Hmachalam, 1990). It was as early as in 1950 that the need for entrepreneurial behaviour development was first felt and since then substantial amount of research has been done in this sphere.

Concept of Entrepreneur

An overview of literature pertaining to concept of entrepreneur is absolutely essential. The word entrepreneur is derived from **French** word **'Entreprendre'** and the **German** word **"Unternehmen"**, both mean **"To undertake"**. For a long time there was no equivalent for the term 'entrepreneur' in the English language. Three words were commonly used to connote the sense the French tenn carried: adventurer, undertaker and projector; these were used interchangeably and lacked the precision and characteristics of a scientific expression (Gopakumar, 1995). Hence the term 'entrepreneur' did not find any prominence in the history of economic thought. The earliest attempt to invest the concept with some economic content could be traced to the works of an 18th century French writer, Bernard F. de Belidor, who defmes entrepreneurship as buying labour and materials at uncertain prices and selling the resultant output at contracted prices (Hoselitz, 1960). Entrepreneurship as a concept gathered prominence in economic literature mainly through the writings of Richard Cantillon (1680-1734), who gave the concept some analytical treatment and assigned the entrepreneur an economic role by emphasizing on 'risk' as a prominent entrepreneurial function (Gopakumar, 1995). According to Encyclopaedia Britanica entrepreneur means individual response for the operation of a business, including the choice of a product, the mobilization of necessary capital, decisions on product prices and quantities, the employment of labour and expanding or reducing the productive facilities.

Sara (1998) concluded that farm household productivity was a major factor in enabling the survival of small scale agricultural production. Many farm businesses combine agricultural production with other income generating activities and such multiplicity of activities has always been an important and distinctive feature of the farming sector. That's why the farmer must be treated as a part of rural entrepreneurial class.

Prakasam (1998) concluded that the entrepreneur, being an economic man aims to optimise his profits through innovative means. He has a will to act, is ready to assume risk and to organize human resources for production of foods and services which in turn generate and multiply economic and social activities that induce change in the area and in the society.

Ramana (1999) entrepreneurs are defined as those people who work for themselves are called "entrepreneurs". The.word entrepreneur is derived from French word "Entreprendre" meaning to "Undertake". The entrepreneur is thus a person who organizes and manages in an activity/organisation, undertaking the risks for fulfilling some of his needs. His job involves the quality of boldness, courage, dynamism and risks taking in sufficient measure. Therefore entrepreneur in this context is defined as one who could start a new activity or a new entreprise which is a deviation from his traditional family occupation or profession.

Concept of Entrepreneurship

An overview of literature pertaining to concept of ; entrepreneurship is absolutely essential. -Cole (1949) stated that entrepreneurship comprises any purposeful activity that initiates maintains or develops a profit-oriented business in interaction with the internal situation of the business or with the economic, political and social circumstances surrounding the business.

According to Diamond (1957) entrepreneurship is equivalent to "enterprise" which involves the willingness to assume risks in undertaking an economic activity particularly new one it may involve an innovation but not necessary so. It always involves risk taking and decision-making, although neither risk nor decision making may be of great significance.

Hostelits (1957) opined that entrepreneurship was associated with personality pattern in which achievement motivation was strong. But the presence of strong achievement motivation in a group of individuals did not necessarily produce an abundance of entrepreneurs unless certain other conditions of social structure and culture strongly favour achievement oriented individuals.

Hagen (1968) stated that entrepreneurship is not only the conceiving of the idea behind a venture but also designing and maintaining the organisation for carrying it out.

According to Schumpeter (1970) entrepreneurship is

- ✰ a function of group level pattern.
- ✰ a function of managerial skill and leadership.
- ✰ an organisational building function.
- ✰ a function of high achievement.
- ✰ input-completing and gap filling.
- ✰ a function of status withdrawl, function.
- ✰ a function of social, political and economic structure.

Wick (1998) entrepreneurship refers to the intentional creation or transformation of an organization for the purpose of creating or adding value through organization of resources values added by entrepreneurial behaviour result in increases in personal wealth, increases in cash flow for a region, new or better jobs and new or better products or services.

Khanka (1995) states that making the entrepreneurship is a complex phenomenon. This is a crystallisation of social milieu from which entrepreneurs come, family imbib~' make up of their minds, personal attitudes, caste system, educational background, family occupation and so on. In fact, several factors go in to the making of an entrepreneur and thereby entrepreneurial society, these are grouped under three major headings

- ✰ Stimulation
- ✰ Support
- ✰ Sustaining.

Manimekalai (1998) opined that entrepreneurship development among the rural labour force would strengthen the village economy, promote regional development, bring job diversification, relieve the dependence on agriculture. The small scale agricultural production in Trichy district was becoming increasingly unprofitable due to the continuing drought and large scale agricultural production has been transforming it self into an agro business in Trichy district. All these suggest that, there has been a need for self employed industrial activity or non farm activity as an alternative to agricultural employment as agriculture fails to support the dependent population.

According to Ranlana (1999) entrepreneurship is a purposeful activity indulged in initiating, promoting and maintaining economic activities for the production and distribution of wealth. The individual as an entrepreneur is a critical factor in economic development and an integral part of socio-economic transformation. Therefore, the basic concept of entrepreneurship connotes effectiveness, an urge to take risks in the face of uncertainties and intuition.

Concept of Entrepreneurial Behaviour

Parsons and Shills (1951) proposed a theory of action which assumes that the actor strives to achieve goals. The theory of action is conceptual scheme for the analysis of behaviour of living organisms. There are four points to be noted in the conceptualization of behaviour.

- ✰ Behaviour is oriented to the attainments of ends or goals or other anticipated states of affair.
- ✰ It takes place in a situation.
- ✰ It is normatively regulated.
- ✰ It involves expenditure of energy or effort or motivation. This means that any behaviour of a living organism might be anticipated in terms of the anticipated state of affairs towards which it is directed.

According to Legans (1961) behaviour is response to stimuli refers to what an individual knows (knowledge), what he can do (skill), mental and physical, what he thinks (attitude) and what he actually does (action).

Calder and Ross (1976) stated that behaviour was a function of many personal factors -motives, habits, attitudes, and so on - and of many environmental factors, such as norms law, rewards, and punishment.

According to Fisher (1982) behaviour is a function of continuous process of multi directional interaction between the person and the situations, including other persons, that he or she encounters.

Robbins (1996) reported that behaviour was a function of it's consequences. and found that people would most likely engage in desired behaviour if they were rewarded for doing so, these rewards were most effective if they immediately followed the desired response, and behaviour that was not rewarded, or was punished, was less likely to be repeated.

Entrepreneurial Behaviour

Fishbein (1965) reported that entrepreneurial behaviour was a function of the surrounding social structure both past and present, which could be readily influencing by manipulable economic and social incentives.

Mcclelland (1965) stated that casual sequences of entrepreneurial behaviour were as follows:

Ideological values ⟶ Family socialisation need for Acluevement ⟶ E.B.

Javillionar and Peters (1973) listed three dimensions of, entrepreneurial behaviour *viz.*

- ☆ Risk taking ability.
- ☆ Novel or energetic instrumental activity and
- ☆ Individual responsibility as indicators of entrepreneurial performance.

According to Rao (1985) entrepreneurial behaviour is the result of an interaction of individual, situational, psychological, social and experiential factors.

Singh (1986) stated that Entreprenerurial behaviour (EB) was a function of an individuals personality characteristics and environmental factors. This could be represented as

$$EB = f\,(PE)$$

where,

P: Personality characteristics

E: Environmental factors.

Manjula (1995) stated that entrepreneurial behaviour was the change in knowledge, skill and attitude of entrepreneurs towards the selected enterprise.

Patel and Sanoria (1997) stated that the activity of an individual to decide for adopting certain enterprise to make profit was regarded as entrepreneurial behaviour.

Paul (1998) revealed that majority of respondents in all the groups of small (65 per cent), medium (50 per cent), and big farmers (55 per cent) had medium entreprenurial behaviour whereas low entrepreneurial behaviour constitutes more among small farmers (22.5 per cent) and medium farmers (32.5 per cent). The high entrepreneurial behaviour (32.5 per cent) of big farmers might be due to their sound financial condition to take risk and adopt new technologies and also more innovativeness and leadership abilities.

Against this backdrop Entrepreneurial Behaviour has been defined as package of personality characteristics and environmental factors related to dynamic agent of change for transforming physical, natural and human resources into corresponding production possibilities.

Phases of Entrepreneurial Venture

- ☆ **PHASE – I Identification of enterprise** by an entrepreneur may come from unsatisfied need of people as well as unsatisfied personal need. This is reflected by the alertness of entrepreneur him/herself, his ability to identify **opportunity,** family and educational background, professional experience, formal and informal networks etc.
- ☆ **PHASE – II Initiation/Creation of Enterprise** The entrepreneur needs to create an organization for transforming the concept into a marketable product by accumulating and combining physical and other resources. The ideas of entrepreneurs and environment interact to turn the ideas into reality. The new venture takes away large portion of time and attention of

entrepreneur from the management of environment to the management of organization. Interaction with several components like bankers, regulatory agencies, experts and advisors, suppliers and so on entrepreneur alone has to constantly interact with them to have the control over the enterprise that he/she created.

- **PHASE – III Nurturing of Enterprise** Organiosation translates the business concept into marketable product and offers it to the customer. The entrepreneur gets feedback on market response in terms of profitability, sales etc. The feedback is measured in terms of efficiency, competitiveness, effectiveness, innovativeness, flexibility etc.
- **PHASE – IV Concluding/Transition Phase** Entrepreneurial process does not end with achieving stability and reaches success. At this stage organizations require different managerial style as venture competition gradually builds up more and more.

Typology of Entrepreneur

Ramachandran and Ray (1998) described the typology of entrepreneurs based on the outcome of entrepreneurial ventures. Empirical evidence suggests the existence of following four types of entrepreneurs.

- **Mobile entrepreneurs**: They are the persons who leave the ventures as soon as venture is created. They are the true Schumpeterian entrepreneurs as according to Schumpeter (1934) the entrepreneurial activity ends as soon as the venture is created. Ted Nieren berg, founder of Dansk designs in the US, Mohan of Good knight and Vikram Sarabhai both from India comes closer to this type.
- **Managerial entrepreneurs**: They are the persons who prefer to continue in the same venture by transforming themselves to fit the changing demands. As for example persons like Henry Ford of Ford Motors (Lacey, 1986; Pollarq, 1996), Pierre S. Du Pont (Chandler and Salsburry, 1971), and George Eastman of Kodak (Brooke ball, 1996).
- **Innovative entrepreneurs**: They are the persons who create an organization and remain engaged in their pursuits of innovation and creation of novel product's and technology. Walt Disney, Ibuka and Akio Morita of Sony Corporation (Morita, 1987), Sochiro llinda of llinda motor Company (Pewins, 1967)"and Bill Gates of Microsoft (Marshall, 1996) are some of the notable entrepreneurs belonging to this type.
- **Empire Builders**: They are the persons engaged in creating chain of new ventures having an ownership. They have the qualities of vast vision, flair of innovation al1d managerial capability to build an empire for themselves. John D. Rockfeller of the US (Abels, 1965), J.N. Tata (Harris, 1958) and Ghanshyam Birla (Jaju, 1985) both from India, Konosuke Matsushita of Japan (and wins, 1967) -and Chung Ju Yung of Hyundai and Kim Woo Chong of Daewoo both from South Korea are some of examples of this type.

Typology of Entrepreneurial Firms

Typology of Entrepreneurial Firms

Business Innovation Rate		
High	Resource Constrained	Glamorous
Low	Economic Core	Ambitious
	Low	High

Business Growth Rate

This is based on rates of innovation and growth, developed by Kirchhoff (1994). He has developed a matrix having four classes of entrepreneurial firms, *viz.*, 1. Economic core firms 2. Constrained growth firms 3. Ambitious firms and 4. Glamorous firms.

1. **Economic Core Firms:** These firms are low innovation and low growth firms. These firms are most common form of entrepreneurship. These firms primarily satisfy the owner's desire for independence. It fulfills a specific need in a small market and it does not obtain a significant growth. Initially it achieves a degree of growth. But once it reaches a size that can satisfy the owner's needs, growth stops.
2. **Constrained Growth Firms:** These firms are having high rates of innovation, they lack adequate resources for growth, therefore it fails to grow. These firms can be classified into two classes, i. Firms unable to acquire the needed resources and ii. Firms internally make decision to constrain growth. The owner of resource constrained firms are willing to pay the cost of resources but are unable to attract the capital needed to sustain high growth. Where as the owner of internally constrained firm make decisions not to acquire the resources needed to support high growth. These firms are easy prey to better financed competitors. They find their product copied entrepreneurs having adequate resources devour the market.
3. **Ambitious:** These firms are having high growth rate with limited number of innovations in a large market a single successful product can sustain high growth for many years. These firms growth rate will eventually decline if new products are not introduced because markets are not stable.
4. **Glamorous:** These firms are having high growth rate with high rates of innovation. These firms are technology branded firms which require regular updating. Kirchoff calls these firms glamorous firms because they often

attract high media attention and receive local and national awards for their success.

Conclusion

The conceptual framework presented here identifies several key components of entrepreneurs. Entrepreneurship process starts when the entrepreneur recognizes an opportunity in the environment. Entrepreneurship Development process goes through i. identification of enterprise ii. Creation of enterprise iii. Nurturing of enterprise and iv. Concluding/transition phase of enterprise. The creation is 1. Economic Core Firms 2. Constrained Growth Firms 3. Ambitious and 4. Glamorous firms.

Refernces

Agarwal, R. (1997). Innovations in entrepreneurship: major issues in new era of gloalisation. Abhigyan. Vol (15), pp. 65-71.

Aitken, H.G.J. (1963). The future of entrepreneurial research, explorations in entrepreneurial history. Second series, Vol (1), pp. 3-9.

Akhouri, M.M.P. (1975). Planning and monitoring entrepreneurship. Developing entrepreneurship. A hand book (Ed. Rao, T. V and Pareek, U). Learning system, New Delhi. pp. 336-347.

Akhouri, M.M. (1996). Conceptual background and rationale for entrepreneurship development training programmes. Paper presented in orientation course on entrepreneurship development in agriculture. December 17-31.IARI, New Delhi. pp. 10.

Ali, H. (1995). Gateways to market entry. The Journal of entrepreneurship. Vol (4), No. (I), pp. 49 69.

Appley, L.A. (1956). Management in action. American management association, New York. pp.19.

Atkinson, J.W. and Litwin, G.H. (1960). Achievement motive and test anxiety conceived motive to approach success and pp. 105- motive to avoid failure. Journal of abnormal\ and social psychology, Vol (60), No. (1), pp. 52-63.

Banerjee, M. and Talukdar, R.K. (1997). Variables influencing entrepreneurship of women entrepreneurs. Indian Journal of Extension Education. Vol. (33), No. (1 and 2). pp. 25-36.

Bankston, C.L. and Zhou, M. Q 996). Go fish: The Louisiana vietnamese and ethnic\ entrepreneurship in extra active industry. National Journal of Sociology. Vol (10), No. (1), pp. 37-55.

Basavanna, M. (1971). A study of self-confidence as an attitude of self-concept. Ph.D. Thesis (Unpublished). Srivenkateswara University, Tirupathi, Andhra Pradesh.

Basu, S.K. and Moulik, T.K. (1979). Preference oriented selection criteria for rural entrepreneurs, identification and selection of small scale entrepreneurs. (Ed.

Rao, T.V. and Moulik, T.K.). Indian Institute of Management, Ahmadabad, Gujarat.

Battacharjee, S.K. and Akhouri, M.M.P. (1975). Profile of small industry entrepreneurs. SEDME. Vol. (2), No. (1), pp. 73-85.

Battacharya, S.N. (1983). Entrepreneurship development in India and the South East Asian countries in agricultural and industrial sectors. Metropolian book company, New Delhi.

Bema, J. (1960). Industrial entrepreneurship in Madras State. Asia Publishing House, London.

Bethal, L.L. Atwar,F.S. Smith, G.B.E. and Stockman, H.A. (1930). Industrial Organization and Management. Mc Graw Hill Book Company, New York, pp. 126.

Bhagat, G.R. and Singh, Y.P. (1995). Effectiveness of entrepreneurial development training, on the knowledge level of farm entrepreneurs. Indian Journal of Extension Education. Vol. (31), No. (1-4), pp. 102-104.

Bhatt, R.S. (1974). Growth of entrepreneurship in small and medium sectors. The Indian Journal of Public Administration. Vol (20), No. (3), pp. 453-465.

Bhattacharya, D. Sarkar, S. and Gangopadhyay, P. (1997). Decision making role of women in rural development. A quantitative study. Social science perspectives in agriculture. Edited by Das Gupta, D. and Mukhopadhyay, S.D. Santineketan, W.B. pp : 143-156.

Bisht, N.S. and Sharma, P.K. (1991). Entrepreneurship expectations and experience. Himalaya publishing house, New Delhi. pp. 3-4.

Bradford, L.A. and Glenn, L. J. (1953). Farm management analysis. John wiley and sons, Inc. New York.

Bull, I. and Willard, G. E. (1993). Towards a theory of entrepreneurship. Journal of Business Venturing. Vol. (8), pp: 183-195.

Busenitz, L.L. (1996). Research on entrepreneurial alertness: Sampling measurement and theoretical issues. Journal of Small business management. Vol. (34), No. (4), pp. 35 44.

Bygrave, W.D. and Hofer, C.F. (1991). Theorizing about entrepreneurship. Entrepreneurship Theory and Practice. Vol. (6), No. (2), pp. 13-22.

Cantillon, R. (1955). International encyclopedia of social sciences. Quoted by Cochran, T.C. (1968) under entrepreneurship. Mac millan, New York. Vol. (5), pp. 88-90.

Carland, J. W. (1984). Differentiating entrepreneurs from small business owners. Academy of Management Review. Vol. (9), No. (2), pp. 354-359.

Carrier, C. (1996). Entrepreneurship in small business: An exploratory study. Entrepreneurship Theory and Practice. Vol. (2), No. (1), pp. 5-20.

Christopher, K.J. (1969). Socio psychological factors influencing the adoption of innovations of starting a small industry unit. SIET Institute, Hyderabad, pp. 68-71.

Christopher, K.J. (1971). Identification and detection of latent entrepreneurship. Productivity. Vol. (12), No. (1), pp. 78- 83.

Cohen, N. (1980)"The Five Ages of the Entrepreneur." *Venture*: 40– 42.

Cole, A.H. (1949). Entrepreneurship and entrepreneurial history. In change and the entrepreneur. Prepared by the research centre in entrepreneurial history. Harward University Press, Cambridge.

Cole, A.H. (1959). Business enterprise in it social setting. Harward University Press, Cambridge.

Cole, A.H. (1968). Meso-economics: A contribution from entrepreneurial history. Explorations in Entrepreneurial History. Vol. (6), No. (1), pp. 78-86.

Cooper, A.C. Timothy, B and Woo, C. (1995). Entrepreneurial information search. Journal of business venturing. Vol. (10), No. (2), pp. 107-120.

Covin, J.G. and Slevin, D.P. (1991). A conceptual model of entrepreneurship as firm behaviour. Entrepreneurship Theory and Practice. Vol. (16), No. (1), pp. 7-25.

Dasgupta, A. (1969). Indian business management. Deptt. Of business management and industrial relations, University of Delhi. pp. 60.

De, D. (1985). Status symbol and innovative entrepreneurship are predictors of farmers progressivism. Journal of Extension System. Vol. (1), No. (1), pp. 81-84.

De,D. (1986). Factors affecting entrepreneur characteristics of farmers. Indian Journal of Social Work. Vol. (XLVI), No. (4), pp. 541-546.

De D and Rao M. S. (2001) Entrepreneurial behaviour of farmers: An Axiomatic theory. Ganga Kaveri Publishing House, D.35/77 Jangamwadi Math Varanasi.

Deivasena Pathy, P. (1986). "Entrepreneurial success: Influence of certain personal variables". The Indian Journal of Social Work. Vol. (XLVI), No. (4), pp. 547-553.

Derossai, F. (1971). The maxican entrepreneur. Development centre of the organisation for economic co-operation and development. Paris.

Desai, G.R. (1981). A critical analysis of the contribution of Extension guidance to economic performance of cotton farmers of Karnataka State. Thesis (Unpublished), U.A.S., Banglore.

Downing, J. (1991). Gender and growth of enterprises: Small Enterprise Development. Vol. (2), No. (1), Drucker, P. (1954). The practice of management. Harper and Row, New York. pp. 17.

Drucker, F.P. (1985). Innovation and Entrepreneurship –Practice and Principles. Heinemann, London. 31.

East West Centre, (1977). Entrepreneurial discovery and development progress of action research. East –West technology and development institute, Honolulu, Hawai. pp. 82 83.

Encyclopedia Britanica : Micropedia Vol. III : Ready reference and index.

Encyclopedia Britanica, Inc. Benton, W. Publishers, 1943-1973. Helen Hemingway, Chicago.

Encyclopedia of psychology, Search press, Ed. Eysenck, H.J. and Arnold, W. and Meili, R. London. Vol (2).

Encyclopedia of social sciences, (1933). Mc Millan Company, New York. Vol. (20), pp. 76 77.

English,H.B. and English, A. (1958). A comprehensive dictionary of psychological and psychoanalytical terms. Longmans. Green and Company, New York.

Fishbein, M. (1965). Attitude and the prediction of behaviour. In Fishbein, M (Ed.), Readings in attitude theory' and measurement. Jolm wiley and sons, New York. pp. 477-492.

Fraser, T.M. (1961). Achievement motivation as a factor in rural development, a report on research in Western Orissa. Haverford College, Haverford, Paris.

Gaikwad, J.G. Inamke, N.M. and Kadam, K. R. (1997). Involvement of farm women in decision making. Maharashtra.J. Extn. Edn. Vol. (16), pp. 359-361.

Gaikwad, V.R. (1978). Entrepreneurship: The concept and social context in developing entrepreneurship. A hand book of Rao, T. V. and Pareek, U. in collaboration with Deshpande, M. V Mehta, P. and Nadkarni. M.S. Learning systems, New Delhi.

Gartner, W.B., Bird, B.J. and Starr, J.A. (1992). Acting as if Differentiating entrepreneurial from rganizational behaviour. Entrepreneurship theory and Practice. Vol. 16 (3). pp. 13-31.

Ghosh, R. Gupta, M. and Dhar, S.N. (1998). Women and entrepreneurship in India. Entrepreneurship innovation edited by Kanungo, R.N. Sage Publications, New Delhi. pp.156-175.

Gogoi, M. and Talukdar, R.K. (1997). Determining achievement motivation of agricultural scientists. Indian Journal of Extension Education. Vol. (33), No. (1 and 2), pp. 64-67.

Gopakumar K (1995) Entrepreneurship in economic thought: A thematic review. Journal of Entrepreneurship, 4 (1) Pp 1-17.

Gordon, L. (1961). Private enterprise and international development. The American Review. Vol. (6), No. (1), pp. 5-13.

Gupta, C.B. and Srinivasan, N. P. (1992). Entrepreneurial development: Second edition. Su1tanchand and sons educational publishers, New Delhi.

Harbison, F.N. and Hyers, C.A. (1964). Education man power and economic growth. In Nodder, B.W. (Eds.). Economic development in the topics, Meuthen, London.

Haredero, J.M. (1979). Agricultural entrepreneurship. Identification and selection of small scale entrepreneurs. (Ed. Rao. T.V. and Moulik, T.K.). Indian Institute of Management, Ahmadabad. pp. 151

Harper, M. and Vyakaranam, S. (1988). Rural enterprise: Case studies for developing countries. Intermediate technology publications, London.

Hawley, F.B. (1882). Entrepreneurship in interaction. In Encyclopedia o/the social sciences (1968). Mac Millan, New York. Vol. (5), pp. 87-90.

Hegan (1968). The economics of development. Irwin RD Inc. Illinois, New York

Hepple, L.M. (1959). Group organization and Leadership in rural life. Lucas bros, Columbia, Missouri.

Himachalam, A (1990). Developing entrepreneurial industrialist - An alternate approach. Productivity. Vol. 34 (2). pp.281-287.

Homans, G.C. (1970). The relevance of psychology to the explanation of social phenomena. Borger, R. Cioffi, F. (Ed.) Explanation in the behavioural sciences. Cambridge University Press, Cambridge. Hornaday, J.A. and Aboud, J. (1971). Characteristics of successful entrepreneurs. Personal psychology. Vol. (14), No. (6), pp. 66-78.

Hoselitz, B.F. (1957). Non-Economic factors in economic development in Tandoil, B.C. (Eds.). Environmental and entrepreneur. Cheigh publications, Allahabad, India.

Hundal, P.S. and Singh, S. (1975). Structure of personality characteristics and motive pattern of farmers. Indian Journal o/Psychology: Vol. (50), No. (1), pp. 33-43.

Jaiswal, N.K. (1965). A study of factors associated with low level of adoption of improved agricultural ractices. Ph.D. thesis, Division of agricultural extension, IARI, New Delhi.

Jenkins; J. C. (1983). A resource mobilization theory and the study of social movements. Annual review of sociology. Vol. (9), pp. 527-533.

Jhamtani, A. (1996). Dynamics of entrepreneurship. Orientation course on entrepreneurship development in agriculture. December 17-31, JARI, New Delhi, pp 5-6.

Jones, S. and Cohen, M.B. (1990). The emerging business. Ronald Press, New York.

Joshi, S.S. and Kapur, T.R. (1973). Fundamentals of farm business management. Kalyani publishers, ND

Jyothi, V. (1990). Entrepreneurship development among women - study of entrepreneurs in Guntur district. M. Phil dissertation submitted to Nagarjuna University. Guntur, Andhra Pradesh.

Jyothi, V. and Prasad, G. (1993). A profile of potential rural women entrepreneurs. Small enterprises development management and extension Journal (SEDME). Vol. (20), No. (1), pp. 23-38.

Kanungo, R.N. and Bhatnagar, J. (1978). Achievement orientation and occupational values: A comparative study of Young French and English Canadians. Canadian Journal of Behavioural Science. Vol. (10), No. (3), pp. 202-213.

Kanungo, R.N. and Conger, J.A. (1995). Model orientation in leadership research and their implications for developing countries. New approaches to employee management. Vol (3), pp. 155-170.

Kanungo, R.N. and Mendonca, M. (1994). Fundamentals of organizational behaviour. Kendall Hunt publishing company, Iowa.

Keshari, P.K. and Kota, R. (1995). Determinants of the propensity to export: The case of small firms in the diesel engine and chemical industries in Gujarat. The Journal of entrepreneurship. Vol. (4), No. (1),

Khan, N.A. (1998). Agricultural marketing: Present position and future strategy, Kurukshetra. Vol. (47), No. (3), pp. 11-13.

Khan, W, M. (1998). Technological innovation in SMEs in Pakistan. Entrepreneurship and innovation edited by Kanungo, R.N. Sage publications, New Delhi. pp. 287-309.

Khanka, S.S. (1998). Making the entrepreneurial society. Yojana. Vol. (42), No. (2), pp. 9 11.

Khemmani, M. (1983). Turning thoughts towards training, Manita, phillippines, Asia and Pacific programme for development training and communication planning (DTCP), UNDP. Training tips No.1 Editor: Andrew P. Bartlett.

Kilby, P. (1971). Entrepreneurship and economic development. Free press, New York.

Kirzner, I.M. (1979). Perception, opportunity and profit: Studies.in the theory of entrepreneurship. The University Chicago press, Chicago.

Kittur, M.M. (1976). A study of adoption behaviour of marginal farmers in relation to their characteristics and value orientation in Bizapur district of Karnataka state. M.Sc. (Ag.) Thesis. Division of Agricultural Extension, U.A.S. Dharwar.

Knudson, W., Wysocki A., Champagne J. and H. Christopher Peterson, (2004). Entrepreneurship and innovation in the Agri-food System, Amer. J. Agr. Econ. 86:1330-1336.

Kogan, N. and Wallach, M.A. (1964). Risk taking. A study in cognition and personality. Holt, Rinehart and winston, New York.

Kotler, P and Levy, S.J. (1969). Broadening the concept of marketing. Journal of marketing. Vol (16), No. (8), pp.63-64.

Koontz, H. (1994). Essentials of management. (I edition). Mc. Graw. Hill publishing company, New York. pp. 39.

Kourilsky, M. (1980). Predictors of entrepreneurship in stimulated economy. Journal of creative behavioul: Vol. (14), No.3. pp. 76-86.

Krishnan, S.B. Kumar, K. G. and Kanungo, R.N. (1998). Entrepreneurship development concept and context. Entrepreneurship and innovation. Sage publications, New Delhi. pp. 21.

Kumar, E.S. (1.998). Theorizing in entrepreneurship: A review, critique and future directions.Entrepreneurship innovation. Edited by Kanungo, R.N. Sage publications, New Delhi, pp.90-.

Kumar, G.K. (1995). The entrepreneur in economic thought: A thematic overview. The Journal of entrepreneurship. Vol. (4), No. (1), pp. 1-15.

Kumar, K.G. (1998). Management skills for rural enterprises. Entrepreneurship innovation. Edited by Kanungo, R.N. Sage publications, New Delhi, pp. 338-357.

Kumar, V.S. (1998). Small business marketing: Need for specilized skills. Entrepreneurship innovation. Edited by Kanungo, R.N. Sage publications, New Delhi, pp. 325-337.

Lamb, R.K. (19652). The entrepreneur and the community in men in business. In Baumback,C.M. and Mancuso, JoR. (Eds.). 'Entrepreneurship and venture management. Prentice Hall inc, Engle wood cliffs, New Jersey.

Learned, K.E. (1992).What happened before the organization? A model of organization formation. Entrepreneurship theory and Practice. Vol. 17(1). pp. 39-48.

Leeds, C.N. and Stainton, R.S. (1978). Management and business studies (Second edition). Macdonald and Evans..

Leibenstein, H. (1976). Beyond economic man: A new foundation for micro economics. Harward University, Cambridge, Massachusetts.

Lekha, K.C. Kalyan, W and Lavoie, D. (1998). Micro enterprise management by women in India. Entrepreneurship innovation. Edited by Kanungo, R.N. Sage publications, New Delhi, pp. 176-199.

Lim, C. (1977). Small industry in Malaysia: A socio economic profile of entrepreneurs. Social action. Vol. (27), No. (2), pp. 22-36.

Lumpkin, G.T. and Dess, G.G. (1996). Clarifying the entrepreneurial orientation construct and linking it to performance. Academy of management review. Vol. (21), No. (1), pp. 135-172.

Ma, Hao and Tan, Justin, (2006). Key components and implications of entrepreneurship: A 4P framework, Journal of Business Venturing. 21:704-725.

Mall, C.P. (1993). Impact of financial incentives on entrepreneurial development in backward regions: A case study of Deoria district of V.P. The Journal of entrepreneurship. Vol. (2), 1 No. (2), pp. 199-207.

Mancuso, J. and Mascolo, M. (1987). Recognising achievement motivation. Motivation and emotion. Vol. (11), No. (3). pp. 323-330.

Manimala, M.J. (1998). Net working for innovation: Anecdotal evidences from a large sample study of innovative enterprises. The Journal of entrepreneurship. Vol. (7), No. (2), pp. 153-169.

Manimekalai, N. (1998). Gem cutting and rural entrepreneurship development. Kurukshetra. Vol. XLVI, No. (5), pp. 22-26.

Manimekalai, N. and Abdullah, M. (1998). Charcoal production and rural entrepreneurship development. The Indian economic journal. Vol. (45), No. (2), pp. 159-169.

Manjula, S. (1995). A study of entrepreneurial behaviour of rural women in Ranga Reddy district of Andhra Pradesh. M. Sc. (Home Science). Thesis. ANGRAV, Rajendra Nagar, Hyderabad.

Massie, J.L. (1982). Essentials of management (Third edition). Prentice Hall of India private limited, ND

Mcclelland, D.C. (1953). Achievement motive. Appleton, New York.

Mcclelland, D.C. (1965). Need achievement and entrepreneurship: A longitudinal study. Journal of personality and social psychology. Vol. (1), pp. 389-392.

Mcclelland, D.C. (1969). Motivating economic achievement. Free Press, New York.

Mishra, S.P. (1979). A study of farm entrepreneurship in a backward district of Bihar, Ph.D.Thesis (Unpublished), IARI, New Delhi.

Mishra, S.P. (1996). Training in entrepreneurship -A conceptual background. Paper presented in orientation course on entrepreneurship development in agriculture. December 17-31. IARI, New Delhi.

Mokry, B.W. (1988). Entrepreneurship and public policy. Quorum books, New York. Nair,K.B. Sivaraman, M. and Kanungo, R.N. (1998). Rural entrepreneurship in a developing economy: A contingency approach. Entrepreneurship innovation. Edited by Kanungo, R.N. Sage publications, New Delhi. Nandapurkar, G.G. (1980). A study of the entrepreneurial behaviour of small farmers. Unpublished Ph.D. thesis. U.A.S., Bangalore.

Nandi, A. (1973). Motives, modernity and entrepreneurial competence. Journal of social psychology. Vol. (91), No. (3), pp. 127-136.

Neill, R.E. and Rogers, E.M. (1963). Measuring achievement motivation among farmers. Department series A.E. 346. Ohio Agricul. Expt. Stn. Columbus.

Niles, M.C. (1956). The essence of management. Orient longmans, Bombay. pp. 20.

Palmer, B.l. Ahiamadu, R.K. (1997). Leadership style and performance of public enterprises. A Nigerian study. Productivity. Vol. (38), No. (1), pp. 54-64.

Pandey, VK. and Sharma, K.C. (1996). Crop diversification and self sufficiency in food grains. Indian Journal of agricultural economics. Vol. (5), No. (4), pp. 644-651.

Pareek, U. and Nadkarni. M. (1978). Development of entrepreneurship. A conceptual model. In Rao, T.V and Pareek, U. (Eds.), Developing entrepreneurship: A hand book, learning systems, New Delhi.

Pareek, U. and Rao, T.V. (1995). Counseling and helping entrepreneurs. The Journal of entrepreneurship. Vol. (4), No. (1), pp. 19-34.

Patel, D.R. (1995). Entrepreneurship and small business development for women. Kurukshetra. Vol. (XLIII), No. (11), pp. 65-68.

Patel, G. V. (1970). A successful ventures in developing entrepreneurs. Lessons for Gujarat.A symposium paper on development of entrepreneurial talent. New Delhi.

Patel, M.M. Sanoria, Y.C. (1997). Correlates of entrepreneurial behaviour of sugarcane growers. Maharashtra J Extn. Edn. Vol. (16), pp. 344-346.

Paul, K.C. (1998). A study on entrepreneurial behaviour of vegetable growers in Krishna district of Andhra Pradesh. M.Sc. (Ag.) thesis, ANGRAU, Rajendra Nagar, Hyderabad.

Pearce, I.A. Kramer, T.R. and Robbins, K. (1997). 'Effect of managers' entrepreneurial behaviour on subordinates. Journal of business venturing. Vol. (12), No. (2), pp. 147 160.

Pollo, L.V. (1996). On economic theory of entrepreneurship : The strategic role of the entrepreneurial function. International review of sociology. Vol. (6), No. (2), pp. 191-213. PRADAN (2000) System setting document. Self Help Groups Pradan Programme, New Delhi

Prakasam, D.S. (1998). Entrepreneur: An important change agent in every society. Kurukshetra. Vol, (46), No. (10), pp. 51- 52.

Prasad, R.M. (1983). Comparative analysis of achievement motivation of rice growers in three states in India. Ph.D. thesis (unpublished), U.A.S., Banglore.

Prasad, S.K. (1996). Promoting an entrepreneurship development programme. Laghu Udyog. Vol. (21), No. (2), pp. 3-5.

Rajak, M.P. (1998). Craft technology and entrepreneur. Kurukshetra. Vol. (XLVI), No. (5), pp. 17-19.

Ramachandran K and Ray Sougata (1998) A framework and entrepreneurial typology for developing a comprehensive theory of entrepreneurship. In Entrepreneurship and innovation: Models for development. Ed Kanungo R. N. Sage Publications New Delhi. Pp.40-63.

Ramana, A. V. and Papaiah, R. (1998). Nuances of entrepreneurship development. Indian management. Vol. (37), No. (2), pp. 48-53.

Ramakrishnan, K. (1979). Interview techniques as a method of selection in identification and selection of small entrepreneurs (Eds. Rao, T.V. and Moulik, T.K.) Indian Institute of Management, Ahmedabad.

Ramana, A. V. (1999). Entrepreneurship and economic development. Kurukshetra. Vol. (48), No. (2), pp. (8 to 10 and 45).

Rani, C. (1986). Potential women entrepreneurs. A study. SEDME. Vol. (13), No. (3), pp. 13 32.

Rao, G.K.V. (1985). A prediction analysis of farming performance of farmers through their entrepreneurial behaviour factors. Ph.D. thesis, A.P.A.U. Rajendra Nagar, Hyderabad.

Rao, M.S. (2000). Axiomatic approach to theory building in entrepreneurial behaviour of farmers. Ph.D. thesis, Institute of Agricultural Sciences, B.H.U., Varanasi.

Rao, M.S. and Mishra, S.K. (1997). Farmers perceptions of environmental risk in pesticide use. Maharashtra Journal of Extension Education. Vol. (16), pp. 214-219. Rao, T. V (1975). Development of an entrepreneur: A behaviouristic model. SEDME. Vol. (1), No. (4), pp. 325-345.

Rao, T.V. (1975). Characteristics of entrepreneurs as perceived by entrepreneurial development agents. Paper presented at the work shop on identification and selection of entrepreneurs. Indian Institute of Management, Ahmedabad.

Rao, T.V. and Mehta, P. (1978). Psychological factors in entrepreneurship. In Rao, T.V. and Pareek, U. (Eds.). Developing entrepreneurship. A hand book. Learning systems, New Delhi.

Rao, T.V.S. (1989). A study on entrepreneurial characteristics of vegetable growers in Guntur district of Andhra Pradesh. Unpublished M.Sc. (Ag.) thesis, A.P.A.U., Rajendra Nagar, Hyderabad.

Ray, S. and Ramachandran, K. (1996). Towards a frame work for a comprehensive theory of entrepreneurship. The Journal of entrepreneurship. Vol. (5), No. (1), pp. 1-22.

Razin, E. and Langlois, A. (1996). Metropolitan characteristics and entrepreneurship among immigrants and Ethnic groups in Canada. International Migration Review. Vol. (30), No. (3), pp. 703-727.

Reddy, Y.V.R. (1992). Economics of Agro forestry system with particular reference to agro horticultural systems on dry land region. Int. sym. on natural resource management for a sustainable agriculture, New Delhi. Feb. 6-10.

Richart, C.B. Maurer, J.G. (1997). The moderating effect of environmental variables on the entrepreneurial and marketing orientation on entrepreneur -led firms. Entrepreneurship Theory and Practice. Vol. (22), No. (1), pp. 47-58.

Ricky, W.G. (1992). Essentials of management prentice Hall of India Private Lmited, New Delhi. pp. 22-24.

Ripsas, S. (1998). Towards on interdisciplinary theory of entrepreneurship. Small business economics. Vol. (10), No. (2),pp.103-115.

Rogers, E.M. and Svenning, L. (1969). Modernization among peasants: The impact of communication. Hort, Rinehart and Winston, New York.

Rogers, E.M. and Shoe maker, F.F. (1971). Communications of innovations, A cross cultural approach. The Free Press, A division of Macmillan publication company, Inc. New York.

Rogers, F.M. and Neill, R.E. (1966). Achievement motivation among columbian peasants. Michigan state University, East I Lansing, Michigan..

Rousmasset, J.A. (1977). Risk and uncertainty in agricultural development. Seminar report No. 15. Agricultural Development Council, New York.

Samanta, R.K. (1977). A study of some agro economic, socio psychological and communication variables associated with repayment behaviour of agricultural credit users of nationalised banks. Ph.D. thesis, Bidanchandra Krishi Vidyalaya, Nadia, West Bengal.

Samanta, R.K. {1989). Management in agriculture and rural development. UDH publishing house, Naisarak, New Delhi.

Samuelson, P.A. (1963). Discussion -problems of methodology. American economic review. Vol. (53),

Sara, C. (1998). Portfolio entrepreneurship in the farm sector: Indigenous growth in rural areas? Entrepreneurship and, regional development. Vol. (10), No. (1), pp. 17-32.

Say, J.B. (1815). Quoted under entrepreneurship in international encyclopedia of the social sciences (1968). Macmillan, New York. Vol. (5), pp. 87-90.

Schumpter, J.A. (1961). The theory of economic development: An enquiry into profits, capital, interest and the business cycle. Cambridge Harward University Press. Cambridge.

Schumpter, J.A. (1970). The entrepreneur as innovator. Reading in management (Second edition).Mc. Graw Hill, New York. Schwartz, E.A. (1979). Entrepreneurship: New female frontier. Journal of Contemporary business winter issue. Vol. (15), pp.150 163.

Scott, C.E. (1986). Why more women are becoming entrepreneurs. Journal of small business management. Vol. (24), No. (4), pp. 37-44.

Scott, M. (1997). Reengineering the rapidly growing entrepreneurial company. Industrial management. Vol. (39), No. (3), pp. 21-23.

Sengupta N (1998) Empowerment: A socio-psychological approach to Self Help Group formation. Prajnan. XXVI (4): 525-531

Shailendra, V. (1.990). When the harvest is in developing rural entrepreneurship. Oxford and IBH publishing Co. Pvt. Ltd. \ New Delhi

Sharma, A Mehta: P. and Sha S.K.(1996) Evaluation of task m diversified crop farming of Himanchal Pradesh,. A case study of vegetable farmers; Indian Journal of Agricultural Economics; Vol.. (51), No. (4..

Sharma, K.L. (1973). Growth of small scale entrepreneurship in India: A theoretical frame of analysis. Interdisciplinary. Vol.. (10), No. (3), pp. 97-108.

Sharma, S. V.S. (1980). Characteristics of entrepreneurs. In developing entrepreneurship: Issue and problems. SIET (Small industry extension training). Institute, Hyderabad.

Singh, D.N. (1986). Developing entrepreneurship for economic development. Vikas publishing house private limited, New Delhi. pp. 83-92.

Singh, K. (1992). Women entrepreneurs. Ashish publishing house, New Delhi.

Singh, K.N. and Pal, K. (1974). A study on achievement motivation of farmers. Interdiscipline. Vol. (II), No. (I), pp. 31-40.

Singh, N.P. (1979). Achievement motivation among agricultural and business entrepreneurs of Delhi. Journal of Social Psychology. Vol. (81), pp. 146-149.

Singh, N.P. (1979). Tribal entrepreneurship, efforts and achievements. In Rao, T.V. and Moulik, T.K. (Eds.), identification and selection of small scale entrepreneurs. Indian Institute of Management, Ahmedabad.

Singh, N.P. and Gupta,,R.S. (1985). Promotion women entrepreneurs. NIESBUD, New Delhi.

Singh, N.P. and Singh, K. (1971). Motivational components of agricultural and business entrepreneurs in India. Indian Journal of Industrial relations. Vol. (7), No. (1), pp.31 52.

Singh, P:N. (1986). Factors influencing entrepreneurship. Developing entrepreneurship for economic development. Vikas publishing house private limited, New Delhi. pp. 46 92.

Singh, 'S. (1978). Achievement motivation, decision making orientation and work values of fast and slow progressing famlers in India. The Journal of social psychology. Vol.:(.106), pp. 153-160.

Singh, S. (1979). Motives, Personality characteristics and work values of fast and slow progressing farmers. Behavioural science and rural development. Vol. (2), No. (2), pp.71-77.

Singh, T..P. and Kumar, K. (1975). Achievement motivation, socio- economic status and green revolution. Indian Journal of Extension Education. Vol. (1.1), No. (10), pp. 1-7.

Sinha,B.P. and Mehta, P. (1972). Size and land holding and farmers achievement motivation and change proneness. Behaviour science and community development. Vol. (6), No. (2), pp.244-25.1.

Sinha, T.N. (1996). Human factors in entrepreneurship effectiveness.. The Journal of entrepreneurship. Vol. (5), No. (1), pp. 23-39.

Sisodia, G.S. (1976). A study of the achievement motivation of farmers for mechanization. Indian Journal of Extension Education. Vol. (13), No. (3 and 4), pp. 72-75.

Stevenson, H.H. and Jari110, I.C. (1990).. A paradigm of entrepreneurship: Entrepreneurial management. Strategic management Journal, 10th summer special issue. Vol (11),pp. 17-27.

Strayton, R. (1986). The organization of course and conferences., in ILO (1986b) pp. 1-27. Sullivan, A.O. Munir, K. and Dougherty, D. (1998). Building organizational capacities for product innovation. Entrepreneurship innovation edited by Kanungo, R.N. Sage publications, New Delhi.

Sundari, S.S.K. (1995). Entrepreneurship development. for rural women. SDEM. Vol. (1)., pp. 15-30.

Suresh, B. Kumar, G.K. and Kanungo, R.N. (1998). Entrepreneurship development: Concept and context. Entrepreneurship innovation. Edited by Kanungo, R.N. Sage publications, New Delhi.

Sutton, R.I. and Staw, B.M. (1965). What theory is not. Administrative science, Vol. (40), No. (3), pp. 371-384.

Swamy, B.A. (1988). Agricultural entrepreneurship in India. Ph.D. thesis, Andhra University, Waltair, Andhra Pradesh.

Terry, G.R. (1968). Principles of management. Irwin, R.D. Inc., Homewood, Illinois.

Thagamuthu, and Manimekhalai, N. (1989). Generation of employment for women through DWACRA. Journal of rural development. Vol. (8), No. (40), pp. 431-438.

Thomas, T. T. and Padmakar, K. (1993). Developing entrepreneurial industrialists. An alternative approach. Productivity. Vol. (34), No. (2), pp. 281-287.

Timmons,l.A. (1996). Characteristics and role demands of entrepreneurship. Paper presented in orientation course on entrepreneurship development in agriculture. December 17 31, IAR!, New Delhi.

Tripathi, D. (1985). An integrated view of entrepreneurship. Economic and political weekly review of management. Vol. (20), No. (48), pp. 62-69. Administrative science. Vol. (31), No. (3), pp. 439-465.

Tyson, S. and York, A. (1989). Personal management. Made simple books, an imprint of Heinemann rofessional publishing limited. Oxford, England.

Vplankar, A. T. (1977). A study of self employed industrial entrepreneurs. The Indian Journal of Social Work. Vol. (38), No. (2), 137-147.

Vijaya, V. Kamalanabhan, T.J. (1998). A scale to assess f entrepreneurial motivation. The Journal of; entrepreneurship. Vol. (7), No. (2), pp. 183-198.

Vinze, M.D. (1987). Women entrepreneurs in India. Mittal publications, New Delhi.

Vyas, V.S. (1996). Diversification in agriculture: Concept, rationale and approaches. Indian Journal of Agricultural Economics. Vol. (51), No. (4), 635-643.

Weber, M. (1930). Entrepreneurship and economic development. The Free Press, New York.

Webster, S. (1968). New world dictionary of the American language. Cleveland.

Westermark, N. (1951). The human factor and success in farming. Acta agriculture scandanavica. Vol. (1), pp. 146.

Westermark, N. (1961). The human factors and economic progress. Indian Journal of Agricultural Economics. Vol. (16), pp. 12-17.

Wick, J.H. (1998). Understanding entrepreneurial intentions: The attitudinal determinants. Entrepreneurship and innovation. Edited by Kanungo, R.N. Sage Publications, New Delhi, pp. 119-138.

Wilken, P.H. (1979). Entrepreneurship. A comparative and historical study. Ablex publishing, Norvood. pp. 60.

Yasunobu, K. (1997). Entrepreneur fanners in peninsular Malaysia : profile of Peladang Jaya. JIRCAS News letter. (December). No. (13). pp. 4-5.

Zahir, M.A. (1994). Entrepreneurship in Punjab. A case study. Yojana. Vol. (38), No. (8), pp. 24-28.

Chapter 34

Orchard or an Enterprise in Generation of Income and Livelihood

Abu Hassain

Globalization, trade liberalization and change in consumer demand are creating new market opportunities for farmers and land less laborers especially through horticultural system in the tropics and subtropics. Many cereal farmers are already converting a portion of their land into production of high value horticultural crops. A strong horticulture sector is an engine for economic growth. It creates jobs, supports agribusiness, and generates income to a greater degree than staple crops. Furthermore, horticultural crops can provide micronutrients that are essential yet lacking in the diet of half of the world's population. Horticultural crop production diversifies crop production and when integrated into cereal banned cropping systems, breaks decrease cycles and conserves soil and water. In horticulture fruit crops occupies a major position. India is the second largest producer of fruits (49 million tones) and contributes 10 per cent share in global fruit production. World wide production of fruit crops has grown faster than that of cereal crops, aebe it from a much lower bone. In a commercial orchard farmer grow and market fruit for profit. In crease in fruit traded world wide have also been dramatic and the total value of horticultural crops traded at present is more than double that of cereal crops, reaching nearly 21 per cent of total value exported from developing countries. Likewise the consumption of fruits has also increased. Annual growth in domestic consumption trends of fruits and vegetable is reported to have gone up from 2.5 per cent in 1980-81 to 7 per cent in the recent years.

Creation of Income

The generation of additional employment opportunities in rural areas where labour is abundant is critical for achieving wide spread and equitable growth. Cultivation of fruits allows for productive employment where labour/land ratio is high, since fruit production is mually labour intensive. Depending on the crop, production of horticulture crops requires at least twice the labour and up to five times the labour days per hac on compared to cereal crops. Increasing fruit production contributes to commercialization of the rural economy and creates many off farm jobs. The market integration of producers of fruits and vegetables is usually higher than that of stopic crop producers. Commercialization benefits the poor both through generation of employment and increasing agricultural productivity. As the production of fruit crop is often more risky, became these crops are much more costly to produce per hectare than traditional crops, and became yields and prices are more variable than for staple crops. Resource poor farmers must then be supported by an enabling institutional environment such an access to credit and capital, and meant be provided with access to market price information.Value addition to fruit crops generates further employment in the associated agri business and further down the commodity chain from the producers to the consumer. But the value addition is only 7 per cent in India an against 23 per cent in China and 88 per cent in U.K. development of new products like juice punches, banana chips and fingers, mango neetar and fruit kernel derived cocoa substitute, and inertial oils from citrus, fruit wines, dehydrated products from grape, pomegranate, mango, apricot are getting popular day by day. In each season due to in ability to met quality requirements a considerable share of fresh fruits cannot be sold in the market. These fruits can either be sold to fruit processing industry or can be processed directly on the farm. Prices achieved for home made fruit products such as brandy, dry fruits, vinegar and cider are much higher than prices of fruits sold to the fruit processing industry.

Nutrition and Human Health

Low levels of fruit and vegetable consumption in some parts of the world have devastating health effects. It is estimated that insufficient fruit and vegetable intake, and the micronutrient deficiencies that often result from such shortages, came some 2.7 million deaths each year, making it one of the top ten risk factors contributing to mortality. The diverse array of micronutrients offered by horticultural products is an answer to these deficiencies. Then more research is needed to analyze functional properties of fruits and vegetables, particularly those already available in high risk areas, and to develop new varieties that add value for poor, malnourished populations. Analysis of the nutritional properties of selected indigenous and traditional crops and varieties is especially important for delivering benefit to the poor. The bioavailability of specific nutrients from mineral rich foods can help to determine which crops should be promoted and marketed for their health benefits in zones of known micronutrient deficiencies. Fruits and its processed products are enjoying increasing domestic and international demand. The expansion of markets and the liberalization of trade policies are providing new opportunities for rural people escape poverty through production and exchange of non staple crops. How ever, much more focus efforts needs to be undertaken to bring resource poor farmers and land less labours into this development. Today, 1.1 billion people continue to live in extreme poverty on less than U.S $ 1 a day. Another 1.6 billion live on between U.S $ 1-2 per days. Opportunities in horticulture and food industries must increase especially for women, to enable their escape from poverty and live long, healthy lives. However growth in the horticultural sector does not always favour the poor. Most growth and development in the horticultural sector which presents opportunities for small farmers and rural economic occurs in local and regional markets. The rapidly changing dynamics and criteria of global horticultural markets often act barriers to their participation of the global value chain.

The research agenda for agriculture must be broadened and more emphasis must be put on the growing importance of the fruit sector and how these crops can contribute to poverty alleviation, economic development and human health. Such investments will allow a significant portion of the world's poor nations, farmers and land less laborers to gain from horticulture and propel societies toward health's and more diverse diets. Demand for fruits is rising, both in domestic and international markets. This market is changing rapidly, partially fueled by the spread of supermarkets. In developed countries a derives for year-round availability and increased diversity of foods, as well as growing awareness of the relationship between diet and health all contribute to the increased consumption of these commodities. Many consumers today purchase a broad range of relatively expensive commodities such an off season produce exotic fruits and organic produce. Increasing participation of women in the labour market of developed economics has created demand for processed, ready to cost convenient products. Developing countries are taking advantage of this trend, and over the past decade the increase of their processed food experts has exceeded that form the developed regions. Many work tanks, such as chopping washing, labeling, and bar coding, are being transferred to developing countries and are generating new jobs, especially for women.

Chapter 35

Scented Rices: Value Addition, Trade and Livelihood Development

Kaushik Brahmachari

Introduction

Earliest reference to the existence of aromatic rices in India can be traced to the documents of Susrutha (circa 400 BC), the great Indian pioneer in medicine and surgery. The first usage of the word "Basmati" is found in the Punjabi classic Heer Ranjah (1766) by the Punjabi poet Waris Shah. The names of rice cultivars such as Basmati, Begumi and Satthi (60-day rice) suggest that "Basmati" rices might have been in the cultivation scene long before 1700 AD.

Subsequent reference to "Basmati" can be found in "A Dictionary of the Economic Products of India" compiled by George Watt and published in 1891. The word "Basmati" appears to have been derived from two Sanskrit words *i.e.,* vaas (fragrance) and matup (possessing). Vaasmati pronounced now as Basmati appears to mean the one that "possesses fragrance." In North India "va" is often pronounced as "ba", and that is how the word "vasmati" must have become "Basmati".

Basmati rice was the preferred rice for preparation of Biryani and Pulao during festive season and special occasions. It was patronised by emperors, while poets have been composing poems admiring its qualities. The credit for conserving and improving Basmati rices to suit various socioeconomic and agroclimatic conditions should go to farmers of the geographical region in the Indian subcontinent. The names of Basmati varieties that exist now are quite often suggestive of places, where they are predominantly grown. Haryana Basmati, for instance, is from Haryana and Dehradun Basmati (Type 3) from Dehradun area in Uttaranchal.

One of the key characteristics of Basmati for which it is commanding high premium price is its unique aroma. The Basmati aroma is found at all stages of crop growth in almost all aerial parts of plant except roots. Development and retention of aroma is influenced by genetic and environmental factors. Aroma develops from a cocktail of more than 100 compounds which include hydrocarbons, alcohols, aldehydes and esters. Among many compounds the most prominent fragrance causing chemical compound in Basmati rices, Pandanus leaves, and jasmine rice has been found to be 2-acetyl-1-pyrroline (2-AP).

Study of genetics of aroma has been an attractive research topic and many researchers studied employing various sensory tests. But for a few like Reddy and Reddy who reported two to three recessive or dominant genes to determine the fragrance, most believe Basmati fragrance is controlled by a single recessive gene.

Rice, the second most widely grown cereal crop, is the staple food for more than half of the world's population. Rice is endowed with amazing genetic diversity with more than one hundred thousand landraces and improved cultivars maintained in the germplasm collections spread worldover. Varietal diversification appeared to have occurred as a function of adaptation to diverse agro-climatic conditions coupled with conscious and continuous selection subsequently by man for his diverse quality preferences. A unique varietal group that has distinguished itself as a result of natural and human selection, which found wider acceptance all over the world as a speciality rice is called "Basmati rice". The unique features of Basmati rices such as extra long slender grain, lengthwise excessive elongation on cooking, soft and fluffy texture of cooked rice, and pleasant aroma which together determine uniqueness of Basmati rice. Aside their unique cooking quality, Basmati rices are also reported to have low glycemic index and are micronutrient rich especially for iron and zinc. Unlike other aromatic rices, the unique quality traits of Basmati rice found their expression only when they are grown in the north-western foot hills of the Himalayas in the Indian subcontinent. And, because of its geographic specific manifestation of quality features, Basmati is now a Geographical Indication (GI) belonging to a specific geographical area in the Indian subcontinent, and it is like "champagne" among wines and "scotch" among whiskies. Basmati has attained 'heritage rice" status as it is considered as "farmers cultivar" being maintained and grown by farmers of Punjab region of India and Pakistan and Haryana and Uttar Pradesh regions of India for more than 250 years. The special quality of Basmati rice is attributed to unique combination of soil, water, climate and cultural practices under which it is grown, besides the inherent genetics governing the features.

History

Earliest reference to the existence of aromatic rices in India can be traced to the documents of Susrutha (circa 400 BC), the great Indian pioneer in medicine and surgery. The first usage of the word "Basmati" is found in the Punjabi classic Heer Ranjah (1766) by the Punjabi poet Waris Shah The names of rice cultivars such as Basmati, Begumi and Satthi (60-day rice) suggest that "Basmati" rices might have been in the cultivation scene long before 1700 AD. Subsequent reference to "Basmati" can be found in "A Dictionary of the Economic Products of India" compiled by George Watt and published in 1891. The word "Basmati" appears to have been derived from two Sanskrit words *i.e.*, vaas (fragrance) and matup (possessing).

Vaasmati pronounced now as Basmati appears to mean the one that "possesses fragrance." In North India "va" is often pronounced as "ba", and that is how the word "vasmati" must have become "Basmati". Basmati rice was the preferred rice for preparation of Biryani and Pulao during festive season and special occasions. It was patronised by emperors, while poets have been composing poems admiring its qualities. The credit for conserving and improving Basmati rices to suit various socioeconomic and agro-climatic conditions should go to farmers of the geographical region in the Indian subcontinent. The names of Basmati varieties that exist now are quite often suggestive of places, where they are predominantly grown. Haryana Basmati, for instance, is from Haryana and Dehradun Basmati (Type 3) from Dehradun area in Uttaranchal.

One of the key characteristics of Basmati for which it is commanding high premium price is its unique aroma. The Basmati aroma is found at all stages of crop growth in almost all aerial parts of plant except roots. Development and retention of aroma is influenced by genetic and environmental factors. Aroma develops from a cocktail of more than 100 compounds which include hydrocarbons, alcohols, aldehydes and esters. Among many compounds the most prominent fragrance causing chemical compound in Basmati rice, Pandanus leaves, and jasmine rice has been found to be 2-acetyl-1-pyrroline (2-AP). Study of genetics of aroma has been an attractive research topic and many researchers studied employing various sensory tests. But for a few like Reddy and Reddy who reported two to three recessive or dominant genes to determine the fragrance, most believe Basmati fragrance is controlled by a single recessive gene.

In Brief

Scented rice have been known in the Indian subcontinent since the times of Charaka (1600 BC to 700BC – eds) and Susruta (200 BC and 400 BC – eds). These rice have played an important role in many regional economics, and have been the favorites

Table 35.1: Scented Rice Identified at Various Times during the Course of History

Group of rice variety	Literal meaning	First referred by	Referred as scented by	Notes
Mahasali	Large	Charaka	Charaka [600 BC (c. 700 BC – eds.)] Hiuen-Tsang (640 AD)	
Kalama	Stiff stem	Charaka	Kashyapa (800 AD)	
Sugandhika	Fragrant	Charaka	Charaka [600 BC (c. 700 BC – eds.)]	
Promodaka	Fragrant	Susruta	Susruta [200 BC (c. 400 BC – eds.)]	
Pundrika	Softness, color, and the fragrance of a white lotus	Susruta	Susruta [200 BC (c. 400 BC – eds.)]	
Pushpandaka	Grains with the fragrance of flowers	Susruta	Susruta [200 BC (c. 400 BC – eds.)]	
Black sambhaka	Black	Kashyapa	Kashyapa (800 AD)	
Palasa vrihi	Red, full of flavor	Kashyapa	Kashyapa (800 AD)	
Sukhdas		Ain-i-Akbari	Ain-i-Akbari (1590 AD)	Deep water rice
Madhkar		Ain-i-Akbari	Ain-i-Akbari (1590 AD)	Deep water rice
Jhanwar		Ain-i-Akbari	Ain-i-Akbari (1590 AD)	Deep water rice
Mushkin	*Musk* means scent	Ain-i-Akbari	Ain-i-Akbari (1590 AD)	
Deergasali	Long with sweet flavor	Kashyapa	Kashyapa (800 AD)	
Gandhsali	Scented	Someshvardeva	Someshvardeva (1126–1138 AD)	Possibly, *sugandhika* of Charaka

Table 35.2: Scented Rice Varities in India and their Distribution

State	Varieties (district)
Assam (65)	**Ranga joha I, Joha** (47 types). **Bongali, Bor, Bhabeli, Kanjoha, Kanku, Khorika kala, Kopausali, Manki, Ranga, Rampal, Bagri bhog, Tulsi bhog, Govind bhog, Badshahbhog, Prasad bhog, Malbhog, Kalajira**
Bihar (42)	**Basmati 3** (Patna); **Katarani** (Bhagalpur, Champaran); **Kari bank** (Patna, Bhojpur, Munger, northern Bihar); **Mohin dhan**, **Sagarbhog**, **Hansraj** (Patna, northern Bihar); **Sonachur** (Bhojpur, Rohtas, northern Bihar); ***Badshahbhog*** (Bhojpur, Bhagalpur); **Kanakjira** (Bhojpur, northern Bihar); **Shamjira** (Rohtas, Aurangabad, northern Bihar); **Shapasand** (Rohtas, northern Bihar); **Tulsi phul** (Rohtas, northern Bihar); **Kanehonehur** (Gaya); **Mehijawain** (Aurangabad, northern Bihar); **Tulsimanjri** (Bhagalpur, Munger, northern Bihar); **BR 9, BR 10** (Bhagalpur, northern Bihar); **Marueya, Lakhisar** (Munger, northern Bihar); **Badshahpasand, Bahraini, Bhuri C. basmati, Chenaur, Devtabhog, Kamod, *Kali* Champaran basmati, Kesarbani, *Lal C. basmati*, Marcha, Malbhog, Ramjawain, Sonalari, Tulsipasand** (northern Bihar); **Mircha, Malida, Satari** (Muzzafarnagar); **Amad, Abdul, Ramjain** (western Champaran); **Brama bhusi** (Semra, Ramgarh, western Champaran); **Deobhog** (Darbangha); **Kamini** (Bhagalpur)
Gujarat (5)	**Pankhali, Kamod** (Kheda); **Krishnakamod** (Ahmedabad); **Kolhapur scented** (Saurashtra); **Zeersal**
Haryana (2)	**Basmati 370** (Rohtak, Kaithal); **Karnal local** (Karnal, Kurukshetra, Panipat)
Himachal Pradesh (9)	**Mushkan, Ramjawain, Achhoo, Seond basmati, Baldhar basmati, Madhumalati, Chetru basmati** (Kangra valley); **Pansara local** (Kullu); **Hathkoti basmati** (Shimla)
Jammu and Kashmir (7)	High hills: **Gul zag, Zagir, Muskkanti, Tumlazag;** Mid-hills: **Musk budji, Qadir baig, Ranbir basmati** (R.S. Pura, Katua, Jammu)
Karnataka (19)	**Ambemohor** (Belagoan, Dharwar); **Devamallig** (north Kanara). **Gumsali** (Haveri); **Gandhsali**, **Gulvadi, Gamanasanna** (south Kanara); **Huggibatta** (Belagoan, Dharwar); **Jeerigesanna** (Mysore, Bangalore, Kodagu, Chikmanglur); **Kagisali** (Belagoan, Dharwar, Haveri); **Kumudh** (Haveri); **Karigajavile** (Belagoan, Dharwar, Haveri); **Krishnapasangi** (Raichur, Gulbarga, Bellary); **Kunsum Kesari** (north Kanara); **Kalabatta** (Tumkur, Bangalore); **Kavali** (Bidar); **Rattansagar** (Bidar); **Sindhagi local** (Bijapur); **Vasane Sanna Batta** (north Kanara); **Yalakkisali** (Haveri)
Kerala (7)	**Gandhakasala, Jeerakasala, Velumbala, Chomala, Kayama** (Wyanad); **Kothampalari** (Kannur); **Pookkilathari** (Palakkad)
Madhya Pradesh (33)	**Amarjyoti** (Mandalla); **Adamchini, Antraved** (Damoh, Panna); **Badshahbhog** (Bastar); **Batanphul** (Sidhi); **Chakarbhatta** (Chattarpur); **Chhatri** (Jabalpur); **Chindi kapur** (Raigarh); **Chinoor** (Balaghat); **Chirna khai** (Bastar); **Dilbaxa** (Tikamgarh, Satna, Reva); **Dubraj** (Raipur, Durg, Rajnandgaon, Bilaspur, Mahasamund, Dhamtari, Janjgir, Korba, Kanker); **Gangaprasad** (Rajnandgaon);

of Kings, religions heads, royalty, and the elite of the society. Most of these rice are highly area specific: hence each Indian State has its own special scented rice(s). Scented rice of short and medium grain size are grown in most States, but long grained basmati rice of Northeastern State of India has gained popularly all over the country. It is known that scent is present only in the handful of rice varieties and that it is conspicuously absent in wild rice. Biotechnological studies have revealed that scent originated as a mutation in normal rice in the BAD 2nd generation. Even in the ancient times, the existence of several groups of Scented rice was known. Like other rice, these have shown a spread from one area to another, revealing important links with important people and even in the history. This proper attempt to bring together information on the history, diversity and spread of scented rice, and patronage offered by the royalty. It traces basmati from Vedic period and its association with mahasali, the well known variety of ancient and medieval times.

Table 35.3: Notified Basmati Varieties

Cultivar	*Pedigree*	*Method of Development*	*Notified by*	*Type*
Basmati 370	Selection	Pure line selection	India, Pakistan	Traditional Basmati
Dehraduni Basmati	Selection from Type30	Pure line selection	India	Traditional Basmati
Taraori Basmati	Selection from Basmati370	Pure line selection	India	Traditional Basmati
Basmati 386	Selection from Pakistani Basmati	Pure line selection	India	Traditional Basmati
Ranbir Basmati	Selection from Basmati 370	Pure line selection	India	Traditional Basmati
Basmati 217	Punjab (Indian sub-continent)	Pure line selection	India	Traditional Basmati
Pusa Basmati-I	Pusa 150/Karnal Local	Convergent Breeding	India	Evolved Basmati
Haryana Basmati	Sona/Basmati370	Pedigree Method	India	Evolved Basmati
Kasturi	Basmati370/CR 88-17-1-5	Pedigree Method	India	Evolved Basmati
Mahi Sugandha	BK-79/Basmati370	Pedigree Method	India	Evolved Basmati
Punjab Basmati	TN1/Basmati370	Pedigree Method	India	Evolved Basmati
Pusa 1121	Sister line of Pusa Basmati-1	Convergent Breeding	India	Evolved Basmati
Kernel Basmati	CM7/Basmati370	Pedigree Method	Pakistan	Evolved Basmati
Super Basmati	Basmati320/IR661	Pedigree Method	Pakistan, India	Evolved Basmati
Basmati385	TN1/Basmati370	Pedigree Method	Pakistan	Evolved Basmati
Basmati 198	Basmati370/TN1	Pedigree Method	Pakistan	Evolved Basmati

Table 35.4: Basmati Quality Rice Varieties Released in India till 2008

IET no.	Variety	Parentage	Grain type	Flowering duration (days)	Year
Central release					
10367	Haryana Basmati 1	Sona/Basmati 370	LS	110	1991
18990	Improved Pusa Basmati 1	PB 1/PB 1/IRBB 55	LS	105	2007
8580	Kasturi	Basmati 370/CR 88-17-1-5	LS	100	1989
10364	Pusa Basmati 1	Pusa 150/Karnal Local	LS	105	1989
17251	Pusa RH 10	Pusa 6A × PRR 78	LS	95	2001
16310	Pusa Sugandh 2	Pusa 1238-1/Pusa 1238-81-6	LS	105	2001
16313	Pusa Sugandh 3	Pusa 1238-1/Pusa 1238-81-6	LS	105	2001
17021	Pusa Sugandh 5	Pusa 3 A × Haryana Basmati	LS	102	2004
16775	Sugandhamati	Pusa Basmati 1 × IET 12603	LS	114	2004
15391	Vasumathi	PR 109/Pak.Basmati	LS	113	2001
14720	Yamini	BR4-10/Pak.Basmati	LS	101	2001
18990	Improved Pusa Basmati (Pusa 1460)	Pusa Basmati-II/Pusa Basmati-1/IRBB55	LS	108	2007
18005	Pusa Basmati 6 (Pusa 1401)	Pusa Basmati-1/Pusa 1121-92-8-2-7-1	LS	119	2008
State release					
Andhra pradesh					
15824	Sumati	Chandan/Pak. Basmati	LS	105	2001
Bihar					
	Kamini	Pure line selection from Katarni rice	LS	105	1991
Gujrat					
	GR 101	IR 8/Pankhari 203	LS	100	1984
Himachal Pradesh					
8580	Kasturi	Basmati 370/CR 88-17-1-5	LS	100	1989
15060	Hasan Sarai	Introduction from Iran	LS	95	2000
Haryana					
	Taroari Basmati	Pure line selection from local Basmati	LS	105	1996
Jammu & Kashmir					
11348	Ranbir Basmati	Selection from Basmati 370	LS	85	1995
Jharkhand					
18028	Birsamati	IR 36/BR 9	LS	130	2003
Karnataka					
13549	Mugad Sugandha	Selection from Basmati composite	LS	100	2001
Madhya Pradesh					
3785	Madhuri	7aya/RII	LS	84	1980
Maharashtra					
	Ambica	MAU Sel-1	LS	88	1984
12897	Indrayani	Ambemohar 157/IR 8	LS	102	1987
11338	Pawana	IR8/Pusa 33	LS	90	1988
10651	Sakoli-7	TN 1/Basmati 370	LS	100	1988
9296	SYE-ER 1	Sonal/SYE 44-3	LS	185	1990
	SYE-2		LS	105	1985
13549	Bhogavati	Selection from Basmati composite	LS	100	2004
New Delhi					
18004	Pusa Sugandh 4 Renamed as Pusa Basmati 1121	P 614-1-2/P 614-2-4-3	LS	105	2003
Orissa					
	Geetanjali	Mutant of Basmati 370	LS	100	2005
Punjab					
13185	Basmati 385	TN 1/Basmati 370 (introduction from Pakistan)	LS	105	1992
14710	Basmati 386	Set from Pakistan Basmati	LS	105	1994
7313	Punjab Basmati 1	Sona/Basmati 370	LS	95	1982
18031	Super Basmati	IR 662/Basmati 320	LS	110	2003
	Basmati 217[a]	Selection from Local	LS	115	1962
Rajasthan					
7028	BK 79	TN1/NP 130//Basmati 370	LS	100	1981
	Khoshboo	Baran Basmati/Pusa 150	LS	100	1994
12601	Mahisugandha	BK 79/Basmati 370	LS	100	1994
Tamil Nadu					
	Pusa 33[a]	Improved Sabarmati/Ratna	LS	80	1975
13544	JJ 92	Selection from Basmati 370	LS	80	1993
Uttar Pradesh					
	Type 3	Selection from Basmati Deharadun	LS	115	1973
19493	MAUB 21 (Vallabh 21)		LS	130	2007
Uttarakhand					
14132	Pant Sugandh Dhan 15	Basmati 370/Sadri//Baharl/Muskan 41	LS	114	2002
17263	Pant Sugandh Dhan 17	Pusa Basmati 1/UPRI 95-154	LS	105	2004

Source: Shobarani et al. [28]

Rice that possess a scent in their plant parts and grains are known as scented rice. These rice emit the scent in the field and during milling, and retain the scent in storage, and cooking (Gibsion, 1976; Jafferson 1985). Scent is not restricted to any specific type and occurs in indica and japonica, glutinous and non-glutinous rices, long and short, course and fine grain types and in black, red and white rice. Scent is however, conspicuously absent in all wild species. Scented rice were know since the ancient times, and were considered the most among the specialty rice. Through out the World, they have been the choicest food of kings royalty, the elite of society, as well as the common man. Scented rice have always had a special place in countries where rice is the staple food. For example Japan, home grown Koshihikart (scented rice) is considered worth its weight in gold. Black scented rice were a favorite of the royalty in China.

Scented rice were known since the ancient times and were considered the best among the specialty rice. Throughout the world, they have been the choicest food of kings, royalty, the elite of society, as well as the common man. Some important scented varieties of Indian include the basmati of northeastern India, Kalanamak of Uttar Pradesh, Dubraj and Chinoor of MP, Ambemohor of Maharashtra, Radhunigopal of West Bengal, Jeerakasambha of Tamil Nadu, Gandhakasala of Kerala and Kalajira and Badshabhog of Odisha.

Conclusion

- ✰ Basmati rice of the Indian sub-continent, despite being the most expensive, is much in demand in the domestic and international markets. Taking advantage of the growing demand, premium price offered and the tariff concession available from the European Union, there is ample scope to expand the area under Basmati cultivation and thereby increase the export volume and forex earnings. Such strategy, however, warrants priority attention to certain research and development and policy related issues and associated problems.
- ✰ Given the kind of hassles the country had to pass through in challenging the patent protection obtained by an American Company on the ground that the name 'Basmati' is generic, the foremost issue to be addressed is legally protecting 'Basmati' rice as a GI. Evolved Basmati varieties meeting all the key quality traits can be designated as "Basmati".
- ✰ This being a policy issue, the Ministry of Commerce, in consultation with Indian Council of Agricultural Research (ICAR), should take appropriate decision in the interests of expanding export potentials. They would have, as we experience now, a place of their own as long as the consumer world looking for relatively less expensive Basmati rice exists.
- ✰ Further, to sustain the farmer's profit of margin, production, and export, Government needs to enhance the minimum support price of Basmati rice. As for the expansion of area under the traditional Basmati rice, no matter how high commercial value they enjoy and forex earn because of certain trait(s) of aesthetic value, policy makers should exercise caution in

promoting such a strategy, keeping in view the great challenging goal of making the country food and rice secure.

- ✰ While restricting the area under very low yielding traditional Basmati varieties to what is planted to now, we should rationally plan our future rice research and development strategies giving emphasis to evolve stress tolerant and nutritionally rich Basmati quality varieties and hybrids in high yielding backgrounds. Genomic tools especially molecular marker technology now available is used increasingly in breeding such value added varieties.
- ✰ Again, area under such cross-bred Basmati quality varieties be based on demand position in the domestic and international markets. The foresaid research priority should not mean that we neglect improvement of traditional Basmati rices. Absence apparently of needed variability for yield enhancement, defending against biotic stresses etc. could be overcome by strengthening diversity survey and screening for still uncovered traits of value using state of the art genomic tools available today.

Chapter 36

Potentials of Enterprise Generation from Plant Biodiversity and Farmers' Patent Rights

Samir Kumar Samanta

Plant genetic resources (PGR) and its management in appropriate way indicate the status of sustainable development in agriculture particularly staple food production system in any country. In India major initiatives were taken during implementation of National Agricultural Technology Project (NATP) through NBPGR. Besides government initiatives public awareness and continuous fight by different organizations in international arena gained momentum in preserving our plant diversity. In this context many aspects of genetic diversity important for every people living in India are discussed.

1. Genetic Diversity Helps to Ensure the Survival of Species

Biological diversity (also known as biodiversity) is the total variability within all living organisms and the environments they inhabit. It includes diversity among ecosystems, among species, and within species (known as "genetic diversity").

The genetic diversity that exists in a staple food crop or a population of trees in a forest is the result of processes of mutation and recombination of genes over long periods of time, influenced by generations of farmers and breeders. As a source of genes for natural and human selection, this diversity is one of humanity's best weapons against poverty and food insecurity. In the field, genetic diversity also gives species the ability to adapt to changing climates and environments, and to new pests and diseases. Many plants are susceptible to environmental stresses. But if a plant population is genetically mixed, some individuals may be able to withstand these stresses, allowing the species as a whole to survive - in effect, Darwin's survival of the fittest. Agriculture depends on the diversity of relatively few plant and animal species. Of the approximately 250 000 identified and described plant species, some 7000 can be used for food but just three (maize, wheat and rice) supply almost 60 per cent of the calories and protein in the human diet. Given its enormous dependence on so few food species, humanity depends very heavily on the diversity within these species for its survival. In many cases, this diversity is considerable. For example, farmers have developed about 100 000 different varieties of rice over time. Nevertheless, the world's genetic diversity is in danger of being lost. With the advent of modern agriculture, untold numbers of locally adapted crop landraces and traditional varieties have been replaced by the widespread use of genetically uniform high-yielding modern varieties. In China, for example, between 1949 and 1970 the number of wheat varieties cultivated by farmers dropped from approximately 10 000 to 1 000. Activities such as overgrazing and changes in land use, including the conversion of natural ecosystems to farmland, are also taking a toll on the diversity still available in the wild relatives of crop plants and other wild species.

Extinction is a natural process. Species have been created, flourished and have died out over the aeons. What is alarming is that, owing largely to human activity, today's rate of extinction is thousands of times greater than the rate at which new species appear. This threatens the life-sustaining systems of our planet. As human populations increase, environmental problems are likely to intensify. Climate change, deforestation, desertification and other processes could have serious consequences for crop productivity. By conserving genetic diversity where it occurs naturally (*in situ*), we can facilitate the natural processes of evolution and adaptation, thus mitigating the effect of these problems. By conserving genetic diversity in genebanks (*ex situ*), we can ensure the continuing availability of a broad range of genes for crop improvement or for reintroduction in case of loss. Using this diversity, farmers and professional breeders can create crop varieties that are better able to adapt to changing environmental conditions, pests, and diseases and thus ensure that we are able to meet the future food needs of the world's rapidly growing population.

2.Plant Genetic Resources Increase Options for Environmentally Sound Agriculture

Present decades have observed rapid developments in agricultural practices in order to meet the increasing world population for increased earnings as well as for food, feed, fibre and related basic needs. In developing countries like India it has been possible only due to expansion of cultivable area, increased cropping intensity through increased facility of irrigation and most importantly genetic improvement and release of high yielding varieties combined with fertilizer use and increased nitrogen use efficiency meaning thereby more yield.

3. Plant Genetic Resources are Needed to Meet Challenges to Human Survival

Considering the projected population growth in the country about 60 per cent more food would be required during next 30 years. Alternatively as much food will be needed in the next three decades as has been produced in the past 10 000 years.

Biotechnology is opening up exciting new possibilities for meeting the challenges of food production. New molecular techniques can help to tap the valuable genetic potential in crop wild relatives and landraces. Such techniques allow scientists to search plant genomes for the genes responsible for desirable traits, such as drought and pest resistance, yield and nutritional characteristics. Genetic engineering may help solve problems that have proved intractable using conventional breeding approaches. These include the development of crop varieties with resistance or tolerance to pests and diseases. Using genetic mapping techniques, it is possible to pinpoint individual genes on their particular chromosomal location. These techniques have already been used to locate genes controlling such properties as fruit weight in tomatoes and yield in maize.

A limitation of conventional breeding is the time needed for repeated back-crossing and testing in order to identify material containing the trait of interest and to weed out offspring containing unwanted "extra" genes. The use of DNA markers linked to characters can speed this process considerably, by enabling the quick selection of the offspring containing the desired genes. Such "marker assisted breeding" has been used to introduce malting quality and disease resistance genes into barley, and scientists predict that it will be especially useful in future for "pyramiding" multiple resistance genes into crop lines to generate more broad and durable resistance.

Advances in molecular genetics have also provided plant conservationists with a range of new techniques for their work. Genebank managers and conservationists concerned with both *in situ* and *ex situ* management endeavour to conserve the greatest possible genetic diversity of their target species. Molecular markers can provide detailed genetic information, which will significantly improve the effectiveness of these efforts. Such techniques will greatly improve procedures for locating, sampling, identifying, characterizing, monitoring and maintaining genetic diversity.

4. Plant Genetic Resources are a Vital Element of a Nation's Cultural Heritage

The diversity found in indigenous plants and traditional crops is a result of many years of adaptation to the environment, combined with careful management by local communities. The multipurpose Neem tree (*Azadirachta indica*), native to south and southeast Asia, is cultivated and revered in all tropical regions of the world. It yields a huge range of products including traditional medicines, a natural insecticide, fuelwood, animal fodder, cosmetics and fertilizer. In India alone, over 40 neem-based products are traded commercially. Rural people have spread the tree throughout Asia and beyond through deliberate planting and by creating conditions suitable for its regeneration and protection. There is a vast traditional knowledge and folklore about neem but systematic study and exploration of its genetic resources has only recently begun.

5. Coordinated Efforts are Essential for the Sustainable Conservation and Effective Use of Plant Genetic Resources

Both the Convention on Biological Diversity and Agenda 21 emphasize that the strategic needs of countries can best be met if national plant genetic resources programmes bring together the full range of experience and knowledge about genetic resources that exists in the country. National coordination is very important in order to make the best use of available human, financial and technical resources and to maximize the overall effectiveness of efforts throughout the country.

A key objective of this coordination should be to help promote the complementarily of local, national and international efforts as well as of strategies and plans of the formal and informal, public and private sectors. In many cases, this will require reconciling environmental concerns with development priorities. There is no one ideal model for a coordinated national plant genetic resources programme but the common elements of strategic planning, coordination and communication are likely to apply in all cases. The adjustment or even fundamental reshaping of decision-making mechanisms may be needed if environment, agriculture and development concerns are to be fully integrated in national economic and political decision making. The sustainability of all of these efforts will depend upon long-term financial commitment by governments. Financial commitment in turn needs to be matched by commitments from stakeholder groups to ensure that policies and plans are put into action.

6. Exchange of Information on Plant Genetic Resources is Necessary Both to Raise Awareness and to Allow the Transfer of Technology

The effective coordination of plant genetic resources efforts requires regular communication between stakeholders. Yet unfortunately, the flow of information among the many groups associated with conservation and use is often irregular and *ad hoc*. The weakness or, in some cases, complete absence of communication channels between genetic resources conservationists and users can forestall the creation of coherent national policies and strategies. It also makes it more difficult for them to learn about each other's priorities and needs, and limits the collaboration and the interdisciplinary activities that do so much to support the implementation of national policies and plans.

7. For Plant Genetic Resources to Contribute to Future Agricultural Development, they Must be Used as well as Conserved

Some national conservation programmes actively promote the use of plant genetic resources and have strong links to breeders. Unfortunately, the management of gene bank collections and protected areas in many countries is often not conducive to the use of the genetic resources they hold. Collections are often maintained in isolation from breeding programmes and other activities that promote use. Both Agenda 21, adopted in 1993, and the Global Plan of Action for plant genetic resources, adopted in 1996, recognize the importance of strengthening the links between the conservation and use of plant genetic resources and urge countries to develop strategies and plans that integrate both activities.

Many curatorial and research activities related to plant genetic resources collections (*e.g.* genetic diversity studies, characterization, evaluation, the establishment of core collections, pre-breeding) add value to the collections by making the conserved material, whether *ex situ*, *in situ* or on-farm, more accessible and easier to use. Sustained government support for these activities should be regarded as a long-term investment in a country's future.

8. National Policy and Legislation Affecting Plant Genetic Resources should Support both International Obligations and National Objectives

In recent years, numerous international policy fora, with different interests and objectives, have emerged or have expanded their mandates to include matters affecting the conservation and use of plant genetic resources. Yet, despite a significant overlap in the issues treated by these diverse bodies, there has been little effort to coordinate their efforts. The closely linked issues of access and benefit-sharing, for example, are currently under negotiation by the Conference of the Parties to the Convention on Biological Diversity, the Commission on Genetic Resources for Food and Agriculture, the World Trade Organisation, GATT/TRIPS and others. Only rarely do countries ensure continuity in representation to the various fora, nor have the different sectors been particularly successful in sharing with each other the developments in these fora.

As a result, national decision-makers are faced with a bewildering array of policy positions in related, but quite distinct fora. Achieving clarity and harmony among these positions is critical to the development of a coherent national framework. Ensuring such harmony is one of the most crucial tasks facing governments.

9. Why Plant Genetic Resources Matter

Plants are vital for the development of human society. They are the key to food security, providing us with cereals and other foods in addition to feeding the livestock that produce milk, meat and eggs and provide farm power. The wise use of plant genetic resources can help eradicate poverty, both by lowering the price of food and other commodities and by raising and diversifying the incomes of producers and processors. They also protect and enhance the environment, for example by preventing erosion and desertification and by absorbing atmospheric carbon.

The multiple uses of plants depend on a crucial characteristic of plant life, its diversity. Diversity exists at three main levels: the combinations of species that make up different ecosystems, the number of different species, and the different combinations of genes in species. All three levels help sustain agricultural systems, as well as ensure their productivity. Genetic diversity, in particular, provides species with the ability to adapt to changing stresses, such as pests and diseases or drought.

Today, the diversity of the Earth's plant life is under threat as never before. In agriculture, the widespread adoption of a few improved varieties has narrowed the genetic base of important food crops and led to the disappearance of hundreds of landraces. In the forests, up to 8 per cent of all plant species are expected to disappear over the next 25 years as deforestation continues.

Conserving and using plant genetic diversity is vital to meeting the world's future development needs. The earth's population is predicted to double, perhaps even triple, before it stabilizes late in the 21st century. This will place enormous pressure on the environment, and will require a huge increase in the production of food and non-food commodities.

10. National Sovereignty over Plant Genetic Resources Implies Government Responsibility for Conservation and Use

Public sector efforts will be most effective if governments concentrate on areas that are not of great interest to the private sector and are thus sometimes overlooked, such as the needs of resource-poor farmers in marginal farming ecosystems. In addition to conservation in genebanks, specific government actions might include support for conservation efforts by local farming communities who maintain genetic resources with valuable adaptive characteristics.

Efforts on Combating Biopiracy: 06 May 2005

India's 10-year-long battle against the grant of a patent on the use of neem as a fungicide has finally been won at the European Patent Office (EPO). Neem's fungicidal properties are part of the 'Traditional Knowledge' (TK) base of this country. Nonetheless, the patent rights on a method for controlling fungi on plants by the aid of hydrophobic extracted neem oil had originally been granted by the EPO in September 1994 to the United States Department of Agriculture (USDA) and the New York-based multinational agribusiness corporation WR Grace.

In June 1995, a legal opposition to the grant of this patent was filed jointly by Dr Vandana Shiva, director of Delhi-based Research Foundation for Science, Technology and Ecology (RFSTE), Ms Magda Aelvoet of the Green Group in European Parliament and Ms Linda Bullard of International Federation of Organic Agriculture Movements (IFOAM). Their joint legal opposition claimed that the fungicidal properties of the Neem tree had been public knowledge in India for centuries thereby nullifying the novelty (one of the criteria for patentability) claimed to be involved in the aforesaid patented subject matter. Consequently, the EPO revoked the patent in May 2000. The MNC, however, appealed the patent revocation. It is against this appeal that the EPO has finally upheld its earlier decision, thereby revoking in its entirety the patent rights once and for all on March 8, 2005.

Unfortunately, this is not the only instance of such biopiracy of traditional Indian knowledge. Instead, several cases of biopiracy have come to the fore in the past as well. The patents granted for instance, on wound-healing properties of 'haldi', or on anti-diabetic properties of `karela', `jamun', and brinjal, are some such examples of blatant biopiracy of Indian TK that have generated considerable hue and cry in not-too-distant past.

— Part V —

Motivation and Communication: The Lifeline for Extension Manager

Chapter 37

Theories of Motivation: Planning on Entrepreneurial Behaviour

Manas Mohan Adhikary and Sankar Kumar Acharya

Key Theories of Motivation

Motivation is the driving force behind human behaviour.

Motivation and human management

Motivated people are those who have made a conscious decision to devote considerable effort in achieving something that they value. What they value will differ greatly from one individual to another. There are a variety of ways to motivate people, including the fear of losing a job, financial incentives, self fulfillment goals and goals for the organization or groups within the organization.

The traditional view - finds some of its origins in the work of Taylor and the school of scientific management. At its most extreme, this view postulates the following:

- people dislike work
- people will only work for money
- people are not capable of controlling their work or directing themselves
- simple, repetitive tasks will produce the best results
- workers should be closely supervised and tightly controlled
- extra effort must lead to greater reward
- people will meet standards if they are closely controlled
- firm but fair supervision will be respected

Taylor took the view that there is a right (meaning best) way to perform any task.

It is management's job to determine the right way. Workers gain from this approach because the 'right way' is easier and pay is enhanced as a result of increased productivity.

The human relation view - originates in the work of Mayo, known as the Hawthorne Studies. The series of studies essentially concluded that the strongest motivational force behind most employees' behavior at work was the preservation and nurturing of social relationships with their colleagues. The main tenets of this view are as follows: people want to be made to feel valued and important people want recognition for their work people want to be controlled sensibly managers must discuss the plans they make for staff they must take any objections on board they must encourage self-regulation on routine tasks.

Mayo's work leads to an approach towards people which encourages contribution and self-direction, advocating full participation on matters of significance in order to improve the quality of decisions made and the nature of supervision.

Theory X and Theory Y Managers

- Douglas McGregor proposed a Theory X and Theory Y model to explain basic human traits. Theory X assumptions are:
- The average human being has an inherent dislike of work and will avoid it if he can.
- Most people must be coerced, controlled, directed or threatened with punishment to get them to expend adequate effort towards the achievement of organizational objectives.
- The average human being prefers to be directed, wishes to avoid responsibility, has relatively little ambition, and wants security above all.
- On the other hand, Theory Y makes different assumptions about the nature of people:
- Expenditure of physical and mental effort in work is as natural as play or rest.
- The average human being does not inherently dislike work, which can be a source of satisfaction.
- External control and the threat of punishment are not the only means of bringing about effort. People can exercise self-direction to achieve objectives to which they are committed.
- Commitment to objectives is a result of the rewards associated with their achievement. The most significant of those rewards is satisfaction of the self actualization needs.
- The average human being learns, under proper conditions, not only to accept, but to seek, responsibility. Avoidance of responsibility, emphasis on security and low ambition are the result of experience and are not inherent in man's nature.

- ☆ Capacity to exercise a relatively high degree of imagination, ingenuity and creativity in the solution of organizational problems is widely, not narrowly, distributed in the population.
- ☆ Under conditions of modern industrial life, the intellectual potential of the average human being is only partially utilized.

Needs and Goals

These various needs can be categorized in a number of ways eg, physiological and social motives or intrinsic and extrinsic motivation.

Extrinsic motivation is related to tangible rewards such as salary and fringe benefits, security, promotion, contract of service, the work environment and conditions of work.

Intrinsic motivation is related to 'psychological' rewards such as the opportunity to use one's ability, a sense of challenge and achievement, receiving appreciation, positive recognition and being treated in a caring and considerate manner.

Goals also influence motivation.

Classical and Modern Theories of Motivation

Content theories focus on the question of what arouses, sustains and regulates goal directed behaviour *i.e.*, the particular things that motivate people. They offer ways to profile or analyze individuals to identify their needs.

Often criticized as being static and descriptive they appear to be linked more to job satisfaction than to work effort. Maslow, Herzberg and McGregor take a universal approach whereas McClelland and Argyris list forces and drives that will vary in relation to different individuals.

Process theories attempt to explain and describe how people start, sustain and direct behaviour aimed at the satisfaction of needs or the reduction of inner tension. The major variables in process models are incentive, drive, reinforcement and expectancy. The best-known work in this area has been concerned with Vroom's expectancy theory, Handy's motivation calculus and Adam's equity theory.

Maslow's Hierarchy of Needs

Maslow advanced the following propositions about human behaviour:

Man is a wanting being.

A satisfied need is not a motivator of behaviour, only unsatisfied needs motivate.

Man's needs are arranged in a series of levels - a hierarchy of importance. As soon as needs on a lower level are met those on the next, higher level will demand satisfaction. Maslow believed the underlying needs for all human motivation to be on five general levels from lowest to highest, shown below.

Within those levels, there could be many specific needs, from lowest to highest.

- ☆ Self
- ☆ Ego

- Social
- Safety/Security
- Basic/Physiological fulfillment

Maslow's hierarchy of needs Related aspects at work

- Challenging job
- Creative task demands
- Advancement opportunities

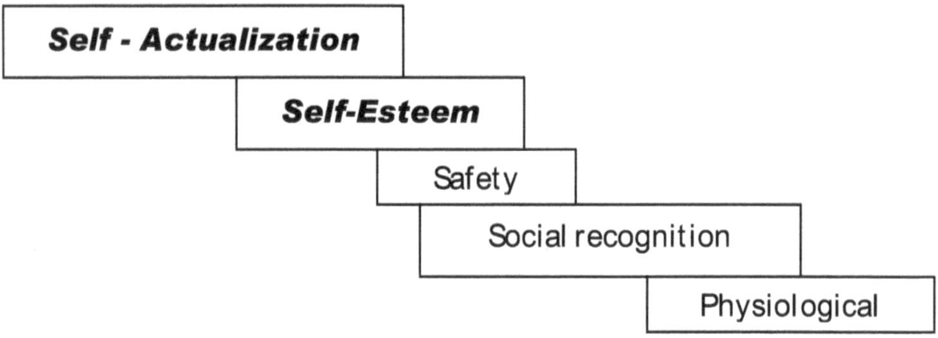

Figure 37.1: Maslow's Hierarchy of Needs Related Aspects at Work.

- Achievement in work
- Merit pay increase
- High status job title
- Compatible work group
- Friendships at work
- Fringe benefits
- Job security
- Basic salary
- Safe working conditions

Physiological - the need for food, drink, shelter and relief from pain.

Safety and security – once the physical needs of the moment are satisfied, man concerns himself with protection from physical dangers with economic security, preference for the familiar and the desire for an orderly, predictable world.

Social - become important motivators of his behaviour.

Esteem or egoistic - a need both for self-esteem and the esteem of others, which involves self-confidence, achievement, competence, knowledge, autonomy, reputation, status and respect.

Self-fulfillment or self-actualization – is the highest level in the hierarchy; these are the individual's needs for realizing his or her own potential, for continued self-development and creativity in its broadest sense.

Remember the assumptions of Maslow's hierarchy: individuals have multiple needs are ordered into levels, creating a 'hierarchy' a need, once satisfied, is no longer a need.

To be of use, **Maslow's basic theory** needs qualification to include the individual as a determining factor in motivation and behaviour. These include:

Levels in the hierarchy are not rigidly fixed; boundaries between them are indistinct and overlap.

There are individual exceptions to the general ranking of the hierarchy. Some people never progress beyond the first or second level (for example, many inhabitants of the third world), others are so obsessed with the higher needs that lower ones may go largely unnoticed.

Variables apart from individual needs may motivate eg, social standards and a sense of duty.

An act is seldom motivated by a single need; any act is more likely to be caused by several needs.

The same need will not give rise to the same response in all individuals.

Substitute goals may take the place of a need that is blocked.

Herzberg's Theory of Motivation

The major finding of the study was that the events that led to satisfaction were, not surprisingly, of a quite different kind from those that led to dissatisfaction.

One set of factors are those which, if absent, cause dissatisfaction. These factors are related to job context. They are concerned with job environment and are extrinsic to the job itself.

They are called 'hygiene' factors and include such elements as: company policies and administration supervision working conditions interpersonal relations money, status and security.

The other set of factors are those which, if present, serve to motivate the individual to superior effort and performance. These factors are related to the job content of work. They are 'motivators' or growth factors. Motivation factors include: achievement increased responsibility challenging work recognition for achievements growth and development

David McClelland

David McClelland (Studies in Motivation, 1955) identified three basic types of motivating needs present in people. He shows that all three needs can be present in a person but the weight attached to each can vary. The three needs are:

(a) **Need for achievement** - where this is high then people have an intense desire to succeed and an equally intense fear of failure.

(b) **Need for affiliation** - where this is high people tend to seek acceptance by others, need to feel loved and are concerned with maintaining pleasant social relationships.

(c) **Need for power** - people with a high need for power seek opportunities to influence and control others, seek leadership positions and are often articulate, outspoken and stubborn.

C Argyris

Even though management based on the assumptions of **Theory X** is perhaps no longer appropriate in the opinion of McGregor and others, it is still widely practiced. Consequently a large majority of the people in organizations today are treated as immature human beings in their working environment.

In attempting to analyse the situation, Argyris compared bureaucratic/pyramidal values (the organisation's counterpart to Theory X assumptions) that still dominate most organisations with a more humanistic/democratic value system (**Theory Y** assumptions about people).

Expectancy Theory

The common themes in expectancy theories are: conscious decisions by individuals to behave in certain ways individual values with regard to choosing desired outcomes individual expectations concerning the amount of effort required to achieve specific outcome individual expectations concerning the probability of being rewarded for achieving a desired outcome.

Motivation is a function of the relationship between: effort expended and perceived level of performance the expectation that rewards (desired outcomes) will be related to performance. There must also be the expectation that rewards are available. These relationships determine the strength of the motivational link. the strength of the individual's preference for an outcome the belief in the likelihood that particular actions will achieve the required goal

This process may be illustrated in the following way:

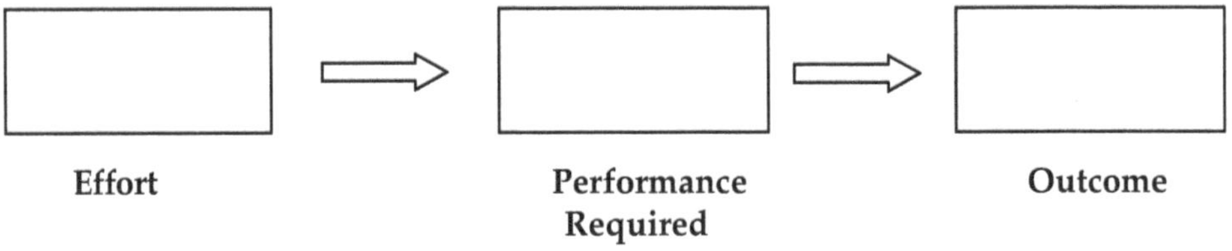

Force = Valence × Expectancy

Force is Strength of Motivation

- ☆ Valence is strength of preference for an outcome.
- ☆ Expectancy is the level of belief that changes in behaviour will achieve the required outcome.
- ☆ The recommendations to management that go with this model are outlined as follows:
- ☆ Discover what outcome each employee values most.

- ☆ Define for employees the kinds of performance that are desired or required, ie explain what constitutes a 'goal' and 'adequate performance'.
- ☆ Ensure that the desired levels of performance are achievable.
- ☆ Link the outcomes desired by employees to the specific performance desired by management.
- ☆ Ensure that the overall motivation strategy avoids conflict between the positive expectations it seeks to create and other factors in the work situation.
- ☆ Make sure that outcome or rewards are sufficiently attractive to motivate the desired level of performance.

Porter and Lawler's Model

Vroom's theory has been extended by the findings of Porter and Lawler (Managerial Attitudes and Performance, 1968) in devising a more complete model of motivation for management.

Basically, Porter and Lawler's model (shown below) shows that the amount of effort generated depends upon: the value of the reward the amount of effort seen to be necessary the probability of receiving the reward. The amount of effort deemed necessary and the probability of receiving the reward are in turn influenced by the individual's record of performance to date, and range of skills, personality, perception of his role, and any number of other environment factors.

- ☆ Value of rewards
- ☆ Probability of achieving reward
- ☆ Effort
- ☆ Role perception
- ☆ Performance
- ☆ Skills, abilities, personality

Remember that rewards may be:

(a) External rewards that are given by others and form part of the job situation (eg wages, status, security)

(b) Intrinsic rewards which the individual manager awards himself. These arise from the performance of the tasks (eg feelings of self-esteem, accomplishment)

Handy's Motivational Calculus

The idea put forward here is that each person has a specific 'motivation calculus' in respect of every decision taken – and this can be on a conscious or a subconscious basis. This assesses three factors:

- ☆ **Needs** - these may be defined in accordance with the ideas of Maslow or any other researchers, and they are the person's needs at that time.
- ☆ **Desired Results** - these are what a person is expected to accomplish in the work.

- **Expenditure (E) factors** - these relate to the expenditure of effort, energy, and excitement in attaining the desired results. Handy suggests that motivational theories have been too preoccupied with 'effort'. He notes that there seems to be a set of words (coincidentally beginning with 'e') that might be more helpful. As a result of this assessment we have a 'motivation decision'. This is the strength of the motivation to achieve the desired results. It will have to be dependent upon: the strength of the needs of the individual the expectancy that the expenditure will achieve the desired results the expectancy that these desired results will assist in the satisfaction of the needs. Charles Handy therefore suggests that: each individual must know exactly what is expected he or she should participate in the setting of the targets in accordance with personal objectives feedback, on a regular basis, is necessary to inform individuals of their performance in relation to the targets

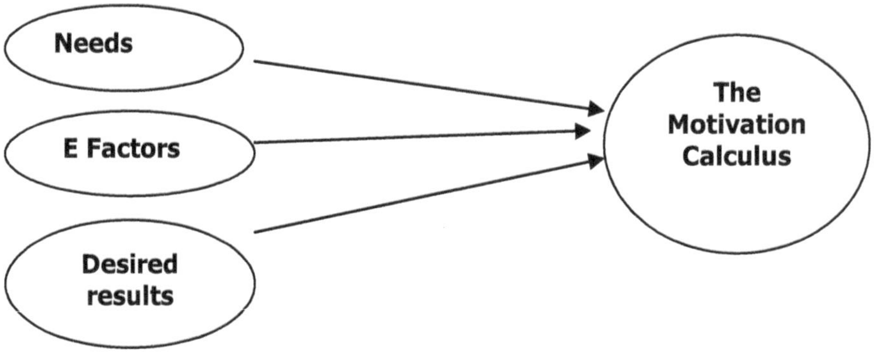

Adams' Equity Theory

When people sense inequities in their work they will be aroused to remove the discomfort and restore a state of felt equity to the situation by: changing work inputs changing rewards received leaving the situation quantity or quality of their work, whilst those who are underpaid (feel negative inequity) do the opposite. Feelings of inequity are determined solely by the individual's interpretation of the situation - the fact that a manager feels that the annual pay review is fair is immaterial.

Content and Process Theories of Motivation

Content and Process Theories

Content theories - such as those of Maslow and Herzberg, stress the satisfaction of needs.

Process theories - such as those of Vroom, emphasize the importance of rewards. Content theories attempt to explain those specific things that actually motivate motivates. There is the assumption that everyone responds in much the same way to motivating pressures and that there is, therefore, one best way to motivate everybody. These theories provide a prescriptive list which managers can follow in an attempt to increase productivity.

Process theories (expectancy and goal) change the emphasis from needs to the goals and processes by which workers are motivated. They attempt to explain and describe how people start, sustain and direct behavior aimed at the satisfaction of needs or reduction of inner tension. They place emphasis on the actual process of motivation. Process theories also attempt to identify major variables that explain behavior, but the focus is on the dynamics of how the variables are interrelated in explaining the direction, degree and persistence of effort. The major variables in process models are incentive, drive, reinforcement and expectancy.

Ways to Motivate Staff

Job Satisfaction

Drucker suggested that employee satisfaction comes about through encouraging - if need be, by 'pushing' - employees to accept responsibility. There are four ingredients to this: careful placement of people in jobs high standards of performance in the job providing the worker with the information needed to control his/her own performance, opportunities for participation in decisions that will give the employee managerial vision designing rewarding jobs etc.

Herzberg Defines Three Avenues to Improve Staff Satisfaction and Motivation

Job enrichment - is a deliberate, planned process to improve the responsibility and challenge of a job. A job may be enriched by: giving it greater variety (although this could also be described as job enlargement) allowing the employee greater freedom to decide how the job should be done encouraging employees to participate in the planning decisions of their managers ensuring that the employee receives regular feedback on his/her performance.

Koontz and Weihrich identify four elements that are necessary to make job enrichment a practical and worthwhile exercise:

There has to be a clear understanding of people's needs. If the purpose of enrichment is to increase productivity then it must be shown, at the earliest stage that workers will share in the benefits. People need to be treated as individuals and involved in the planning and introduction of the schemes. Staff like to feel that managers are committed to the success of the programme. So managers must be seen to be enthusiastic and take part in all stages. Job enlargement - is an attempt to widen jobs by increasing the number of operations in which a job-holder is involved. Job enlargement is a 'horizontal' extension of an individual's work, whereas job enrichment is a 'vertical' extension.

A well-designed job should involve the following:

- ✰ Giving the individual scope for setting his/her own work standards and targets
- ✰ Giving the individual control over the pace and methods of working
- ✰ Varying the work to encourage efficiency

- Giving the individual a chance to add his/her comments about the design of the product or of the job
- Providing feedback to the individual about his/her performance.

Job rotation - is the planned rotating of staff between jobs to alleviate monotony and provide a fresh job challenge. It can take two forms: An employee might be transferred to another job after a longish period (2-4 years) in an existing job to give a new interest and challenge. Job rotation might be regarded as a form of training where employees learned about a number of different jobs by spending six months or one year in each job before being moved on.

Participation

There are certain guidelines that must prevail if participation is to be effective in raising motivation:

- The participation must be seen as part of a continuing approach not just a 'one-off' exercise.
- Staff must be fed the results of their involvement as quickly and fully as possible.
- The participation must be genuine.
- People must have the ability, equipment and will to be involved.

Quality of Work Life

The intention is to improve all aspects of work life, especially job design, work environment, leadership attitudes, work planning and industrial relations. It is an all-embracing systems approach, which usually starts with a joint management and staff group looking at the dignity, interest and productivity of jobs.

The Importance of the Reward System

Money

Whether in the form of wages, piecework, incentive pay, bonuses, stock options, company paid insurance or any of the other things given to people for performance, money is important. Economists, accountants and many managers tend to regard money as a prime motivator. However, behavioral scientists tend to place it low on the scale of motivators. Probably, neither view is correct as an all-embracing approach; it depends on the individual. Money in the form of pay is a powerful motivator, which can be related to the motivational theories that we have examined.

In discussing money as a motivator it is necessary to recognize its effects at two levels.

- Money in absolute terms, as an exact amount, is important because of its purchasing power. It is what money can buy, not money itself that gives it value.
- Money is also important as an indication of status.

Incentive Schemes

There are several features which are unique to money as a motivating force and which can affect workers in different ways:

- ☆ Money is more important to people who are seeking to establish an initial standard of living rather than those who have arrived.
- ☆ Schumacher defined his principle of motivation in which he states that if all efforts by the company are devoted to doing away with work by automation and computerization, then work is a devalued activity which people put up with because no other way has been found of doing it. People would therefore be working just for money.
- ☆ Most organizations use money not as a motivator, but simply as a means of ensuring adequate staffing.
- ☆ The need for apparent fairness encourages the use of salary grades and hence comparability of earnings.
- ☆ Financial incentives operate with varying strengths for different people in different situations.
- ☆ Much research has shown that money is not a single motivator or even a prime motivator.

Chapter 38

Communication, Enterprise Management and Value Addition

Jayanta Kumar Das

A large majority of farm youth across the country are losing confidence in farming. Among several reasons, the single most reason for this mindset is continuous decline in income derived from farming and consequently farming is no more considered as profitable profession.

Agripreneurship can contribute to the sustainable transformation of Indian agricultural sector from primarily food self-sufficiency to more of a market orientation in support of poverty alleviation and income generation.

Developing value chains involve all stakeholders (farmers, processors and traders) targeting loss prevention, value addition, rural employment and quality raw material and processed products. It has been found that entrepreneurship plays a crucial role in the development and well being of the society, as it creates jobs, drives and shapes innovation and promotes competition which in turn improves productivity. Agripreneurship works as a growth catalyst for agriculture.

Communication

- ☆ Communication is the life blood of the business. No business can develop in the absence of effective communication system.
- ☆ Communication is the mortar that holds an organization together, whatever its business or its size.
- ☆ When people within the organization communicate with each other, it is internal communication and when people in the organization communication with anyone outside the organization it is called external communication.

☆ Ability to work well in teams, to manage your subordinates and your relationship with seniors, customers and colleagues depends on your communication skill.

American management association defines," communication is any behaviour that results in an exchange of meaning.'

Peter Little defines," communication is the process by which information is transmitted between individuals and/or organizations so that an understanding response results".

Newman and Summer Jr. state that, "communication is an exchange of facts, ideas, opinions or emotions by two or more persons."

According to Keith Davis, "The process of passing the information and understanding from one person to another. It is essentially a bridge of meaning between the people. By using the bridge a person can safely across the river of misunderstanding."

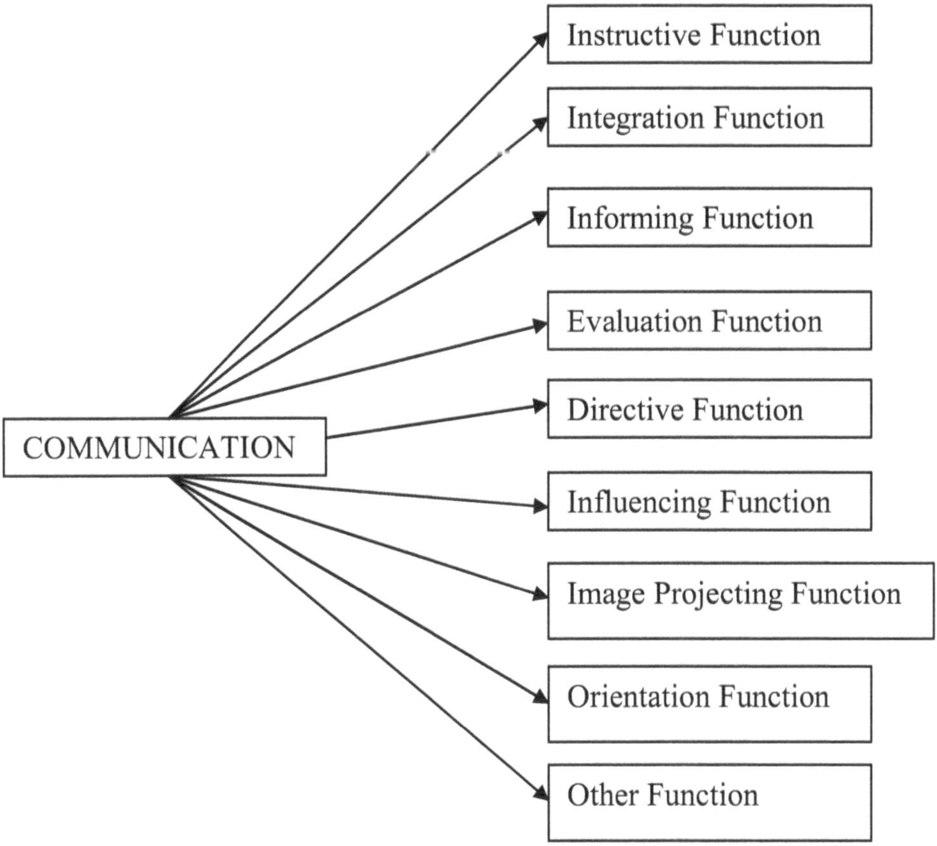

Figure 38.1: Purpose of Communication.

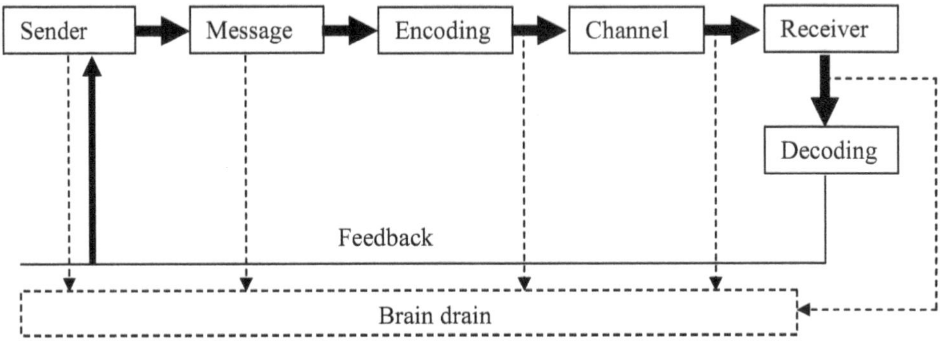

Figure 38.2: The Communication Process/Cycle.

Louis, A. Allen defines," Communication is the sum total of all the things that a person does, when he wants to create an understanding in the mind of another. It involves a systematic and continuous process of telling, listening and understanding."

Value Addition

Value added agriculture's potential lies in creating long-term solutions rather than short-term fixes. It allows for an increase in the amount of income flowing into the pockets of farmers and rural residents. By doing so, production agriculture will rely on market forces. Value added agriculture provides an alternative growth strategy that allows the opportunity to expand by moving upward in the food chain rather than expanding horizontally. This opportunity is available because the food supply chain is becoming more integrated. By moving aggressively, farmers have the opportunity to play a role in the development of this new integrated system. This opportunity will provide farmers a greater role in the ownership and control of the new system. However, because of its long-range focus, it takes time to implement. If you are considering any of the value-added opportunities that are available, you must remember that it may take a considerable amount of time before the benefits from these value-added activities are received.

In addition to new production skills, new business skills are also needed. Creating and operating a value added business venture is different than most farm businesses. For example, marketing a food product is much different than marketing a commodity. Product marketing involves identifying customers, designing and implementing a promotional plan, pricing your product, identifying and analyzing competitors and a host of other activities. You also will be involved in conducting feasibility studies and designing and implementing business plans. Other skills involve human relationships. As you move from commodity production to value added activities, relationships with others become more important. Value added activities often involve groups who work together to accomplish business objectives. Teamwork, communications and conflict management skills become more important. Working with customers, employees, producers, distributors, processors and others are important to your success.

- ☆ Leadership
- ☆ Entrepreneurial
- ☆ Decision making
- ☆ Business management
- ☆ Teamwork
- ☆ Strategic management
- ☆ Negotiation
- ☆ Financial and risk management
- ☆ Time management
- ☆ Marketing/selling
- ☆ Communication
- ☆ Organization and planning
- ☆ Conflict management
- ☆ Public relations
- ☆ Inter-personal
- ☆ Personnel management
- ☆ Stress management

Table 38.1: The Relationship between Agriculture and Micro-enterprise

100 per cent Farm Income	*Strong Farm Production Supplemented by non-farm business*	*Balance between Farm and Non-Farm*
Small holder agriculture with no or little business	Mixture of small scale agriculture	
No or Little land and no income generating activities	Viable-stable non-farm business	100 per cent income farm business

Income-Generating Activities Viable Stable Growth

Increasing Concentration on Business and Diversification

Figure provides a matrix for analyzing the relationship between agriculture and micro-enterprise. This tool developed out of research conducted in the southern region of Africa from 1995-2001, reflects the different options that households face in combining farm and non-farm activities and links these two different levels of livelihood security. While in country specific, the matrix provides useful insights in the needs of the rural poor. The vertical Y-axis of the matrix shows the level of household income from agriculture, and the horizontal X-axis shows the level of income from micro enterprise. The household's position on the matrix reflects the level of income from each of these two livelihood strategies. Households in the bottom left-hand corner are subsistence farmers with limited income from both agriculture and micro-enterprise. Depending on their objective, households can move up, along or diagonally across the matrix. Households that move up the Y-axis specialize in commercial agriculture at the expense of micro-enterprise. Households that move along the X-axis specialize in micro-enterprise at the expense of agriculture. Households along the diagonal tend to balance agriculture with micro-enterprise.

As the diagram illustrates, multiple approaches are required to meet the needs of producers in rural areas. At one end of the spectrum some farmers with little or no land and no income-generating activities will be simply surviving from day to day, and small farmers with mixed cropping and livestock arranging will just be coping. Others with strong farm production and off-farm income will adapt to new circumstances and show potential for immediate development. Still others will accumulate capital as a result of balance between farm and non-farm income. Whom to serve first, what capabilities to be developed, and what programmes to provide, as well as how to fund particular projects – these are hard questions that precede hard choices.

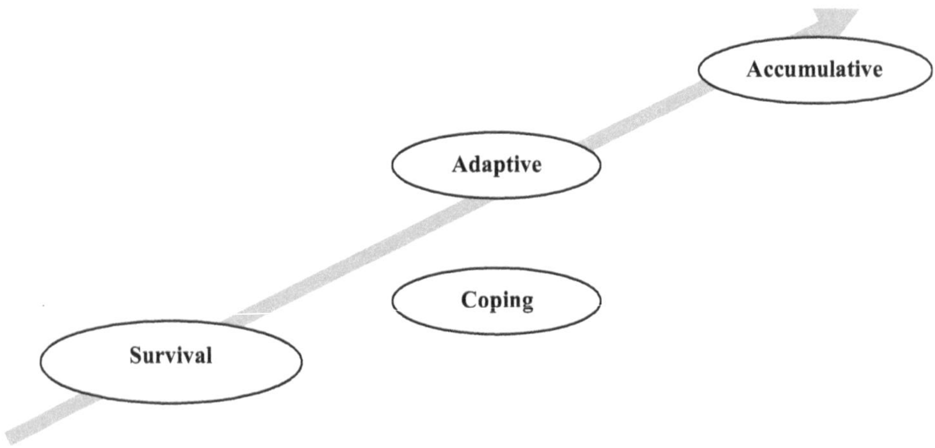

Business Communication – Mind Map

The Message

- ☆ Aim
- ☆ Purpose
- ☆ Type
    - Good news
    - Bad news
    - Information
    - Instruction
- ☆ Clear/confused
- ☆ Importance

The Medium

- ☆ Verbal (face to face)
- ☆ Paper based
- ☆ Electronic
- ☆ Image/Visual

Role of ICT

- ☆ Speed
- ☆ Accuracy
- ☆ Clarity
- ☆ Personal/impersonal
- ☆ Cost vs. Benefits

Value

- ☆ Motivation
- ☆ Clarity of line management
- ☆ Reflection
- ☆ Co-ordination and organization of the business

To Whom

- ☆ Formal
- ☆ Informal
- ☆ Staff
- ☆ Customers
- ☆ Shareholders
- ☆ Suppliers
- ☆ Local Community

Barriers

- ☆ Language
- ☆ Technical content
- ☆ Not recognizing the receivers needs
- ☆ Inadequate feedback
- ☆ Emotional interference
    - Emotions-anger, frustration, etc.
    - Body Language
- ☆ Degree of knowledge and expertise of both sender and receiver
- ☆ Quantity of information
- ☆ In appropriate medium
- ☆ Lack of trust/Honesty

- Cultural differences
- Poor listening skills
- Hierarchy

Developing a Strategic Communications Plan

In creating your strategic communications plan, you may wish to consider the following 10 sections:

1. Situation Analysis
2. Stakeholder Analysis
3. Strategic Business Objectives
4. Communications Objectives
5. Key Messages
6. Communication Strategies
7. Communications Tactics
8. Communications Tools
9. Budget
10. Timeline

References

NAIP (ICAR), News : Vol. 1 No.5, Aug.2012

Business communication An Introduction.www.newagepublisher.com

Should You Participate in Value-added Agriculture? Don Hofstrand, retired extension value added agriculture specialist.www.extension.iastate.edu

FAO Corporate Document Repository, Rural Development.html

Biz-Business Communication-Mind Map Biz.ed.html.

Developing a strategic communications plan, www.socialedge.org

Chapter 39

Knowledge Management: The New Age Approach

Sankar Kumar Acharya and Manas Mohan Adhikary

Introduction

What do we mean by knowledge?

- Drucker (1994): "The knowledge society will be more competitive than anything that we have seen so far."
- Why? With knowledge being universally accessible there will be no reason for poor performance.
- Cyert (1991): "The most crucial variable in economic development is Knowledge."
- Leonard-Barton (1995): "Organizations that are successful innovators are those that build and manage knowledge effectively through activities as developing shared problem-solving skill, experimentation, integrating knowledge across functional boundaries, and importing expertise from external sources."

Knowledge Management

- Ancient
- Collaboration at the organizational level
- Could revolutionize collaboration and computing

Opening Vignette: Knowledge Management Gives Mitre a Sharper Edge

- Mitre - knowledge management system (KMS) to leverage organizational knowledge effectively throughout the organization

- ☆ Internal marketing during development
- ☆ Supported at the highest level
- ☆ Provided an important application
- ☆ Organizational culture shift was critical
- ☆ Saved $54.91 million/invested $7.19 million

Knowledge Management

- ☆ Leverages intellectual assets
- ☆ Delivers appropriate solutions to anyone, anywhere
- ☆ Good managers have always done this
- ☆ Ancient concept

DSS Insights- GEM: A DSS for Workload-Planning Decisions Overview

- ☆ GEM a large stevedoring company
- ☆ Schedules developed a week ahead
- ☆ Each ship is expected to arrive within 10 days
- ☆ Unexpected conditions cause the schedule to be re- written

DSS Insights- GEM: A DSS for Workload-Planning Decisions

System Description

* means very important variable

- ☆ Ships
- ☆ ETA
- ☆ Cargo Information
- ☆ Ship's workload per location
- ☆ DWT
- ☆ Permitted berths
- ☆ Maximum number of elevators
- ☆ ETD*

DSS Insights- GEM: A DSS for Workload-Planning Decisions Berth

- ☆ Equipment information
- ☆ Availability of equipment
- ☆ Maximum permitted length
- ☆ Maximum permitted draught
- ☆ Maximum permitted DWT

DSS Insights- GEM: A DSS for Workload-Planning Decisions

- Other characteristics
- The planner can override the system
- Each ship has a max number of elevators which can be set by the planner

System Operation

- Run planning scenario with no penalties
- Study results
- If there are ships in an unfavorable position (ETD) - manipulate penalties to improve ships position
- Repeat until satisfactory

DSS IN FOCUS 9.1

THE IMPORTANCE OF MANAGING KNOWLEDGE

Knowledge or intellectual capital management is critically important because

- Intellectual capital is a firm's only appreciable asset. Most assets depreciate from the day of acquisition.
- Knowledge work is increasing, not decreasing. The service economy is growing, and so intellectual capital is growing in importance.
- Employees with the most intellectual capital have become volunteers because the best ones are relatively portable.
- Many managers ignore intellectual capital and lose out on the benefits of its capture and use.
- Employees with the most intellectual capital are often the least appreciated.
- Many current investments in intellectual capital are misfocused (e.g., knowledge is mismanaged).

Source: Adapted from Ulrich (1998).

DSS IN FOCUS 9.2

HOW TO ENCOURAGE KNOWLEDGE SHARING

Though knowledge sharing not difficult, many people do not understand what it is all about, do not see the value in it for their organization, or have been discouraged by organizational culture barriers. Here are some tips on how to convince people to share knowledge.

- Educate people on the value of knowledge.
- Revamp reward and recognition systems.
- Show people what knowledge sharing looks like.
- Let people know it's alright to make a mistake and not know something.
- Make knowledge sharing a requirement of the job.
- Educate people about the kinds of knowledge that are valuable and how they can be used.
- Make the technology work for people; don't expect the people to work for the technology.

Source: Adapted from Holloway (2000).

DSS IN FOCUS 9.3

WHY PEOPLE DON'T SHARE KNOWLEDGE

Even though it can cripple an organization, sometimes people refuse to share what they know. Usually organizational culture barriers create this situation. Here are some reasons of why people hoard their knowledge.

- People believe that knowledge is power and that hoarding it guarantees job security.
- People won't get credit for sharing knowledge or won't be able to *own* it anymore.
- People don't have time to share knowledge.
- People are afraid of making mistakes and looking bad or being reprimanded.
- The technology for sharing knowledge does not really meet their needs.
- People don't know how much they know.
- People don't realize that their knowledge is valuable to the organization.
- People don't know how to share their knowledge.

Source: Adapted from Holloway (2000).

Knowledge Management

Helps Organizations

- Identify
- Select
- Organize
- Disseminate
- Transfer

Important information and expertise within the organizational memory in an unstructured manner

Knowledge

As a form of capital, must be exchangeable among persons, and must be able to grow

Intellectual Capital- as the competence of an individual and the commitment of the individual to the organization's goals (competence * commitment)

Knowledge Management

Requires a major transformation in organizational culture to create a desire to share

Knowledge

- Information that is contextual, relevant, and actionable
- Knowledge is INFORMATION IN ACTION
- Higher than data and information

Knowledge Types

- Advantaged knowledge
- Base knowledge

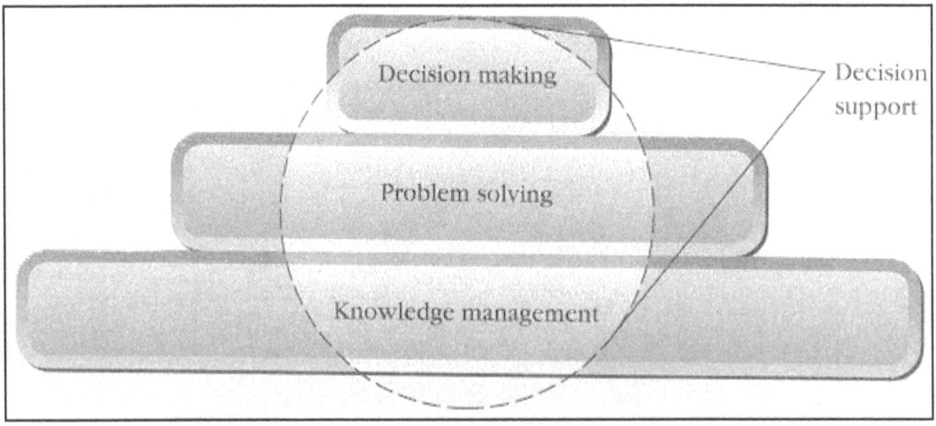

Figure: 39.1 The Knowledge-Management Foundation.

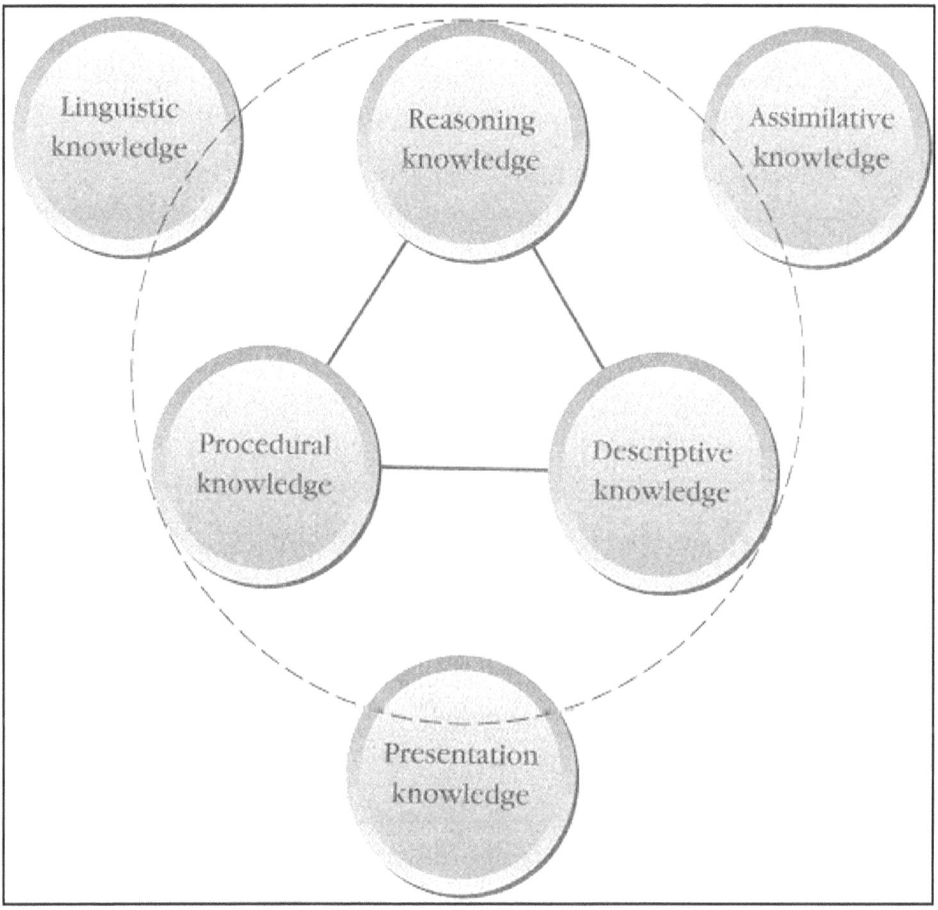

Figure: 39.2 Six Types of Knowledge.

- ☆ Trivial knowledge
- ☆ Explicit knowledge
- ☆ Tacit knowledge
- ☆ Advantaged Knowledge- Knowledge that provides competitive advantage
- ☆ Base Knowledge- Knowledge that is integral to an organization, providing it with short-term solutions (*i.e.* best practices)
- ☆ Trivial knowledge- knowledge that has no impact on the organization

Explicit Knowledge

- ☆ Objective, rational, technical
- ☆ Easily documented
- ☆ Easily transferred/taught/learned

Tacit Knowledge

- Subjective, cognitive, experiential learning
- Hard to document
- Hard to transfer/teach/learn
- Involves a lot of human interpretation

THREE VIEWS OF KNOWLEDGE

Knowledge is:

- A usable representation of something;
- A set of states ranging from raw data to decisions;
- The result of a production activity (i.e., learning) involving acquisition and/or derivation.

KNOWLEDGE QUALITIES

Knowledge Validity
- Accuracy
- Consistency
- Certainty

Knowledge Utility
- Clarity
- Meaning
- Relevance
- Importance

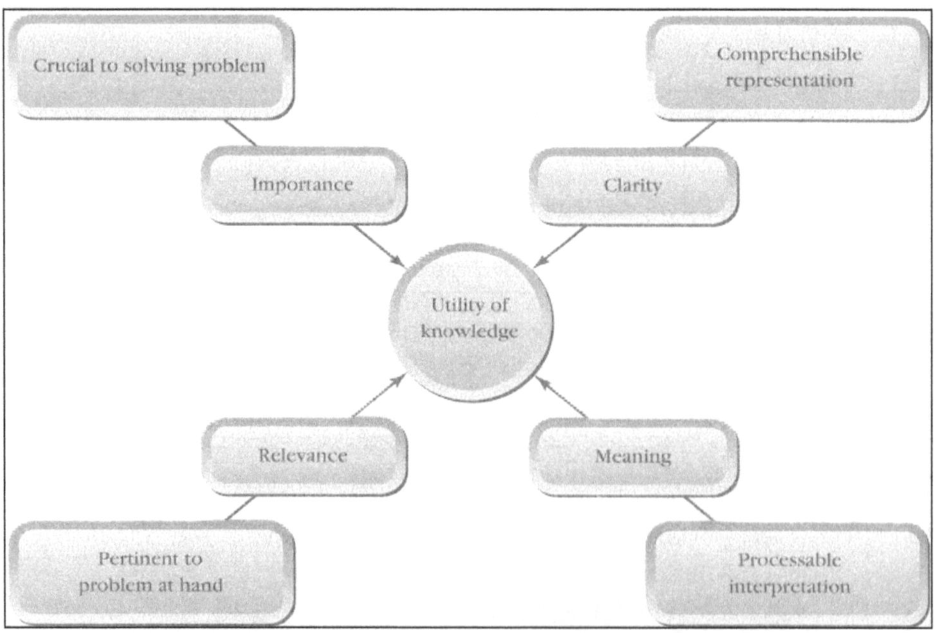

Figure: 39.3 Factors Inflencing the Usefulness of Knowledge.

Knowledge Has

- Extraordinary leverage and increasing returns
- Fragmentation, leakage, and the need to refresh
- Uncertain value
- Uncertain value sharing

Organizational Memory

- Individual wells
- Information well
- Culture well
- Transformation well
- Structural well
- Ecology well

Organizational Learning Focuses

- Knowledge source
- Product-process focus
- Documentation mode
- Dissemination mode
- Learning focus
- Value chain focus
- Skill development focus

Organizational Culture

- Culture is a pattern of shared basic assumptions
- Most important aspect of KM success
- Why don't people share knowledge?

Knowledge Management (KM)

- A process of elicitation, transformation, and diffusion of knowledge throughout an enterprise so that it can be shared and thus REUSED
- Helps organizations find, select, organize, disseminate, and transfer important information and expertise
- Transforms data/information into actionable knowledge to be used effectively anywhere in the organization by anyone

How Core Competency is Linked to Explicit and Tacit Knowledge

KM Objectives

- Create knowledge repositories
- Improve knowledge access

- ☆ Enhance the knowledge environment
- ☆ Manage knowledge as an asset

KMS Manage

- ☆ Knowledge creation through learning
- ☆ Knowledge capture and explication
- ☆ Knowledge sharing and communication through collaboration
- ☆ Knowledge access
- ☆ Knowledge use and reuse
- ☆ Knowledge archiving

Knowledge Repository

- ☆ Not a database
- ☆ Not a knowledge base (like for ES)
- ☆ A collection of internal and external knowledge

Knowledge Repository Types

- ☆ External
- ☆ Structured internal knowledge
- ☆ Informal internal knowledge

KM Activities

- ☆ Externalization
- ☆ Internalization
- ☆ Intermediation
- ☆ Cognition

KM Features

- ☆ Create a knowledge culture
- ☆ Capture knowledge
- ☆ Generate knowledge
- ☆ Explicate (and digitize) knowledge
- ☆ Share and reuse knowledge
- ☆ Renew knowledge

Cyclic Model of KM

- ☆ Create knowledge
- ☆ Capture knowledge
- ☆ Refine knowledge
- ☆ Store knowledge
- ☆ Manage knowledge
- ☆ Disseminate knowledge

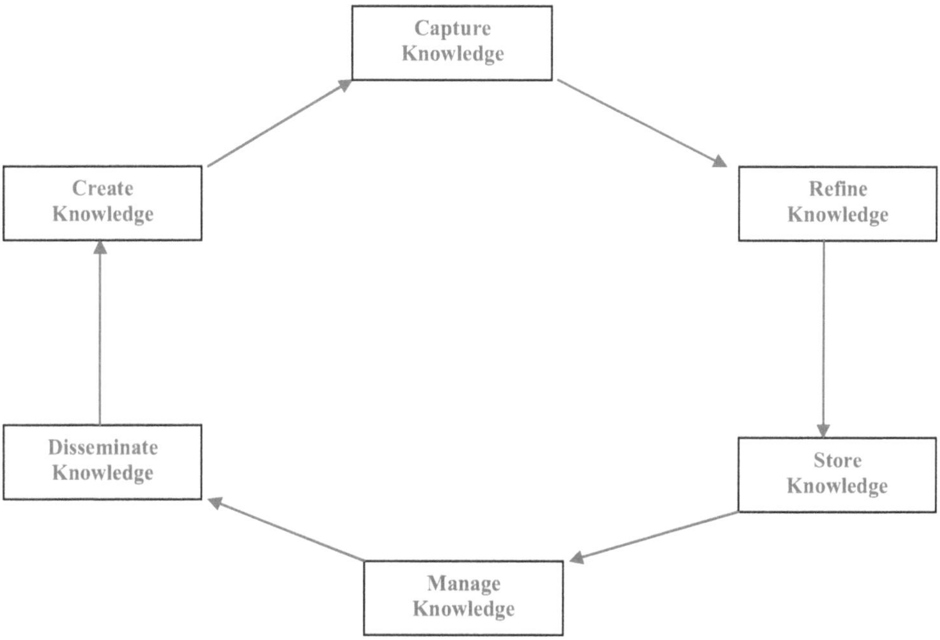

Cyclic Model of Knowledge M

KM Examples

- Mitre
- Dow Chemical Company
- Xerox
- Chrysler
- Monsanto
- Chevron
- Buckman Laboratories
- KPMG
- Ernst and Young
- Arthur Andersen
- Andersen Consulting

Why Adopt KM

- Cost savings
- Better performance
- Demonstrated success
- Share *Best Practices*
- Competitive advantage

Chief Knowledge Officer (CKO)

- ☆ Maximize firm's knowledge assets
- ☆ Design and implement KM strategies
- ☆ Effectively exchange knowledge assets
- ☆ Promote system use

KM Development

- ☆ Need a knowledge strategy
- ☆ Identify knowledge assets

KM Development

1. Identify the problem
2. Prepare for change
3. Create the team
4. Map out the knowledge
5. Create a feedback mechanism
6. Define the building blocks
7. Integrate existing information systems

Strategies for Successful KM Implementation

1. Establish a KM methodology
2. Designate a point person
3. Empower knowledge workers
4. Manage customer-centric knowledge
5. Manage core competencies

More Strategies

1. Foster collaboration and innovation
2. Learn from best practices
3. Extend knowledge sourcing
4. Interconnect communities of expertise (communities of practice)
5. Report the measured value of knowledge assets

KM Methods, Technologies, and Tools

- ☆ Email or messaging
- ☆ Document management
- ☆ Search engines
- ☆ Enterprise information portal
- ☆ Data warehouse
- ☆ Groupware
- ☆ Workflow management

- Web-based training
- Others

How to KM

- Integrate the technologies to manage knowledge effectively

KM Tool Categories

- Information architecture
- Technical architecture
- Application architecture

KM Software

- Know ware still developing but…
- Decision Suite
- Wincite
- Data Ware
- Knowledge X
- Knowledge Share

KM Success

- Economic performance
- Technical and organizational infrastructure
- Standard, flexible knowledge structure
- Knowledge-friendly culture
- Clear purpose and language
- Change in motivational practices
- Multiple channels for knowledge transfer
- Worthwhile level of process orientation
- Nontrivial motivational encouragement
- Senior management support

Measuring Success

- Balanced Scorecard
- Skandia Navigator
- Economic Value Added
- Inclusive Valuation Methodology

KM Failure Causes

1. Unclear definition of knowledge
2. Overemphasis on knowledge stock, not flow
3. Belief that knowledge exists outside people's heads

4. Not recognizing the importance of managing knowledge
5. Failure to manage tacit knowledge

More Failure Causes

1. Failure to disentangle knowledge from its uses
2. Downplaying reason and thinking
3. Focusing on the past and present, not the future
4. Failure to recognize the importance of experimentation
5. Substituting technology contact for human interface
6. Overemphasis on measuring knowledge, not its outcomes

KM and AI

- ☆ Can use AI in KM
- ☆ Can use KM in AI
- ☆ Data mining can create knowledge

Electronic Document Management

- ☆ A KM for documents
- ☆ Everyone is on the *same page*
- ☆ Documents are *up to date*
- ☆ Simple example: corporate phonebook
- ☆ Lower costs
- ☆ Better performance

The Knowledge-Based View of Decision Making

Accepting Messages: see next slide

- ☆ A decision maker (human-being) can accept stimuli from the environment
- ☆ The stimuli are messages that carry knowledge (information)
- ☆ Some messages have a direct and immediate impact on the decisions being manufactured
- ☆ Other messages can be:
  - Discarded
  - passed along to others and or other places
  - Stored for future use

The Knowledge-Based View of Decision Making

Issuing Messages

- ☆ The decision maker can issue messages to the surroundings:
  1. Other people
  2. Documents/storage vessels
- ☆ The message may also be the Manufactured Decision

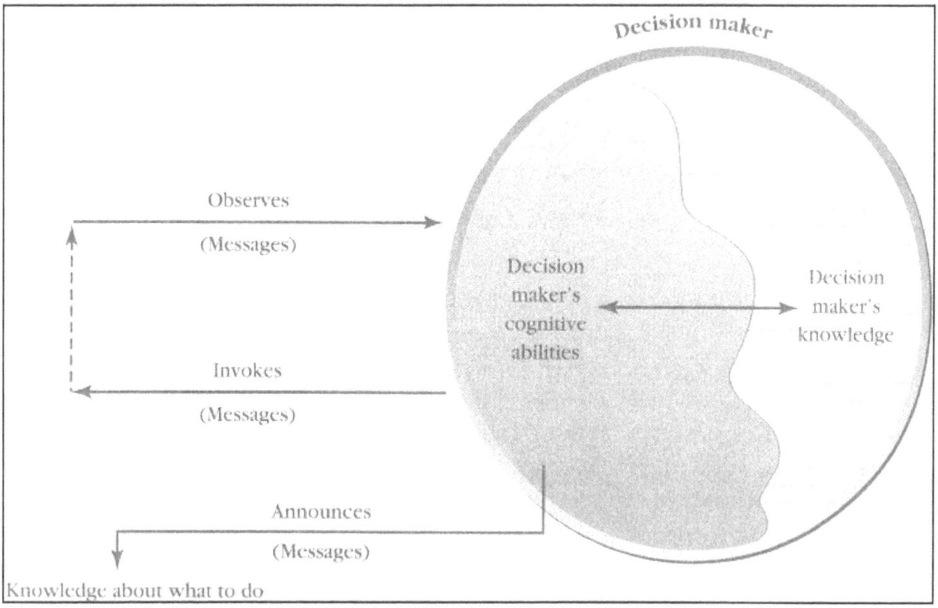

Figure: 39.4 Bird-Eye View of a Decision Maker

The Knowledge-Based View of Decision Making

Assimilating Knowledge

- ✰ Once the Decision maker has established the meaning of an incoming message it can be assimilated with the DM's knowledge store
- ✰ When new knowledge is assimilated, it alters the knowledge store:
  - Just be added
  - Cause existing knowledge to be altered,
  - dDiscarded, or marked as being obsolete
  - It may cause fundamental alterations of the knowledge store

The Knowledge-Based View of Decision Making

Recognizing the Need for a Decision

- ✰ May be very obvious:
  - Highly structured
  - Happens frequently
- ✰ May be take many repetitions of the event/stimuli to initiate action, thus it is:
  - unstructured
  - novel

- by observing conditions (economic, political, mechanical) we may come to recognize that:

  - a problem exits

  - a solution is required

The Knowledge-Based View of Decision Making

Manufacturing a Decision: The manufacturing process produces new knowledge from knowledge.

☆ The sources of raw materials are the decision maker's storehouse of knowledge (experience, facts, rules, etc).

☆ Knowledge is extracted on an as needed basis and manipulated by Cognitive abilities to produce solutions for the flow of problems that constitutes the **KNOWLEDGE Manufacturing Process**.

☆ The solution that is the product of the process is the **NEW KNOWLEDGE**.

A Manufacturing Analogy	
Material Product	Decision
1. The process begins in reaction to a customer order or anticipated order.	The process begins in reaction to a recognized need or opportunity.
2. The process draws on an inventory of raw materials.	The process draws on an inventory of KNOWLEDGE.
3. Items entering inventory are subject to quality testing controls.	Knowledge is assimilated into inventory only if it is expected to be usable.
4. Abilities for manipulating materials transform/assemble raw materials into final products.	Abilities for manipulating knowledge transformation/assemble existing knowledge into new knowledge about what to do.
5. During the process may yield material by-products that are stored in inventory or discarded.	During the process there are intermediate pieces of knowledge called problem solutions.
6. The process may yield material by-products that are stored in inventory or are discarded.	The process may yield knowledge by-products that are stored in inventory or are discarded.
7. The manufacture may be an individual or have multiple participants.	The decision maker may be an individual or consist of multiple participants
8. The finished product is packaged for distribution.	The decision is packaged for distribution.

Source: From Holsapple & Whinston, 1996

Three Views of Knowledge

☆ Knowledge Representations: If a system has and can use a representation of something then the system itself can also be said to have KNOWLEDGE.

The textbook can be a representation of knowledge if it can be read.

☆ Representation is pattern of Symbols => an abstraction

☆ It embodies knowledge:

Defining Knowledge

☆ A useable representation of something

- From a DSS point of view we must be concerned with
  - the computer memory and how it processes knowledge represents knowledge
  - Clean data and defined objects are required

Defining Knowledge

- One state of knowledge can be used to generate different states of knowledge
- DSS help in:
  - Acquiring knowledge
  - deriving knowledge

Knowledge Production

- The result of a productivity activity (*i.e.* LEARNING) involving acquisition and/or derivation
  - The flows and stock of knowledge
  - Stocks are the inventories of knowledge
  - The flows are the messages that tell the stock to do something

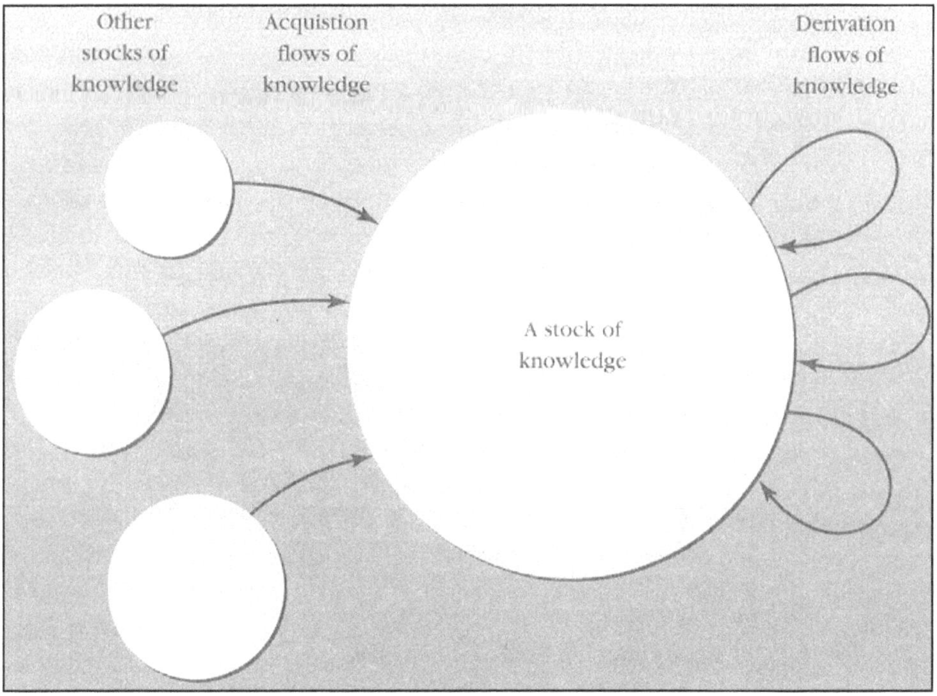

Figure: 39.5 A Production View of Knowledge.

Knowledge Sources

The decision makers store house of Knowledge: Internal and External

- ✰ The DM can be active or Passive about acquiring knowledge
- ✰ Active: Message can be emitted to invoke a response
- ✰ Passive the DM observes without invoking reactions

The Decision to Acquire/Derive Knowledge

- ✰ General a mixture of acquiring and derivation of knowledge
- ✰ Acquiring knowledge may tax: cognitive abilities, time, economic limits
- ✰ There are tradeoffs
- ✰ DSS tens to promote greater reliance on internal production of knowledge

Knowledge Sources

Reliability of Knowledge:

- ✰ Do we get the same knowledge from internal and external sources?
- ✰ If there are multiple external sources- to they yield the same result
- ✰ DSS

  Without a DSS it may be infeasible to produce it internally

  - Use the DSS in parallel with the knowledge acquisition to check the source reliability

Knowledge Sources

- ✰ Knowledge Qualities
  - DSS
  - accurately retaining knowledge
  - flagging inconsistencies
  - analyzing uncertainties
  - tracking multiple sets of knowledge

Knowledge Sources

Utility of Knowledge: Usefulness

- ✰ Knowledge can be useful to different people
- ✰ To a history professor knowledge about particle physics is probably not useful
- ✰ DSS:
  - Present what is relevant to a specific DM
  - Provide high quality knowledge

Knowledge Management

- ✰ Techniques
- ✰ Text Management
- ✰ Forms Management

- ✰ Business Graphics
- ✰ Solver Management
- ✰ Rule Management
- ✰ Database Management
- ✰ Report Generation
- ✰ Spreadsheet Analysis
- ✰ Programming
- ✰ Message Management
- ✰ Knowledge Management

Reasons for Understanding Knowledge Management

- ✰ Position or integrate Knowledge into a decision
- ✰ Extend the role of supporting participants "from mere production to the processing, storage, retrieval, dissemination, utilization and general management of knowledge."
- ✰ Facilitate and develop a philosophy and methodologies for handling knowledge
- ✰ Shift the role of supporting participants "from producing certainty and complete knowledge to structuring ignorance and managing uncertainty

Knowledge Management

Five types of Knowledge

1. Practical
2. Intellectual
3. Pastime
4. Spiritual
5. Unwanted

Knowledge Management

Three Primary Types of Knowledge

1. **Descriptive:** Includes descriptions of past, present, future, and hypothetical situations. DATA and INFORMATION- To Know What
2. **Procedural Knowledge**: The how to do; Step-by-Step
3. **Reasoning Knowledge**: To know Why- Approaches to problem solving

Knowledge Management

Three Secondary Types of Knowledge

1. **Linguistic**: Vocabulary and grammar, body language, meaning of gestures.
2. **Assimilation Knowledge**: The basis for controlling changes to the knowledge store.

A filtering mechanism

3. **Presentation Knowledge**: The basis for packaging outgoing responses

Knowledge Management

- ☆ The Decision Maker possess knowledge
- ☆ The DSS has processing abilities that can supplement the Decision Maker

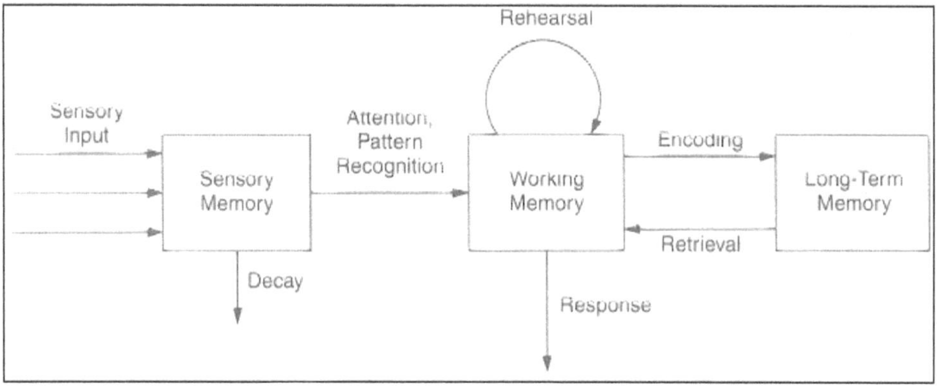

Figure: 39.6 Overview of Information Processing System.

The Cognitive Basis for Knowledge

- ☆ Declarative Knowledge
  - factual information that is static in nature
  - it is usually describable to us
  - history- events, facts
  - flexible- it can be reorganized to suite our purposes
  - Knowing That

Cognitive- Knowledge

- ☆ Some knowledge can be encoded in a declarative format which can
- ☆ Later be transformed into a procedural format as we become familiar with the information.
- ☆ Examples: Reading Windows for Dummies
  Reading a Golf technique book then truing on the PC/Play golf

Cognitive Knowledge

- ☆ Attention: the concentration and focusing of mental activity
- ☆ Paying attention seems to accentuate, or enhance, sensory input that has been focused on

The Cognitive Basis for Knowledge

- ☆ Procedural Knowledge
  - the underlying skillful actions we possess
  - it is dynamic
  - it is not very describable
  - the acquisition of a skill involves making and detecting errors (skiing, bike riding, ballet)
  - with additional experience we improve
  - Knowing How

Table: 39.1 Relationship between Cognitive Codes and Cognitive Processing

	TYPE OF COGNITIVE CODE	
Processing	Declarative	Procedural
Automatic	Some of my lectures	Driving on the interstate
Controlled	Solving simultaneous equations by hand	Driving downtown in rush-hour traffic

Information Processing

- ☆ Sensory system: where specific aspects of the environment are detected and organized-into cognitive code
- ☆ The code is passed into memory
- ☆ Memory
  - Working memory: a workbench for cognitive codes (short term memory)
  - Permanent memory: long term storage of declarative and procedural knowledge

Memory

- ☆ External Input
- ☆ Sensory Register
- ☆ Short-term Storage
- ☆ Long-term Storage

Sensory Register

- ☆ Where our feature detection and pattern recognition process produce a cognitive code that can be stored for a short time.
- ☆ The Sensory register does not depend on resource allocation- we do not have to pay attention to incoming stimuli in order to have this cognitive code created.

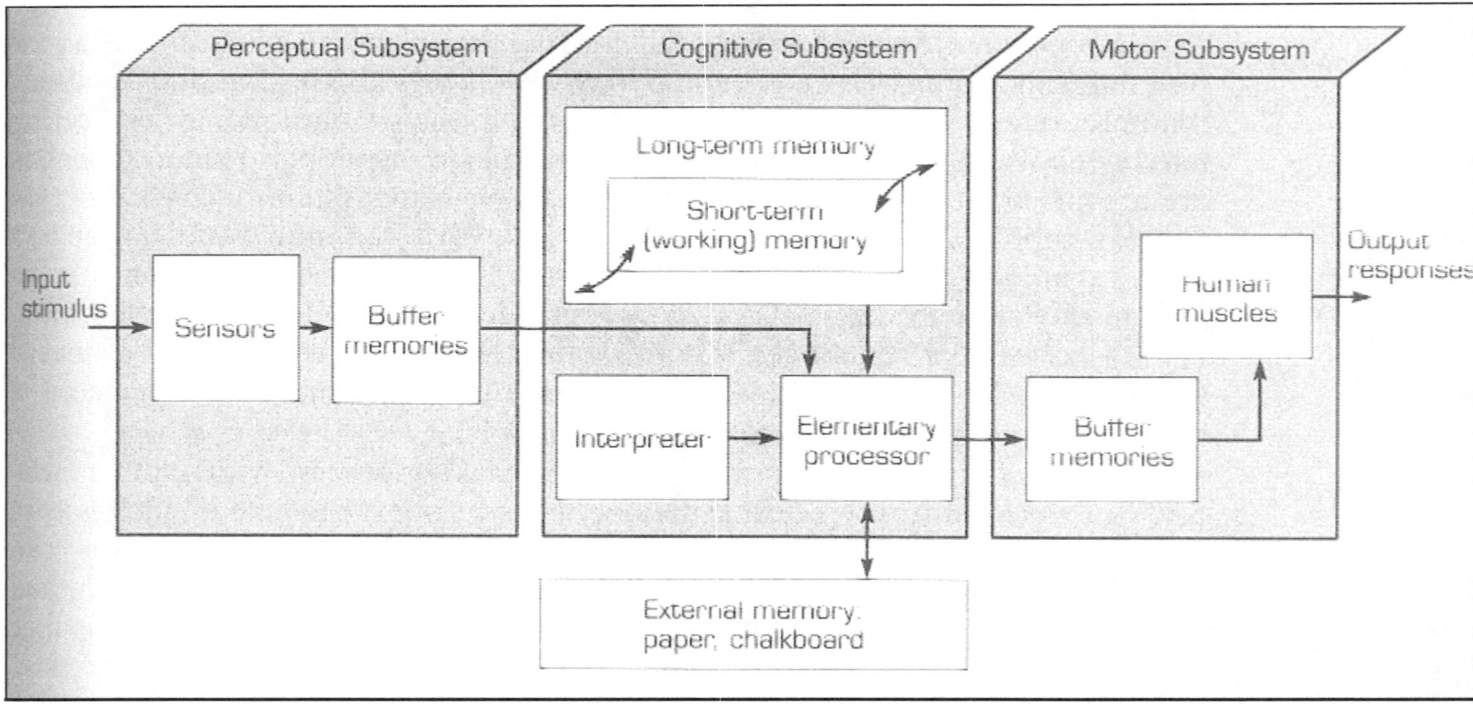

Figure: 39.7 Newell-Simon Model of Human Information Processing.

Source: Adapted from P. Harmon and D. King. *Expert Systems*, ©1985, John Wiley & Sons, Inc., New York, Reprinted by Permission of John Wiley & Sons, Inc.)

Sensory Register

- ☆ It must have a large storage capacity
- ☆ It is modality specific:
  - has difference storages for audio, visual
- ☆ The code in storage Decays over time
- ☆ Resources must be allocated to transfer the code to STM or LTM

Short-term Memory

- ☆ Limited capacity (RAM)
- ☆ Storage is organized by sensory component:
  - acoustic, verbal, linguistic

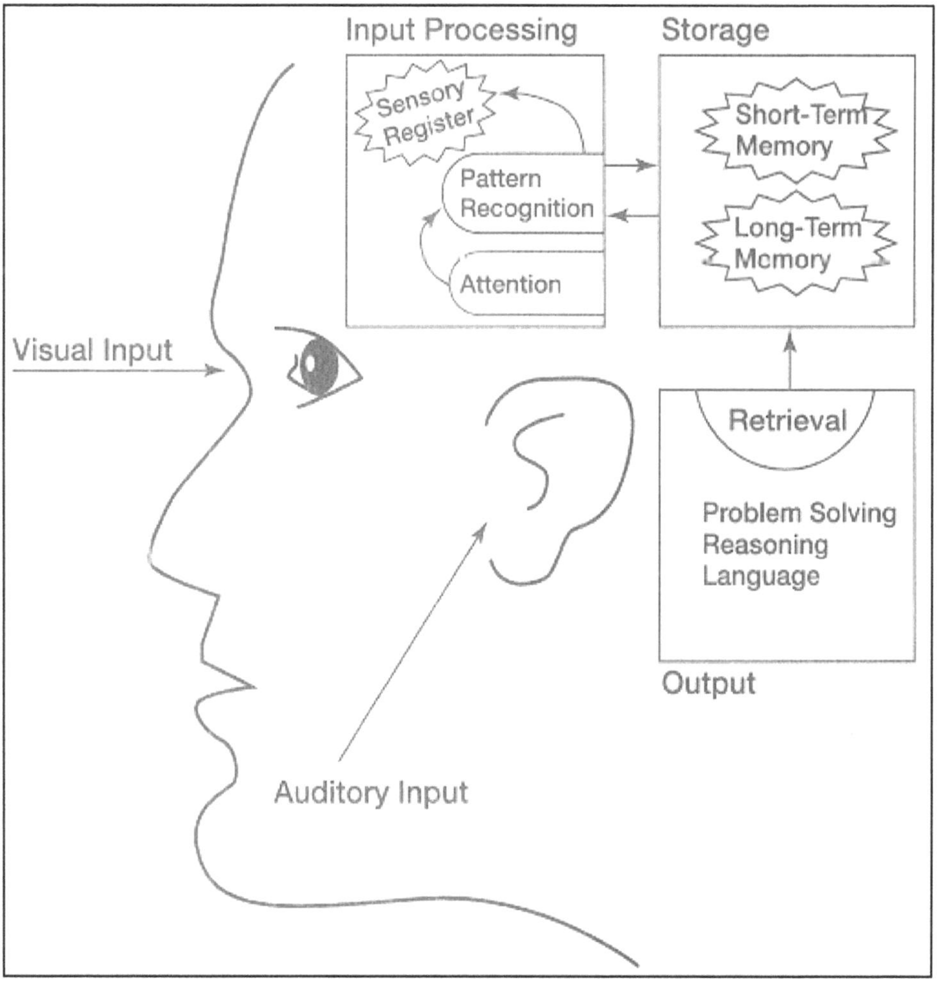

Figure: 39.8 Sequence of Information Processing.

- Storage duration of unrehearsed material is about 30 seconds
- Material that is not elaborated or transferred decays.

Long-term Memory

- To go from STM to LTM requires rehearsal

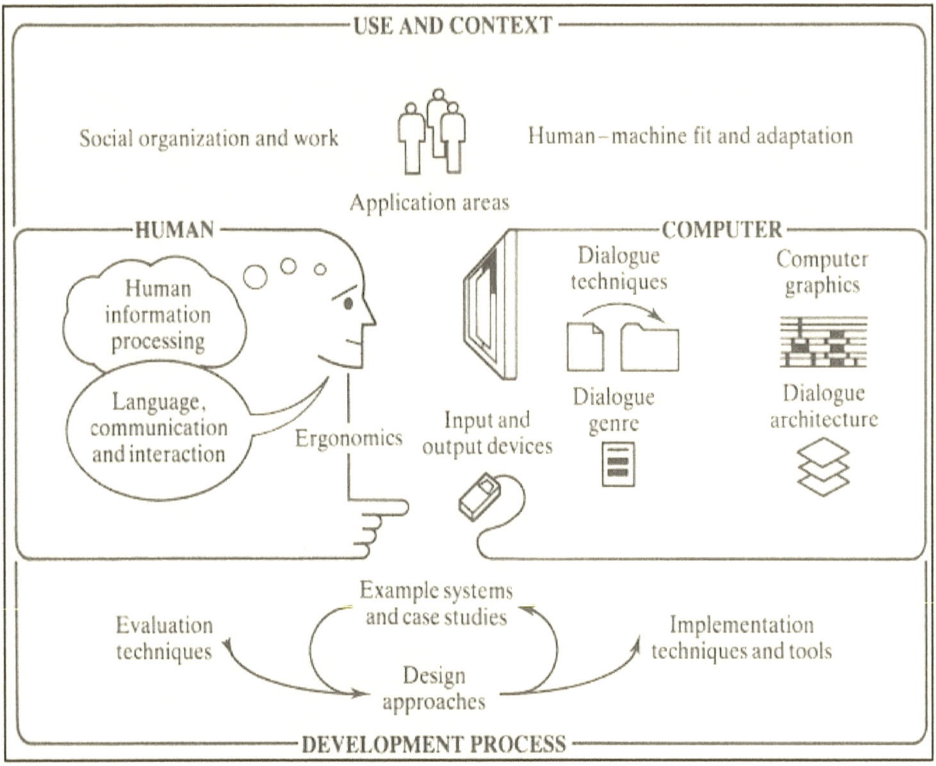

Figure: 39.9 Human-Computer Interaction as Described by the SIGCHI Committee (ACM SIGCHI, 1992).

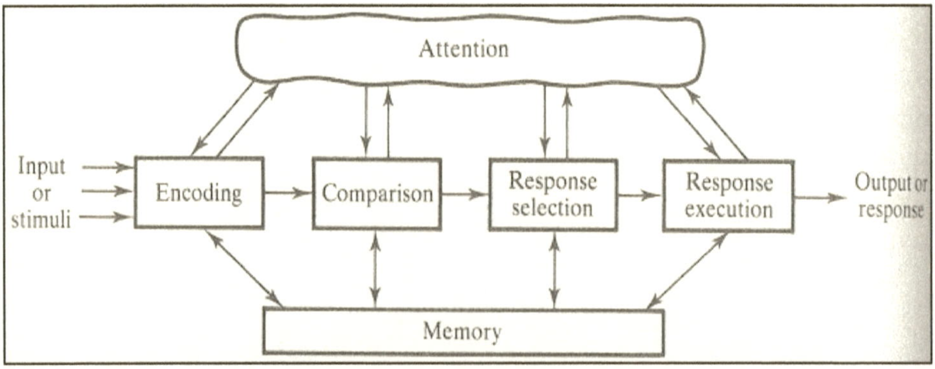

Figure: 39.10 Extended Stages of the Information Processing Model (Adapted from Barb, 1988).

☆ Rehearsal:

- procedures that maintain the vitality of the code
- STS code will last indefinitely if it is occasionally refreshed by rehearsal.
- Rehearsal duplicates and augments the code for long-term storage (associations/links are created).

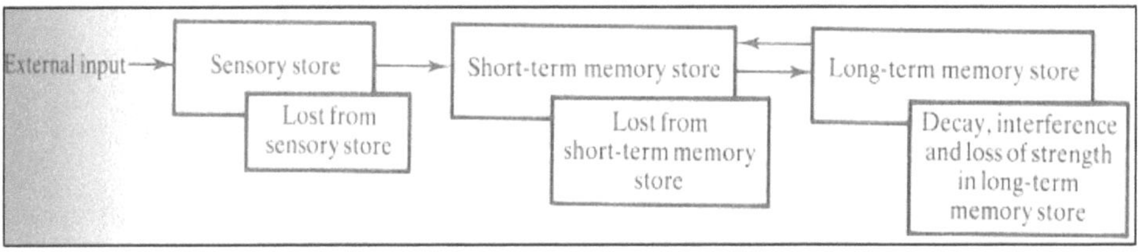

Figure: 39.11 The Multi-Store Model of Memory (Adapted from Atkinson and Shiffrin, 1968).

Figure: 39.12 The Human Processor Model (Adapted from Card *et al.*, 1983).

Knowledge Engineering and Acquisition

Taxonomy of Knowledge Types

- **Primary**
  - *Descriptive:* data, information, descriptions of past, present, and future situations
  - *Procedural:* how to do something
  - *Reasoning:* codes of conduct, regulations, policies, diagnostic rules
- **Secondary**
  - *Linguistic:* vocabulary, grammar, knowledge of gestures
  - *Assimilative*: permissible contents, retention cycles, relevancy filters
  - *Presentation*: modes of communication, graphing, messaging, inverse of Linguistic knowledge

Basic Pre-Interview Checklist

- Decide what you need to know.
- Ask yourself why this information is needed.
- Determine that an interview is the best method for obtaining this information.
- Determine the appropriate degree of structure for the interview.
- Consider the method in which the answers to your questions will be coded and analyzed.

Necessary Task Conditions for Successful Concurrent Protocols

- The sample of cases employed must be highly representative of the task under study.
- Each task must have a clearly defined conclusion or point of completion.
- The task must be able to be completed in one protocol session.
- All data must be presented to the expert in a familiar form.
- A test case should be given to the expert prior to the collection of protocols so that he or she may become familiar and comfortable with the verbalization process.

Knowledge Base Validation Measures and Techniques

- **Accuracy:** How well does the system reflect reality. How correct is the knowledge in the knowledge base.
- **Adaptability:** Possibilities for future development or changes.
- **Adequacy:** The portion of the necessary knowledge that is included in the knowledge base.
- **Appeal:** How well the knowledge base matches intuition and stimulates thought and practicability
- **Breadth:** How well is the domain covered?

- **Depth:** The degree of the detailed knowledge.
- **Face Validity:** How credible is the knowledge.
- **Generality:** Capability of a knowledge base to be used with a broad range of similar problems.
- **Precision:** Capability of the system to replicate particular system parameters. Consistency of advice and coverage of variables in the knowledge base.
- **Realism:** Accounting for the relevant variables and relations. Similarity to reality.
- **Reliability:** The frequency of system predictions that are correct.
- **Robustness:** Sensitivity of conclusions to model structure.
- **Sensitivity:** The impact of changes in the knowledge base on the quality of outputs.
- **Technical/Operational:** Goodness of the assumptions, context, constraints, and conditions.
- **Turing test:** Ability of a human evaluator to identify if a given conclusion is made by a real expert or a computer.
- **Usefulness:** How adequate the knowledge is (in terms of parameters and relationships) for solving correctly.
- **Validity:** The capability of the knowledge base for producing empirically correct predictions.

KM – The Future

- Not a fad
- Impact is immense
- Research on organizational culture
- How to do each step
- Are they the right steps?

Knowledge Management

- The definition is clear
- The concepts are clear
- The challenges are
- Clear
- Surmountable
- The benefits are clear (and can be huge)
- The tools and technologies are viable

Knowledge Management–Key Issues

- Organizational culture
- Executive sponsorship

- Measuring success
- The future: Comprehensive standardized KM packages

Knowledge Management

- "The wise see knowledge and action as one" (Bhagvad Gita)
- Intelligent organizations recognize that knowledge is an asset, perhaps the only one that grows over time, and when harnessed effectively can sustain the ability to continuously compete and innovate.

Chapter 40

Rural Tourism: A Fast Simmering New Enterprise

Manashi Saha and Amitava Biswas

Tourism is currently the world's largest industry and the fastest growing sector of the market. Tourism is usually viewed as being multidimensional, possessing physical, social, cultural, economic and political characteristics. "Definitions of tourism share a range of common elements" (Dowling 2001, p24). Mathieson and Wall (1982) opine that tourism is the temporary movement of people to destinations out of their normal home and workplace, the activities undertaken during the stay, and the facilities created to cater for their needs. Tourism is the fastest growing industry in the world. According to the World Travel and Tourism Council (WTTC) tourism is the world's largest industry generating 12 per cent of the global gross national product and it employs around 200 million people worldwide (WTTC, 1995). Worldwide tourism is ranked second highest revenue-generating industry next to the oil industry. It is necessary to differentiate between different type of tourists to understand and analyses their purpose of visit. There are different ways to attract domestic and foreign tourists, we need to understand what types of services are required to attract and retain the customers. There is a large potential market for rural tourism especially for foreign tourists, which has not yet developed because government has not taken up any systematic approach to attract foreign tourists.

Tourism: The Concept

Tourism is the travel for recreational, leisure, or business purposes. The **World Tourism Organization** defines **tourists** as people "travelling to and staying in places outside their usual environment for not more than one consecutive year for leisure, business and other purposes".

The word *tour* is derived from the Latin, 'tornare' and the Greek, 'tornos', meaning 'a lathe or circle; the movement around a central point or axis'. This meaning changed in modern English to represent 'one's turn'. The suffix *–ism* is defined as 'an action or process; typical behaviour or quality', while the suffix, *–ist* denotes 'one that performs a given action'. When the word *tour* and the suffixes *–ism* and *–ist* are combined, they suggest the action of movement around a circle. One can argue that a circle represents a starting point, which ultimately returns back to its beginning. Therefore, like a circle, a tour represents a journey in that it is a round-trip, *i.e.*, the act of leaving and then returning to the original starting point, and therefore, one who takes such a journey can be called a tourist."

Forms of Tourism

Sustainable Tourism

"Sustainable tourism is envisaged as leading to management of all resources in such a way that economic, social and aesthetic needs can be fulfilled while maintaining cultural integrity, essential ecological processes, biological diversity and life support systems." (World Tourism Organization). Sustainable development implies "meeting the needs of the present without compromising the ability of future generations to meet their own needs." Sustainable tourism can be seen as having regard to ecological and socio-cultural carrying capacities and includes involving the community of the destination in tourism development planning. It also involves integrating tourism to match current economic and growth policies so as to mitigate some of the negative economic and social impacts of 'mass tourism'.

Ecotourism

Ecotourism, also known as ecological tourism, is responsible to travel to fragile, pristine, and usually protected areas that strives to be low impact and (often) small scale. It helps to educate the traveller; provides funds for conservation; directly benefits the economic development and political empowerment of local communities; and fosters respect for different cultures and for human rights.

Pro-poor Tourism

Pro-poor tourism, which seeks to help the poorest people in developing countries, has been receiving increasing attention by those involved in development; the issue has been addressed through small-scale projects in local communities and through attempts by Ministries of Tourism to attract large numbers of tourists. Research by the Overseas Development Institute suggests that neither is the best way to encourage tourists' money to reach the poorest as only 25 per cent or less (far less in some cases) ever reaches the poor; successful examples of money reaching the poor include mountain-climbing in Tanzania and cultural tourism in Luang Prabang, Laos.[27]

Recession Tourism

Recession tourism is a travel trend, which evolved by way of the world economic crisis. Identified by American entrepreneur Matt Landau (2007), recession tourism is defined by low-cost, high-value experiences taking place of once-popular generic retreats. Various recession tourism hotspots have seen business boom during the recession thanks to comparatively low costs of living and a slow world job market suggesting travellers are elongating trips where their money travels further.

Medical Tourism

When there is a significant price difference between countries for a given medical procedure, particularly in Southeast Asia, India,Eastern Europe and where there are different regulatory regimes, in relation to particular medical procedures (*e.g.* dentistry), travelling to take advantage of the price or regulatory differences is often referred to as "medical tourism".

Educational Tourism

Educational tourism developed, because of the growing popularity of teaching and learning of knowledge and the enhancing of technical competency outside of the classroom environment. In educational tourism, the main focus of the tour or leisure activity includes visiting another country to learn about the culture, such as in Student Exchange Programmes and Study Tours, or to work and apply skills learned inside the classroom in a different environment, such as in the International Practicum Training Program.

Creative Tourism

Creative tourism has existed as a form of cultural tourism, since the early beginnings of tourism itself. Its European roots date back to the time of the Grand Tour, which saw the sons of aristocratic families traveling for the purpose of mostly interactive, educational experiences. "Creative tourism" as tourism related to the active participation of travellers in the culture of the host community, through interactive workshops and informal learning experiences. More recently, creative tourism has gained popularity as a form of cultural tourism, drawing on active participation by travellers in the culture of the host communities they visit. Several countries offer examples of this type of tourism development, including the United Kingdom, Austria, France, the Bahamas, Jamaica, Spain, Italy and New Zealand.

Dark Tourism

This type of tourism involves visits to "dark" sites, such as battlegrounds, scenes of horrific crimes or acts of genocide, for example: concentration camps. Dark tourism remains a small niche market, driven by varied motivations, such as mourning, remembrance, education, macabre curiosity or even entertainment. Its early origins are rooted in fairgrounds and medieval fairs.

Doom Tourism

Also known as "Tourism of Doom" or "Last Chance Tourism" this emerging trend involves travelling to places that are environmentally or otherwise threatened before it is too late. This type of tourism is believed to be on the rise. Some see the trend as related to sustainable tourism or ecotourism due to the fact that a number of these tourist destinations are considered threatened by environmental factors such as global warming, over population or climate change. Others worry that travel to many of these threatened locations increases an individual's carbon footprint and only hastens problems threatened locations are already facing.

Sports Tourism

Since the late 1980s, sports tourism has become increasingly popular. Events such as rugby, Olympics, Commonwealth games, Asian Games and football World Cups have enabled specialist travel companies to gain official ticket allocation and then sell them in packages that include flights, hotels and excursions.

Tourism Growth in India

- ✰ According to *World Tourism Organization* estimates, India will lead in South Asia with 8.9 million arrivals by 2020
- ✰ India is poised to emerge as the 2nd fastest growing (8.8 per cent) tourism economy in the world over 2005-14 according to the *World Travel and Tourism*

Foreign Earnings

- ✰ There has been a growth of more than 13 per cent in foreign tourist arrivals at 3.9 million during 2005, up from 3.4 million foreign tourists who visited India during previous year.
- ✰ Foreign exchange earnings from foreign tourists were up by more than 20 per cent at $5,730.86 million in 2005, up from $4,769 million earned the previous year.

Background: Rural Tourism

Tourism is one of the largest and fastest growing industries in the world. In both developed and developing countries, tourism is frequently supposed to be a viable means of raising the economic activity of regions. Additionally, the development of a tourism industry has been noted to promote the destination's image, enabling the region to achieve other objectives, such as business recruitment and retention. According to the World Travel and Tourism Council, tourism generates 12 per cent of the global gross national product and it employs around 200 million people worldwide. Tourist arrivals are estimated to reach 1 billion by 2010 and 1.6 by 2020 worldwide.

The tourism and hospitality industry is an important contributor to Alberta's economy and is identified as a priority value-added sector in the province's economic development strategy. In 2006, tourism revenues were estimated to be in the order of $5.3 billion. Approximately 10 per cent of Alberta's workforces – over 180,000 people – work in the tourism and hospitality industry. Alberta's tourism and hospitality industry is expected to experience strong growth over the next 10 years. Employment in the accommodation and food services segment of the industry is expected to grow on average by 2.2 per cent annually over the next five years and generate almost 10,913 new jobs by 20116.One of Travel Alberta's reports show that Alberta's current tourism revenue is $5.8 billion with visits in Alberta numbering 23.6 million. The goal is to enable industry to grow revenues to $6.2 billion and visits to 25.1 million by 2013. Given the expected growth of tourism there is a great optimism for rural tourism.

Numerous agencies and academic researchers have identified tourism as a potential economic development tool, particularly for rural communities. Rural Tourism is increasingly being used as a development strategy to improve the social and economic well being of rural areas.

Rural Tourism encompasses a huge range of activities, natural or manmade attractions, amenities and facilities, transportation, marketing and information systems9. Rural tourism is very diverse and fragmented in terms of operational structures, activities, markets and operating environments. Rural tourism is not just farm-based tourism. It includes farm-based holidays but also comprises special interest nature holidays and ecotourism, walking, climbing and riding holidays,

What is Rural Tourism?

Rural tourism can be defined as the country experience which encompasses a wide range of attractions and activities that take place in agricultural or non-urban areas. Its essential characteristics include wide-open spaces, low levels of tourism development, and opportunities for visitors to directly experience agricultural and/or natural environments. Consequently, rural tourism in its purest form should be Located in rural areas. Functionally rural – built upon the rural world's special features of small-scale enterprise, open space, contact with nature and the natural world, heritage, — traditional societies and Traditional practices. Rural in scale — both in terms of buildings and settlements — and, therefore, usually small-scale. Traditional in character, growing slowly and organically, and connected with local families. It will often be very largely controlled locally and developed for the long term good of the area.

Rural Tourism: The Values and Assets

- ✰ *Scenic value* including mountains, seashores, lakes, islands, rivers, and special interest scenery such as wetlands, native bush, geological features;
- ✰ *Special wildlife assets* - flora and fauna both native and exotic:
- ✰ *Cultural assets* - including historic buildings, towns, settlements, historic sites and other cultural experience opportunities, other ethnic heritage;
- ✰ *Agricultural/horticultural/forestry assets* - farm systems and activities *e.g.* sheep rearing,cattle rearing, interesting crops, flowers etc;
- ✰ *Special facilities for sporting activities* - including hunting, fishing, skiing, trekking, walking, etc.

Rural tourism could help in boosting the local performing arts and help conserve the local culture,and can prevent rural-urban migration.Thus rural tourism could attract tourists by providing excellent glimpse of the village ambience with local cuisine. Moderate, but clean, accommodations for tourists should be constructed by the villagers in traditional design and architecture. Bank finance Different festival of India like Holi, Diwali, Durga Puja, Ganesh Utsav, Navaratri, etc are some of the festivals of India should be more organize to attract Domestic as well as International tourist. Melas and festivals are needed to be promoted to attract tourists.

Types of Rural Tourism

Tourism is synthesized from mass and alternative tourism. Mass tourism is characterized by large numbers of people seeking culture holidays in popular resort destinations. **Alternative tourism** is sometimes referred to as ¯special interest tourism or responsible tourism and it's usually taken to mean alternative forms of tourism which give emphasis on the contact and understanding of inhabitants way of living and the local natural environment. The diversity of attractions included within rural tourism includes heritage tourism (sometimes referred to as cultural heritage tourism), nature-based tourism/ecotourism, agritourism, as well as partnership-based approaches, such as scenic byways and heritage areas. Heritage tourism refers to leisure travel that has as its primary purpose the experiencing of places and activities that represent the past.

A second major type of rural tourism activity is nature-based tourism/ **ecotourism** (sometimes called recreation-based tourism), which refers to the process of visiting natural areas for the purpose of enjoying the scenery, including plant and animal wildlife. Nature-based tourism may be either passive, in which observers tend to be strictly observers of nature, or active (increasingly popular in recent years), where participants take part in outdoor recreation or adventure travel activities.

A third major form of tourism is **agritourism**, which refers to, the act of visiting a working farm or any agricultural, horticultural or agribusiness operation for the purpose of enjoyment, education, or active involvement in the activities of the farm or operation. It includes taking part in a broad range of farm-based activities, including farmers markets¯petting farms, roadside stands, and pick-your-own operations; engaging in overnight farm or ranch stays and other farm visits; and visiting agriculture-related festivals, museums, and other such attractions¯Tourism alternatives, potentials and problems in the development of tourism.

Why Rural Tourism

- ✰ Sustainable Livelihood
- ✰ Poverty Alleviation
- ✰ Uplift of Rural Artists
- ✰ Uplift of Rural Artisans
- ✰ Empowerment of Women
- ✰ Improvement in Urban and Rural Infrastructure
- ✰ Better Image, Quality of Life and Attitude of People
- ✰ Revival of Traditional Crafts
- ✰ Heritage Conservation
- ✰ Conservation of Natural Resources

Driving Forces in Rural Tourism

These forces have been identified by many authors and can be summarized as:

- Tourism generating regions for rural tourism are highly developed and urbanized – the stresses of urban living and the remoteness from the natural environment has created a desire for escape from the monoculture of city living. Rural locations offer an idealized release from stress and the opportunity to re-engage with a simpler, quieter way of life that offers rest and relaxation.
- Demand fuelled by media, over-familiarity and congestion with traditional tourist resorts and increased interest in alternative attractions – with its voracious appetite for content and the resultant over-exposure of many traditional tourist destinations, the media have sought out new and interesting tourism experiences for their lifestyle productions.
- Increasing environmental awareness and interest in the relationship between humans and the environment. Green issues have raised the attractiveness of rural experiences as ecologically sustainable tourism.
- Transport, communications, and the removal of political and economic barriers to travel have facilitated accessibility of rural areas.
- Increasing numbers of Free Independent Travellers and world-wide long-haul travel – many more travelers are FIT than in the past due to the increased capacity, especially in long-haul transport modes. When combined with increasing discretionary incomes, greater awareness of the range of experiences on offer, and greater mobility through private transport, the accessibility and attractiveness of rural destinations has been dramatically improved.
- A move toward short-break holidays - income and leisure time have changed so that shorter breaks with greater choice of leisure activities are sought. Changing work patterns have increased the popularity of shorter breaks that minimize the absence from work and the effect of absences on work flow and involvement.
- Better-educated travelers have increased interest in outdoor recreation, eco-tourism and special interest tourism - individualism drives a need for unique experiences and rural tourism, because of its fragmented nature and diversity of offerings, can satisfy this need.
- An increased interest in heritage can be satisfied through rural tourism as rural areas are often the repositories of remnant heritage.
- Rural areas are perceived as healthier, offering fresher air, cleaner water and the opportunity for outdoor recreation. Rural areas offer fresh, and sometimes, specialty foods.
- An increasing desire for authentic experiences including interaction with local people - Rural tourism is REAL (Rewarding, Enriches the spirit, provides Adventure and Learning); authenticity is believed to be found in genuine country experiences and lifestyles

Essential Elements for Development of Rural Tourism

- ✰ Creation of infrastructure
- ✰ Restructuring and liberalization of policies
- ✰ Encouragement for investment
- ✰ Law and Order
- ✰ Tourist Police
- ✰ Complaints Handling
- ✰ Standardization of goods and services
- ✰ Government support

Rural tourism can help in creating sustainable development in some of our villages in rural areas.

1. To develop of rural areas
2. Provide more jobs in rural sector
3. Improve the standard of living of rural people
4. Bring rural people to main stream
5. Totally reduce migration of people to cities in search of employment tourists to India

Ministry of tourism has identified many spots which are being developed as tourist destinations in rural area. These sites are unexplored and incredibly beautiful. The rural tourism in India is will also help to promote the skills and arts of the villagers along with natural, cultural and historical heritage of India.

Benefits of Rural Tourism

Rural tourism, while still only a minority tourism market, is making a valuable contribution to rural economies. Its contribution can be expressed not only in financial terms, but also in terms of jobs, contributions towards funding conservation, encouragement to the adoption of new working practices, and the injection of a new vitality into sometimes weakened economies. Potentially rural tourism promises some of the following benefits to rural development:

1. Job Retention

Rural tourism cash flows can assist job retention in services such as retailing, transport, hospitality and medical care. It can also provide additional income for farmers, and, in some cases, for foresters and fisherman. Job retention is not as politically glamorous as job creation, but, by helping the viability of small communities, it is critical to the survival of marginal areas. Studies of rural Austria, Sweden and Ireland have documented the role of tourism in job retention.

2. Job Creation

Job creation typically occurs in the hotel and catering trades, but can also take place in transport, retailing, and in information/heritage interpretation. Studies in

Britain suggest that job creation varies by enterprise type. Farmhouse accommodation and bed-and-breakfast can create up to 23 jobs per £ 100 000 of tourism revenue. Job creation effects are less marked in hotels and caravan/campsites, yielding approximately six jobs per £ 100 000 of revenue.

3. New Business Opportunities

Tourism generates new opportunities for industry. Even those rural businesses not directly involved in tourism can benefit from tourist activity through developing close relationships with tourist facilities where **local foods** can be used as part of the tourism offering in a locality. Rural tourism facilitates expansion of complementary businesses such as service stations and new businesses are created to cater to tourist needs for hospitality services, recreational activities and arts/crafts

4. Opportunities for Youth

The tourism industry is often promoted as an exciting and growing industry suited to the energies and enthusiasm of young people. Career options are enhanced with the opportunities for training and direct involvement in running tourism businesses, especially those within small communities

5. Service retention

Visitor information services can be provided by existing outlets, such as shops, thus increasing income flows if payment is made for acting as information outlets. Services can also benefit by the additional customers which visitors provide. Finally, tourism's importance to national economies can strengthen the political case for subsides to help retain services.

6. Community diversification

Community diversification is an important activity in many upland and climatically marginal regions. Forest regions have suffered serious socio-economic problems in recent years, partly because of the mechanization of tree felling and processing, and partly because of falling prices following reduced timber demand. Rural tourism can assist forestry by diversifying income sources for forest communities if the special qualities of the forest environment for recreational use are realized and developed.

7. Rural Tourism Enhances and Revitalizes Community Pride

Tourism encourages conformity to an ideal image of community which can result in growth of personal ties and community solidarity. Thus the basis for community solidarity shifts from shared cultural background to shared image. Amenities play a fundamental role in shaping a community's identity and pride and so the potential of tourism for improvements to facilities and amenities has positive implications for community pride, particularly rural museums as an important repository of rural culture.

8. Preservation of Rural Culture and Heritage

In rural tourism the sense of place is a fundamental element in both the tourists' and host community's feelings of what makes the area attractive to visit and live in.

This sense of place is maintained partly through rural museums which play a vital role in preserving heritage.

9. Increase Arts and Crafts Sale

Arts and crafts have a special place in the cultural heritage of regions and nations. Many commentators have noted that tourism can assist arts and crafts, both by recognizing their importance, and by purchasing craft products. Income flows from these activities are well documented. Support between the arts and tourism can be a two-way process. Many communities now use arts and crafts festivals as a marketing mechanism to encourage visitors to come to their areas.

10. Landscape Conservation

Landscape conservation has become an increasingly important form of heritage protection. Landscape is of crucial importance to rural tourism but, equally, visitor use is vital to the landscape conservation industry.

11. Environmental Improvements

Environmental improvements such as village paving and traffic regulation schemes, sewage and litter disposal can be assisted by tourism revenues and political pressures from tourism authorities. These help develop pride of place, important in retaining existing population and businesses, and in attracting new enterprises and families.

12. The Historic Built Environment

The historic built environment can benefit from rural tourism in two ways. Many historic properties now charge for admission in order to maintain their fabrics and surrounding gardens and parklands. Secondly, there are important buildings from the past which have become redundant. Churches have lost their congregations, castles have lost their wars, farm buildings have become too small for modern equipment, railway stations have lost their trains, and canal warehouses no longer have barge traffic.

The tourist industry can usually use these redundant buildings profitably and imaginatively: they can become attractions in their own right. The small town of St. Jacobs in Ontario, Canada, has converted grain stores into a craft centre; in Lanarkshire, Scotland, a folly constructed in the shape of a pineapple has become sought after as a cottage for holiday lets; in the Swiss valley of Safiental, a timber farm-house is maintained in its original pre-twentieth century condition by using it as a Youth Hotel.

A wide range of benefits has been identified as potential outcomes associated with promoting the development of agritourism. From an agricultural industry perspective, agritourism is perceived to be a means of:

- ☆ expanding farm operations;
- ☆ using farm based products in new and innovative ways;
- ☆ improving farm revenue streams;

- developing new consumer market niches;
- increasing awareness of local agricultural products;
- increasing appreciation of the importance of maintaining agricultural land uses;
- channeling additional on-farm revenues directly to family members who might otherwise have to work off the farm;
- giving a stimulus to upgrade farm living and working areas, as well as leisure facilities;
- providing opportunities to acquire managerial skill and entrepreneurial spirit;
- increasing the long term sustainability for farm businesses.

Rural Tourism Development: Pre-requisites

- Image/Visibility
- Infrastructure
- Tourism enabled services
- Heritage conservation and management
- Tourism friendly environment (non-intrusive, facilitating, safe and secure)
- Tourism products
- Deepening tourist experience (repeat visits/longer stay)

Issues in Rural Tourism

Understanding the Market

- Understanding and relating to the market
- To identify a body to coordinate a study to ascertain market size Identifying potential tourist
- Study to understand tourists profile, perceptions and expectations
- To focus on outbound tourists.

Concept Development

- Destination Development
- To offer a complete package to tourists
- To design macro-plan for each state and link to the micro-plan for each district
- Seasonality Factor
- Scheduling events to attract different segments during different times of the year

Infrastructure and Other Facilities

- Provision and management of infrastructure
- Primary infrastructure - road, electricity, water etc. by govt.

- ☆ Secondary infrastructure - hotels, transport etc. by public/private players
- ☆ Provision of hygienic food
- ☆ Rural communities to provide local food in hygienic environment
- ☆ Guide services and interpreters
- ☆ To enable tourists to understand them earnings of local culture, traditions and festivals
- ☆ Employment a venue for local youth Community involvement and community- industry cooperation
- ☆ Role of NGOs/CBOs/local govt. and community Partnership arrangement required.

Local Involvement

- ☆ Community involvement and community- industry cooperation
- ☆ Role of NGOs/CBOs/local govt. and community
- ☆ Partnership arrangement required

Impact of Rural Tourism in our Society

Rural tourism can bring various benefits such as foreign exchange, employment, higher government revenues and the transformation of agriculture into a modern service industry, it can also damage many indigenous societies.

1. Economic Impact

Positive Economic Impact

It will create employment for the rural people and generate income for them. The villagers will able to provide better food and education for their children.

- ☆ Create employment especially for the rural youth.
- ☆ Income level will rise.
- ☆ Generate foreign exchange.
- ☆ Demand for other goods and services will increase.
- ☆ Improvement in the public services.
- ☆ Generate revenue or the government.
- ☆ Modernization of agriculture and other rural activities.
- ☆ Local small businessman will be benefited.

Negative Economic Impact

- ☆ The rural people can be exploited.
- ☆ The rural people have to depend on the urban entrepreneur, so the benefit may not reach them.
- ☆ The urban investor will take away most of the profit.
- ☆ Food, drink and necessary products will be imported from outside and not produced locally.

- ☆ Rural people may be under paid.
- ☆ Local artisan may not get benefited.
- ☆ Due to competition the local handicraft and farm produce products will be sold at lower price.
- ☆ Increase in the price of local agro products.

2. Environment Impact

The rural people can learn to develop the healthy environment with proper sanitation, roads, electricity, telecommunication etc for better living on one side on other side the tourist can exploit natural resources and have a heavy impact on the environment.

Positive Environment Impact

- ☆ The rural people learn from the visitors how to lead a healthy and hygienic life.
- ☆ Help in create and maintain the natural park.
- ☆ Learn the importance of preservation of natural resources.
- ☆ They will learn to use modern tools and technology.
- ☆ They will learn to preserve the natural habitats, biodiversity historical monuments.

Negative Environment Impact

The visitors may overexploit natural resources and it can have a heavy impact on the environment. In addition, rural tourism will require infrastructure, transportation and other facilities which can cause environmental distortion.

- ☆ Development of infrastructure may distort the natural beauty.
- ☆ Huge number of visitors may exploit the natural resources
- ☆ Hunting can affect the wild life population.
- ☆ Natural ecology will be disturbed.

3. Socio-Cultural Impact

Since the income from tourism is much higher than what rural people can earn from agriculture, tourism has been accepted willingly in many rural areas in spite of its negative effects. It is widely recognized that such negative impacts on rural communities have become stronger, and that rural tourism must be modified to give rural people its benefits. Poorly planned tourism can mean that villages are invaded by foreign visitors with different values, disrupting rural culture. The higher standards of living in urban tourist destinations have caused emigration from nearby rural neighbours, resulting in changes in the demographic structure and possible culture shock.

Positive Socio-Cultural Impact

The rural people will learn the modern culture. They will learn to come out of their traditional values and beliefs. They will adopt different practice of modern society.

- ✰ Education and health of the rural community will improve.
- ✰ Provide higher standard of living or the rural people.
- ✰ Cultural understanding through fairs and festivals.
- ✰ Exchange of cultural beneficial for both.
- ✰ Foreign tourist will bring cultural change faster.
- ✰ Reduce migration of rural people to urban areas.
- ✰ Market for agro products and handicrafts will develop in rural areas farmers and artisans will develop a direct contact with the customers.

Negative Socio-Cultural Impact

Poorly planned tourism can affect the villagers. It may affect the traditional and cultural practices, agriculture and other allied activities.

- ✰ Can create disharmony in development.
- ✰ Modernization can affect their traditional values and cultural practices.
- ✰ Traditional products will be replace by modern products.
- ✰ Rural people may shift from traditional business to tourism activities.
- ✰ Decline in participation in rural traditional and cultural practices follows.

Challenges in Rural Tourism

The major challenges are need to preserve the environment and natural resources, the need for education, proper understanding for both tourists and local people, and the need to generate a democratic movement which helps people at all levels to participate in tourism development. These are some major challenges are-

1. Lack of Basic Education

According to 65 per cent of the respondents, lack of literacy is a major problem in rural tourism. The rural literacy is 59 per cent as per the Census report 2001 while the literacy rate was 45 per cent in 1991 in rural India. Only 2.7 per cent of the rural population is graduate and above.

2. Language Problems

75 per cent of the respondents have said language can be one of the hindrances in rural tourism. Although Hindi is an official language, but in many parts of India people do not understand it. Either the rural people have upgrade them to communicate with the urbanites otherwise they will not going to get much benefit from the rural tourism.

3. Legislation Problems

48 per cent of the respondents have pointed out there can be legal problems. All hotels, motels and cottage having license are paying taxes to the government. According some respondent, the rural tourism should have a tax holiday or it should be tax free. There could be law and order problem.

4. Insufficient Financial Support

65 per cent of the respondents have suggested that to start a rural tourism, sufficient fund is required to promote this business. Central government and State government should encourage rural tourism by providing financial support to start the project. Because it will create employment in rural areas and it will also help in flow of fund from urban to rural. It can help in preventing the migration of people from rural area to urban areas.

5. Lack of Trained Manpower

61 per cent the respondents said that there will problem of trained manpower from hospitality industry. The trained people in hotel management may not be interested to go to rural area to work. The rural people who will be appointed are required to be trained for discharging their duties. Decorating the cottages or suites and maintain them. The success of rural tourism totally depends on the quality of service provided to the tourist.

6. Underdeveloped People

70 per cent of the respondents said that a vast majority of the rural populations are uneducated and literate; they are bounded by the traditional values and customs. Their culture, religion, superstition have a strong influence on their attitudes and behaviours. But it has been observed that after globalization the rural economy started showing the aspiration to grow, the media is playing an important role in changing the mindset of the rural consumer. They are exposed to different technology by provided through government or non-government initiatives.

7. Lack of Proper Physical Communications

80 per cent of the respondents said physical communication will be a big problem. Nearly half of the villages in the country do not have all-weather roads. For developing in tourism in any rural area we not only need all-weather roads but we also have arrange safe drinking water, electricity, telephone, safety and security, etc.

8. Business Planning Skills

48 per cent of the respondents said that the rural people generally lack in the business acumen. To do any business, they require technical knowledge and skill to organize and maintain it. With the help of government or non-government organization they can prepare business plan. If some outside is preparing a project for rural tourism, the villagers should participate in developing and implementing the project, otherwise it will not going to give much benefit to the rural people. Advertisement and sales promotion will play a very important role in creating awareness and attracting the customers.

9. Trained Tourist Guide

According to 57 per cent of the respondents availability of trained guide will be a problem. The guide plays a very important role in attracting tourists. Some of the respondents pointed out that the guide should have thorough knowledge about the place and he or she should able to generate interest in the mind of tourist to visit the site. The guide can show the album, video film, brochure to give knowledge about the places. The guide should give what he promise during booking.

10. Communication Skills

71 per cent of the respondent has pointed out the communication will be a major problem. Language and education is the basic hindrance in communication. The ability to communicate effectively is essential. Much of success will depend on your ability to give warm welcome to tourists. They have to understand the tourist wants and needs.

Opportunities

75 per cent for the respondents are interested for rural tourism, if the trip can bring pleasure and relaxation of body and mind. Once in a year, urbanites have expressed their interest to go for rural tourism. Especially to those area where there is beauty of natural elements like mountains, forests, sea, lakes etc. Along with that traditional customs, handicraft of the rural flock, traditional foods and their hosting style and the cultural programmes. For an example In Rajasthan, we can see reflection of these types of programmes in Jaipur and Udaipur especially. Jaipur and other cities of Rajasthan have developed some unique selling proposition to attract domestic and foreign tourists. Some of the hotels of Rajasthan are providing traditional food in a very traditional style. As tourists become more and more educated, more aware of facilities available and more experienced, so their expectation has also increased. People are interested to explore new places. Rural tourism in India has great future, since it not only provides natural elements of beauty but also the indigenous local traditions, customs and foods. Direct experience with local people can be a unique selling proposition to attract tourists. Every state in India has some unique handicraft, traditions and foods. The Rural tourism should not go for a mass marketing. Rural tourism should develop different strategy for different segment to make it successful.

Tourism and Biodiversity

Biodiversity is vital for tourism. Coasts, mountains, rivers and forests are major attractions for tourists around the world. Tourism in the Caribbean, Mediterranean and much of Southeast Asia depends strongly on the recreational opportunities provided by their coastal environments. In southern and eastern Africa, tourism based on wildlife safaris is a dominant attraction and source of income for the tourism sector. Wildlife and landscapes are important attractions for tourism in mountain areas.

Biodiversity plays different roles in different types of tourism. All tourism –even in city centers – relies on natural resources for supplies of food, clean water and other 'ecosystem services' that ultimately depend on biodiversity. For most other types of tourism, biodiversity contributes significantly to the attractiveness and quality of destinations, and therefore to their competitiveness: for example, coastal water quality and natural vegetation are both ecosystem services that contribute to destination attractiveness. And biodiversity is a direct attraction at the heart of nature-based tourism products– such as wildlife watching, scuba diving or tourism in protected areas.

However, biodiversity is under pressure worldwide and has suffered severe losses as more land is converted for human use from a natural state, and as these human uses become more intensive.

In 2005, the UN's Millennium Ecosystem Assessment concluded that human activities threatened the Earth's ability to sustain future generations. Clearance of land for tourism developments has contributed to these losses, particularly in coastal and mountain areas. Inappropriately sited developments have damaged natural coastal defenses, making coastal areas more prone to storm damage; or have damaged mountain vegetation and soils, making them more liable to erosion and increasing flooding risks. The physical pressure from the many visitors who are interested in and want to visit sites with rich biodiversity but fragile environments has also created problems. For example, coral reefs are easily damaged at heavily used scuba diving sites, and the effects of trampling by visitors can change and, eventually, destroy vegetation in mountain areas.

Rural Tourism as an Enterprise in West Bengal: Case Study

Rs 70 lakh project will be opened at Ballavpur Danga village near Santiniketan. The project, which has been funded by the West Bengal Tourism department along with Central funds, will primarily be run by self-help groups (SHG) thereby, employing local people, state tourism minster Manabendra Mukherjee said and a another similar project Mukutmanipur in Bankura district will also be opened soon

Ballavpur Danga is 3 km from Shantiniketan, the celebrated home of Nobel laureates Rabindranath Tagore and Amartya Sen. It is a Santhal Adivasi tribal community, framed in the rustic beauty of rural Bengal. The Sonajhuri forest lies to the east and the Ballavpur Avayaranya forest area and bird sanctuary is to its south. Santhali art, craft and culture is woven into the community's life. Attractive designs adorn the walls and floors of their homes, with colours made from red earth and cowdung. Tribal dances unfold in the open air, with fire, to the rhythmic beats of the madal, their traditional musical instrument, while oral Adivasi myths and history remain etched in song and recitation. For all these ventures, while the state government will prepare the physical infrastructure, SHGs will undertake the daily activities, for which specific soft-skills training will be imparted.

Relevance and Impact of Rural Tourism in Agricultural Extension

1. The conventional agriculture has turned classical And mundane in providing income and livelihood for rural population. Thus rural tourism open up a new vista to them.
2. Through the application of proper extension strategy, the function of agro-ecosystem consisting of water body, orchards, animal resources and indigenous technical knowledge can well be characterized into rural tourism components.
3. Rural tourism creates a platform for exchanging and assimilating rural-urbane values, cultures and livelihood technologies.
4. A new goner of extension pursuits will be resultant, can be branded as *Rurban* extension approach.

5. The socio-economic and ecological indicators are going to be the must to have an objective analysis on what kind of impacts are being (positive/ negative) generated from.
6. The economic and ecological analysis of ethno medicines biodiversity, practices and customs, wisdom mentoring can be resultant of upcoming rural tourism strategy.

Recommendation

Governments should recognize importance of rural tourism at priority and help in creating healthy competitive business environment. Government should try to generate data for decision-making bodies investing for developing the human resources, create adequate facilities and suitable infrastructure like accommodation, roads, airport facilities, rail facilities, local transport, communication links and other essential amenities become essential for development of rural tourism.

Some of the essential services required for rural tourism.

- ☆ Build confidence in safety and security.
- ☆ Plan for sustainable growth of rural tourism
- ☆ Invest in new technology
- ☆ Business must balance economics with people, culture and environment.
- ☆ Develop rural tourism protecting natural resources, local heritage and lifestyles.
- ☆ Fill the narrow gaps between the 'haves' and 'have-nots'.
- ☆ Promote traditional tourism products.
- ☆ Improve quality, value of rural tourism.
- ☆ Give quality standards of services at all levels
- ☆ Transfer skills and best practice.
- ☆ Share information to make better business decisions.
- ☆ Communicate more effectively.
- ☆ Share the pursuit of long-term growth and prosperity.
- ☆ Set environmental policy goals.
- ☆ Develop and deploy skills effectively.
- ☆ Collaborate on information requirements.

Conclusion

Rural tourism will emerged as an important instrument for sustainable human development including poverty alleviation, employment generation, environmental regeneration and development of remote areas and advancement of women and other disadvantaged groups in the country apart from promoting social integration and international understanding. It can help inflow to resources from urban to the rural economy. It can prevent migration of rural people to urban. Both short-term and long-term planning, implementing and monitoring are vital in avoiding damage to rural areas. Environmental management, local involvement, sound legislation, sustainable marketing, and realistic planning are crucial for development of rural tourism. The government should promote rural tourism to ensure sustainable economic development and positive social change.

References

Statistics Canada and Alberta Tourism, Parks, Recreation and Culture, *Canadian Business Patterns – Business Register March 2006, Statistics Canada, Total Tourism Sector and Labour Force Survey 2006, Statistics Canada*

Mason, P., and Cheyne, J. (2000). Residents' attitudes to proposed tourism development. *Annals of Tourism Research 27, 391–411.*

"UNWTO technical manual: Collection of Tourism Expenditure Statistics". World Tourism Organization. 1995. p. 10. Retrieved 26 March 2009.

"International tourism receipts surpass US$ 1 trillion in 2011".UNWTO. 7 May 2012. Retrieved 15 June 2012.

"2012 Tourism Highlights". UNWTO. June 2012. Retrieved 17 June 2012.

"International tourism challenged by deteriorating global economy". *UNWTO World Tourism Barometer* (World Tourism Organization) 7 (1). January 2009. Retrieved 17 November 2011.

"Manila Declaration on World Tourism". World Tourism Conference. Manila, Philippines. 10 October 1980. pp. 1–4.

Barre, H., Jafari, J., 1997, Culture, Tourism, Development: Crucial Issues for the Twenty-First

http://albertacountryvacation.com

www.tourismindia.com

www.rajasthantours.allindiaguide.com

www.incredibleindias.org

www.tourismindiaonline.com

www.indiabusiness.nic.in

Chapter 41

Interpersonal Communication in Agril. Entrepreneurship

Manas Mohan Adhikary and Sankar Kumar Acharya

"A spoken word flies, you won't catch it". A Russian proverb.

"I will pay more for the ability to deal with people than for any other ability under the sun". John Rockefeller.

"Trust is the single most important factor in both personal and professional relationships".

Interpersonal communication skills are much more important than technical skills if you want to become a senior manager. Good communication is a very important aspect of interpersonal relationships. Listening is one part of communication which sounds fine in theory but is rare to find in practice. Empathy will help you to understand people and respond to their needs more effectively. Understanding the basic communication model should help to improve your communication skills. Experts maintain that, as much as 65 per cent of the message communicated is non-verbal or body language. There are some practical tips you can apply to help people to like you more.

Similarly, an extension worker fails to achieve results in the absence of interpersonal communication skills though he may be endowed with the best of the technology or even the best of the strategies. The ability to get along with others is immensely helpful for getting success in almost all walks of life. Whether one acts in role of extension worker, farmer, Village Developmental Officer, Scientist, Research worker, the manner in which we interact with people would determine very largely if life remains buoyant or miserable.

Research reveals that there is a positive correlation between good relationship and productivity, profit. Success of an extension worker will largely depends upon an effective communication which in turns depends upon the interpersonal communication skills such as empathy, listening, positive attitudes, perception, presentation skills, written communication skills, courtesy, manners, a sense of humour, etc.

Interpersonal Communication Skills

Communication is two-way

Communication is the art of sharing ideas, information, instruction or feelings. The basic ingredients of good communication are clear thinking, clear speaking and clear writing. We spend about 75 per cent working hours in some form of communication such as talking, listening, reading and writing.

There are 1 million words in the English language (about 200,000 of them are technical). The average person has a recognition vocabulary of 10,000 words and a conversation vocabulary of 2000 words. However, in everyday usage most people manage on a vocabulary of as little as 300 words. In everyday conversation therefore, you should choose words that are most frequently used and understood. It is also wise to keep sentences short and to the point. Avoid jargon and 'in-company' phrases, particularly when dealing with outsiders such as suppliers and customers.

Listen to Learn – Learn to Listen

The average speaking rate is 175 words per minute, but we think at about 400words per minute. This gives each of us a lot of spare capacity for mental doodling. Try to counteract this natural phenomenon by concentrating on what the speaker is saying and meaning. Focus a central ideas rather than details. Use your spare mental capacity for summarizing main points anticipating what is going to be said and observing body language.

Concentrate

Expert in communication estimate that only about one-quarter of all listeners are able to understand the main idea when listening to a speaker. If you find your kind wandering when listening to somebody, you must get back into focus. You can use the spare capacity productively. It makes good sense to summarize in your own mind the key concepts and ideas the speaker is trying to convey. If you are unsure about the message, be brave and repeat it back to the speaker's satisfaction. This is called feedback and ensures that two-way communication has taken place. Remember, communication does not take place until it is understood. Speaking quickly and indistinctly may also prove a barrier to communication.

You are now beginning to realize that communication is not as easy as it seems. We talk, listen, read and write each day but we don't give it much thought. However, we must continually work at each of them if we are to become skilled in that most underrated art, communication.

Ask questions. Ask, don't tell. Use open-ended questions. Questions beginning with whom? What? Where When? How and why? In dealing with enquiries we need to elicit information in order to identify and solve the problem. So use this questioning technique.

Effective Listening

Have you ever been at a meeting when nobody seems to listen to anybody else? Of course, you have. The meeting consists of a series of monologues. The people are not really listening to each other but thinking and planning what to say as soon as the other person pauses for a breath. Experts maintain that the best way to make friends is to become an attentive listener. Many people fail to make a favourable impression on others, simply because they do not take the trouble to listen. Communication is a two way process and listening is very important aspect of it. Somebody once said that the reason we have two ears and only one mouth is because we should listen twice as much as we talk. Apart from anything else, it is just plain good manners to listen. The next time you are in conversation with someone, lean a little forward to show your interest and be seen to be actively listening and eager to help. Consider some of the barriers to effective listening are:

- ☆ **Wishful thinking:** Most people only hear what they like to hear. This phenomenon is known as wishful thinking.
- ☆ **Attention:** Don't permit your thoughts to stray or your attention to wander.
- ☆ **Semantics:** This is the science, which deals with the development of the meaning of words. Do not put an interpretation on words and phrases other than that intended by the speaker.
- ☆ **Talk person to person:** Don't be too formal. Be sincere and sympathetic in your manner. Build trust. Never breach confidence. Make the other person feel welcome and important. Use little courtesies such as 'Good morning', "please" 'thank you', and so on.
- ☆ **Attitude:** Develop a positive attitude to others. Don't close your mind to other people's opinions and viewpoints. There are enough 'know alls' in the world. Show respect for their opinions. Don't consider yourself too good to learn from others. Pride is one of the seven deadly sins. Listen, you may learn something.
- ☆ **Get the facts:** Don't prejudge a person or situation. Many people stereotype others by their nationality, colour, race, religion or appearance. Listen, understand and then judge. Don't jump to exclusions.
- ☆ **Excessive talking:** How many of us are infatuated with the sound of our own voice? We cannot listen and talk at the same time. Comment on what the speaker is saying without interrupting the flow. Occasionally paraphrase what you hear. This provides vital feedback and ensures that the message is being received and understood.

Empathy

Sensitivity is probably the most important factor in successful communication. Sensitivity in a nutshell is the ability to see things from the other person's point of view. Empathy, which means listening with understanding, is the modern expression for the same idea. People have feelings, opinions, fears and prejudices. Empathize with them and show concern and respect. People like to be appreciated and made to feel important.

Self Awareness

Self-awareness is a key factor in good personal relationships. Be aware of your strengths and weaknesses. Optimise your strengths and minimize your weaknesses. You must be come aware of your personal biases, likes and dislikes and shortcomings. If they are interfering with your social skills, then you must eliminate them from your behaviour. Before you develop the ability to cope with others, you must first have the ability to live with yourself. Keep your emotions under control. Be cool, calm and collected. Never lose your temper with others. Remember, the best way to win an argument is to avoid it and the best way to cook your goose is with a boiling temper.

There is an old Chinese proverb: 'Man with sour face should not open shop'. Yet how many times do we go into shops to be met by unsmiling, uncaring and disinterested faces? It does not cost anything to smile. A smile says, 'I like you', I am glad you are here. When you meet people, do you smile?. Develop a sense of humor. Don't take yourself or life too seriously. Learn to laugh at yourself. However, don't laugh at others and avoid sarcasm at all costs.

Are you responsive to others? Listen and accept the feelings content as well as the word content of the message. Listen and question for facts, feelings and opinions so that you have a total understanding of the message. Be positive. Tell the person how you propose to solve the problem and the action you will take with a specific comeback date.

The Communication Model and Interpersonal Interaction

In the basic communication model there is a sender, a message and a receiver. In other words, a speaker, words and a listener.

Sender

First the sender conceives an idea which must be translated into words. The effectiveness with which the sender conveys the message will depend on choice and appropriateness of words, vocabulary, tone and gestures. All these factors help to project the message with precision and clarity. The message may be distorted, ignored or misunderstood, if any of these is mishandled by the sender.

Message

The message consists of words and non-verbal cues, what are words? Words are symbols which represent concrete objects and concepts. It is important to rember that words can create understanding and inspire people to great deeds or they can create misunderstanding, distract, annoyance and bad feeling. So pick your words with care and discretion.

Words are not necessarily the precision tools of description that they are made out to be.

For Example

- ✰ Words mean different things in different contexts. In the context of a training, programme, 'course' means curriculum, but in the context of a horse or a greyhound race it means route. Therefore meaning of words lies in the people and not in the words.

- ☆ Words mean different things in different cultures.
- ☆ Words mean different things to different people. A bull to a farmer means livestock. On the other hand, to a stockbroker, it means a person who buys shares.

Receiver

The third ingredient in the communication model is the receiver. The involvement of the receiver can be considered in three ways. Firstly, the receiver recognizes what the speaker is saying and tunes in. secondly, the receiver interprets the meaning of what is being said by mentally summarizing and reorganizing the idea. Thirdly, the receiver understands the message, which has been assimilated, associates and consolidates with exiting knowledge and only then responds to what has been said. You can now see just how complicated communication is. Mishaps can occur at any stage and interfere with the message. This is the communication cycle.

Verbal and Non-Verbal Communication

The experts tell us that only 35 per cent of what we mean is in the verbal message. The other 65 per cent of the meaning is contained in the non-verbal communication. Facial expressions, eyes and other body movements must be considered when interpreting meaning. In any social interaction with other people you must be aware of the non-verbal cues and signals. Psychologists have been paying increasing attention to these non-verbal aspects of communication. The following are some of their findings.

- ☆ **Boredom:** The hand over mouth to cover a yawn can eman boredom. Glances at watches, frequent recrossing of legs, constant repositioning on seat, fidgeting and eyes wandering can also indicate boredom, disinterest, impatience or tiredness.
- ☆ **Arms folded:** Folded arms usually indicate resistance; the listener is forming a protective barrier against you and your ideas. But beware, it can also mean other things.
- ☆ **Frown:** The frown may suggest disagreement, lack of understanding or annoyance. Several responses are possible. You can back up and restate your last point. You can pause and ask for questions, or ask you can clarify your point.
- ☆ **Upward glance:** Watch peoples eyes. The eyes are the minor to the soul. The sky ward glance may mean "I have heard that one before". Eye contact is essential for good communication.

Memory

The average person has a poorly trained memory. Both the long-term and short-term memories are important factors in communication. Research has shown that we forget 50 per cent of what we read immediately unless we review. We forget 80 per cent of what we hear within 24 hours. We forget 90 per cent within 48 hours. Don't rely on your over worked memory that we tend to remember what we like and forget what we dislike.

Encouraging People to Like You

There are some practical steps we can take to encourage other people to like us better. We can put the following tips into practice straight away:

Positive Thinking

Many books have been written about the power of positive thinking. The problem with negative thinking is that it becomes a self-fulfilling prophecy. The good news is that positive thinking is likewise a self-fulfilling prophecy. Shakespeare said: There is nothing good or bad except that thinking makes it so. Believe in yourself. Think positive experiences. Think about your successes rather than your failures. Develop a mindset for success. Positive expectations are more likely to lead to positive outcomes. Positive thoughts precede positive actions and positive actions produce positive results. Negative expectations are more likely to lead to negative outcomes. Positive thinking is about developing a favourable image of yourself. Remember there is often little correlation between success and a positive self-image. Many successful, Hollywood stars have a very negative self-image and commit suicide. Unless you have a favourable image of yourself you cannot impress others favourably. You must be able to like yourself before you can like other people. Develop faith in your own ability to handle colleagues and bring work task to completion. The more successful you are the more confidants you will become in the future. Don't underestimate yourself and never sell yourself short.

Show Interest

Take a genuine interest in the people around you and in the work environment. To be interesting, be interested. Ask questions that the other person will enjoy answering. Encourage them to talk about themselves and their achievements. Disraeli said : Talk to a man about himself and he will listen for hours. Develop a friendly and helpful attitude. Always try and be sincere. If you are artificial, people will eventually see through you. Be positive. If you don't know something, say: 'I don't know, but I'll certainly find out and let you know', rather than a curt: 'Don't know'. Develop a reputation for reliability. If you say you're going to do something, do it!

Remember to Smile

A smile is like a piece of sunshine. You'll be surprised at the friendly way people will react. Smiling facilitates positive thinking by giving you a psychological lift. You can't smile and entertain negative thoughts at the same time. Develop a cheerful disposition.

- ☆ **Name**: People like to hear the sound of their own names. One of the best ways of crating goodwill is to remember names and use them. To remember names, relate them to the person's features and repeat the name frequently during the conversation.

- **Manners**: Good manners are a sign of good breeding. Manners help make living pleasant. Like smiling, good manners cost nothing but reap many benefits.
- **Learning is a lifelong continuous process**: Seek out opportunities for self-development, not only in your work life but also in your personal life. Action cures fear. Don't procrastinate. Undertake that educational programme non-whether formal, such as certificate, diploma or degree programme, or informal, such as challenging recreational pursuits. It will make you a more interesting person.
- **Practice**: Practice the previous advice. It's like driving a car, if you practice these, they will become automatic confident responses. Bernard Sha once remarked: 'If you teach a man anything, he will never learn'. Learning is an active process. We learn by doing. Only knowledge that is used sticks in your mind.
- Some experts maintain that the ability to handle people is three times more important than technical expertise in determining the suitability of people for senior managerial positions. So all the technical knowledge (or job knowledge) in the world is of relatively little use if you aspire to senior management, without the social skills to handle people.

Handling People

The following guidelines will help you to handle people.

- **Tact**: Don't argue: The best way to win an argument is to avoid it. Let the other person do most of the talking. Use the feedback technique. Ask questions in a friendly voice. Use the open-ended questioning technique – Why ? What ? When ? Where ? How? And Who? Arguing encourages emotional responses such as anger. Anger leads to irrational behavior and anything that encourages anger should be avoided. If, despite everything, a person gets angry, don't get angry back. 'Anger blows out the light of reason, but, after a roaring hurricane expends its fury, the full of calm sets in'. Let the angry person blow themselves out.
- **Respect**: Show respect for the other person's feelings and opinions. Tact and diplomacy are the watchwords here. Never say directly, or suggest to a person, 'You're wrong'. You could say you may be right, but let's look at the facts'. Nobody likes being told directly that they're wrong especially in front of a third party. Lord Chesterfield said to his son : 'be wiser than other people, if you can; but do not tell them so'. You may praise but never criticize in front of others. If you use diplomacy and tact, they may admit to themselves that they are wrong. But never, if you crudely try to ram the fact down their throats.
- **Empathize**: See things from the other person's point of view. How would you feel if you were in their shoes? That is what empathy means. Think ahead and try and anticipate reactions. Henry Ford said : 'If there is any one secret of success it lies in the ability to get the other person's point of view and see things from his angle as well as from your own'.

- **Simple**: Speak in simple language. Remember the KISS technique-Keep If Short and Simple. Avoid jargon or technical terms. Don't try to impress others by using big words. Always use the simple work in preference to the long work. Keep your sentences short. Big words and long-winded sentences amount to woolly thinking, confusing others as well as yourself.
- **Praise**: Most people are very reluctant to praise others, even when it is due. We spend most of our time knocking others. Treat people as winners and they will live up to your expectations. A sincere compliment for work well done is a boost to morale and an incentive for sustained excellence in the future. 'You can always catch more bees with honey than you can with vinegar'. However, insincerity is counterproductive. If you criticize for whatever reason, always soften the criticism with praise first criticize the act not the persons.
- **Appeal to senses**: Dramatise your ideas and involve the other person. Appeal to as many of the senses as possible – hearing, sight, taste, smell and touch. When making a presentation to a group use visuals-a picture speaks a thousand words. Remember the old proverb: 'I hear, I forget;; I see, I remember; I do, I understand'.
- **Synchronize with your listener**: Watch for body language and non-verbal communication. Try and synchronize your words with the speed of receptivity of the listener (you can gauge this from the actions and expressions of the listener). Mirror your body language to that of the other person.
- **Sell** : Make other person feel the idea is theirs. This is the best way to implement your ideas. Use the open-ended questioning technique to obtain viewpoints. Concentrate and develop those on which there is common ground and agreement. Show what's in it for them. Most people are motivated by self-interest.

Making an Impression on Others

The critical points about making an impression on other people are

- Posture
- Look, listen and learn
- Expression
- Appearance
- Speech
- Eagerness to help

Posture

Your posture can indicate clearly whether or not you are going to be friendly and helpful. Indifference, nerves, restlessness and lack of confidence can all be indicated by posture.

Look, Listen and Learn

For the total message listen for the words and observe the accompanying body language. Pay attention to what they say and show that you are really interested in them. People prefer to talk about themselves, rather than listen to you. Don't interrupt the speaker or change the subject or you will show lack of interest and may create resentment.

Expression

Most people look at your face, and particularly your eyes, at some point during your conversation. Shakespeare said that the eyes are the mirror of the soul. Remember, your face and eyes reveal your feelings. They show the other person not only how your yourself feel – tired, interested, uninterested – but also how you feel about the other person. It's polite and a sign of attentiveness to look at the person who is talking to you (but avoid staring). Looking at a person also gives you and opportunity to notice their expression. Use your mouth. A smile, especially when greeting someone, can be the biggest icebreaker of all. Your mouth can show friendliness as easily as it reveals boredom and hostility.

Appearance

When somebody comes up to you, both you and your work area are on view. Do you give the impression that you are friendly, neat, well-groomed, smart and organized? Remember, if you look untidy, your work may be untidy too. People act on impressions. Make sure that you are neatly turned out and look after your appearance and personal hygiene.

Speech

Your first words create an impression, which colour the other person's reply. If they are friendly and positive, they invite a smile and a thank you from the other person. Nothing is more annoying than unhelpful, negative remarks. Remember to use the person/s name. There is nothing sweeter to a person's ear than the sound of their own name. Use the name frequently during the conversation. Link some outstanding feature with the person's name to remember it better. To be heard and understood, you need to speak clearly and look at the person as you speak. Do not use bad language or slang in conversation. If you use bad language you will let yourself down and people will judge you by the way you speak.

Eagerness to Help

Adopt a positive attitude and helpful manner in your dealings with people. Show in a positive way that you are enthusiastic, eager and willing to help solve problems. If you want to make friends you must be seen to be obliging, unselfish and thoughtful.

Consequences of Poor Personal Inter-Communication Skills

- ☆ Stress
- ☆ Lack of communication

- ☆ Irritation
- ☆ Close mindedness
- ☆ No team spirit
- ☆ Lack of credibility
- ☆ Poor self-esteem
- ☆ Suspicion
- ☆ Loss of productivity
- ☆ Isolation
- ☆ Poor health
- ☆ Distrust
- ☆ Anger
- ☆ Prejudice
- ☆ Un co-operative behaviour
- ☆ Conflict
- ☆ Frustration
- ☆ Unhappiness

Interpersonal communication skills are much more important than technical skills if you want to become a senior manager. Good communication is a very important aspect of interpersonal relationships. Listening is one part of communication, which sounds fine in theory but is rare to find in practice. Empathy will help you understand people and respond to their needs more effectively. Understanding the basic communication model should help to improve your communication skills. Experts maintain that as much as 65 per cent of the message communication is non-verbal or body language.

Some important guidelines for handling people were tact, respect; empathize; simple; praise, synchronize with your listener.

The critical points in making a good impression on others can be posture; look, listen and learn, expression; appearance; speech, eagerness to help.

Consequences of poor interpersonal communication skills are stress; lack of communication; irritation; close mindedness; no team spirit; lack of credibility; poor self esteem; suspicion; isolation, poor health etc.

Chapter 42

Organisational Communication for Entrepreneurship and Value Addition

Sankar Kumar Acharya and Manas Mohan Adhikary

Communication is the most vital ingredient of an organization. In fact, an organization cannot be conceived of without communication. An organization is a group of persons constituted to achieve certain specific objectives. The achievement of these objectives largely depends upon a proper Co-ordination and integration of human effort in an organization. The people working in an organizaton are interrelated; their activities are also interrelated because all activities are performed only to achieve the organizational objectives. Co-ordination and integration of various human activities are possible only if there is an effective system of communication in the organization which provides for exchange of information and sharing of various ideas. The more effective the system of communication is, the better are the relations between the workers, and between workers and the management.

Communication is the nervous system of an organization. It keeps the member of the organization informed about the internal and external happenings relevant to a task and of interest to the organization. It coordinates the efforts of the member to achieving organizational objectives. It is the process of influencing the action person or a group. It is a process of meaningful interaction among human beings to initiate, execute, accomplish, or prevent certain actions. Communication is, thus, the life food of an organization. Without communication, an organization is lifeless and its very existence is in danger.

Concept of Organisation

Organization is the backbone of management. Without efficient organization, no management can perform its functions smoothly. Hence let us define what an organizations is

- ☆ According to Allen, organiztion may be defined "as the process of identifying and grouping the work to be performed, defining and delegating responsibility and authority and establishing relationship for the purpose of enabling people to work most effectively together in accomplishing objectives".
- ☆ According to Harold Koontz and o'Donnell – "organizing involves the grouping of activities necessary to accomplish goals and plans, the assignment of these activities to appropriate departments and the provision for authority delegation and co-ordination".
- ☆ According to Prof. Haiman. "organization is a process of defining and grouping the activities of the enterprise and establishing the authority relationship among them".
- ☆ According to Mooney and Reiply – "organization is the form of every human association for the attainment of a common purpose".
- ☆ In the words of chester Barnard – "A system of co-operative activities of two or more persons is called organization.

Concept of Communication

Communication is the nervous system of an organization. It is the life blood of an organization. The following are the some of the definitions given by different authors.

- ☆ Communication is the sum of all things, one person does when he wants to create understanding in the minds of another. It involves a systematic and continuous process of telling, listening and understanding". Allen louis.
- ☆ Communication has been defined "As the transfer of information from one person to another whether or not it elicits confidence". Koontz and o'Donell.
- ☆ "Communication is an exchange of facts, ideas, opinions or emotions by two or more persons". George Terry
- ☆ Communication is defined as "The process of passing information and understanding from one person to another. It is essentially a bridge of meaning between people. By using the bridge of meaning a person can safely cross the river of misundersanding". Keith Davis.

Concept of Organizational Communication

"Organizational Communication is an evolutionary, culturally dependent process of sharing information and creating relationships in environments designed for manageable, goal oriented behaviour"

Types of Organisational Communication

Communication in an organization carries innumerable kinds of messages which may be difficult to map out; but it may be possible to classify communications in regard to how to transmit, or who communicates to whom, or what kinds of relationships communication develops. Thus, Communication may be grouped on the following basis.

According to Organizational Structure

Formal Communications

Such communications are those communications, which are associated with the formal organization structure. The travel through the formal channels – officially recognized positions in the organization chart. They are established mainly by the organization structure. Formal communications are mostly in black and white. We generally hear the phrase "through proper channel". It explains the essence of formal channels. Such communications include orders, instructions, decisions or intensions etc., of the superior.

Informal Communications

Informal Communications are also known as "Grapevine Communications". They are free from all sorts of formalities, because they are based on the informal relationship between the parties, such as friendship, membership of the same club or association or origin from the same place. Such communication includes comments, suggestions of any other informal reaction also. They may be conveyed by a simple glance, gesture, nod, smile or mere silence too.

According to Direction of Communication

Downward Communication

Communications which flow from the superiors to subordinates are known as downward Communications. They include orders, rules instructions and policy directives etc. Their nature is directive. It would be impossible to manage an enterprise without downward Communications.

Upward Communications

Upward Communications are just reverse of the downward Communications. It flows from the sub-ordinates to their superiors. Such communications include reactions and suggestions from workers, their grievance etc. Contents of the upward Communication are reports, reactions, suggestions, statements and proposals prepared for the submission to the boss etc. Upward Communication is considered to be a main source of motivation in employees.

Lateral/Horizontal Communication

This type of communication refers to communication between various departments or units representing the same level, or people within the same or different departments, without having a superior – sub-ordinate relationship *viz.*, peer groups,

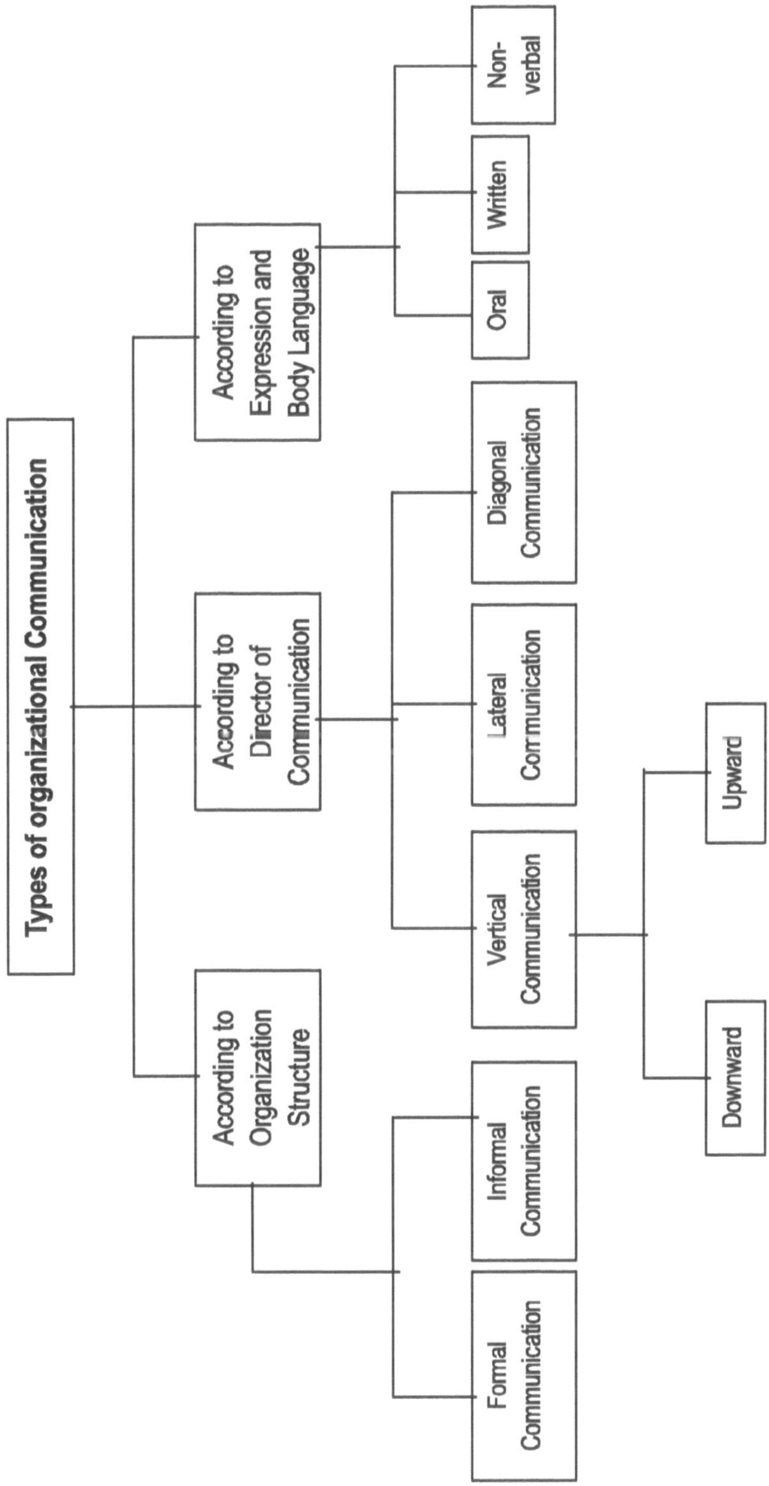
Types of organizational Communication
According to Organization Structure
According to Director of Communication
According to Expression and Body Language
Formal Communication
Informal Communication
Vertical Communication
Lateral Communication
Diagonal Communication
Oral
Written
Non-verbal
Downward
Upward

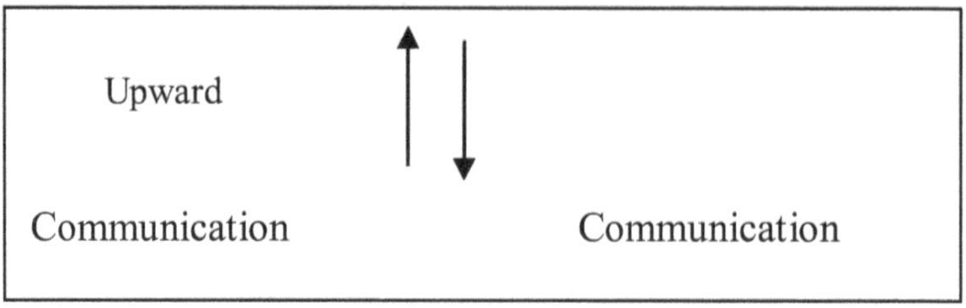

Exhibit : Vertical Communication

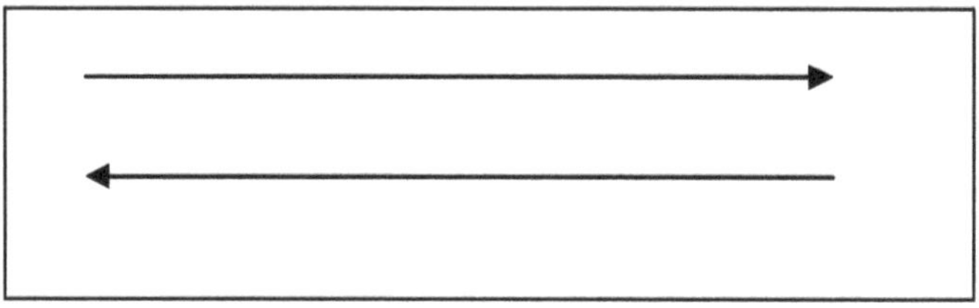

Exhibit : Lateral Communication

friends and trainees. It flow between persons at the same hierarchical level. The main object of this type of communication is to co-ordinate the efforts of different departments performing different but related activities. Such communications may be oral or written.

Diagonal Communication

In an organization, communication does not necessarily traverse along a stipulated path. While vertical and lateral forms for transmission of messages are important, there is yet another mode to be taken cognizance of, that is, the diagonal. In this type of communications, there is no direct path chalked out for information to travel. It could, at certain stages, take on the upward path, then a lateral direction and finally, move downward, or it could even skip certain stages.

This channel proves to be very affective as hierarchical bindings are done away with and communication flows irrespective of position or status. It also helps in building relationships and binding ties between the superior and the subordinate.

According to Expression and Body Language

According to way of expression, and body language, the communication may be oral or written and Non-verbal.

1. Oral or Verbal Communication

In oral communications both parties to the process of communication exchange their ideas through oral words either in face-to-face communication or through any mechanical device such as telephone, etc. Meetings and conference, lectures and interviews are other media of such communications.

2. Written Communication

They are communications on black and white. They include written words, graphs, diagrams, pictures etc. They may take the form of circulars, notes, manuals, reports, posters or memos etc.

3. Non-Verbal Communication

Non-Verbal Communication is an integral part of us and helps in communicating effectively. The way an individual positions him, holds his hands, tilts his head, all transmit volumes about the individual.

A receiver observes non-verbal communication 55 per cent of the time, which is much more than listening merely to the words and the voice articulation. Lack of emphasis in this area is due to paucity of material and lack of expertise.

Barriers to Organizational Communication

When a communication is sent to the receiver, it must be effective and it is effective only when it has been understood by the receiver in the same sense in which the sender intended it. If it is not properly understood by the receiver, the very purpose of communication is lost and the communication proves ineffective. There may be so many obstructions in the way of its being effective. The message may not reach the receiver; there may be some flaws in encoding and decoding the message; the channel used may be defective or there may be some noise in the channel, or the channel used may be wrong, there may also be personal problems. Thus, for various reasons, the message may not be received by the receiver as was intended by the sender.

Types of Barriers

Barriers, obstructions, and interruptions in communication may broadly be categorized into the following groups:

- ✰ External Barriers
- ✰ Organisational Barriers
- ✰ Personal Barriers

However, this classification does not suggest that these categories are mutually exclusive.

External Barriers

External barriers are those caused by factors other than organizational and personal factors. Such external barriers may be (a) Semantic barriers, (b) emotional or psychological barriers.

(a) Semantic Barriers

These barriers are obstructions caused in the process of receiving or understanding a message during the process of encoding or decoding it into words and ideas. The linguistic capacity of the two parties may have some limitations, or the symbols used may be ambiguous. Symbols may have several meanings and, unless the context is known to the receiver, he is likely to take the meaning of the symbol according to his preconceived notion and misunderstand the communication. For this purpose, a meaningful distinction should be made between inferences and facts. Inferences are meaning taken out of the context of the communication and at times cannot be avoided in communication process. Since inferences can give a wrong signal, one should be aware of them and analyze them carefully. In case of any doubt, more feed back may be sought. Symbols may be classified as language, picture, or action.

(i) Language

In written or verbal communication, words used are important. A word used in the communication may have several meanings. In a face-to-face communication, it is easy to seek clarification of words used, if any doubt is encountered. In case of doubt feedback is required. Many words which we use informally may be taken literally in other contexts, non-friendly situations, or in written communication. Thus, effective communication is idea-centered rather than word-centered. The communication may be decoded correctly by the receiver only if the context is known to him; otherwise, it may be incorrectly interpreted. Without context, language is just like an eyesore that initiates our senses and interferes with our perceptions.

Sometimes, lack of clarity and precision in a message makes it difficult to understand. Poorly chosen and incorrect words and phrases, careless omissions, lack of coherence, bad organization of ideas, awkward sentence structure, inadequate vocabulary, platitudes, numbing repetitions etc. are some of the faults found in many cases of poor communication.

Communication with technical personnel and special groups may pose another type of language problem. Such persons or groups use technical words in their messages. It creates hindrance in communication with persons not in their specialty, area because of the receiver's ignorance of this type of jargon.

(ii) Picture

Picture is another type of symbol. Pictures are visual aids worth-thousands of words. An organization makes extensive use of pictures like blueprints, charts, maps, graphs, films, three dimensional models, and other similar devices. A viewer may come to understand the whole story when he sees them.

Sometimes, a picture creates confusion in the mind of the observer: it may be ambiguous if it is not supplemented by words or actions.

(iii) Action

Action is another type of symbol. We communicate by both – by action or by lack of it. To do or not to do, both have a meaning for the receiver. For example, if a subordinate does a good job, patting and non-patting on his back by the superior, both have a meaning. Patting may inspire him to do a better job again, and non-patting may make him disappointed. In this sense, we communicate all the times on the job whether we intend to do so or not. Action or non-action may influence the perception of the receiver.

Action speaks louder than words: if a person says something and does another, there is a credibility gap. The bigger the gap, the less seriously people take him, and he will lose the confidence of his subordinates. They will not take seriously what he says.

Body language is a type of action in which a message is communicated through the movement of the whole body or a part of it. Mostly face and hands are used in body language. Examples are eye contact, eye movement, smiles and frowns, touching a furrowed brow, closeness, lip movement, breathing rate etc.

These different types of symbols are used in communication. If these are interpreted incorrectly, the communication may not carry the sense intended by the source but something altogether different.

(b) Emotional or Psychological Barriers

Personal or emotional or psychological barriers arise from motives, attitudes, judgement, sentiments, emotions, and social values of participants. These create a psychological distance that hinders the communication, or partly filters it out, or causes misinterpretation, thereby making the communication inadequate.

The following are some emotional barriers:

(i) Premature Evaluation

Premature evaluation is a tendency to evaluate a communication prematurely, rather than keeping an open mind during the interchange. Such evaluation interferes with the transfer of information and begets a sense of futility in the sender. This barrier can be remedied by empathy and non-evaluative listening.

(ii) Loss in Transmission and Retention

When communication passes through various levels in a organization, successive transmissions of the same message are decreasingly accurate. A part of information is lost in transit: it is said that about 30 per cent of the information is lost in each transmission. Poor retention of the information is again a malady. Research reveals that workers retain only 50 per cent of information, and the supervisors retain 60 per cent of it.

(iii) Distrust of Communicator

The communicator is sometimes distrusted by his own subordinates. It happens when he lacks self-confidence, or is less competent in his position. He frequently makes ill-considered judgements or illogical decisions, and then reviews his own decisions when he fails to implement them. Repeated experience of this kind gradually conditions the receiver to delay action, or to act unenthusiastically, hence making the communication ineffective, though apparently it is complete.

(iv) Failure to Communicate

Sometimes, managers do not communicate the needed messages to their subordinates. This might be because of laziness or procrastination on their part, or they arbitrarily assume that everybody has got the information, or they may hide information deliberately to embarrass the subordinate.

(v) Undue Reliance on the Written Word

In cases where the employees have undue reliance on the organisation's written policies and viewpoints and hesitate in accepting the face to face verbal communication of their superiors, communication cannot be effective. This is because they will be disinclined to accept oral orders of their superiors, which are not consistent with the written policies of the organization. Such a situation embarrasses both the subordinates and the superiors, unless there is a high degree of mutual confidence between the organization and its employees. A written communication might fail to explain the purpose of the order, procedure, or directive. Written communication often tells what is to be done bur not why it should be done and, thus, it lacks persuasion. Hence, written communication should not be relied upon very rigidly, rather it should be used as supplementary to productive face to face relationships.

(vi) Inattention

The preoccupied mind of the receiver and the resultant non-listening is one of the major chronic psychological barriers. It is a common phenomenon that people simply fail to react to bulletins, notices, minutes, and reports.

Organisational Barriers

An organization is a deliberate creation of management for the attainment of certain specific objectives. The day-to-day functioning of the organization is regulated in such a way as to contribute to the attainment of these objectives in the most effective manner. For this purpose, a variety of official measures are adopted such as designing of the structure, arrangement of activities, formulation of various policies, rules and regulations, and procedures, laying down of norms of behaviour, instituting a reward and punishment system etc. All these variables markedly affect the organisation's functioning. As such major organizational barriers are:

i. Organisation Policy

The general organizational policy regarding communication provides overall guidelines in this matter. This policy might be in the form of a written document, or it has to be inferred from organizational practice, particularly at the top level. If the policy creates hindrance in the free flow of communication in different directions, communication would not be smooth and effective.

ii. Organisational Rules and Regulations

More often, different activities of an organization are governed by specific rules and regulations, such rules and regulations prescribe the subject-matter to be communicated as also the channel through it is to be communicated. The rules may restrict the flow of certain messages and may omit many important ones. For example, the rules may prescribe that upward communication shall be only through proper channel. Such restrictions may delay the message and may deter employees from conveying any message. An employee may give up the idea of conveying a message to the top executive to avoid the observance of rules. The message may be important to the organization. It may also be the case that the superior may not allow him to convey the message. This barrier is strongly operative in India public sector undertakings where observance of rules and regulations is more rigid.

iii. Status Relationship

The placing of people in superior-subordinate relationship in a formal organization structure also blocks the flow of communication and more particularly, in the upward direction the grater the difference in hierarchical positions in terms of their status, the greater would be the worry of middle managers about what their senior bosses might think; this leads to their paying little attention to the needs and demands of their subordinates. The subordinates are reluctant to communicate, or the managers in the middle of hierarchy may be reluctant to pass on their juniors' communication to the next higher executive, which their bosses do not like or which may have a negative effect on their own relationships. Dislike, distrust, dissatisfaction with job and work environment are other reasons for reluctance to tell anything to their bosses. Such obstruction may be overcome by crating an atmosphere of trust and confidence in the organization.

iv. Complexity in Organisation Structure

In an organization where there are a number of managerial levels, communication gets delayed and it moves along the hierarchical line. Also, chances of the communication getting distorted are greater as the number of filtering points is higher. This is particularly true in upward communication because people at intermediate levels do not like to pass on negative remarks either of themselves or of their superiors.

v. Organisational Facilities

Certain organizations provide certain facilities for smooth, adequate, clear, and timely flow of communication such as meetings, conferences, complaint or suggestion boxes, open door system etc. If these facilities are not properly emphasized, people generally fail to communicate effectively.

Personal Barriers

As communication is basically an interpersonal process, many personal factors inherent in the two parties to communication, the sender and the receiver, influence the flow of communication and present many hurdles in the way of effective communication.

1. Barriers in Superiors

Superiors play an important role in communication. Because of their hierarchical position, they act as barriers in a number of ways as follows:

(i) Attitude of Superiors

General attitude of the superiors about communication, or attitude towards a particular communication, affects the flow of messages in different directions. If the attitude is unfavourable, there is greater possibility of filtering or colouring of the information. Any information received from the top may not reach the bottom in the same form, or even the reverse may happen. Managers at intermediate levels may colour the information, sometimes intentionally, with a view to twist the situation in their favour. In some cases the superior quotes his subordinate incorrectly or may say something against him before his boss just to spoil his career, or his chance of promotion, or his image in the eyes of the boss.

(ii) Fear of Challenge to Authority

The superiors in an organization generally try to withhold the information coming down the line or going up as frequent passing of information may disclose their own weaknesses; thus what happens generally happens when the superior lacks self-confidence and is afraid that someone else might be promoted in his place if his weaknesses were to come to light.

(iii) Insistence on Proper Channel

There are channels of communication in an organization along which information passes upward or downward. Some officers insist too much on communication through proper channel. They do not like any bypassing in communication. But, sometimes, by passing becomes necessary in the interest of the organization; however, the superiors think bypassing as thwarting of their authority and block the flow of communication.

(iv) Lack of Confidence in Subordinates

The superiors generally perceive that their subordinates are less competent and, they are not capable of advising their superiors. Therefore they feel, whether correctly or otherwise, that they are overburdened and have no time to talk to their subordinates.

(v) Ignoring Communication

Sometimes, the superiors ignore a communication or a part of it, to and from their subordinates, to maintain their importance. In some cases, information does not reach the receiver in the same form as it was received from the sender: the superiors filter the information.

2. Barriers regarding Subordinates

There are certain factors in the subordinates, which adversely affect their participation in the communication process. Some factors like attitude, lack of time, applicable to the superiors are also applicable here. Two more factors in the subordinates need special attention, which are responsible for blocking communication in the upward direction.

(i) Unwillingness to Communicate

The subordinates generally are not willing to communicate upward any information, which is likely to affect them adversely. If they feel that supply of such adverse information is necessary for control purposes they would modify it in such a way so as not to harm their interest.

(ii) Lack of Proper Incentive

Lack of incentives to communicate also prevents the subordinates from communicating upward. They are punished when they are wrong but may not be rewarded when they work marvelously well and offer a novel suggestion.

The superiors assume that better performance on the part of subordinates is their duty towards the organization or it is their own interest. The rewards and punishment system and the attitude of the superiors towards their subordinates are responsible for this agony. If a novel suggestion by a subordinate does not evoke any interest in the superior he would not convey it.

7 C's and 4 S's in Organisational Communication

In any business environment, adherence to the 7C's and the 4 S's help the sender in transmitting his message with ease and accuracy. Let us first take a look at the 7 C's.

1 **Credibility**: If the sender can establish his credibility, the receiver has no problems in accepting his statement. Establishing credibility is not the outcome of a one-shot statement. It is a long-drawn out process in which the receiver through constant interaction with the sender understands his credible nature and is willing to accept his statements as being truthful and honest.

2. **Courtesy**: Once the credibility of the sender has been established, attempts should be made at being courteous in expression. In the business world, almost everything starts with and ends in courtesy. Much can be accomplished if tact, diplomacy and appreciation of people are woven in the message.

3. **Clarity**: Absolute clarity of ideas adds much to the meaning of the message. The first stage is clarity in the mind of the sender. The next stage is the transmission of the message in a manner which makes it simple for the receiver comprehends. As far as possible, simple language and easy sentence constructions, which are not difficult for the receiver to grasp should be used.

4. **Correctness**: At the time of encoding, the sender should ensure that his knowledge of the receiver is comprehensive. The level of knowledge, educational background and status of the decoder help the encoder in formulating his message. In case there is nay discrepancy between the usage and comprehension of terms, miscommunication can arise. If the sender decides to back up his communication with facts and figures, there should be accuracy in stating the same. A situation in which the listener is forced to check the presented facts and figures should not arise. Finally, the usage of terms should be nondiscriminatory, *e.g.* the general concept is that women should be addressed for their physical appearance whereas men for their mental abilities. This, however, is a stereotype and the time of addressing for praising members of both the sexes, the attributes assigned should be the same. Similarly for occupational references. In the business

world almost all professions are treated with respect. Addressing one individual for competence in his profession but neglecting the other on this score because of a so-called 'inferior' profession alienates the listener from the sender.

5. **Consistency**: The approach to communication should, as far as possible, be consistent. There should not be too many ups and downs that might lead to confusion in the mind of the receiver. If a certain stand has been taken, it should be observed without there being situations in which the sender desires to bring about a change in his understanding of the situation; he should ensure that the shift is gradual and not hard for the receiver to comprehend.
6. **Concreteness**. Concrete and specific expressions are to be preferred in favour of vague and abstract expressions. In continuation of the point on correctness, the facts and figures presented should be specific. Abstractions or abstract statements can cloud the mind of the sender. Instead of stating. "There has been a tremendous escalation in the sale figure". Suppose the sender made the following statement: "There has been an escalation in the sales figures by almost 50 per cent as compared to last year. The receiver is more apt to listen and comprehend the details.
7. **Conciseness**. The message to be communicated should be as brief and concise as possible. Weighty language definitely sounds impressive but people would be suitably impressed into doing precisely nothing. As far as possible, only simple and brief statements should be made. Excessive information can also sway the receiver into either a wrong direction or into inaction. Quantum of information should be just right, neither too much nor too little, *e.g.*

7 C's

C's	*Relevance*
Credibility	Builds trust
Courtesy	Improves relationships
Clarity	Makes comprehension easier
Correctness	Builds confidence
Consistency	Introduces stability
Concreteness	Reinforces confidence
Conciseness	Saves time

4 S's: An Understanding of the 4 S's is Equally Important

1. **Shortness**: "Brevity is the soul of wit, "it is said. The same can be said about communication. If the message can be made brief, and verbosity done away with, then transmission and comprehension of messages is going to be faster and more effective. Flooding messages with high sounding words does not create an impact. Many people harbour a misconception that they

can actually impress the receiver, if they carry on their expeditious travels. Little do they realize how much they have lost as the receiver has spent a major chunk of his time in trying to decipher the actual meaning of the message.

2. **Simplicity**: Simplicity both in the usage of words and ideas reveals a clarity in the thinking process. It is normally a tendency that when an individual is himself confused that he tries to use equally confusing strategies to lead the receiver in a maze. Reveal clarity in the thinking process by using simple terminology and equally simple concepts.
3. **Strength**: The strength of a message emanates from the credibility of the sender. If the sender himself believes in a message that he is about to transmit, there is bound to be strength and conviction in whatever he tries to state. Half-hearted statements or utterances that the sender himself does not believe in add a touch of falsehood to the entire communication process.
4. **Sincerity**: A sincere approach to an issue is clearly evident to the receiver. If the sender is genuine, it will be reflected in the manner in which he communicates. Suppose there is a small element of deceit involved in the interaction or on the part of the sender. If the receiver is keen and observant, he would be able to sense the make-believe situation and, business transactions, event if going full swing, would not materialize.

4 S's

C'c	*Relevance*
Shortness	Economizes
Simplicity	Impresses
Strength	Convinces
Sincerity	Appeals

Informal Communication

Communications are not planned or deliberately created channel of communication. It is free from all formalities.

Grapevine Communication

The informal channel of communication is also known as 'Grapevine'. It is the result not of any official action, but of the operation of social forces at work-place. The term 'grapevine' arose during the days of U.S. Civil War. At that time, intelligence telephone lines were strung loosely from tree to tree in the manner of a grapevine, and the message thereon was often distorted; hence any rumour was said to be from the grapevine. Today the term applies to all informal communication. While formal communication exists to meet the utilitarian needs of the organization, informal communication is the method by which people carry on social, non-programmed activities within the formal boundaries of the system. It, thus, exists outside the official network. Though continuously interacting with it, this informal channel is generally multiple in nature; same person having social relationships with a number of people working in the same organization.

Many authors on this subject opine that the communication through informal channel generally performs a positive service to the organization. It operates with much greater speed. Moreover there might be certain subject matter of communication which do not require their transmission through the formal channel. This channel also satisfies the communication needs of the various persons in the organization, more particularly those persons who freely mix up with others and rely upon informal relationships. Informal communication is thus a need fulfillment device generated when it is found that the existing formal communication is inadequated or insufficient. The requirement of speedier communication creates conditions for informal communication which is supplemental to the formal one. Chester Bernard opines that 'communication function of the executives includes the maintenance of informal executive organization as an essential means of communication. The functions of informal executive organization as the communication of intangible facts, suggestions, suspicious that cannot pass through formal channels without raising issues calling for decisions without dissipating dignity and objective authority.

Informal or grapevine communication has certain basic limitations also. It is less orderly and less static. Sometimes messages communicated through the channel are so erratic that any action based on these may lead to difficult situation in the organization. In this case, the irresponsibility of the persons communicating through the informal channel is the most important factor. Since origin and direction of the flow of information is hard to pinpoint, it is difficult to assign responsibility for false information or morale lowering rumours. Moreover, as each person conveying the message may add, subtract or change the original message according to his motive, informal communication problems multiply. There is a change that by the time a communication complex journey it may be completely distorted.

The informal communication is a part and parcel of the organizational process. The only thing management can do in this respect is to take suitable action to minimize the adverse effect of this channel. Management must be aware, through the understanding of the informal communication, the utility and the positive contributions to the operation of the organization that the informal communication can make. As such, proper analysis of informal communication and suitable clarification in this respect will be helpful in making its use towards organizational efficiency.

Types of Grapevine

Four types of informal communication channels have been identified so far : single strand, gossip, probability and cluster.

In single standard network, the individual communicates with other individuals through intervening persons. In the gossip network, the individual communicates non-selectively. In probability network, the individual communicates randomly with other individuals according to the law of probability. In the cluster network the individual communicates with only those individuals whom he trusts. Out of these informal networks, the cluster is most popular.

Chapter 43

Information Communication Technology for Agro-Processing and Value Addition

Manas Mohan Adhikary and Sankar Kumar Acharya

ICT: The Definition and Relevance

ICT or Information and Communications Technology in simple terms, can be defined as the basket of technologies, which assist or support in storage, processing of Data/Information, or in dissemination/communication of Data/Information, or both. ICT thus includes technologies such as desktop and laptop computers, software, peripherals and connection to the Internet that are intended to fulfill information processing and communication functions. According to Wikipedia (2008), the term ICT is the broader term of Information Technology (IT), to explicitly include the field of electronic communication, in addition to IT. The term IT is defined as "the study, design, development, implementation, support or management of computer-based information systems, particularly software applications and computer hardware." IT deals with the use of electronic computers and computer software to convert, store, protect, process, transmit and retrieve information, securely.

ICT is thus used as an umbrella term that includes any communication device or application, encompassing radio, television, cellular phones, computer and network hardware and software, satellite systems and so on, as well as the various services and applications associated with them, such as videoconferencing and distance learning. The importance of ICT lies less in the technology itself, than in its ability to create greater access to information and communication among the hitherto un-reached geographies and populations. Appropriate ICT interventions are yielding positive results in developing and underdeveloped economies. The "Grameen Phone" initiative in Bangladesh, Kothamale Radio Project in Srilanka, and ITC's e-Chaupals in India, are examples of such innovations. Many countries around the world have established organizations for the promotion of ICTs, because it is feared that unless less technologically advanced areas have a chance to catch up, the increasing technological advances in developed nations will only serve to aggravate the already-existing economic gap between technological "have" and "have not" areas.

Types of ICT Tools

There are various ICT tools or applications available to simplify the operation for general purpose use to mission critical applications. The general purpose applications can be used to simplify the day-to-day operations in the organization. The following range of ICT tools (applications) is available for the general purpose use.

- Word processing applications - used for preparing text documents.
- Presentation software - used for preparing presentations
- Spreadsheets - used for calculation and preparing tabular data with calculations
- Databases - used to store the information in the form of records and retrieving information using query mechanism.
- Multimedia – used to organize the information more meaningful manner with the help of text, images, animation, audio and videos
- Web browsers and E-mail – used to searching information and sending or receiving mail using Internet.

Presentation of Information Using Word Processing, Publications and Presentation Software

- Enter, cut, copy, paste and move text
- Format text, eg. bold, italic, underline, justify, change font
- Incorporate clip art/graphic images and tables
- Make use of document formatting features, eg. headers, footers, bullet points
- Use word wrapping facilities around images/objects
- Use mail-merge facilities, eg. merging database data into a document

Organisation and Analysis of Numerical Information Using Spreadsheet Software Enter a Range of Data

- Format cells to match data types
- Cut, copy, paste and move data between cells, rows and columns
- Insert and delete rows and columns
- Enter and replicate formulas
- Use simple functions, eg. SUM, AVERAGE
- Produce charts with labels, eg. axis titles, legends

- ☆ Use relative and absolute cell references
- ☆ Print selected areas

Organisation and Analysis of Structured Information Using Database Software

- ☆ Prepare database structure and validation rules for different data types, eg. text, currency, date
- ☆ Enter data including use of data entry forms
- ☆ Establish a relationship between fields in two tables
- ☆ Search and sort including use of related tables
- ☆ Produce reports showing the results of searches and sorts

Organisation and Presentation of Information Using Multimedia Software

- ☆ Establish structure and navigation route through the presentation
- ☆ Create and/or find the separate components of the presentation, eg. text files, images, sound files
- ☆ Produce the individual frames/layers/backgrounds/slides
- ☆ Combine the separate components to create a final presentation
- ☆ Edit the separate components and the final presentation

Communication, Searching and Selection of Information Using the Internet

- ☆ Use e-mail for communication between individuals and groups
- ☆ Understand and apply the main search principles of internet search engines
- ☆ Searches, multiple criteria searches
- ☆ Understand and apply the main features of browser software, eg. forward and back buttons, book marking and organising favourites
- ☆ Navigate purposefully large web sites, eg. locate a specific information resource in a given site.

Some applications are available or developed to solve a specific purpose. For example; capture, manipulate and enhance graphic images, automate and control processes, monitor and record physical and financial data for analysis and interpretation. And, new applications can be developed as per the need of a particular task of organization.

File Management and Standard Ways of Working

There are many reasons for having standard ways of working in ICT. The most important is that information in ICT systems can be easily lost or misused, and that having standard ways of working can help to overcome these problems. In your work with ICT, you must ensure that you:

- ☆ Keep information secure, eg. from theft, loss, viruses, fire

- ✰ Protect confidentiality, eg. prevent un-authorised access to documents or records
- ✰ Respect copyright, eg. not using the work of others without permission
- ✰ Save work regularly and use different filenames
- ✰ Keep dated backup copies of files in another location
- ✰ Manage your work effectively, eg. use appropriate filenames and locations
- ✰ Work safely, eg. use the correct position for the monitor and chair, avoid trailing cables, take regular breaks
- ✰ Take account of relevant legislation and codes of practice

Trends in Agricultural Information Management

Virtual Community

A virtual community, e-community or online community is a group of people that primarily interact via communication media such as letters, telephone, email or Use net rather than face to face. If the mechanism is a computer network, it is called an online community. Virtual and online communities have also become a supplemental form of communication between people who know each other primarily in real life. Many means are used in social software separately or in combination, including chatrooms and forums that use voice, video and/or text.

Virtual Information Networks

A number of virtual information networks with the objective of linking agricultural institutions for facilitating better access to information resources have been initiated successfully. While some of these network agricultural institutions within a country, link to agricultural institutions around the globe. Some of these are profiled here.

International Initiatives

Agriculture Network Information Center (AgNIC) www.agnic.org

AgNIC is an Internet based network of public and private agricultural libraries and information centers, coordinated by the National Agriculture Library (NAL), USA. The network aims to provide global access to agricultural information. AgNIC Members represent 40 land-grant universities and other national and international partners including universities and research institutions, government agencies, and a non-profit organization. Through its website the network provides access to a network of electronic sources on research and teaching in agriculture, food, renewable natural resources, forestry, and physical and social sciences. AgNIC is a distributed discipline-oriented source of agricultural information in electronic form on the Internet. The goals are to: identify major collections of agriculture-related information; provide mechanisms to facilitate access/retrieval from these information resources; create mechanisms to encourage organizations to collaborate in creating/using AgNIC. AgNIC partners select important information sources for inclusion in the system. Services include: Resource database with web sites, image collections, lists of publications, documents, databases, and other resources; Calendar of events which includes meetings, seminars, national and international symposia and conferences, conventions, and workshops in agricultural and related sciences; News items; Specialized Services – such as Plant Disease Announcements; discussions on emerging plant diseases around the world; Expertise where specialists respond to individual questions (NAL, 2006). There are partnerships between libraries and United States Department of Agriculture (USDA) Cooperative Extension programs; between libraries and academic departments within colleges; between states and between technologists and librarians. Member participants take responsibility for small segments of agricultural information and develop Web sites and reference services in their specific subject areas. Nearly all participating AgNIC institutions have developed partnering relationships with a variety of internal and external institutions, groups, and agencies to develop content and tools for their respective Web sites.

Agricultural Libraries Network (AGLINET)

The Agricultural Libraries Network (AGLINET- www.fao.org/library/ info_services EN/aglinet EN.htm) coordinated by the Food and Agriculture Organization of the United Nation (FAO), is a world-wide, voluntary network of international agricultural libraries with regional/country coverage and other comprehensive or specialized subject resource collections. All member libraries provide, upon request, access to the literature originating in the country or region or for a given specialization. AGLINET aims at combining resources for mutual and rational use through delivery of primary documents, by means of Inter library loan provision of reproductions, bibliographic information; with appropriate regional and subject specialization. Member libraries include Agricultural libraries of national or regional importance with comprehensive collections and strong regional coverage; Libraries in special subject fields within the broad domain of agriculture with world wide coverage. AGLINET Network members include Libraries from Argentina, Australia, Belarus, Belgium, Benin, Botswana, Brazil, Bulgaria, Brazil, Canada, China Costa Rica, Czech republic, Denmark, Egypt, Estonia, Ethiopia, Finland, France, Germany, Hungary, India, Indonesia, Italy, Japan, Jordan, Kenya, Korea, Latvia, Lithuania Malawi Malaysia, Mexico, Netherlands, Nigeria, Norway Philippines, Poland Portugal, Russia, Slovenia, south Korea, sprain, Sri Lanka, Sweden Syria, Thailand, UK, Uruguay, USA, West Indies and Zambia.

International Information System for Agricultural Sciences and Technology (AGRIS) http://www.fao.org/agris

The Food and Agriculture Organization (FAO) of the United Nations initiated agricultural information networking through AGRIS. Established in 1974, AGRIS facilitates information exchange and brings together world literature dealing with all aspects of agriculture and related subjects. AGRIS is a cooperative system in which 240 national, international and intergovernmental centers in participating countries input references to the literature produced within their countries and draw on the information provided by the other participants. The new vision of AGRIS, in response to the needs includes: a decentralized approach with greater emphasis on national partnerships, improved linkages between the AGRIS network and other FAO initiatives; focus on management of documents in agricultural science and technology in full text; information about activities, organizations, and people in agricultural science and technology; etc.

Agrigate (www.agrigate.edu.au)

Coordinated by the National Library of Australia is a project of the libraries of the Universities of Melbourne, Adelaide and Queensland, and the Commonwealth Scientific and Industrial Research Organization (CSIRO) in Australia. This is a subject information gateway for resources, both online and offline, in agricultural research. The purpose is to support the identification and dissemination of quality research materials selected by an editorial review process consisting of members of the agricultural research community. The content covers agriculture and related areas of value to the Australian agricultural research community. The resources selected have been reviewed by subject specialist librarians. The majority of resources identified in the database are available online.

AGROWEB CEE network (http://www.iaaldcee.hu/awnt.html)

AGROWEB CEE network has been established by organizations and individuals involved in agricultural and rural development in Central and Eastern Europe. This collaborative network is ICTs: Concepts and Relevance maintained by representatives of participating countries, and facilitated by International Association of Agricultural Information Specialists (IAALD) and FAO. Twenty five countries from the Balkan region, New Independent States, Baltic states, Caucasus region, Central Asia, and European Union are participating in the network. National web pages have been established as portals for the countries in the region, which provide access to information about agriculture-related institutions - ministries, libraries, information centres, universities, research centres, NGOs, agro-marketing organizations and other organizations. The AgroWeb national portals provide Internet links, to these national institutions and to other relevant national web sites. In addition to the national portal pages, cross-cutting subject areas have been identified and a number of thematic sectors of the network are being coordinated by regional focal points.

Asia-Pacific Association of Agricultural Research Institutions(APAARI) (www.apaari.org)

Information exchange among the Asia-Pacific national agricultural research systems is one of the primary objectives of the Asia-Pacific Agricultural Research Information System (APARIS). Components of APARIS include Management Information System (MIS) Tools *viz.*: Regional Research Networks (RRN) Database, NARS Database, etc.; information on regional events; access to scientific publications generated by agricultural research in the region; electronic Forums to facilitate dialogue among stakeholders of Agricultural Research for Development (ARD) in the region on issues of strategic importance; gateway/portal service - portal to Regional Research Networks (RRNs), websites of NARS Institutions in the Asia/Pacific region, web-enabled information on key topics/themes of ARD; Knowledge Networks on ARD in the APAARI region; information dissemination through APAARI web site and publications. The web site has linkages to various national, regional and international institutions, networks, and other organizations.

The Consultative Group on International Agricultural Research (CGIAR) (www.cgiar.org)

This is a research network of fifteen international agricultural research centers. The 15 centers supported by the CGIAR are independent institutions, each with its own charter, international board of trustees, director general, and staff. The institutions are Africa Rice Center (WARDA), Biodiversity International, Centro Internacional de Agricultura Tropical (CIAT), Center for International Forestry Research (CIFOR), International Maize and Wheat Improvement centre (CIMMYT), International Potato center (CIP), International Center for Agricultural Research in the Dry Areas (ICARDA), International Crops Research Institute for the Semi-Arid Tropics (ICRISAT), International Food Policy Research Institute (IFPRI), International Institute of Tropical Agriculture (IITA), International Livestock Research Institute (ILRI), International Rice Research Institute (IRRI), International Water Management Institute (IWMI), World Agro-forestry Center (ICRAF) and World Fish Center (WFC). Thirteen of the 15 CGIAR Centers are located in developing countries and the knowledge gathered is shared and disseminated across the world. The CGIAR website gives access to a Virtual Information Center and Library. One can tap into agricultural information databases, including online libraries of the CGIAR Centers and the Core Collection Database. The CG Library gives access to databases and e-journals and facility to go directly to the full text of publications. One can search CGIAR libraries/other agricultural libraries/by specific topics. The Virtual Information Center provides information on various topics mentioning the source against each topic. Each institutional library is the source for/and responsible for providing information on specific subject areas *viz.,*: Agriculture in the Dry Areas- ICARDA library; Agriculture in semi arid tropics- ICRISAT library; Agro-forestry- ICRAF library; Aquaculture and fisheries -WFC library; Food policy- IFPRI library, Agricultural biodiversity – IPGRI library; Water Management-IWMI library; Wheat - CIMMYT library, etc.

DAINet - German Agricultural Information Network

The German Agricultural Information Network DAINet is a catalogue of Internet sources in the field of nutrition, agriculture and forestry. DAINet was established in 1995 to guide people interested in subject specific information to sources on the Internet. DAINet has about 7,500 references to information sources and about 800 weblinks to agriculture-related organizations. Three sections cover, 'information', 'service' and 'dialogue'. The 'information section' organized by subjects, structures worldwide existing agricultural information under 'subject fields', 'user groups' and 'topics'. The 19 'subject fields' cover various topics of agriculture. The division into five user groups caters to the specific needs of farmers, economists, scientists, educators and journalists. The 'service section' is structured according to types of data. The 'dialog section' is used for professional communication through Internet. 'Information' and 'service' sections offer links to databases and web pages and also give access to newsgroups and mailing lists.

ELIN - Entomology Library and Information Network

ELIN project hosted by the Natural History Museum, London, at http://www.nhm.ac.uk/hosted_sites/elin/aims to establish a global Entomology Library and Information Network to provide entomologists with a co-ordinated and structured electronic information resource including a gateway to a wide range of entomological information. A mailing list was launched in 1998 and has over 100 members from libraries and information centres from around the world, which has encouraged international communication between specialists working with entomological information. The ELIN project provides up-to-date, detailed information about entomological libraries around the world.

Information Network on Post-harvest Operations (INPhO) www.fao.org/inpho/

INPhO is an FAO databank project of the Post-harvest Management Group. It is an international collaborative effort by the FAO, GTZ and CIRAD which aims to support the collection and dissemination of information on proven technologies and products in post harvest systems. Components of the network include a comprehensive collection of information on post harvest issues, communication/interactive services; links with other databases. The website includes full text documents of training and technical publications, country profiles, crop profiles (main focus is on cereals and grains, fruits and vegetables, oil seeds, roots and tubers). Recipients of this information include people working in the agricultural production and marketing sector *viz.*, producers, researchers, policy makers, private investors and donors (FAO, 2006).

Philippine Agricultural Libraries and Information Services Network (PhilAgriNet) http://www.philagrinet.org.ph/

PhilAgriNet aims to create and maintain a central electronic database of Philippine technical agricultural literature and make this accessible to agricultural scientists worldwide. Membership is open to agricultural institutions. The database intends to cover all technical publications on agriculture, generated by member institutions, both published and unpublished, and written in any language or dialect, regardless of format. Affiliation is open to libraries of: Government agencies and corporations engaged in agricultural research; state colleges and universities, whose curriculum includes agriculture and independent or private agricultural agencies.

Virtual Extension Research Communication Network (VERCON)

Virtual Extension and Research Communication Network (VERCON) established by FAO aims to use the potential of the Internet and apply it to strengthening and enabling linkages among the research and extension components of the national agricultural knowledge and information system. VERCON aims at improving linkages between and within agricultural research and extension institutions through human and technological components. The human component is a network of staff of research and extension institutions, faculty of agricultural education, NGO workers, in some cases agro producers committed to strengthening collaboration, communicating, sharing information and supporting improved agricultural production. In order to link the human component there is the technological component, which allows members of the network to communicate and develop, share, store and retrieve information (FAO, 2006). Network members can engage in two-way horizontal communication to address problems, discuss solutions and coordinate local, regional and national agricultural activities. An illustration of how VERCON functions: A farmer visits an extension office with an insect that is causing heavy damage to his cotton crop. The extension worker identifies the insect by consulting a database of digital photographs of various cotton pests that have been photographed in the area and scanned. Using a chat room, a number of integrated pest management specialists are consulted simultaneously about how to accurately identify and control the pest. Relevant information is quickly compiled from research results, pest distribution maps, publications and the photo file into a Fact sheet using a template. A message with the Fact sheet attached is sent to all extension officers in the region, warning about the presence of the pest and providing recommended integrated pest management measures. This is communicated to farmers through IPM farmer field schools, facilitated by extension. The Egyptian Government has established a pilot VERCON to support research and extension interaction and the flow of information between research, extension and farmers in Egypt.

Web-based Information Service for Agricultural Research for Development (WISARD) http://www.wisard.org/wisard/home.html

WISARD is a web-based information platform that provides searchable information on experts, organizations, outputs and projects in the fields of Agricultural Research for Development (ARD), Natural Resource Management (NRM) and Sustainable Development (SD) from the mid-nineties till date. The system can be used at organisation, network, national and international levels. WISARD allows decentralized data input and management through focal points at organization, network or national level. A module has been added for depositing outputs of research and another is under development for 'Technologies and Best Practices'. Present partners include among others, CGIAR, IPM-Europe, Rice-Wheat Consortium for the Indo-Gangetic Plains, FAO, donor agencies, international and national NGOs and European agencies and national agricultural research systems of India, Pakistan, Nepal and Bangladesh.

Electronic Publishing in Agriculture

Electronic Publishing (e-Publishing) is quickly becoming an important part of publishing mainstream. Recent innovations in this area have made it possible to publish on the information super highway. E-document can be accessed at the computer. It makes much easier for readers to search the information. It is very quick and easy for a reader to browse the table of the contents of previous issues, to jump directly to a particular section of a document or even particular section of an article.

Advantages of e-Publishing

e-Publishing (EP) difference lies in the new levels of value it provides through features not possible in traditional media. EP products may differ to an even greater degree than print products. Nevertheless, there are some common features to distinguish EP from print Publishing in terms of value to end-users. ICTs: Concepts and Relevance EP products create additional value for the user with regard to following three dimensions: content availability, content transparency and interactivity and content format.

Content availability means that EP products can be delivered and accessed with more independence of time and place than can be traditional print products and that their delivery is less limited with regard to quantity. Content availability includes: time of delivery - available any time; location of delivery - consumption anywhere; amount of information - end of traditional boundaries imposed by paper volume and price. Content transparency and interactivity refers to new tools and opportunities concerning information navigation. The main features are: Interactivity- contextual hyperlinks open new dimensions of information retrieval and lead to new types of information behavior: browsing, etc. the possibility to integrate content and services; and search tools across one or thousands of documents - interactive information processing.

Electronic Publishing in Agriculture in India

The Directorate of Information and Publications of Agriculture, (DIPA), New Delhi, is the official publication wing of the ICAR through which the research and other activities are revealed to the world. DIPA brings out a variety of publications in English and Hindi languages for the use of scientists, researchers, students, policy planners, extension personnel, farmers and the general public. The e-publications of DIPA include - "Handbook of Horticulture", ICAR Research Projects Information – Research Project Files (RPF) Database, ICAR Vision 2020 Document etc. Some other important institutions bringing out e-publications in Agriculture in India include, National Institute of Agricultural Extension Management, MANAGE, Hyderabad (www.manage.gov.in), Indian Agricultural Research Institute, New Delhi (www.iaripusa.org), Indian Farmers Fertilizer Cooperative Limited (IFFCO), (www.iffco.nic.in), Krishak Bharati Co-operative Limited (KRIBHCO), (www.kribhco.net), National Bank for Agricultural and Rural Development (NABARD) (www.nabard.org), National Agricultural Co-operative Marketing Federation of India Ltd. (NAFED) (www.nafed-india.com), Indian Agricultural Statistics Research Institute (IASRI), New Delhi (www.iasri.res.in) and State Agriculture Universities of Tamilnadu, Uttarakhand, Punjab, Haryana and Andhra Pradesh. Two major open universities namely Indira Gandhi National Open University (IGNOU), Delhi and Yeshwantrao Chavan Maharashtra Open University (YCMOU), Nashik and National Institute of Agricultural Extension Management, MANAGE, Hyderabad have taken open and distance learning in Agriculture on a national scale, and all these institutions are in the process of bringing out a number of e-publications in Agriculture.

ICT Indicators for Infrastructure and Access

The ICT Indicators for Infrastructure and Access take account of Telephone (including Mobile Phones) connectivity, Internet Connectivity, Telephone and Internet access and Radio and Television infrastructure. These Indicators are further categorized as basic core and extended core indicators:

Basic Core

- Fixed telephone lines per 100 inhabitants
- Mobile cellular subscribers per 100 inhabitants
- Computers per 100 inhabitants
- Internet subscribers per 100 Inhabitants
- Broadband Internet subscribers per 100 inhabitants
- International Internet bandwidth per inhabitant
- Percentage of population covered by mobile cellular telephone
- Internet access tariffs (20 hours per month), in US$, and as a per cent of per capita income
- Mobile cellular tariffs100 minutes of use per month, in US$, and as a per cent of per capita income
- Percentage of localities with public Internet access centres (PIACs) by number of inhabitants (rural/urban)

Extended Core

- Radio sets per 100 inhabitants
- Television sets per 100 inhabitants

Indicators for Access to and Use of ICT by Households and Individuals

Basic core Indicators for Access to and Use of ICT by Households and Individuals include the Indicators on Radio, Television, Telephones and Internet availability at Household level. These Indicators are:

Basic Core

- Proportion of households with a radio
- Proportion of households with a TV
- Proportion of households with a fixed line telephone
- Proportion of households with a mobile cellular telephone
- Proportion of households with a computer
- Proportion of individuals who used a computer in the last 12 months
- Proportion of households with Internet access at home
- Proportion of individuals who used the Internet in the last 12 months
- Location of individual use of the Internet in the last 12 months
- Internet activities undertaken by individuals in the last 12 months

Extended Core

- Proportion of individuals with use of a mobile telephone
- Proportion of households with access to the Internet by type of access
- Frequency of individual access to the Internet in the last 12 months

Reference Indicator

- ☆ HHR1 Proportion of households with electricity

Indicators for Use of ICT by Business

Basic Core

- ☆ Proportion of businesses using computers
- ☆ Proportion of employees using computers
- ☆ Proportion of businesses using the Internet
- ☆ Proportion of employees using the Internet
- ☆ Proportion of businesses with a Web presence
- ☆ Proportion of businesses with an intranet
- ☆ Proportion of businesses receiving orders over the Internet
- ☆ Proportion of businesses placing orders over the Internet

Extended Core

- ☆ Proportion of businesses using the Internet by type of access
- ☆ Proportion of businesses with a Local Area Network (LAN)
- ☆ Proportion of businesses with an extranet
- ☆ Proportion of businesses using the Internet by type of activity

Indicators for ICT Sector and Trade in ICT Goods

Basic Core

- ☆ Proportion of total business sector workforce involved in the ICT sector
- ☆ Value added in the ICT sector (as per cent of total business sector value added)
- ☆ ICT goods imports as a percentage of total imports
- ☆ ICT goods exports as a percentage of total exports

ICT Indicators and India

Twenty years ago, India faced tremendous challenges when it set on its information journey. The PC revolution was yet to encompass the country, the telecom infrastructure was low and there was virtually no indigenous software or hardware development to talk about. The ICT industry, at a very nascent stage, appeared far behind its Western counterpart. Today, in 2008, the scenario has undergone an amazing transformation. The Indian ICT Industry in particular the IT software and services and ITES (IT Enabled Services) sectors, have not only managed to catch up with their more technology savvy global leaders, but they are also being actively sought by companies worldwide for their onsite, offshore expertise and wealth of manpower resources. Indian ICT organizations are now counted among the well known and reputed ICT solutions and services providers across the world and scores of global ICT leaders have invested in India, making the country their hub for software development, offshore outsourcing and R and D.

The Use of Personal Computers has tremendously increased from 5.4 million PCs in 2001 to 14.5 million in 2005. In 2005, only one in every hundred persons had a personal computer, which was much less compared with any developed country.

Internet Users per 100 Populations

Though we have a rapid positive trend for this indicator, compared to the developed countries, we are still at the infant stage. However, every 35th person is using internet in India. The Government vision is to use Information Technology as a tool for raising the living standards of the common man and enriching their lives. Towards this end an ambitious programme of PC and Internet penetration to the rural and under-served urban areas has been taken up. The Department of Information Technology has initiated a programme of establish State Wide Area Network (SWAN) up to the block level with a minimum Bandwidth of 2 MBPS to provide reliable backbone connectivity for E-Governance. The National Policy of the government recognizes the potential of E-Governance not only to improve governance but also to facilitate people's access to government services. Ministry of Communications and IT is working on a National E-Governance action plan that seeks to lay the foundation and provide impetus for a far more pervasive spread of E-Governance to reach the Common man particularly in far-flung areas. Seeking to do so Government of India is putting together various elements that are needed for leveraging the enormous power of ICT for the economic development of our country and enable the common man to access Government services in an efficient, convenient and cost effective manner.

Internet for Information and Communication

We are in the midst of Information and Communication revolution. The world is rapidly shrinking to a 'global village', which some courageously calls a global family. In the merger of telephony, television and computers a new world of Communications is evolving. What triggered this undreamt of merger is the emergence of Internet-millions of computers and computer networks connected with each other exchanging information. The word Internet flashes many images upon the canvas of the mind. The applications aspect of Internet is the multitude of different services it offers, example email, searching information over web, discussion groups etc. The Internet was started as a military network in USA and has undergone tremendous change over a period of time, offering variety of services to the users. The Internet is a huge resource of information that was accessed by millions of users every day, accounts to 100 terry bits of data passes through Internet backbone for every minute. As per the ITU (International Telecommunications Union, www.itu.int) website information, there were 1.13 billion subscribers to Internet in December 2006. India had 6.93 million subscribers and 60 million users of Internet. The Internet penetration figures (per cent of population using internet) for selected countries are given in the table. The services and information that Internet provides are increasing at a very fast pace.

Internet

Internet is a worldwide network of computer networks. It is an open inter connection of networks that enables connected computers to communicate with each other. These networks are scattered over the globe, yet are inter connected making it possible to communicate with each other in a few seconds. Internet is not owned by any individual organization or the country; it is a free for all open service facility. It is governed by INTERNIC (Internet Network Information Centre). The use of Internet and the ease of use of Internet have been growing in parallel. Till the early 80's, using the Internet was a complex process of issuing text commands and remembering the complex numeric addresses of the communicating sites. However, its power was obvious. There was no other method to connect up universities and research labs around the world that was so fast, convenient and flexible. The Internet users at universities came up with the software to participate in discussions over the network. They created documents and software libraries on the network, which were accessible to all users. During this period the Internet remained within the narrow confines of the academic and research-lab world.

Table 43.1: The Internet Penetration Figures (Per cent of population using internet) for Selected Countries

Sl.No.	*Name of Country*	*NRI Rank*	*Sl.No.*	*Name of Country*	*NRI Rank*
1.	Denmark	1	2.	Sweden	2
3.	Singapore	3	4.	Finland	4
5.	Switzerland	5	6.	Netherlands	6
7.	US	7	8.	Iceland	8
9.	UK	9	10.	Norway	10
11	India	44	12.	China	59

Source: www.itu.int.

Terms Used with Internet

World Wide Web (WWW): World Wide Web is a wide area, hypermedia information retrieval initiative aiming to give universal access to a large universe of documents. Hypermedia is a natural extension of hypertext, in that the contents of each document not only include text but also, images, sounds and video. WWW provides a consistent means to access a variety of information in a simplified manner to the users on computer networks. WWW contains a vast storehouse of hypertext documents written using the Hypertext Markup Language (HTML). Hypertext is a method for presenting text, images, sound, and videos that are linked together in non-sequential web of associations. Hypertext format allows the user to browse through topics in any order.

WWW enables the users to view variety of information on any subject in the form of textual material, pictures, audio and videos. The information is also in the form of e-magazines, archives, public and university library resources, current world and business news. It provides a web of interactive documents that contain text, pictures, graphics, multimedia, animations, etc. The hyperlinks provide the links to the resources of the same page, other pages of the web site or the pages belongs to other web sites. The user can navigate through the information by pointing to special designated text or other objects on the screen. These objects link to the other WWW pages on the same server or any other WWW server on the network.

Web Browser

Web Browser is a software programme, which facilitates to access the information and presents it on the screen and helps in navigation on the internet. The browser provide with powerful, and easy to use features that allow taking full advantage of web contents. The browser presents the formatted text, images, sound or other objects such as links in the form of web page on the computer screen. The web browsers also called as "Client" programmes, which takes commands from user and sends requests to "web server" to get information from it and presents it on the browser window. Web browsers give access to special multimedia contents that provide audio, video and interactive web pages. Web browsers were initially designed to interact with the content of World Wide Internet for Web. Most browsers now also interact directly with Gopher servers, FTP-sites and other internet tools and systems, thus providing a uniform, easy-to-use interface with many services of the Internet.

Browsers can be divided into two basic groups: text mode and Graphic User Interface (GUI). Text-mode browsers are often faster and usable with a variety of hardware and software systems, but they have limitations as they can handle only text *i.e.* words. The Lynx is the most popular text-mode browser available on UNIX based platforms. GUI browsers are easier to learn, faster to control and use. The GUI browsers can perform the same tasks as text-mode browsers. They are accomplished largely through mouse point-and-click operations in keeping with the native interface be it Windows, Macintosh, or X-Windows. The GUI browsers generally have some page-handling features like – the ability to save the page being viewed currently; print the page being viewed currently. Some browsers let you e-mail a page to yourself or someone else and some will even let you create and mail messages from within the browser. The leading browsers are Microsoft's Internet Explorer, Netscape Corporation's Netscape Navigator and Fire box.

Web Page

A web page is a single unit of information called a hypertext document. A web page may consist of multimedia content such as text, images, sound and videos. A group of web pages created by one person or a company or organization is referred as web site. A hyper link can be used to link other documents, sounds, images, databases, e-mail addresses etc. The links contained in web pages can point to areas within the same page, to other pages residing on the same web server, or to pages sitting on a computer on the other side of the world. Hyperlinks are usually underlined and are referred as URL. There is no need to know or type the URL. Each time the mouse moves over these links, the mouse pointer changes to a hand.

How Does the Internet Work?

There is no single central server or Computer or organization to make Internet work. All the Computers working independently, at various locations across the globe, are connected by the Internet. The information can be exchanged between these computers over the Internet, irrespective of the architecture of the hardware system, and operating system software working on the computers. It works because they follow the rules framed by the TCP/IP protocol (Transmission Control Protocol/

Internet Protocol). Because of heterogeneous computers and operating systems, a specific protocol is required, which can connect them into a common platform over Internet to exchange the information easily. The TCP/IP is a standard protocol, which works with Internet. TCP/IP performs an important role when the data was send by a computer. It breaks the data into smaller packets; each packet has three parts, the address where the packet is meant to go, the data and error control information. The data packets will move to the destination in different paths with the help of the address. At the receiving end the packets were reassembled to get back to the original shape of data. There is no central computer or authority. Instead of having the data go to a central computer and then to its destination. Internet is dependent on the existing infrastructure developed by the telephone companies and Inter Service Providers (ISPs) to transmit the data. Internet service providers lease data circuits from the telephone networks and have dedicated computers at their data centers, network devices such as routers, firewalls etc. These rely on the distributed intelligence of networking equipment known as "routers". The content of Internet is hosted on a computer known as "web server". The web servers of data center of ISP may be owned by the organizations, called as "co-location" or certain space on the web server may be given to the organizations on lease to host the content. When a request is made of these servers for the information, they bundle the requested information in small packets, with address as to where it is to be sent, and send them to the nearest connection on the Internet. On the Internet, the packets are received by the router, which is nothing more than a traffic controller, and sent it down in the same general direction of the address. A similar thing happens at the next junction on the Internet. This goes on till the packet is delivered to the right address, where it is put together again with other packets, to make up the original information. Say for example you are sending a message from ICAR Head Quarters in Delhi to a server named Google.com in USA. The message will be broken up into packets of approximately 1500 bytes, and some may travel from MTNL, Delhi ISP to the Google router in the US, some may travel to Hyderabad ISP and then to the Google router and so forth. There is no predetermined path and even individual packets of the same message may follow different paths. It all depends on the traffic at that node, at that moment in time. As the packets reach google.com, they are all put together as in the original message and delivered to the given address.

Domain Names and Addresses

Networks and computer systems on the Internet can communicate with each other. In order to communicate with each other, every computer on the Internet will be identified as a unique system like telephone numbers. All computers on the Internet have been assigned with an address system called IP address, which is unique number on Internet. These addresses are made up of a sequence of four three digit (decimal) numbers separated by periods. *E.g.* 164.100.140.2. Each number is in the range of 0 to 255. For example www.google.com is a domain name, which will be identified by the IP address 64.233.189.104. Because IP addresses are not easy to remember, computers are also identified by a name called domain name. A domain name server translates a domain name into an IP address.

This numeric scheme of IP addresses works well for computer systems, but it is difficult for people to remember and type correctly for every Internet site they need to contact. Therefore, Internet sites also have names associated with them. For example-ori.nic.in. Like the IP addresses, domain names are a sequence of words separated by periods. There are at least two words and can have three or more. The collection of networks making up the internet is divided into groups called domains. The domains represent type of organization and geographical location. For example a site in the domain 'edu' would be an educational institution. The domain name will give meaningful information to the users. For example www.nic.gov.in, which will tell that a web site named as "NIC", owned by the government and belongs to India? An address specified as a domain name is automatically converted to the IP address.

List of Domains by Type of Organization

A list of Domain name types used internationally is given here under:

Domain	*Type of Organization*
.com	Commercial Origination
.edu	Educational Institution
.gov	Government (United States)
.org	Non-profit Organizations
.net	Networks

In India, Centre for Development of Advanced Computing, Mumbai (formerly known as National Centre for Software Technology) is one of the Internet Domain Name Registrars. It regulates issue of domain names.

co.in – for registered commercial organizations

ac.in – for academic community

res.in – for research institutes

gov.in – for government organizations

net.in – for network service providers

mil.in – for military establishments

org.in – for miscellaneous organizations and the indicator.in at the end of all the domain names above indicates that they are registered in India. For other countries there are different (unique) identifiers. A list of some well-known countries domain name indicators is given here under:

List of Geographical Domains

The two letter country code for all the countries is available at anonymous ftp site rtfm.mit.edu in the directory/pub/usenet/news.answers/mail/and on www at http://www.ee.ic.ac.uk/misc/countrycode.html. On the Internet, it is assumed that if there is no geographical code used, then the domain is located within US.

Domain	*Country Name*
.in	India
.au	Australia
.ca	Canada
.jp	Japan
.uk	United Kingdom

Internet Connection

There are number of ways one can connect to the Internet. An Internet Service Provider (ISP) is a company that provides access to Internet. In general, there are two types of connections offered by Internet Service Provider. They are 1) Dial-up connection 2) direct Internet link (leased line or ISDN line).

Dial-UP Connectivity

For smaller organizations, establish link through a dial-up connection that is computer call to ISP over a telephone line (PSTN line) to access the Internet. For instance, you might have a communications server on your network that calls the service provider to send and receive any Internet communications. The obvious advantage of service provider is that they are much less expensive than establishing your own direct link with the Internet. The drawback is that the bandwidth is limited and the speed also depends on the number of connections accessing the same ISP.

Direct Internet Link

The direct Internet link can be provided in the form of leased line connectivity or ISDN connectivity by the ISP. In this type of connectivity, the organization router directly connects to the service provider with a higher bandwidth limit (64 kbps or 128 kbps or 256 kbps or 512 kbps or 1 mbps etc.). This type link is generally used by a larger institutions, corporations and government agencies. It involves establishing internet gateway with a full-time link with the Internet for 24 hours/day. This type connectivity is beneficial to have the maximum traffic and throughput *i.e.* amount of data transferred with the Internet. The drawback, however, is the cost for band width.

Internet Services

Today the Internet is growing tremendously and is known mainly for the services it provides. Some of the best-known services available on the Internet include following:

- ☆ World Wide Web (www)
- ☆ File Transfer Protocol (FTP) service
- ☆ Electronic mail
- ☆ Discussion groups and News Groups
- ☆ Telnet service

World Wide Web

The World Wide Web (www) is the Internet's multimedia service that contains a vast storehouse of hypertext documents written using the Hypertext Markup

Language (HTML). Hypertext is a method for presenting text, images, sound, and videos that are linked together in a non-sequential web of associations. The hypertext format allows the user to browse through topics in any order. There are tools and protocols to explore the Internet. These tools help to locate and transport resources between computers.

File Transfer Protocol (FTP)

File Transfer Protocol (FTP) support is one method of supporting remote networks. It is a protocol, which allows simple file transfer of documents. There are FTP servers, which provides vast amount of information stored as files. The data in these files cannot be accessed directly; rather the entire file must be transferred from the FTP server to the local computer. The most common protocol used for sending files between computers is the FTP. FTP allows for transferring both text and binary files. Both Microsoft operating systems and UNIX system include the traditional character based FTP client. This is one of the utilities that is copied onto the system when the TCP/IP protocol suit is installed. In addition, most Internet browsers such as Microsoft Internet Explorer, Netscape support FTP and use it behind the scenes when transferring files.

E-Mail

E-mail or the electronic mail is the most widely used application on the Internet for sending and receiving electronic messages. It is currently one of the most popular activities on the Internet. For most of the Internet users, it has practically replaced other traditional methods such as telephones, faxes etc. Technically E-mail is a system of delivery of messages on the computers connected via communication networks. E-mail is electronic version of the paper mail or letters used to deliver personal and official messages. E-mail is used to communicate all types of messages- text, graphics, audio, and also visual clips as long as these can be digitized. Hence for all our communication needs e-mail that offers a quicker, cheaper and convenient option. To send e-mail, you must know the recipients' e-mail address.

In government offices and research organizations most of the communication with the international organizations- the World Bank, Food and Agriculture Organization, United Nations Development Programmes, is in form of e-mail. The largest users of e-mail, however, are the students of graduate and post-graduate programmes in the universities. The students use e-mail as most efficient method of keeping in touch with their friends (in some other university in India or abroad), getting information on career and academic opportunities and also for seeking information on academic needs.

Discussion Groups and News Groups

Discussion groups are the virtual networks of Scientists and other stake holders having email interactions/message postings on a common subject. Discussion groups undertake in-depth discussion on email mode. The emerging subject, issue is flagged by one of the group members and then an email alert is sent to all the members. An agreed time frame of one week to 10 days is decided for getting inputs from all the group members and the responses are shared among all. Thus, highly focused discussions take place on the internet, without having any physical meeting.

Discussion groups are emerging as one of the very effective scientific discussion forums on the internet. The solution exchange supported by United Nations Organizations (UN) (website address : www.solutionexchange-un.net.in), has proved to be an excellent enabler of focused group discussions on highly topical issues like "Sustainable Agricultural Extension Systems", "Spreading the ICT Revolution in Rural India- Experiences and Examples", "Establishing Rural Business Hubs" during last two years. Over 4000 experts and field managers have participated and contributed/benefited from the discussions. The consolidated responses on all these topics were later published for wider circulation. The solution exchange has organized its discussion forums in the following groups: Food and Nutrition Security, Education, Environment, Gender, Health, Poverty, Aids, Decentralization, Disaster Management and ICT for Development. Each community (group) has over 1000 members and most of them contribute to make the discussions highly valuable and problem solving in nature. The community also shares latest developments in the concerned area and news about the emerging trends/issues in national and international arena. Discussion forums are also known as web forums, message boards, discussion boards, (electronic) discussion groups, discussion forums, bulletin boards with a little variation in information sharing mechanism. Essentially all these are tools for information-sharing on electronic-platform.

Telnet

Telnet was one of the first Internet protocol. Telnet is used to act as a remote terminal to an Internet host. When a computer connected to an Internet host, it acts as if it was attached to the remote computer. The application settings, programmes can be run on remote terminal using this service. The main use of this service is providing a secured connection to the remote computer and performs the computer operations on remote computer.

Search Engines and Searching

With over a thousand million pages and continuously increasing information in audio and video form on the World Wide Web, the task of finding precisely what you are looking for is very difficult. Search tools available on the internet make your search tasks easier. Many web based search engines are available, the search return the result of an internet search in a matter of seconds.

Search Engine

Search Engines are powerful tools as they do the searching for you by following the instructions you give. Search engines search information on the World Wide Web. You need to supply the key words to the search engine and the search engine returns the index of pages, websites, where it finds match with your key words. The more detailed use of keywords, phrases with the combination of Boolean logic + (and), - (or) in your instructions, the more accurate the results will be. Some of the popular search engines are: google, yahoo, altavista, ask.com, gigablast.com, etc. There are some region or country specific search engines, these include:

- ☆ Ansearch, Australia/US/UK/NZ
- ☆ Araby, Middle East

- Baidu, China
- Daum, Korea
- Guruji.com, India
- Miner.hu, Hungary
- Najdi.si, Slovenia
- Naver, Korea
- Rambler, Russia
- Rediff, India
- SAPO, Portugal

With Web contents increasing at a phenomenal pace there are a number of Domain based search Engines. For example there are "Job Search" search engines, which include:

- Naukri.com (India)
- Bixee.com (India)
- Craigslist (by city)
- Eluta.ca (Canada)
- CareerBuilder.com (USA)
- Hotjobs.com (USA)
- Indeed.com (USA)
- Monster.com (USA)
- Recruit.net (International)
- SimplyHired.com (USA)

Other major categories of Domain based search engines are: legal, blog, news, multimedia, medical, property, geographic, social, kids, agriculture etc.

Agricultural Search Engines

The major search engines for Agriculture domain are:

- www.agfind.com
- www.agriculture.com
- www.agricultureinformation.com
- www.usagnet.com
- www.farms.com
- www.agcareers.com
- www.produceindustry.com
- www.usdareports.com
- www.fruitsearch.com
- www.producelinks.com

Almost all the above agricultural search engines are developed by US based companies and their focus is accordingly to serve their clientele. Hence, most of the information they search is US based/hosted. In India as well as in other developing countries majority of the Agricultural scientists, extension officials use generic search engines like Yahoo, Google, Altavista, Khoj etc.

Important Indian Agricultural Web Sites and Portals

Addresses of some of the important Indian Agricultural Websites are given below; you can start browsing these web sites for accessing information about the indicated institutions/agencies. You can also use search engines to find sites of your interest indicating specific key words for searching. The important Indian Agricultural sites include:

1. www.agricoop.nic.in
2. www.dare.nic.in
3. www.dacnet.nic.in
4. www.agmarknet.in
5. www.indiastat.com
6. www.manage.gov.in
7. www.icar.org.in
9. www.cazri.res.in
10. www.caie.nic.in
11. www.cifa.in
12. www.cife.edu.in
13. www.cpcri.ernet.in
14. www.dryland.ap.nic.in
15. www.crri.nic.in
16. www.iasri.res.in
17. www.iihr.res.in
18. www.spices.res.in
19. www.iisr.nic.in
20. www.nianp.nic.in
21. www.nbagr.ernet.in
22. www.nbfgr.res.in
23. www.nbpgr.ernet.in
24. www.nbsslup.nic.in
25. www.ncap.res.in
26. www.nrcaf.ernet.in
27. www.nrccashew.org
28. www.nrce.nic.in

29. www.iari.res.in
30. www.nrcgrapes.mah.nic.in
31. www.nrcipm.org.in
32. www.nrc-map.org
33. www.nrcmashroom.org
34. www.nrcog.mah.nic.in
35. www.nrcjowar.res.in
36. www.nrcsoya.com
37. www.nrcws.org
38. www.iimahd.ernet.in
39. www.itcibd.com
40. www.esagu.in
41. www.agri.mah.nic.in
42. www.nird.ap.nic.in
43. www.emandi.mla.iitk.ac.in

It is advised to browse some of the above web sites to have better appreciation of information sharing by the concerned departments/institutes.

Expert System

Concepts and Importance of Expert Systems

The branch of computer science known as Artificial Intelligence (AI), covers a number of different fields of application. Expert system is one such field that has attracted significant attention in recent years. Expert systems have been developed and applied to many fields such as office automation, medicine and agriculture. Knowledge representation is an issue that arises in both cognitive science and artificial intelligence. In cognitive science, it is concerned with how people store and process information. In AI, the primary aim is to store knowledge so that programs can process it and achieve the verisimilitude of human intelligence. AI researchers have borrowed representation theories from cognitive science. Thus, there are representation techniques such as frames, rules and semantic networks, which have originated from theories of information processing. Since knowledge is used to achieve intelligent behavior, the fundamental goal of knowledge representation is to represent knowledge in a manner as to facilitate inferencing, *i.e.* drawing conclusions from knowledge. Knowledge engineers are concerned with the representation chosen for the expert's knowledge declarations and with the inference engine used to process that knowledge. A user can use the knowledge acquisition component of the expert system to input the several characteristics known to be appropriate to a good inference technique. They are:

- A good inference technique is independent of the problem domain.
- In order to realize the benefits of explanation, knowledge transparency,

and reusability of the programs in a new problem domain, the inference engine must not contain domain specific expertise.

- ✰ Inference techniques may be specific to a particular task, such as diagnosis of hardware configuration. Other techniques may be committed only to a particular processing technique.
- ✰ Inference techniques are always specific to the knowledge structures.
- ✰ Successful examples of rule processing techniques are forward chaining and backward chaining.

Importance of Expert Systems

The complexity of problems faced by farmers are yield loses, soil erosion, selection of crop, increasing chemical pesticides cost, pest resistance, diminishing market prices from international competition, and economic barriers hindering adoption of farming strategies. The farmer may not become expert manager of all these aspects of farming operations. On the other hand, agricultural researchers need to address problems of farm management and discover new management strategies to promote farm success. Numerical methods have failed to provide better solution because understanding about crop systems is qualitative based on experience and cannot be mathematically represented. Expert system are computer programs that are different from conventional computer programs as they solve problems by mimicking human reasoning processes, relying on logic, belief, rules of thumb opinion and experience. The experience and knowledge of scientists will be used to develop expert system on various issues of agriculture, which in turn will provide advisory support to the farmers. In agriculture, expert systems are capable of integrating the perspectives of individual disciplines such as plant pathology, entomology, horticulture and agricultural meteorology into a framework that best addresses the type of ad hoc decision-making required of modern farmers. Expert systems can be one of the most useful tools for accomplishing the task of providing growers with the day-today integrated decision support needed to grow their crops.

Components of Expert Systems

Expert systems are composed of several basic components such as a user interface, a database, a knowledge base, and an inference mechanism. Moreover, expert system development usually proceeds through several phases including problem selection, knowledge acquisition, knowledge representation, programming, testing and evaluation.

User Interface

The function of the user interface is to present questions and information to the user and supply the user's responses to the inference engine. The questions are mostly in the form visuals that are developed as images, animation clips, and video clips. Any values entered by the user must be received and interpreted by the user interface. Some responses are restricted to a set of possible legal answers, others are not. The user interface checks all responses to ensure that they are of the correct data type. Whenever the user enters an illegal response, the user interface informs the user that his input was invalid and prompts him to correct it.

Knowledge-base

The knowledge the expert uses to solve a problem must be represented in a fashion that can be used to code into the computer and then be available for decision making by the expert system. There are various formal methods for representing knowledge and usually the characteristics of a particular problem will determine the appropriate representation technique employed. The knowledge base is a collection of rules or other information structures derived from the human expert. Knowledge bases can be represented by production rules. These rules consist of a condition or premise followed by an action or conclusion (IF condition...THEN action). Production rules permit the relationships that makeup the knowledge base to be broken down into manageable units. Having a knowledge base that consists of hundreds or thousands of rules can cause a problem with management and organization of the rules. Organizing rules and visualizing their interconnectedness can be accomplished through dependency networks. The knowledgebase can be used to good relational database management systems (DBMS) like Oracle, SQL Serer, MySQL, Access databases to develop the rule base and the query system can be used to retrieve the knowledge from DBMS systems.

Inference Mechanism

The inference mechanism will be integrated as a software program (inference engine) that the part of the program containing reasoning capability. It interacts with a knowledge base (IF…THEN…ELSE statements), which contains information about how to solve problems within the problem domain. This is the global memory where the knowledge based system records information relating to a specific problem that it is trying to solve. Much of this information comes from the user but the memory is also used by the inference engine to record its own conclusions and to remember its chain of reasoning. By comparing what it knows about the problem domain in general with what it knows about the specific problem, the inference engine tries to proceed logically towards a better solution.

Inference Rule

An understanding of the "inference rule" concept is important to understand expert systems. An inference rule is a statement that has two parts, an if-clause and a then-clause. This rule is what gives expert systems the ability to find solutions to diagnostic and prescriptive problems. An example of an inference rule is:

If the symptom of crop is X, Then the nutrition deficiency is Y.

An expert system's rule base is made up of many such inference rules. They are entered as separate rules and it is the inference engine that uses them together to draw conclusions. Because each rule is a unit, rules may be deleted or added without affecting other rules. One advantage of inference rules over traditional programming is that inference rules use reasoning, which more closely resembles human reasoning. Thus, when a conclusion is drawn, it is possible to understand how this conclusion was reached. Furthermore, because the expert system uses knowledge in a form similar to the human expert, it may be easier to retrieve this information from the expert. The knowledge that is represented in the system appears in the rule base. In the rule base described in the cross-referenced applications, there are basically four different types of objects, with associated information present.

Designing Expert Systems for Problem Solving

The architecture of expert systems consists of following steps:

1. The sequence of steps taken to reach a conclusion is dynamically synthesized with each new case. It is not explicitly programmed when the system is built.
2. Expert systems can process multiple values for any problem parameter. This permits more than one line of reasoning to be pursued and the results of incomplete reasoning to be presented.
3. Problem solving is accomplished by applying specific knowledge rather than specific technique. This is a key idea in expert systems technology. It reflects the belief that human experts do not process their knowledge differently from others, but they do possess different knowledge. With this philosophy, when one finds that their expert system does not produce the desired results, work begins to expand the knowledge base, not to re-program the procedures.

There are various expert systems in which a rule base and an inference engine cooperate to simulate the reasoning process that a human expert pursues in analyzing a problem and arriving at a conclusion. In these systems, in order to simulate the human reasoning process, a vast amount of knowledge needed to be stored in the knowledge base. Generally, the knowledge base of such an expert system consist of a relatively large number of "if then" type of statements that were interrelated in a manner that, in theory at least, resembled the sequence of mental steps that were involved in the human reasoning process.

Most expert systems, in the past, have been run only on large information handling systems. Recently, the capacity of personal computers has increased to a point where it is becoming possible to consider running some types of simple expert systems on them. In some applications of expert systems, the nature of the application and the amount of stored information necessary to simulate the human reasoning process for that application is just too vast to store in the active memory of a computer. In other applications of expert systems, the nature of the application is such that not all of the information is always needed in the reasoning process. An example of this latter type application would be the use of an expert system to diagnose a data processing system comprising many separate components, some of which are optional. When that type of expert system employs a single integrated rule base to diagnose the minimum system configuration of the data processing system, much of the rule base is not required since many of the components which are optional units of the system will not be present in the system. The user interface and the procedure interface are two important functions in the information collection process.

End User

The end-user usually sees an expert system through an interactive dialog, an example of which follows:

Q. Do you know which restaurant you want to go to? A. No
Q. Is there any kind of food you would particularly like? A. No
Q. Do you like spicy food? A. No
Q. Do you usually take soft drink with meals? A. Yes

As can be seen from this dialog, the system is leading the user through a set of questions, the purpose of which is to determine a suitable set of restaurants to recommend. This dialogue begins with the system asking if the user already knows the restaurant choice (a common feature of expert systems) and immediately illustrates a characteristic of expert systems; users may choose not to respond to any question. In expert systems, dialogues are not pre-planned. There is no fixed control structure. Dialogues are synthesized from the current information and the contents of the knowledge base. Because of this, not being able to supply the answer to a particular question does not stop the consultation.

During the consultation, the rule base is searched for conditions that can be satisfied by facts supplied by the user. This operation is performed by the inference engine. Once all of the conditions (*i.e.* IF parts of rules) of a rule are matched, the rule is executed and the appropriate conclusion is drawn. Based upon the conclusions drawn and the facts obtained during consultation, the inference mechanism determines which questions will be asked and in what order. There are various inferencing methods available to perform the tasks of searching, matching and execution.

A distinctive characteristic of expert systems that distinguishes them from conventional programs is their ability to utilize incomplete or incorrect data. Given only a partial data set, an expert is likely to have less than absolute certainty in the conclusion. The degree of certainty can be quantified in relative terms and included in the knowledge base. The certainty values are assigned by the expert during the knowledge acquisition phase of developing the system. By incorporating rules in the knowledge base with different certainty values, the system will be able to offer solutions to problems without a complete set of data. The capacity to deal with uncertainty is available in development software.

Advantages and Disadvantages of Expert Systems

Advantages

- Expert Systems are useful in many aspects and ready to use by end user as advisory system.
- Provides consistent answers for repetitive decisions, processes and tasks.
- Holds and maintains significant levels of information.

- ☆ Encourages human expert to clarify and finalise the logic of their decision-making.
- ☆ Never "forgets" to ask a question, as a human might.

Disadvantages

- ☆ Lacks common sense needed in some decision making.
- ☆ Cannot make creative responses as human expert would in unusual circumstances.
- ☆ Domain experts not always able to explain their logic and reasoning.
- ☆ Cannot adopt to changing environments, unless knowledge base is changed

Cases of Expert Systems in Agriculture

Rice-Crop Doctor

National Institute of Agricultural Extension Management has developed an expert system to diagnose pests and diseases for rice crop and suggest preventive or curative measures. The subject matter expert knowledge on rice pathology and entomology has been obtained from Scientists of Directorate of Rice Research, Hyderabad and A.P. Agricultural University. The rice crop doctor illustrates the use of expert systems broadly in the area of rice production through development of a prototype, taking into consideration a few major pests and diseases and some deficiency problems limiting rice yield. The following diseases and pests have been included in the system for identification and suggesting preventive and curative measures. The diseases included are rice blast, brown spots, sheath blight, rice tungro virus, false smut fungi, bacterial leaf blight, sheath rot and zinc deficiency disease. The pests included are stem borers, rice gall midge, brown plant hopper, rice leaf folder, green leafhopper and Gundhi bug.

The brief logic flow of the expert system is as follows:

- ☆ The part of the plant where symptoms have been observed is given by the extension officer.
- ☆ The basic symptoms are given as input:
- ☆ Considering these symptoms the user is expected to give further information based on other visual symptoms.
- ☆ At this step the disease and pest are identified.
- ☆ The user is then given the option to either stop or further diagnose and other disease/pest or get preventive or curative measures on these.

CALEX

This is a blackboard based integrated expert decision support system for agricultural management, developed at University of California. CALEX can be used by growers, pest control advisors, consultants and other managers.

LEY Expert Systems

This is a RF-telemetry based computer-controlled, automated, remote, real-time weather data acquisition and reporting system in Washington State. Cooperation among Washington State University, the National Weather Service (NWS) the U.S. Bureau of Reclamation and several private grower organizations have made this system possible. Data is collected, processed and transmitted to the NWS hourly. These hourly updates of actual conditions are broadcast on the NOAA weather band during the spring frost season to assist fruit growers with frost protection. Real-time weather data are also being used in applications such as irrigation scheduling, crop protection and pest management.

Prototype Decision Support System

A prototype decision support system by P.H. Heinemann is being used to assist managers with the development, implementation, and management of frost protection. The program is divided into four main modules: frost protection strategies, operational management, forecasting, and instrumentation. E. Satellite downlink facility using a unique satellite downlink facility at by R.R. Getz Auburn University, real time meteorological data are being processed on a network of advanced computer workstations. This data feeds a series of computer models that generate site specific predictions of temperature, dew point, wet bulb temperature, and other parameters used to alert Alabama fruit, vegetable, and nursery growers. Model outputs are distributed to Extension specialists and others over a campus-wide fiber optic communications system and to all county Extension offices over the ACENET computer network. During critical freeze situations, information can be disseminated live via satellite to media and rural satellite dish owners using the Auburn University uplink facility. A pilot network of automated weather stations incorporating a 10-m tower for monitoring inversions has been installed to enhance the program.

Application of ICT

E-Governance

Empowering the citizen via electronic governance has been gaining momentum across liberal economies in the world where democratic principles underlie the government constitutional structures. This has been further enabled by the growing use of new information and communication technologies (ICTs), particularly designed for strengthening democratic governance around the world. E-governance is widely acknowledged as the application of new ICTs, especially the Internet and World Wide Web (WWW), as channels for engaging citizens in government as well as facilitating delivery of government information and services.

ICT has a vital role to play in all transactions that the government undertakes. It helps the government cut red-tapism, avoid corruption, and reach citizens directly. This realization is wide spread among all the states of the country. Almost every state has an IT policy in place with the aim of evolving itself from being an IT-aware to an IT-enabled government. State governments are fast recognizing the benefits of an IT-enabled working environment and IT-enabled service delivery. Effective use of ICTs in government or E-Governance is perceived as a facilitator. Governments, at central and state level, in general feel necessity of that being accessible online 24/7 to the average citizen. Why e-Governance? The basic objective of e-Governance is to support and simplify governance for government, citizens and businesses. The use of ICTs can connect these three segments and support processes and activities. Other objectives are to make government administration more transparent, speedy and accountable, while addressing the society's needs and expectations through efficient public services and effective interaction between the people, businesses and government.

Concept and Definition of e-Governance

Definitions of E-governance vary and efforts to incorporate all of them in one single statement seem impossible. However, a common thing is that e-governance refers to the use of Information and Communication Technology (ICT), particularly web based applications, to provide access to and deliver information/service to the public, business, other agencies and governmental entities faster, cheaper, easier and more efficiently. This in turn enhances relationships, enlarges the overall customer base and improves core business operations through reexamination of internal processes. Other definitions go beyond this main core. Wimmer and Traunmuller (2000) consider E-government as a guiding vision towards modern administration and democracy. For them it is concerned with the transformation that government and public administration have to undergo in the coming decades. Hence, e-government is not an option. Instead it is a must for the governments to use the benefits of the information technology in order not to fall behind. Lawson sees e-government as "transferring power to people" by operating by a one stop, non-stop way and doing more for less. e-Governance is defined by M. Backus (Backus, M., 2001), as the "application of electronic means in (a) interaction between government and citizen and between government and businesses as well as (b) in internal government operations to simplify and improve democratic, government and business aspects of governance".

The term *'electronic means'* above stands for Internet and other ICT applications;

The term '*interaction* stands' for delivery of government products and services, exchange of information, communication, transactions, and system integration; Government consists of levels and branches. Government levels include- Central, State, District, Block (Taluka), Mandal level government agencies/departments/ institutions. Government branches include- Parliament, Judiciary, Police, Civil Services and all branches of administration.

Government operations include all back-office processes and inter-governmental interactions within the total government body.

e-Governance and e-Government

e-Governance and e-Government are two very distinct terms. Whereas e-Governance is a broader topic that deals with the whole spectrum of the relationship and networks within government regarding the usage and application of ICTs, e-Government is a narrower discipline dealing with the development of online services to the citizen, *i.e.* government services – such as e-agriculture, e-passport, e-tax, e-transportation, e-health, etc. E-Governance is a wider concept that defines and assesses the impact that technologies are having on the practice and administration of governments and the relationships between public servants and the civil society, and dealings with the elected bodies or outside groups, such as not-for-profit organizations, NGOs or private sector corporate entities. E-Governance encompasses a series of necessary steps for government agencies to develop and administer, in order to ensure successful implementation of e-government services to the public at large.

E-Government

E-Government or Electronic Government is the delivery of more convenient, customer oriented and cost effective public services and sharing of information through electronic media. Specifically, e-Government harnesses information and communication technologies (such as Wide Area Networks, the Internet, and mobile computing) to transform relations with citizens, businesses, and other arms of government. E-government is an administration where citizens can avail of government services such as land records, filing of tax, etc, E-Government are also defined as the use of Information and Communication Technologies to provide better government services. This definition usually refers to more efficient, more transparent and potentially two-way information flow between the government and citizens. But the fact e-Governance is that in most of the cases e-Government is not more than digitalization of the operational functions of the government. The Global e-Government Readiness Report says the role of government has to be reoriented and redefined from "that of managerial authority to that of leadership in a multi-centered government, the culture of governance needs to change from being bureaucratic to participatory from authoritative to accountable, from being monopolistic to being competitive and innovative, from being closed to being participatory, from being autocratic to democratic, and from being exclusive to inclusive". E-Governance is socially inclusive governance with the promise of good governance and effective participation and inclusion of citizens.

E-Governance

E-Governance is the development, deployment and enforcement of the policies, laws and regulations, necessary to support the functioning of a Knowledge Society as well as of e-Government. E-Governance brings about 'anytime anywhere access' to the citizen. E-governance is about transforming the relationship between governments and their citizens through the use of information technologies, and that E-government is about delivering government services using information technologies.

Benefits of e-Governance

E-Governance was initiated worldwide to improve speed, transparency, efficiency and reliability of the Government services. The basic premise to undertake e-Governance initiatives was to harness the advantages of ever-enlarging network and reach of Internet for improving reach, timeliness, credibility and consistency of Government services. E-governance offers the potential to bring citizens closer to their governments. Regardless of the type of political system that a country has, the public benefits from interactive features that facilitate communication between citizens and government. The scope as well as potential benefits of e-Governance are increasing with enrichment of e-services by various departments/agencies. The following are direct benefits of e-Governance:

- Faster, easier, more convenient service
- Convenience, Near to Home, more services at one place
- Faster processing, shorter wait, shorter queues at Government offices
- Less number of trips to government offices
- Better interaction environment; No harassment
- Reduced transport cost
- Avoids wage loss for daily-wage earners
- Better quality service
- More accurate and legible documents, easy recovery from errors,
- Lesser corruption, increased transparency
- Improved access to offices (nearer home, 24X7 or at least 12X6, in place of limited number of working hours or only on working days) and functionaries, no intermediaries.

There are some derived benefits of e-Governance services, as the back-end procedures and databases are also linked to services. This streamlining of data-networking helps the concerned departments to handle queries and complaints more effectively. These benefits include:

- Improved complaint handling, complaint tracking
- Introduces transparency in data, decisions/actions, rules, procedures and performance of Govt. agencies
- Automates processes to take away discretion
- Reduces arbitrariness in decision making
- Entry point for simplification of rules and reengineering processes
- Makes decisions traceable- tracks actions
- Builds accountability- greater access to information through web publishing-role of civil society
- Provides documentation to citizens for follow up
- Tracks procedural/human delays
- Easier to fix responsibility/accountability
- Modularizes Tasks Making Outsourcing Possible
- Introduces competition amongst delivery channels and departments
- Standardized documentation of comments/objections leads to effective supervision- through comparative indicators

- Centralizes data for better audit and analysis Integration of data across applications-provides improved intelligence
- Enables unbiased sampling for audit purposes

Further, there are social benefits of e-Governance projects and services as it induces increased skill and knowledge among the young and curious citizenry. E-Governance creates more job opportunities at grass-root level. Just like the PCO (Public Call Offices) revolution in telecommunications created over a million jobs in 90s, the Community Services Centres (CSCs) are potential creator of over two million jobs in next five years, if the 1,00,000 CSCs are made operational by Ministry of Information Technology, as planned.

Critical Success Factors

A large number of e-Governance projects have been taken up within and outside India. Within our country there are over a hundred e-Governance projects currently operational (at various stages). Some of these are highly successful, benefiting a large number of people and also generating resources to make them self-sustainable. The "e-Seva" project of Government of Andhra Pradesh is one such project. More than 50 government services including payments of telephone-bills, property tax and many more services are available at e-Seva centres. Some other e-Governance projects are delivering high quality information services to the clientele and have potential to be replicated country-wide. "EKrishi- Market Driven Agricultural Initiative through IT enabled Agri-Business Centres in Kerala State" project of Kerala and "Digital Mandi" project of IIT Kanpur can be categorized under such projects. Against a handful of highly successful and replicable projects there are numerous examples of e-Governance projects which have failed to grow beyond pilot stage. Studies have indicated that there are certain critical factors for each successful e-Governance project. Studies carried out within and outside the country have indicated the following as some of the critical success factors for e-Governance projects:

- Political will
- Overall Vision and National e-Strategy
- Full ownership and support of Chief Executive Officer and senior staff
- Effective and Appropriate Project Management
- ICT Infrastructure (telephones, broadband connectivity, computers and electricity)
- Access (within reach- distance as well economic)
- Education of Clients
- Participation of Clients
- Culture
- Cooperation and Collaboration with Private Sector
- Technology
- Human Resources

- ✰ Finances
- ✰ Legal Framework

E-Governance: Ecology and Components

The last 2 decades in the history of mankind have seen enormous growth of technology and its utilization for creating a better global society by disseminating the knowledge down to common man, thereby transcending the geographical boundaries. The accessibility of knowledge is not confined to any nation. Thus the whole world has acquired a bigger vision of coexisting with divergent societal patterns. Unlike other sectors, the application of ICTs in Governance has larger implications on interactions with structural systems of politics and bureaucracy, and value systems of society. For success of e-Governance projects, it is important to understand the whole context or ecology of e-Governance. Professor Sangeeta Sharma (2002) has suggested concept of "Ecology of E-Governance". To understand the ecology of E-Governance it would be pertinent to identify the components, then to elaborate upon how these components interact with each other to deliver the output/service to society.

Components of E-governance

The following components can be identified:

- ✰ Political Component with Enactment dimension.
- ✰ Technological Component with Electronic dimension.
- ✰ Social Component with Egalitarian dimension.
- ✰ Cultural Component with Ethical dimension.
- ✰ Psychological Component with Extensional dimension.
- ✰ Service Component with Empowerment dimension.

Political Component

The political system is essential aspect of governance. It holds responsibility of rationalizing various operative frameworks by enacting laws. This helps to maintain and sustain the cohesive force that is required by society to integrate its people and abide them to follow a uniform policy to fulfill their targets. This refers to the importance of "e" of enactment of laws to stop society from disintegration.

Technological Component

This relates to educate people who are in the bureaucratic structure or outside its periphery regarding use of electronic means to develop better connectivity within and with the system. It requires use of computers (a) in developing the data-base, (b) in networking to facilitate the communication and (c) in creating e-knowledge workers so as to increase their potentiality. This focuses on "e" of electronic knowledge and its proper utilization.

Social Component

The fundamental duty of any government is to educate a society which is based on the principles of equality and justice. A society which is classless with no distinction between haves and haves-not, where nobody is marginalized from the main stream, and is committed to provide a respectful essence of life to its people without any discrimination of cast or race must be set up. This is possible when people are aware of their rights and duties on the one hand, and know about the governmental policies made for them on related issues on the other, hence a vigilant society can be evolved where they can raise their voices by questioning the governmental decisions. This would help in attaining the "e" of egalitarian society with thrust on equality.

Cultural Component

With the advent of the era of knowledge explosion there is a need to reorient some of the value patterns without eroding the fundamental normative structure of any social system. The need to create value patterns conducive for e-governance to operate focusing on work ethos therefore cannot be denied. Thus to work out the ethical framework is the key to move further by discarding obsolete set of values that come in the way of potential utilization. Thus "e" of ethical framework has to be the focal point in constructing a morality-based system.

Psychological Component

Developing required psyche so as to facilitate formation and inculcation of right type of attitudes in the people is prerequisite for efficiency. Apart form this; readiness to connect to people, to listen to their queries, to look for solutions, to improve communicative skills etc. will be necessary elements for behavioral modifications. Hence personality adjustments must be carried out to cater to the needs of common man. This specifically relates to "e" of extension of self so as to have constructive collaborative social relationships.

Service Component

The ultimate function of any welfare government is to serve. As a provider of good services to its people it has to assimilate the basic needs and also expectations of common man in to the documentation of policy, which then has to be implemented in the right spirit. This develops a constant pressure on the government to respond to public demands as this leads to the realization of power they can exercise on Government to maintain impartiality, integrity and transparency in its functioning. This explains "e" of empowerment of people in any system.

Interactions among Various Components

Now let us analyze the interactive relationship amongst the various components identified above so as to get an insight in to understanding the ecology of e-Governance.

These components are not only interdependent but are also interrelated with each other in the way that output of one component becomes input of the other component. Thus a relationship of all purpose type emerges which gets substantiated by assimilating the impact of external forces. A schematic representation is self explanatory in this regard. The definitive interrelationships between the various components reciprocate e-Governance in such a way that resultant impact would be the transformation of society in to one that has right knowledge, right morality, and right way of perpetuating practicing ethos.

Stages of e-Governance

'Stages' of e-governance refer to the degree of progress made by a Government towards the ultimate e-governance goal. There are a number of models on maturity stages of government service delivery and available methodologies to measure it. According to Gartner and Christopher (Gartner and Christopher), The e-Government matures in four stages namely

a) Presence

Web site that lists cursory information about an agency, such as hours of operation, mailing address, and/or phone numbers, but has no interactive capabilities;

b) Interaction

Provides simple interactions and generally revolve around information provision, designed to help the customer avoid a trip to an office or make a phone call by making commonly requested information and forms available round the clock;

c) Transaction

Provides citizens the facility for self-service operations by allowing to complete electronically entire tasks such as license renewals, paying taxes and fees, and submitting bids for procurement contracts. The activities still involve primarily a one-way flow of information; and

d) Transformation

The highest order of evolution for e-government initiatives which facilitates seamless flow of information and collaborative decision making between all levels of government and private partners by removing organizational barriers. The most comprehensive and widely used model is the one suggested by Layne and Lee (2001). According to Layne and Lee model e-government is an evolutionary phenomenon and it evolves in four stages of growth. The four stages outline the structural transformation of e-Governance services and interactions as they progress towards fully integrated and amalgamated with the traditional public administration. These stages are:

1. Cataloguing
2. Transaction
3. Vertical integration
4. Horizontal integration

Cataloguing

In stage one of cataloguing, initial efforts of governments are focused on establishing an on-line presence for the government. Many state governments' efforts on web development and forms-on-line initiatives belong to this stage. Examples of functionalities at this stage are mostly limited to on-line presentations of government information. Towards the end of this stage, mostly pushed by citizens' demands, governments begin to establish index pages or a localized portal site in which scattered electronic documents are organized so that citizens can search for and view detailed government related information and download necessary forms. This first stage is called 'cataloguing,' because efforts are focused on cataloguing government information and presenting it on the web.

Transaction

As government websites evolve, officials as well as citizens come to realize the value of the Internet as another service channel for citizens and want to exploit it. Citizens demand to fulfill government requirements on-line instead of having to go to a specific location to complete paperwork. Electronic transactions offer a better hope for improved efficiency for both the customer and the agency than simply "cataloguing information." In addition, such capabilities provide the opportunity for a broader democratic process by holding interactive conversations with constituents who are reluctant or unable to attend public hearings. There is no question that fully functional e-government will make service delivery more efficient and increase savings for both government and the citizen. This second stage is the beginning of the e-government as a revolutionary entity changing the way people interact with their government. This stage empowers citizens to deal with their governments on-line anytime, saving hours of paperwork, the inconvenience of travelling to a government office and time spent waiting in line. At this stage, citizens can be served on-line by 'e-government.' While the cataloguing stage helps citizens' fact-finding process, this transaction stage e-government presents government on the other side of the Internet as an active respondent. It is now a two-way communication.

Vertical Integration

At this stage, the focus is now moving toward transformation of government services, rather than automating and digitizing existing processes. Making government electronic is not simply a matter of putting existing government services on the Internet. What should and will be happening are permanent changes in the government processes themselves and possibly the concept of government itself. Just as electronic commerce is redefining private business and society in terms of processes and product, electronic government initiatives should be accompanied by re-conceptualization of the government service itself. In the long run, the full benefit of e-government will be realized only when organizational changes accompany technological changes. After on-line transaction services become prevalent and mature, citizens' expectations will increase. Most transaction stage systems are localized and fragmented. A natural progression will be the integration of scattered systems at different levels (vertical) and different functions (horizontal) of government services. Agencies often maintain separate databases that are not connected to other governmental agencies at the same level or with similar agencies at the state and national level. For example, a state input-dealer license database is often separate from a local (district level) business license database. Further, that state license system is probably not connected to the state vendor database. It is expected that vertical integration within the similar functional walls but across different levels of government will happen first, because the gap between levels of government is much less than the difference between different functions/ministries. Many state government departments interact more closely with their central government and district counterparts than other departments at the same level of government. Consequently, at stage three central, state and local counterpart systems are expected to connect or, at least, communicate to each other. While some departments' websites currently provide links to other governmental agencies at different levels, vertical integration goes beyond this simple interconnection. If a citizen conducts a transaction with a state agency, the transaction information will be propagated to district and central government counterparts. These various levels of systems are connected and talk to each other so that results of transactions from one system can be interchanged with another system. Physically, this may be integrated as a central database or a connected web of databases communicating with each other.

Horizontal Integration

The full potential of information technology, from the citizen's perspective, can only be achieved by horizontally integrating government services across different functional walls (or "silos"). The limitations of the functional nature of both the public and private sector will become clearer as more public administrators begin to see the vision opened by the Internet. Typically, citizens requiring assistance from governments need more than one service. Those requiring housing also need governmental assistance for education, food, medical attention, etc. To overcome this problem, some state governments provide one stop service centers where, for example, the homeless can come and obtain information about jobs, etc. Governments continually fight the battle of getting services to the people who need them the most. The horizontal integration of the stage four will considerably improve those efforts. Databases across different functional areas will communicate with each other and ideally share information, so that information obtained by one agency will propagate through out all government functions. In addition, citizens could conduct business across a wide variety of requirements. As an example, when a citizen applies for a driver's license after moving to another state, the basic residence record could be propagated to different functional service branches of government such as the health and the district level election department so that the citizen does not have to fill out a personal record form for each governmental agency. Horizontal integration refers to system integration across different functions in that a transaction in one agency can lead to automatic checks against data in other functional agencies.

Index

F

www.ingramcontent.com/pod-product-compliance
Lightning Source LLC
Chambersburg PA
CBHW051356240926
51852CB00001B/43
9789351302544